H A N D B O O K   O F
CONSTRUCTION
CONTRACTING

VOLUME 1
Plans, Specs, Building

by

Jack P. Jones

CRAFTSMAN

Craftsman Book Company
6058 Corte del Cedro, P.O. Box 6500, Carlsbad, CA 92008

Acknowledgements

The author wishes to thank the following companies and organizations for furnishing materials used in the preparation of various portions of this book.

American Plywood Association — Tacoma, Washington

Asphalt Roofing Manufacturers Association — Washington, D.C.

Brick Institute of America — McLean, Virginia

Canadian Wood Council — Ottawa, Ontario, Canada

First Federal Savings and Loan Association — Warner Robins, Georgia

Home Planners, Inc. — Farmington Hills, Michigan

National Concrete Masonry Association — Herndon, Virginia

National Forest Products Association — Columbus, Ohio

National Woodwork Manufacturers Association, Inc. — Oak Ridge, Illinois

Southern Forest Products Association — New Orleans, Louisiana

Standard Homes Plan Service, Inc. — Raleigh, North Carolina

This book is dedicated to Kenneth

Library of Congress Cataloging-in-Publication Data

Jones, Jack Payne, 1928-
    Handbook of construction contracting.

    Includes index.
    Contents: v. 1. Plans, specs, building.
    1. House construction--Handbooks, manuals, etc.
2. Building--Estimates--Handbooks, manuals, etc.
I. Title.
TH4813.J66   1986   692'.8   86-8925
ISBN 0-934041-11-3 (v. 1)

Contents

Chapter 1

The Business of Building

No field offers more opportunity than construction. Where else can someone start with little cash, no special education and no special skills and build a multi-million dollar company in a few years? This isn't fantasy. It happens regularly.

I'm not saying anyone can do it. It takes hard work, skill, and a little luck. Every year thousands of general contractors, carpenters, masons, plumbers and electricians go into business for themselves. Some of their companies will remain small. Others will grow quickly. Many will fail within a year or two.

Many of the 1000 largest building companies in the U.S. are run by men and women with little formal education. They learned the business from square one, working as laborers or helpers until they had the skills needed to succeed as construction contractors. Most of these people got into construction by accident. Few started out with the intent of becoming professional builders. Typically, here's how it happens:

You worked part-time on construction projects as a teenager. Or maybe you have an uncle or cousin in construction. You felt comfortable on a construction site. You've seen others go into business for themselves as builders. Your first job may have been remodeling an old house. Or maybe it was building a weekend cottage in your spare time. On that first job you made mistakes. Some mistakes cost you plenty. But this was the "tuition" you paid to learn the fundamentals.

Gradually you learned how to estimate material quantities, how to price your work, how to handle employees, how to sell your services, and how to work with clients, architects, lenders and the building inspector. You learned what to watch for and what to avoid. You learned what works and what doesn't. You learned the importance of doing quality work and building a reputation for professionalism. And, gradually, you learned what it takes to make a good living in the construction industry.

A friend or neighbor saw your first project and wanted you to handle some work for him. You took the job, made a decent wage for your effort and turned a small profit to boot. Encouraged by those results, you started another job when that second job was finished. Before too long, you had several requests to bid some work. Two or three tradesmen were regulars on your payroll. Before long, you were running a construction company, had an ad in the Yellow Pages, a company phone and a weekly payroll to meet.

Construction contracting isn't all peaches and cream, of course. This country has a regular construction cycle that brings prosperity to thousands of builders and then washes thousands out of the business four or five years later.

But if you're looking for a way to be your own boss and make a decent living doing quality work, read on. This book is for you. We'll cover all of the basics of construction contracting. We'll discover what separates the successful builder from those who fail. And we'll show you how to increase your profits and reduce costs.

In this first chapter, we'll take a bird's-eye view of the business of building. We'll outline the shape of the construction industry and present a few concepts that you should understand at the outset. The details come in later chapters. If you've been working as a tradesman or contractor for several years, you may want to skip right on to Chapter 2.

Before I leave the subject of careers in the construction industry, let me point out one important benefit that shouldn't be ignored. Few builders see their bank accounts swell into the millions. Of course, many make a good living. But there are benefits to being in construction that never show up on a balance sheet.

Look at the **before** and **after** pictures in Figure 1-1. The finished product isn't the Taj Mahal. But the builder who did the work can take pride in turning an old sow's ear into a comfortable silk purse with modern plumbing and electrical systems. He's provided one of the most basic human requirements, the need for shelter. The family that lives in this remodeled house enjoys more comfort, more convenience and a more attractive environment. That makes for a better life.

If you take pride in doing quality work that makes a better life for others as well as for yourself, you're a "natural" for the construction industry.

Contractors and Developers

Some construction contractors are part-time builders. They have another occupation that provides a regular source of income. When the construction cycle is on the upswing, they put up spec or custom houses or do some remodeling, all the while holding another full-time job. When new homes aren't selling and remodeling jobs are scarce, they concentrate on their other occupation.

Part-time builders have to be good at planning — directing and scheduling work. And they need a reliable full-time employee on site who can make decisions while the owner is occupied elsewhere.

There's nothing wrong with building part-time. Many teachers, for example, find time to run a small construction company as a sideline. Short

school workdays and long summer vacations make running a building business on the side entirely possible.

Full-time contractors usually fall into one of three categories: the on-site contractor, the managing contractor and the land developer.

On-site contractors— Many builders have just two or three regular employees. The only office may be the company pickup truck. The owner works alongside the crew. He does any kind of work that comes along: a spec house, a custom house, an add-on, or a remodeling job.

This type of builder seldom goes broke. What he's selling is his own time. When the construction market turns sour, he doesn't have the burdensome overhead and crushing loan charges that sink other builders during a recession. On-site contractors just pull in their horns, shift emphasis to other types of work (or take another job) and wait until the market improves.

Most of the construction companies in this country are run by on-site contractors. Every community has at least a few. They're the backbone of the residential construction and remodeling industry in the U.S.

Managing contractors— Of course, many contractors haven't picked up a hand tool in years. They spend more time in the office than on the job. This type of builder is primarily a manager. He plans and directs the operation, from buying the lot (for a spec building), to getting permits, estimating, ordering materials, and selecting subcontractors.

A managing contractor directs the work flow, keeping the work on schedule and within budget. He's an inspector and facilitator, seldom a tradesman or supervisor. He doesn't need to be an expert in all trades. But he has to know the difference between quality work and work that doesn't measure up to standard. Some call this type of builder a "paper contractor."

The managing contractor probably has several jobs going at the same time. He has to if he wants to carry the overhead of his own unproductive labor. That's why this type of contractor is vulnerable when construction slows down. It's hard to shed overhead and debts fast enough to stay afloat in a disappearing market.

Land developers— These are land merchants more than construction contractors. Building is more a

Sound, older structure in need of repair

Transformed by the competent builder

Construction know-how and hard work
Figure 1-1

sideline than a major focus. Developers buy tracts of land and turn them into residential, commercial or industrial subdivisions. Developers are gamblers. They bring together borrowed capital, land, design and construction services, and prospective buyers, to create subdivisions that they hope to sell for more than they cost.

When a developer miscalculates, his error is likely to be major. But hitting the right market with the right product can create big profits.

In this book, we'll occasionally refer to land development projects. But we'll focus on smaller-scale construction contracting. Even if you intend to develop land on a large scale someday, you'll want to master building smaller projects first. Remember, finishing successful small projects lays the groundwork for bigger jobs.

Speculative Building

Spec builders come in all sizes, from the contractor who's building his first and only spec home to the developer who has a thousand homes under construction on sites in several states. All spec builders buy or subordinate the land, arrange for a construction loan, take out the permits, put up the buildings, and work to find qualified buyers.

The spec builder's nightmare is unsold inventory. The longer a house remains unsold, the smaller the profit. Eventually, the profit may disappear entirely. Very rarely does a delayed sale bring in enough extra money to cover the cost of holding the property vacant for months or years.

Here's a rule of thumb for estimating the total cost to the buyer for a conventional house with only the minimum of modern conveniences:

Multiply the carpenter's hourly union wage by 3. Then multiply this figure by the total area of the house (in square feet).

Here's an example. Let's say the hourly union wage for a carpenter in your area is $20. You want to estimate the selling price of a 1,500 square foot house. Multiply the union wage ($20) by 3 to get a formula wage of $60. Then multiply $60 by 1,500. This gives you a total cost of $90,000, or $60 per square foot. Remember to use a carpenter's hourly *union* wage in your baseline estimate. (This may not be the same as the actual hourly wage paid on your job.) This "baseline estimate" will usually be high enough to include the lot, profit, and operating expenses like taxes, insurance and interest charges.

Here's another guide to use when considering a spec home project. About one manhour will be needed for each square foot of floor in a home without a basement. Two-thirds of this time will be skilled labor and one-third will be semi-skilled labor.

Let's apply this rule to our 1,500 square foot house. Allowing 1 manhour per square foot of floor space, 1,500 manhours will be required. We'll need 1,000 skilled manhours and 500 semi-skilled manhours.

If you're a skilled carpenter and hire another skilled tradesman and a semi-skilled helper, the three of you can complete the job in 500 hours. If you sub out part of the work, you'll reduce your crew's time accordingly.

Keep in mind that this is only a rule of thumb. It has to be adjusted to fit the situation. Never substitute a baseline estimate for the detailed, itemized material and labor estimate. You need a complete estimate on every job. Before submitting any bid, list every unit of material. Then determine the cost of all those materials and estimate the manhours required for installation.

Building a spec house has a major advantage: If costs run over budget, you can usually increase the asking price a little. With a custom house, the situation is different. Your income is set by contract. Omit some cost and you're stuck with the bill.

Custom Building

Your client's primary concern will usually be the cost, at least until construction gets underway. Use extreme care in preparing your list of materials (take-off sheet). Make sure you include *everything*.

Custom building has advantages and disadvantages. The primary advantage is that your client carries the largest risk. But the builder still carries risk that material prices will increase between the bid date and the date of construction. Lumber and plywood prices, to name just two, can change rapidly. Increases of 15 or 20% over two or three months are common for some materials. Protect yourself. Include price escalation clauses in your contract. Get written bids from subcontractors. Make sure the bid prices are guaranteed for a specified time period. Otherwise, the sub's labor and material price increases end up in your lap.

Before you agree to build a custom home on any site, familiarize yourself with soil conditions. Pay special attention to water levels, rock formations, soil type and topography. If large trees, stumps, and rocks have to be removed, your building costs will be higher. And these costs may not be recoverable.

The less money spent on preparing the site, the better. A $100,000 house sitting on a site that cost $100,000 to prepare looks the same as a $100,000 house sitting on a site that cost just $2,000 to prepare — and it has the same selling price. It doesn't look anything like a $200,000 house. Put your construction money where it shows: *in the house, not under it.*

Changes during construction can be a headache or a money machine for the home builder. Most owners request changes by the bushel as their custom home is built. Once the house is framed, the owner will want to relocate a partition, enlarge a closet, or install a larger window in the master bedroom. After construction begins, *any* change in plans will cost more — more manhours, more for materials, and much more for administrative overhead. Make sure your contract provides that changes will be made only by written request and at your "usual selling price" (what you would like to get, not what the competition would bid). And never price extras at your cost plus 10%. Labor and material cost plus 10% is always less than your true cost.

The owner might also decide that your construction materials are substandard and unacceptable. Dissatisfied owners have been known to knock out studs and joists at the end of the workday after the builder and crew have gone home. Have a clear understanding with the owner about the quality of materials. Include in your contract acceptability standards by grade and species.

Few builders have the time to hand pick the lumber used in a house. But make the time to reject materials that don't meet the standards set in your contract. And send back any finish materials that are defective. Reputable builders don't use substandard materials.

Generally you'll be selected to build a custom home because your bid is either the lowest or among the lowest. That puts a premium on careful estimating. Most builders include a little cushion in their estimates to allow for errors. Usually this is called *contingency*. But building in too much slack will usually guarantee that someone else does the work. The best procedure is to compile complete, accurate estimates, (including overhead and profit) so that little or no cushion is needed.

Remodeling

Remodeling is a big industry in this country. And it's growing faster than construction in general. Room additions, porches, patios, redoing bathrooms and kitchens, repairing fire, storm and insect damage are all big ticket items.

Profit margins are higher in remodeling work than in any other type of construction. And it's good fill-in work for your crew during the slack times when staying busy is a problem. But it's a different type of construction that requires special skills and procedures.

A healthy dose of skepticism is an advantage in remodeling work. Never assume that anything in the existing structure is built according to standard. It probably isn't. And don't assume that the structure is square, plumb or level. It seldom will be.

Assume that everything hidden is either not on center or will move out of position when you drive the first nail. A water pipe or an electrical circuit will always be right where you plan to cut a door or window. When you remove that old commode, you'll discover the floor underneath is so rotten it wouldn't even support a bedpan! And when you start digging the footing for that room addition, you're certain to uncover an abandoned septic tank. The owner forgot all about it, of course.

The windows you agreed to replace are odd-size and will have to be special-ordered. If it's an older house, you may discover that the studs are 2 x 5's. It would have been nice to know that before the custom-made windows arrived — and before you broke out the old windows!

If you assume that the existing floor is level, you may be in for a surprise. When you cut the opening for the doorway to the room addition, your finish floor on the addition may be 1½" higher than the existing finish floor. And advising the owner that it's *his* floor that's skewed won't cut much mustard. Optimistic assumptions cause remodelers a lot of grief. As a remodeler, the only assumptions to make are that *it's either too thin, too thick, too short, too long, too small, too large, or too late.*

Successful remodelers generally would make good building inspectors. They know what to look for and what questions to ask before work begins. The more thorough the inspection, the more questions asked and answered, the fewer surprises during construction.

Estimating standards used in new construction seldom apply in remodeling work. In new construction, it may take only four manhours to install eight linear feet of base and wall cabinet and top. In remodeling, the leveling, furring-out, working around furniture and appliances, taking materials outside for sawing, and an inquisitive child running free can easily make installation an all-day task.

Still, many builders prefer remodeling work because it's plentiful and carries better margins. True, remodeling is good business, *but only for those who have learned through experience what's required to earn a profit.*

Keeping Records

No matter what type of construction you do, accurate, detailed records are essential. Your job estimate sheets, manhour reports and lists of materials are important documents. Set up some orderly filing system so you can identify costs on previous jobs when estimating future jobs. There should be two profits in every job. The first is the money that goes into your pocket. The second is what you have learned while doing the work. Much of what you learned is in written form: material costs, manhour records, subcontractor invoices. Be sure you have these documents available when you need them.

Every builder has to keep payroll and accounting records, even if they're seldom or never used by the business owner. Federal and state law require every business to keep records that establish payroll and income tax liability. The fact that you're not making any money is no defense. Every contractor needs a good record-keeping system to meet state and federal requirements, if for no other reason.

Fortunately there are several good books that describe how to keep good cost records and meet state and federal record-keeping requirements. The order form at the back of this book lists several titles that cover record keeping for builders.

Machinery and Equipment

Most builders should avoid buying expensive heavy equipment such as backhoes, front-end loaders and bulldozers. The initial cost of this equipment is high, and the upkeep can be disastrous.

For example, new rings for a D7 Cat tractor can cost over $5,000. If you install the rings yourself, the cost will be lower. But you're a builder, not a mechanic. If you need dozer or backhoe work, sub it out. Don't consider buying heavy equipment until you're keeping an excavation sub busy full-time. Most builders will never reach that point.

Staying small doesn't mean that a builder is any

less successful. Some very profitable and highly efficient construction companies are run by contractors operating out of their pickup trucks. They handle a steady flow of building and remodeling work with small but experienced crews. Many of these builders are successful enough to finance their own operations, making most borrowing unnecessary.

Construction Loans

Custom home builders usually work with loan proceeds arranged by the owner. As work progresses, your client will be given authority to draw on the construction loan on a prescribed percentage basis.

Spec builders have to find their own construction money at banks and savings and loan associations (S&L's). Spec building loans are normally limited to 75% of the appraised value. If the house is presold, the loan may go up to 95% of either the sales price or appraised value (whichever is lower). But the bank or S&L won't usually dispense more than 80% of the loan until the house is completed.

When you prepare to build your first spec house and apply for a construction loan with a local ender, the procedure will be something like this: You get an application, estimate form, specification sheet and energy requirements form. Samples of these forms are shown in Figures 1-2, 1-3, 1-4 and 1-5. Return the completed forms with a full set of plans, your current financial statement, your two most recent tax returns and either a sales contract or a deed on the land. Next, the bank or S&L will appraise the lot, review the plans and specs, and order a credit report. When all the papers are in order, your application is presented to a loan committee for approval.

When the loan is approved, the lender prepares loan documents. You'll sign a construction loan agreement and a preconstruction affidavit in addition to the note, deed to secure debt, and other legal documents.

At closing, the lender will normally disburse no more than 75% of the lot cost and closing fees. (Be sure to include these fees in your cost estimate.) The rest of the money is placed into an account called "loans-in-process." The bank or S&L will disburse it according to a percentage schedule as work progresses. See Figure 1-6.

The builder who has a reputation for doing quality work and finishing his projects will usually have no trouble getting construction loans.

FIRST FEDERAL SAVINGS AND LOAN ASSOCIATION
CONSTRUCTION APPLICATION AND COST ESTIMATE

Date _____ Mortgage Amount $ _____ Int. Rate _____ % Dis. _____ % A.P.R. _____ %

Term _____ Months   Loan to Value _____% (not to exceed 80%)

LEGAL DESCRIPTION

_____

ESTIMATED COST OF CONSTRUCTION

STRUCTURE $ _____   PROPOSED SALES PRICE $ _____

LOT COST $ _____   SELLING COST _____ % _____

TOTAL $ _____   PROFIT $ _____

This is to certify that to the best of my knowledge the above is a true estimate of the cost for the structure according to plans and specifications on file with this association.

                                                         Borrower's Initials _____

Personal endorsement of _____

Title to be vested in _____   Tenancy _____

Business Address _____   Phone No. _____

Credit References:

    1. _____   Age _____ Annual Income $ _____

    2. _____   Have you ever taken bankruptcy _____

    3. _____   Outstanding judgements _____

    4. _____   Foreclosures _____

                                          Law Suit _____

Contractor's Phone No. _____   Purchaser's Phone No. _____

STIPULATIONS

1. Interest to be paid calendar quarter.
2. Upon first draw request, a plat must be furnished.
3. Termite certification must be furnished inspector prior to disbursement for same.
4. Percentage schedule will not be diverted from and only materials in place will be considered for draw request.
5. Draw request should be called in by 12:00 Thursdays. Checks will be ready by 1:00 Fridays. No guarantee of draw if not called in by noon Thursday.
6. The maximum number of draws will be eight. Five is recommended.
7. Good construction practices must be followed. Draws will be held up for poor workmanship until corrected.
8. First Federal requests the right to place a sign on property stating the property financed by First Federal Savings and Loan.
9. Construction must not begin until after construction loan closes.
10. Energy features as per handout will be required on all construction.

Signature Borrower _____   Signature Borrower _____

INFORMATION FOR GOVERNMENT MONITORING PURPOSES

The following information is requested by the Federal Government if this loan is related to a dwelling, in order to monitor the lender's compliance with equal credit opportunity and fair housing laws. You are not required to furnish this information, but are encouraged to do so. The law provides that a lender may neither discriminate on the basis of this information, nor on whether you choose to furnish it. However, if you choose not to furnish it, under Federal regulations this lender is required to note race and sex on the basis of visual observation or surname. If you do not wish to furnish the above information, please initial below.

BORROWER: I do not wish to furnish this information (initials) _____	**BORROWER:** I do not wish to furnish this information (initials) _____
RACE/ ☐American Indian, Alaskan Native ☐Asian, Pacific Islander	**RACE/** ☐American Indian, Alaskan Native ☐Asian, Pacific Islander
NATIONAL ☐Black ☐Hispanic ☐White **SEX** ☐Female	**NATIONAL** ☐Black ☐Hispanic ☐White **SEX** ☐Female
ORIGIN ☐Other (specify) _____ ☐Male	**ORIGIN** ☐Other (specify) _____ ☐Male

Loan application
Figure 1-2

Contractor's Estimate

_____
Contractor

_____
Address

Office _____ **Home** _____
Telephones

Estimate for: _____

Address: _____ **City** _____

Phone: _____ **Office** _____ **Home**

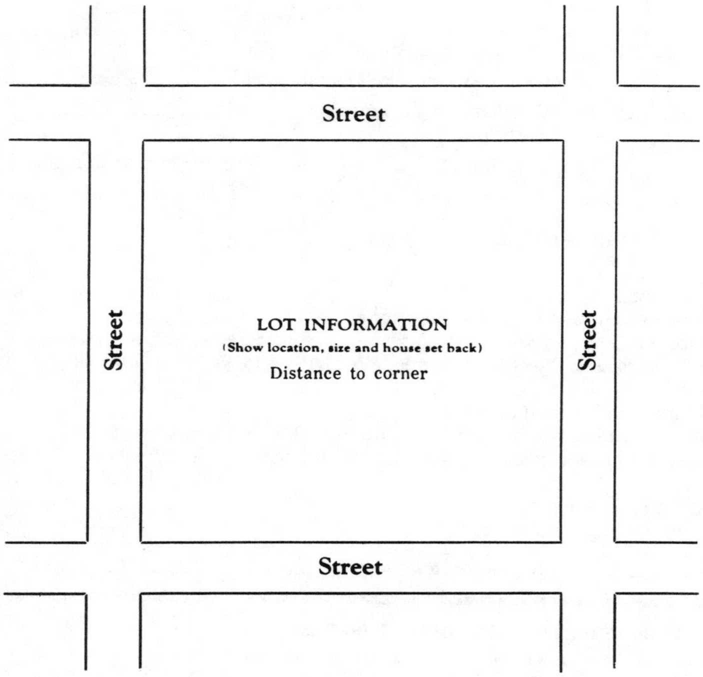

BUILDING SITE

Street _____

Subdivision _____

Lot No. _____ Block _____

UTILITIES AVAILABLE

Water _____

Electricity _____

Sewer _____

Gas _____

Paving _____

Street

Street

Street

Street

LOT INFORMATION
(Show location, size and house set back)
Distance to corner

Front _____ Rear

LOT GRADE

This Estimate Based on Current Cost and Supplies and is good not exceeding

a period of _____ days or until _____, 19___

Estimate form
Figure 1-3

The Business of Building

ITEMS	COST ESTIMATES	Estimated Cost	Actual Cost
PRELIMINARY COST	Plans .... $_____	$	$
	Survey and Laying off House		
	Insurance: Fire — Liability		
	Building Permit		
	Temporary Service Water & Lights		
PREPARATION OF LOT	Removing trees		
	Clearing site		
	Excavation		
	Hauling		
FOOTINGS	Excavation		
	Concrete footings		
	Backfilling		
MASONRY MATERIALS	Block		
	Common brick: Piers		
	Foundation		
	Chimney: Brick		
	Fire brick		
	Damper, ash dump, ...		
	cleanout		
	Tile for hearth		
	Flue lining		
	Face brick		
	Cleaning brick		
	Sand		
	Brixment		
	Waterproofing (where used)		
	Drain tile (where used)		
	Back filling (basement)		
	Scaffold		
STEEL	Lintels		
	Reinforcing rods and mesh		
	Steel sash or vents		
	Pipe piers and columns		
	Flitch plates, bolts, and anchors ...		
MASONRY LABOR	Skilled		
	Common		
CONCRETE WORK	Basement floor		
	Porches		
	Stoops		
	Steps		
	Walks .... Drives		
	Garage or carport floor		
	Forms		
ROUGH LUMBER	Joist, girders and wall plates		
	Stud, plates and purlins		
	Rafters, ridge and valleys		
	Sheathing and Sub-floors		
	Bracing and headers		
	Stringers, ties, and bridging		
ROOFING	Shingles		
	Building paper		
	Flashing chimney and eaves		
ROUGH HARDWARE	Nails		
	Building paper		
	Bolts, anchors and ties		
EXTERIOR MILLWORK	Front doorway		
	Doors and frames		
	Windows and frames		
	Porch columns and rails		
	Moulds		
	Louvres		
	Shutters		
	Garage Door		
FINISH LUMBER	Porch and carport ceiling		
	Porch and carport box		
	Soffit and fascias		
	Moulds		
	Siding or shingles		
SCREENS	Doors .... Windows		
	CARRY FORWARD ☞		

Estimate form
Figure 1-3 (continued)

		Estimated	Actual
	BROUGHT FORWARD ☞	$	$
STORM SASH	Doors .... Windows		
INTERIOR MILLWORK	Doors and frames .... Interior casings .... Mantel .... Flooring (Hardwood) .... Moulds, base and shelving .... Stairway (where shown)		
KITCHEN CABINETS	Shop built .... Built on job		
CARPENTER LABOR	Rough .... Finish		
SHEET METAL WORK	Doors and Window Flashing ...... Termite Shields .... Gutters and Downspouts .... Splash blocks		
INTERIOR WALL FINISH	Lath and Plaster yds........ @........ Sheet-rock sq. feet........... @........ Special Dry Wall finish sq. feet ...... Corner bead		
FINISH HARDWARE	Locks .... Hinges .... Cabinet hardware .... Screen hardware		
WIRING	Electric openings .... Fixtures and installation .... Dishwasher ..... Dryer .... Hot water heater ...... Bell ...... Furnace .... Washer .... Range ..... Air Conditioning		
PAINTING AND PAPERING	Caulking .... Exterior woodwork .... Interior woodwork .... Interior walls and ceiling .... Kitchen .... Bath		
FLOORS	Sanding and finishing (hardwood)		
SPECIAL WALLS AND FLOORS	Kitchen wall............ floor............ Bath wall .... floor.............. Other		
PLUMBING	Cutting street and ditches .... Septic tank........ Well & Pump...... Fixtures .... and piping.......... Hot water heater .... Medicine Cabinets .... Towel, paper, and soap racks ...... Labor		
HEATING	Unit .... Tank................ Ducts		
AIR CONDITIONING	Unit .... Installation............		
ORNAMENTAL IRON	Columns .... Rails		
INSULATING	Walls .... ceilings ...... floors		
WEATHER STRIPPING	Windows .... Doors		
CLEANING AND HAULING	Windows .... House.............		
GRADING, SEEDING AND PLANTING	....		
SHADES OR BLINDS	....		
SPECIAL EQUIPMENT	Refrigerator......... Dishwasher......... Range top & Oven ...... Washer...... Exhause Fan .... Dryer............ Hood & duct .... Disposal...........		
OTHER ITEMS	....		
	Labor and Materials (Total)		
	Taxes and Insurance		
	Contractor's Fee		
	Total Cost		

Estimate form
Figure 1-3 (continued)

ESTIMATING TABLES

BOARD FEET OF LUMBER

2 x 4

	Half for 1 x 4			Double for 2 x 8	
No. Pcs.	10'	12'	14'	16'	18'
1	6⅔	8	9⅓	10⅔	12
2	13⅓	16	18⅔	21⅓	24
3	20	24	28	32	36
4	26⅔	32	37⅓	42⅔	48
5	33⅓	40	46⅔	53⅓	60
6	40	48	56	64	72
7	46⅔	56	65⅓	74⅔	84
8	53⅓	64	74⅔	85⅓	96
9	60	72	84	96	108
10	66⅔	80	93⅓	106⅔	120
20	133⅓	160	186⅔	213⅓	240
30	200	240	280	320	360
40	266⅔	320	373⅓	426⅔	480
50	333⅓	400	466⅔	533⅓	600

2 x 6

	Half for 1 x 6			Double for 2 x 12	
No. Pcs.	10'	12'	14'	16'	18'
1	10	12	14	16	18
2	20	24	28	32	36
3	30	36	42	48	54
4	40	48	56	64	72
5	50	60	70	80	90
6	60	72	84	96	108
7	70	84	98	112	126
8	80	96	112	128	144
9	90	108	126	144	162
10	100	120	140	160	180
20	200	240	280	320	360
30	300	360	420	480	540
40	400	480	560	640	720
50	500	600	700	800	900

2 x 10

No. Pcs.	10'	12'	14'	16'	18'
1	16⅔	20	23⅓	26⅔	30
2	33⅓	40	46⅔	53⅓	60
3	50	60	70	80	90
4	66⅔	80	93⅓	106⅔	120
5	83⅓	100	116⅔	133⅓	150
6	100	120	140	160	180
7	116⅔	140	163⅓	186⅔	210
8	133⅓	160	186⅔	213⅓	240
9	150	180	210	240	270
10	166⅔	200	233⅓	266⅔	300
20	333⅓	400	466⅔	533⅓	600
30	500	600	700	800	900
40	666⅔	800	933⅓	1066⅔	1200
50	833⅓	1000	1166⅔	1333⅓	1500

For lengths over 18'0" Double 10', 12',
or 14' Lengths

NAILS

Use	lbs.	Size	Material
Built-up girders	10	20d Com	100 lf
Floor joist framing	30	20d Com	1M bf
2x6 Ceiling joist-rafters	15	16d Com	1M bf
Frame walls—partitions	10	16d Com	1M bf
Frame walls—partitions	5	10d Com	1M bf
1" Sheathing—sub floor	20	8d Com	1M bf
1" Plaster grounds	10	6d Com	1M lf
1x3" Cross bridging	4	10d Com	100 lf
Bevel siding, wood	20	8d Cas	1M sq ft
T&G beaded ceiling	20	6d Cas	1M sq ft
⅞" Hardwood flooring	50	8d Cut	1M sq ft
Outside trim—moulds	20	8d Cas	1M lf
Outside trim—moulds	5	6d Cas	1M lf
Inside trim—moulds	15	8d Fin	1M lf
Inside trim—moulds	10	6d Fin	1M lf
Asphalt strip shingles	4	1" Gal	100 sq ft
Lath, gypsum	10	1⅛" Lath	1M sq ft

INTERIOR FINISH

LINE A Perimeter of room in lineal feet—Use for base and shoe.
LINE B Square feet in room (Floor or Ceiling)—Use for flooring, ceiling.
LINE C Square feet of wall and ceiling combined for eight foot ceiling. Use for figuring Plaster, Drywall, Painting—For exact amount of material deduct window and door openings. For Walls ONLY deduct amount shown in line. B.

Room Size Feet		2'	3'	4'	5'	6'	7'	8'	9'	10'	11'	12'	13'	14'	15'
2'	A	8	10	12	14	16	18	20	22	24	26	28	30	32	34
	B	4	6	8	10	12	14	16	18	20	22	24	26	28	30
	C	68	86	104	122	140	158	176	194	212	230	248	266	284	302
3'	A	10	12	14	16	18	20	22	24	26	28	30	32	34	36
	B	6	9	12	15	18	21	24	27	30	33	36	39	42	45
	C	86	105	124	143	162	181	200	219	238	257	276	295	314	333
4'	A	12	14	16	18	20	22	24	26	28	30	32	34	36	38
	B	8	12	16	20	24	28	32	36	40	44	48	52	56	60
	C	104	124	144	164	184	204	224	244	264	284	304	324	344	364
5'	A	14	16	18	20	22	24	26	28	30	32	34	36	38	40
	B	10	15	20	25	30	35	40	45	50	55	60	65	70	75
	C	122	143	164	185	206	227	248	269	290	311	332	353	374	395
6'	A	16	18	20	22	24	26	28	30	32	34	36	38	40	42
	B	12	18	24	30	36	42	43	54	60	66	72	78	84	90
	C	140	162	184	206	228	250	272	294	316	338	360	382	404	426
7'	A	18	20	22	24	26	28	30	32	34	36	38	40	42	44
	B	14	21	28	35	42	49	56	63	70	77	84	91	98	105
	C	158	181	204	227	250	273	296	319	342	365	388	411	434	457
8'	A	20	22	24	26	28	30	32	34	36	38	40	42	44	46
	B	16	24	32	40	48	56	64	72	80	88	96	104	112	120
	C	176	200	224	248	272	296	320	344	368	392	416	440	464	488
9'	A	22	24	26	28	30	32	34	36	38	40	42	44	46	48
	B	18	27	36	45	54	63	72	81	90	99	108	117	126	135
	C	194	219	244	269	294	319	344	369	394	419	444	459	494	519
10'	A	24	26	28	30	32	34	36	38	40	42	44	46	48	50
	B	20	30	40	50	60	70	80	90	100	110	120	130	140	150
	C	212	238	264	290	316	342	368	394	420	446	472	498	524	550
11'	A	26	28	30	32	34	36	38	40	42	44	46	48	50	52
	B	22	33	44	55	66	77	88	99	110	121	132	143	154	165
	C	230	257	284	311	338	365	392	419	446	473	500	527	554	581
12'	A	28	30	32	34	36	38	40	42	44	46	48	50	52	54
	B	24	36	48	60	72	84	96	108	120	132	144	156	168	180
	C	248	276	304	332	360	388	416	444	472	500	528	556	584	612
13'	A	30	32	34	36	38	40	42	44	46	48	50	52	54	56
	B	26	39	52	65	78	91	104	127	130	143	156	169	182	195
	C	266	295	324	353	382	411	440	469	498	527	556	585	614	643
14'	A	32	34	36	38	40	42	44	46	48	50	52	54	56	58
	B	28	42	56	70	84	98	112	126	140	154	168	182	196	210
	C	284	314	344	374	404	434	464	494	524	554	584	614	644	674
15'	A	34	36	38	40	42	44	46	48	50	52	54	56	58	60
	B	30	45	60	75	90	105	120	135	150	165	180	195	210	225
	C	302	333	364	395	426	457	488	519	550	581	612	643	674	705
16'	A	36	38	40	42	44	46	48	50	52	54	56	58	60	62
	B	32	48	64	80	96	112	128	144	160	176	192	208	224	240
	C	320	352	384	416	448	480	512	544	576	608	640	672	704	736
17'	A	38	40	42	44	46	48	50	52	54	56	58	60	62	64
	B	34	51	68	85	102	119	136	153	170	187	204	221	238	255
	C	338	371	404	437	470	503	536	569	602	635	668	701	734	767
18'	A	40	42	44	46	48	50	52	54	56	58	60	62	64	66
	B	36	54	72	90	108	126	144	162	180	198	216	234	252	270
	C	356	390	424	458	492	526	560	594	628	662	696	730	764	798
19'	A	42	44	46	48	50	52	54	56	58	60	62	64	66	68
	B	38	57	76	95	114	133	153	171	190	209	228	247	266	285
	C	374	409	444	479	514	549	584	619	654	689	724	759	794	829
20'	A	44	46	48	50	52	54	56	58	60	62	64	66	68	70
	B	40	60	80	100	120	140	160	180	200	220	240	260	280	300
	C	392	428	464	500	536	572	608	644	680	716	752	788	824	860

Estimate form
Figure 1-3 (continued)

FIRST FEDERAL SAVINGS & LOAN ASSOCIATION

PLEASE COMPLETE IN DETAIL

Property Address _____

Owner _____ Address _____ Tel. _____

Contractor _____ Address _____ Tel. _____

Instructions: Describe materials and equipment to be used by entering the information called for in each space. If space is inadequate or if a description is necessary on an item that space is not provided for, describe under the miscellaneous section. All Blanks must be completed. Line through those items or spaces that do not apply.

1. Foundations: Crawl _____   Slab _____
    Foundation Wall: Material _____
    Waterproofing: (Method) _____
    Special Foundations _____
2. Fireplaces and Chimneys: _____
    Material _____ Prefab _____
3. Exterior Walls:
    Siding (Describe) _____
    Masonry Veneer _____ Brick Allowance _____ Stucco _____
4. Floor Framing:
    Joists: Wood, Grade, Species _____
    Concrete Slab, Basement _____ First Floor _____
    Fill Under Slab _____ Membrane _____
5. Finish Floor:
    Room, Grade, Species _____
6. Roofing: Gutters - Yes _____ No _____
    Roofing Type _____ Built Up _____
7. Decoration: (Wall Finish: Paint, Wallpaper, Etc.)
    Kitchen _____ Family Room _____
    Bfst. Room _____ Baths _____
    Living Room _____ Dining Room _____
    Bedrooms _____ Foyer _____
    Hallway _____ Other Rooms (Specify) _____

8. Interior Doors & Trim:
    Doors: Type _____ Material _____
    Special Trim (Specify) _____
    _____
    _____

9. Interior Walls & Ceilings:
    Sheetrock (Thickness) _____ Finish _____
    Insulation (Type & Thickness): Ceiling _____ Walls _____ Roof _____
    Energy package per FFSL _____
10. Windows:
    Type _____ Materials _____
    Special Windows (Describe) _____ _____
11. Cabinets:
    Kitchen: Pre-Manufactured _____ Custom Made _____
    Finish _____ Material _____
    Other Cabinets & Built-in Furniture (Specify) _____
    Allowances _____
12. Special Floors:
    Kitchen _____ Baths _____ Family Room _____
    Foyer _____ Bfst. Room _____ Utility Room _____
    Bedrooms _____ Other Rooms (Specify) _____
13. Plumbing
    Water Supply: Public _____ Community _____ Individual _____
    Sewage: Public _____ Community _____ Individual _____
    Water Heater: Type and Capacity _____

Specification sheet
Figure 1-4

14. Heating and Cooling:
    Type System (Heating) _____
    Central Air Conditioning _____ Yes _____ No _____
    Type and Capacity _____
    Ventilation Equipment (Specify) _____
15. Electrical: _____
    Service Amps: _____
16. Basement:
    Full _____ Partial (Give Sq. Footage) _____ No Basement _____
    Finished _____ Unfinished _____
    If partially finished, give description (i.e. stud in bath, wiring, etc.) _____
    _____
    _____
    _____
    _____
    _____

17. Lighting Fixtures:
    Allowance $ _____ Additional Information _____
18. Hardware:
    Allowance $ _____ Additional Information _____
19. Special Equipment: (Give model description where possible)
    Range _____ Oven _____ Vent Hood _____ Dishwasher _____

    Garbage Disposal _____ Central Vac _____ Inter Com _____ Compactor ___

    Other Equipment (Specify) _____
20. Miscellaneous: (Describe any main dwelling materials, or construction items not shown else-
    where, or use to provide additional information where the space provided was inadequate)
    _____
    _____
    _____
    _____
    _____
    _____

21. Porches, Decks, Patios, Terrances, Etc., (Give Description & Size)
    _____
    _____
    _____

22. Driveway:
    Width _____ Material _____ Approx. Length/Area _____
    Turn Around)approx. dimensions/area) _____
23. Landscaping:
    Lawn (circle one): seeded, sodded or sprigged
    Front yard _____ ft. _____ side yard _____ ft. _____ rear yard _____ ft. ____
    Additional Information _____
    _____
    _____

24. Other On Site Improvements: (Grading, Fences, Walls, Railings, Etc.)
    _____
    _____
    _____

Date _____ Signature _____
                                           (Owner)

                  Signature _____
                                           (Contractor)

Specification sheet
Figure 1-4 (continued)

First
Federal
Savings

and Loan Association

Telephone 721-7329

55 Forestdale • Southgate, Georgia 31093

ENERGY PACKAGE
REQUIREMENTS AND RECOMMENDATIONS

FOUNDATION

UNEXCAVATED OR CRAWL SPACE

Required

4" insulation or R-11 between floor joists.

Recommended

6 mil polyethylene vapor barrier on grade.

SLAB

Recommended

Concrete blocks with 1/2" rigid perimeter insulation between slab and blocks or monolithic slab installed per MPS of VA & FHA.

CRAWL/SLAB

Recommended

Fiberglass sealer or caulking between sills or plates and foundation walls or slabs.

WALL FRAMING

EXTERIOR WALLS

Required

4" insulation or R-11 between 2" x 4"'s. Install 6 mil polyethylene vapor barrier on warm side of walls. Caulk around all windows, pipes, and electrical outlets.

Foam approved type sheathing between exterior framing and finish siding or brick veneer. Minimum R-4 factor.

All the above should equal or exceed a minimum R-19 factor.

Note: On exterior framing consisting of 2" x 6"'s, 24" O.C., use 6" insulation with an R-19 factor. This will increase total R factor on a 6" wall to a minimum of R-27.

Energy requirements form
Figure 1-5

CEILINGS	**REQUIRED**
	6 mil polyethylene vapor barrier on warm side of ceilings. Insulation to have a minimum of R-30 factor. Batts or blown type or a combination of each is acceptable.
	Gable end vents to provide 1 sq. ft. of ventilation for each 300 sq. ft. of ceiling area.*
WINDOWS & EXTERIOR DOORS	**REQUIRED**
	Well weatherstripped units with double insulating glass or well weatherstripped standard wood windows with storm windows.
	Recommended
	Insulated steel entrance doors with double glazing and magnetic weatherstripping.
HEATING & COOLING SYSTEM	Recommended
	Simplified duct system with low inside registers and low central return.
	Heat pump or high efficiency gas heating and cooling system.
OTHER RECOMMENDATIONS	Heavy insulated water heater jacket.
	Hot and cold water pipes insulated to reduce heat loss and to control condensation.
	Energy saving dishwasher.
	Bathroom vent fan with effective damper, exhaust through second damper through exterior wall or roof.
	Special circulation type fireplaces that will use outdoor combustion air.
	Install roof overhang to shade south facing windows in summer.
	*Power roof vents will be accepted.

Energy requirements form
Figure 1-5 (continued)

PERCENTAGE TABLE

Item No.	Completed Construction	%	Item No.	Completed Construction	%
1	Clear Lot - Excavate	1	23	Siding-Brick/Panel	8
2	Footing - Foundation	6	24	A/C Compressor	3
3	Floor Framing, Sub-floor	3	25	Interior Paint/Wallpaper Complete	2
4	Walls	5	26	Electrical Fixtures, Switches, Wall Plugs	3
5	Sheathing	1	27	Interior Trim	1
6	Roof - Sheathing	5	28	Exterior Paint-Complete & Screens	2
7	Roof Shingles	2	29	Driveway, Stoop, Steps, Walk	4
8	Soil Treatment	1	30	Finished Floor: Hardwood, Carpet	3
9	Rough Wiring	3	31	Inlaid Tile	2
10	Rough Plumbing	4	32	Appliances	3
11	Heating & AC Ducts	2	33	Grading, Landscaping, Grass	2
12	Furnace & Coil	2	34	Garage Floor	1
13	Cornice-Carport Ceiling	1	35	Water, Sewer, Gas, Electrical Hook-ups	1
14	Outside Doors & Windows	4	36	Other (Identify)	4
15	Exterior Prime	1		TOTAL	100%
16	Insulation-Wall	2			
17	Sheetrock-Paneling	5			
18	Insulation-Ceiling	1			
19	Inside Doors	1			
20	Bath Tile	2			
21	Cabinets	5			
22	Plumbing Fixtures	4			

(Contractor)   (Date)   Owner-borrower

Owner-borrower

Disbursement schedule
Figure 1-6

Lenders appreciate repeat business just like any merchant. Once you're an established customer, the paperwork will flow much more smoothly.

Getting Started
Almost anyone can become a construction contractor. A lot of us began on a part-time basis with a few tools and a pickup truck. Your attitude is the key. Think quality. Plan your jobs carefully. Don't be satisfied with trial and error methods. Construction is too expensive and too permanent to leave in the hands of amateurs. Think of yourself as a reputable professional builder. If what you're doing isn't what a reputable professional builder would do, don't do it. Instead, do it right and take pride in what you've done. Solve problems before they become disasters.

Work hard. Watch your nickels and dimes. Keep your overhead down. Don't lay down a bundle for a fancy power saw. A modestly priced skilsaw will last through two houses and a couple of remodel-ing jobs. Put a carbide-tipped combination blade in a 7¼'' power saw and you can do it all, from framing to paneling and trim.

A hammer is different. It's the tool builders use the most. Buy a good one. Fiberglass handles last a long time. The 16 oz. hammer is probably the most popular. Most carpenters use the same hammer from the formwork to the roof. But some carpenters keep several types in their tool box.

In this chapter I've used a broad brush to outline the shape of the construction industry. I've included very little here that can help you make a better living in construction. But much of what follows assumes that you know what was in this chapter.

The next two chapters cover the most basic concepts in construction, plan reading and specifications. If you've been working in construction for several years, only a quick skim of Chapters 2 and 3 will be needed. But be sure you understand plan reading and specifications before going on to Chapter 4.

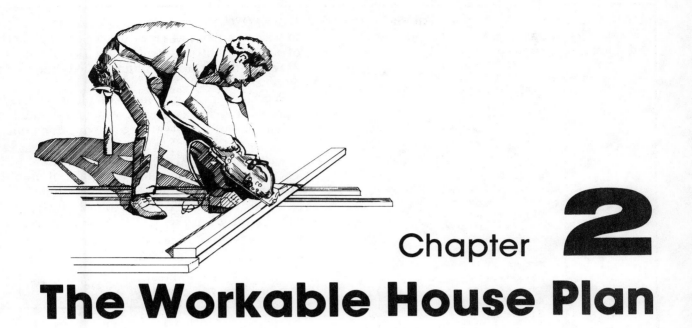

Chapter 2

The Workable House Plan

An experienced contractor can build from a simple pencil sketch on the back of an envelope if he has to. But every builder prefers to work from a complete, clear set of plans. Construction plans are sometimes called *blueprints* because they used to be dark blue. Today they may be blue on white or black on white, depending on the reproduction process. No matter what color, the plans or prints are reproductions of the architect's original working drawings. They're intended to show the size, shape and location of every part of the building under construction.

Whether you build from custom plans designed by an architect or from stock plans, there are common words and symbols that you'll need to know. In this chapter we'll cover all of the important blueprint concepts. We'll look at the different types of construction drawings, hatching and shading, lines and scale. And we'll study each sheet of the Nashville Plan that's bound into this book.

Construction Drawings

A construction drawing consists of lines, figures and notes. A construction plan illustrates and describes a building or some part of it. Let's look at the different types of construction drawings.

Types of Drawings

There are three common types of construction drawings: isometric drawings, perspective drawings, and working drawings.

Isometric drawings— In an isometric drawing the building is tilted up. The sides of the building are positioned at a 30-degree angle to a horizontal line in front of the building. Look at the rectangular block shown in Figure 2-1.

Isometric drawings can help clarify a complicated part of a drawing. Figure 2-2 shows a cutaway isometric drawing of the first floor of a

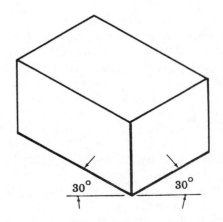

Isometric drawing of a block
Figure 2-1

house. The isometric lets you see the thickness of the walls, and the arrangement of rooms, windows and doors.

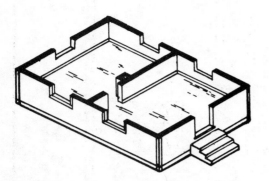

Isometric drawing of a first floor
Figure 2-2

Perspective drawings— A perspective drawing is similar to a photograph. It shows the general appearance of the building but not the exact dimensions. Architects use perspective drawings to give their clients an idea of what the building will look like when it's finished.

Figures 2-3 and 2-4 are examples of perspective drawings.

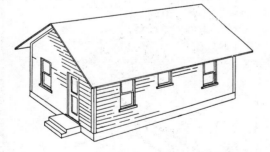

Perpective drawing of a house
Figure 2-3

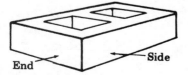

Perspective drawing of a concrete block
Figure 2-4

Working drawings— Working drawings are made to scale so anyone can measure the exact dimension of anything illustrated. Contractors use working drawings to put up a building exactly the way the designer intended.

Working drawings consist of plans, elevations, sections, and details. Here's what you should know about each of these working drawing components:

1) *Plans* show two dimensions: length and width. They show the outline and all details of an object as seen when looking directly down on it.

When the plan shows the top of an object, it's called a *top plan*. Figure 2-5 shows a top plan of a concrete block.

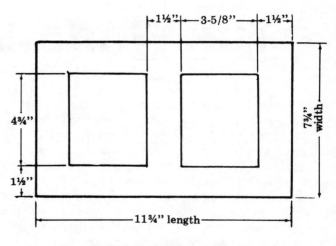

Top plan of concrete block
Figure 2-5

When the plan shows the bottom of an object, it's called a *bottom plan*.

Plans that look up at the ceiling of a room are called *ceiling plans*.

Plans that look down on the floor of a structure are called — you guessed it — *floor plans*. Figure 2-6A shows the floor plans for a two-bedroom cottage. (Compare this to the isometric drawing of the same cottage, Figure 2-6B.)

Plans can show flat, curved or slanting surfaces. Look at the elevations and sections to see which type of surface appears in your plans.

2) *Elevations* show height and length. Two common elevations are side elevations and end elevations.

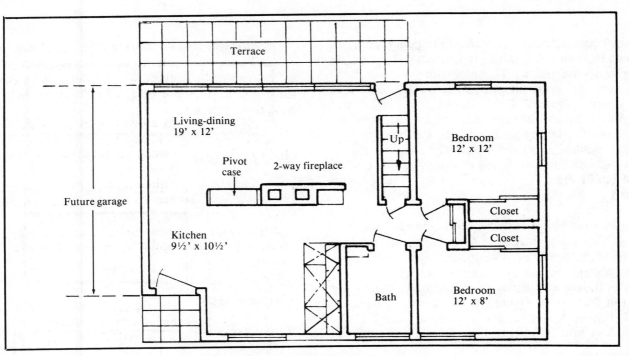

Floor plan for two-bedroom cottage
Figure 2-6A

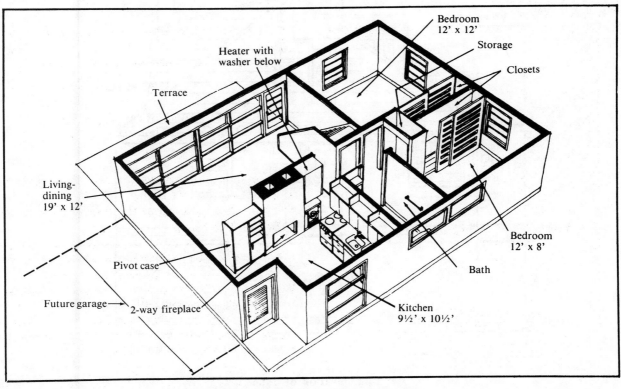

Isometric drawing of two-bedroom cottage
Figure 2-6B

Look again at the top plan of the concrete block shown in Figure 2-5. This view doesn't show us the height of the block. The elevations of the same concrete block are shown in Figure 2-7. The side elevation (Figure 2-7 A) shows the height and length of the block. The end elevation (Figure 2-7 B) shows the height and width of the block.

Use elevations to determine the height of windows, doors and porches. Elevations also show the roof pitch. Figure 2-8 shows the end elevation of a house.

3) *Sections* give a cutaway view of an object. The cut surfaces are marked with hatching or shading to show the inside of the object. Figure 2-9 shows an isometric section of a concrete block.

A double-dot and dash line normally marks the section line on the plan or elevation. In Figure 2-10, line A-A is a section line. In Figure 2-11 A, lines A-A and B-B are section lines.

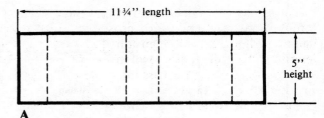

A
Side elevation

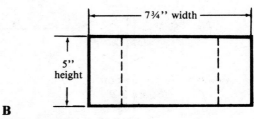

B
End elevation

Elevations of concrete block
Figure 2-7

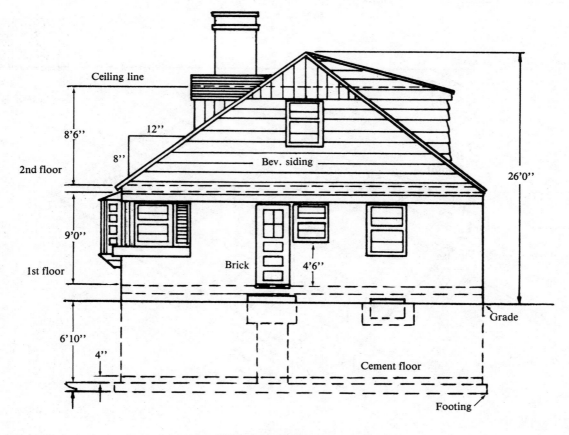

End elevation of house
Figure 2-8

There are two common types of sections: longitudinal sections and cross-sections. Look again at Figure 2-11. Longitudinal sections run the length of the object, as shown in Figure 2-11, B and C. Cross-sections cut across the width of an object. See Figure 2-11, D and E.

But sections are not always taken along a straight line. You can also take a section along a zig-zag line. See Figure 2-12. Figure 2-12 A shows the top plan of a square block with zig-zag section line B-B. Figure 2-12 B shows section B-B. An isometric drawing of the block is shown in Figure 2-12 C. And Figure 2-12 D shows the isometric section.

A third type of section, the main wall section, consists of both longitudinal sections and cross-

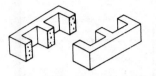

Isometric section of concrete block
Figure 2-9

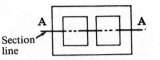

Section line on concrete block
Figure 2-10

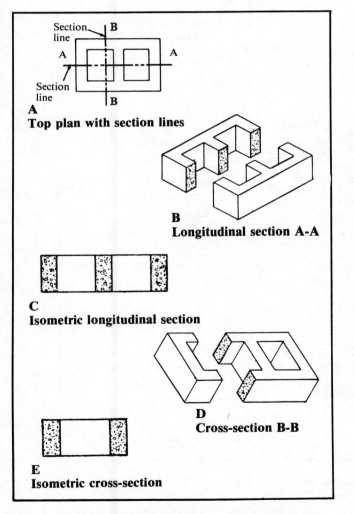

A
Top plan with section lines

B
Longitudinal section A-A

C
Isometric longitudinal section

D
Cross-section B-B

E
Isometric cross-section

Longitudinal and cross-sections of concrete block
Figure 2-11

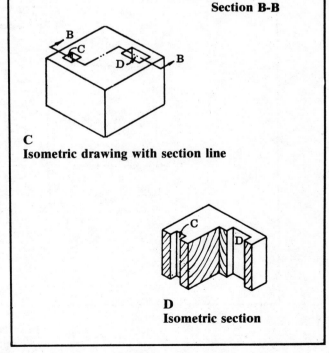

A
Top plan with section line

B
Section B-B

C
Isometric drawing with section line

D
Isometric section

Section along a zig-zag line
Figure 2-12

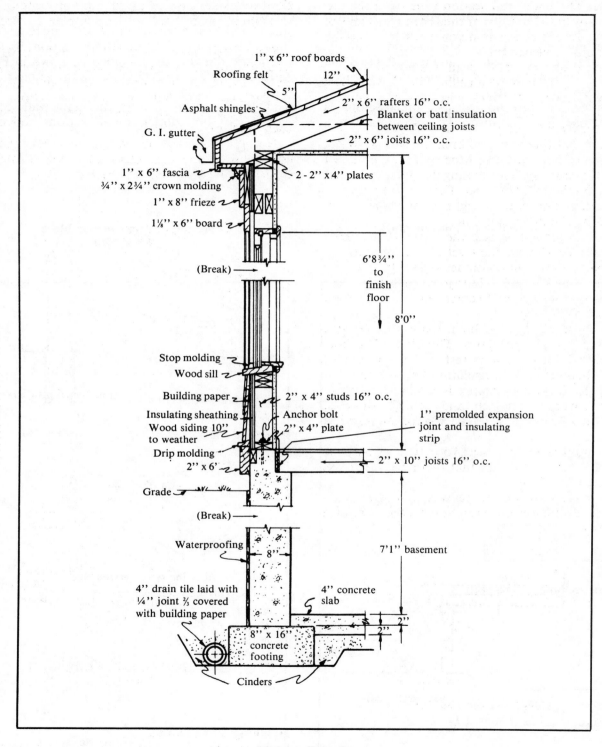

Main wall section
Figure 2-13

sections. The main wall section runs from the top of the roof to the bottom of the footing. Main wall sections provide details that you can't get from the floor plans or elevations.

Let's study the main wall section shown in Figure 2-13. The basement ceiling is 7'1" high. The basement floor is 4" thick with cinders below it. The foundation wall is 8" thick and rests on an 8" x 16" concrete footing.

Notice that there are two breaks in the drawing. This means that portions of the wall have been left out of the drawing. The missing portions are exactly like what is already shown. They're eliminated to save space. The entire height of the wall is indicated by the dimension line.

There's a grade line near the top of the foundation wall. This shows the level of the finished lot. The foundation wall continues on up to the 2" x 4" plate. The 2" x 10" floor joists are placed 16" on center (from center to center) and rest on a shelf or lip on the inside wall.

The house wall is 8'0" high. On top of the wall are two 2" x 4" top plates. The 2" x 6" ceiling joists and 2" x 6" rafters rest on the top plates. There's blanket or batt insulation between the ceiling joists. The roof decking is made of 1" x 6" boards covered with felt and asphalt shingles. There's a galvanized iron (G.I.) gutter attached to the eave.

The small triangle on the roof tells us that the pitch is 5/12. This means that there are 5" of rise for each 12" of run.

The main wall section reveals important construction details for both rough framing and finish carpentry.

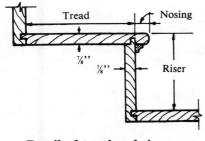

Detail of tread and riser
Figure 2-14

4) *Details* show small portions of a building so you can see the parts clearly. Details of cabinets, fireplaces, stairs and entrances are usually shown in

elevation (side) view. If this doesn't give enough information, there may also be a plan or section view.

Figure 2-14 shows a detail (in section) of the tread and riser of a stair. The detail drawing gives a clearer picture of how to build the stairs.

Compare the plans, elevations, sections and details. The dimensions should be consistent on each sheet of your plans.

Now that we've explored the different types of construction drawings, let's take a look at some of the symbols you'll find on these plans.

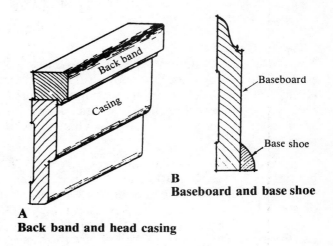

A
Back band and head casing

B
Baseboard and base shoe

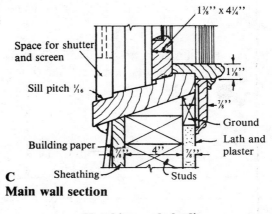

C
Main wall section

Hatching and shading
Figure 2-15

Hatching and Shading
Look at Figure 2-15. Notice that different hatching lines and shading are used to mark different components. For example, Figure 2-15 A shows an isometric section of a back band and head casing

for a door. The hatching lines on the back band are different from those on the casing.

Figure 2-15 B shows a cross-section of a baseboard with a base shoe. The hatching lines on the baseboard are different from those on the base shoe.

Figure 2-15 C shows a portion of a main wall section. Notice again that the varied hatching lines and shading make it easy to see each wall component.

Hatching and shading also indicate the use of different materials in different parts of the building. In Figure 2-15 C, notice that the shading used for lath and plaster is different from the hatching used for sheathing.

Blueprint Lines

There are eight common types of lines shown on construction drawings: the note reference line, outline, dotted line, centerline, section line, dimension line, extension line and break line. These lines are shown in Figure 2-16. Here's what each line means:

Note reference line— This line goes from an explanatory note to the part of the building mentioned in the note. The line can be short, long, straight or curved.

Outline (boundary line)— Outlines are thick, solid lines that mark the visible edges of a building or parts of a building.

Dotted line— When part of a building is marked with a dotted line, it means you can't see that part of the building. A dotted line often marks the base-

ment portion of a house shown in elevation. Notice the dotted line in Figure 2-8.

Centerline— This is a dot-and-dash line drawn through the center of a symmetrical figure, such as a door or window.

Section line— This is a double-dot and dash line drawn on a plan or elevation. It marks where a section is taken.

If there are only a few sections on a drawing, arrows and double letters (such as A-A, B-B) are used to mark the section lines.

When there are several sections on the same drawing, they'll be marked with circles, as shown in Figure 2-17. The top half of the circle identifies the section letter or section number. The bottom half of the circle shows the plan sheet number.

Sections are sometimes taken through parts of the building without showing any section lines. These are called "typical" sections.

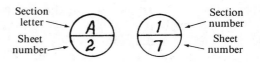

Alternate section markers
Figure 2-17

Dimension line— This line shows a point-to-point measurement for each dimension of the building. There are figures above each dimension line to show the number of feet and inches between the two points.

Extension line— When there isn't room to draw the dimension line along the actual dimension being measured, an extension line will extend the two points out to a space where a dimension line can be drawn.

Look again at Figure 2-8. The 26'0'' dimension line is drawn between two points that have been extended by extension lines.

Break line— When you see a break line, it means that a portion of the construction has been omitted from the drawing. The omitted portion is exactly the same as the construction shown on each side of the break line.

Break lines save space and allow us to leave out

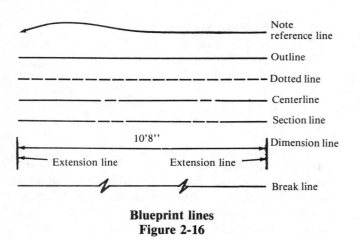

Blueprint lines
Figure 2-16

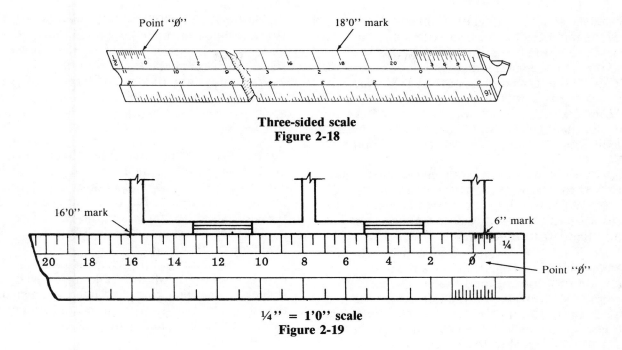

Three-sided scale
Figure 2-18

¼" = 1'0" scale
Figure 2-19

nonessential details. We'll also use them to isolate section details, as shown in Figure 2-15 C.

Scale

Drawings are either full-size or "drawn to scale." When the lines shown on the drawing are the same size as the actual dimensions of the object, then the drawing is full-size. For example, if you draw a hammer, and the lines in your drawing are the same as the actual dimensions of the hammer, then you've made a full-size drawing.

But suppose you need to draw a sledge hammer and you have only a small scrap of paper to draw it on. You'll need to draw the sledge hammer to fit the small piece of paper. You draw it one-half or one-quarter of the actual size, but with every part in exact proportion to the original. In your drawing the relationship between the size of the head and the size of the handle must be the same as in the actual hammer. If you hold all parts in proportion, the drawing is *to scale*. As a contractor, your working drawings will be drawn to scale. A wall that is actually 10 *feet* long may be shown on your drawings as a line only 2½ *inches* long.

House plans and elevations are normally drawn to a 1/4" scale. This means that each 1/4" on the drawing is equal to 1'0" actual size in the finished building. Quarter inch scale is written like this: 1/4" = 1'0".

Sections and detail drawings often use a larger scale, such as 3/8" = 1'0" or 3/4" = 1'0". Look for the scale at the bottom of each sheet of your blueprints.

The amount of information on your working drawings is limited by the size of the paper. There isn't always room to squeeze in all of the dimensions required to build a house. When there isn't enough room to squeeze in some of the dimensions, you have to figure them out yourself. You do this by *scaling* off the dimension.

To do this you need plans that are drawn to scale and an architect's scale. Use the three-sided scale like the one shown in Figure 2-18. It should have all the common scales: 1" = 1'0", 3/4" = 1'0", 1/2" = 1'0" and 1/4" = 1'0".

Place the three-sided scale along the line you want to measure. (We'll use the 1/2" = 1'0" scale shown in Figure 2-18.) Begin measuring your blueprint line at point "0" near the left-hand end of the scale. Now read across the scale from left to right. Each line marker indicates 1'0" of actual wall length.

Suppose your blueprint line extends a fraction of a foot beyond the 18'0" mark. How do you determine the exact length of the line? Look again at the far left end of the scale. Notice that there's a portion of the scale where a foot has been broken down into inches.

Place the 18'0'' mark on the right end of the line. Read the number of inches between point "0" and the left end of the blueprint line. Add this to 18'0'' and you have the total scaled length of your wall.

If a wall is 6' long and a doorway is 3' wide, this relationship of 2 to 1 will also be same on your working drawings. If the wall is 6'' long on your drawing, then the doorway must be 3''. If the wall is shown as 1'' long, then the doorway must be 1/2''.

Let's try scaling another wall length from a plan. The plans for this wall are drawn to a scale of 1/4'' = 1'0''. We'll use the scale shown in Figure 2-19. Notice that you'll read this scale from *right to left*, rather than left to right. Here's how:

Place point "0" on the right end of the line. Reading from right to left, let's say the line extends a fraction of a foot beyond the 16'0'' mark. Now place the 16'0'' marker on the left end of the line. Look again at the far right end of the scale. The inch portion of the scale shows that the fractional part is 6''. The total wall length is 16'6''.

Now you know how to figure the unknown dimensions on your prints. But be careful when scaling dimensions. If the blueprints aren't drawn to scale accurately, your calculations won't be accurate. And perfectly scaled blueprints are hard to find.

Nashville Plan

Figures 2-20A through 2-20H are the Nashville Plan. The plan consists of a foundation plan, floor plan, front and kitchen elevations, rear and bedroom elevations, sections, and details. Notice that the plan has been reduced to one-half the actual size so that it would fit in this book.

The plan includes standard construction details, heating and air conditioning layout, and list of materials.

The house size and design in the Nashville Plan appeal to a broad market. It offers plenty of space for most families. It appeals to young families who might use one of the bedrooms for a study. It's also suitable for elderly couples who need a smaller home with an extra room for weekend guests.

The house is a rectangle with a straight roof. It has separate living and dining areas. The family room and kitchen are on the back of the house and have easy access to the basement stairs, bathroom, and bedrooms. It's a practical and compact design with good use of space.

Now let's apply what we've learned. We've seen the different types of construction drawings. We've looked at hatching and shading, lines and scale. Let's use our knowledge to interpret each sheet of the Nashville Plan.

We'll also take a look at a standard material description form and contract agreement form that are often used with stock plans like the Nashville Plan.

Foundation and Basement Plan (Sheet 1)
Sheet 1 shows the foundation and basement plan. The roof plan and basement stair detail also happen to be on Sheet 1. But they don't have to be on the same sheet as the foundation plan. They're included here because there's some extra space available on this page.

The foundation and basement plan shows the length and width of the building as well as the shape. Notice the two dotted lines on each side of the foundation walls and around the chimney and stoops. These lines mark the position and width of the footings.

The plan calls for an 8/20 concrete footing under 12'' foundation walls, and an 8/16 footing under 8'' foundation walls. This means that the footing must be 8'' thick and 20'' wide under foundation walls that are 12'' thick. The footing must be 8'' thick and 16'' wide under foundation walls that are 8'' thick.

There are five dotted squares along the horizontal center of the basement. These mark the location of the column footings. The column footing size isn't shown on the plan because the footing has to meet local building code requirements. In the center of each square there's a circle, representing a 3'' round steel pipe column.

The basement floor is 4'' concrete over 4'' of sand or gravel. The concrete is reinforced with 6'' x 6'' No. 10 wire mesh. Note that the floor slopes to the floor drain.

Windows are steel basement windows with two sliding lites. Windows measure 2'8'' x 1'10'' and each lite is 15'' x 20''.

Floor joists are 2 x 10's spaced 16'' o.c. The notation *(over)* means that the joists are overhead when you're standing in the basement. Arrows show the direction of the joist run or span. Double joists (D.J.) are shown under partition walls.

Notice the location of access doors in the plan. And there's a 4'' reinforced concrete slab (over) required for stoop floors. Again, "over" means that when you're standing in the basement, the stoop floor will be overhead.

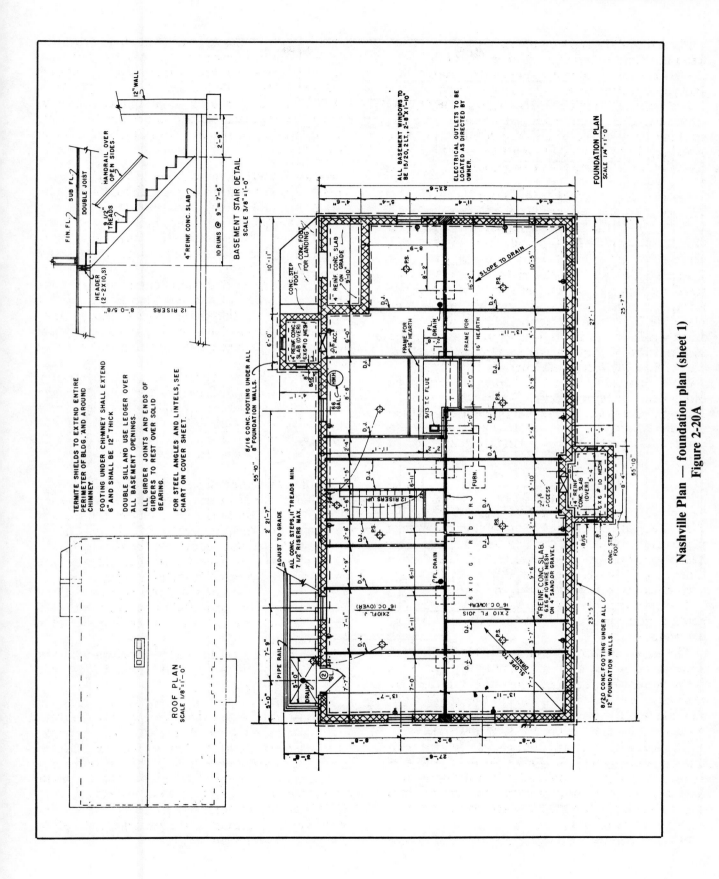

Nashville Plan — foundation plan (sheet 1)
Figure 2-20A

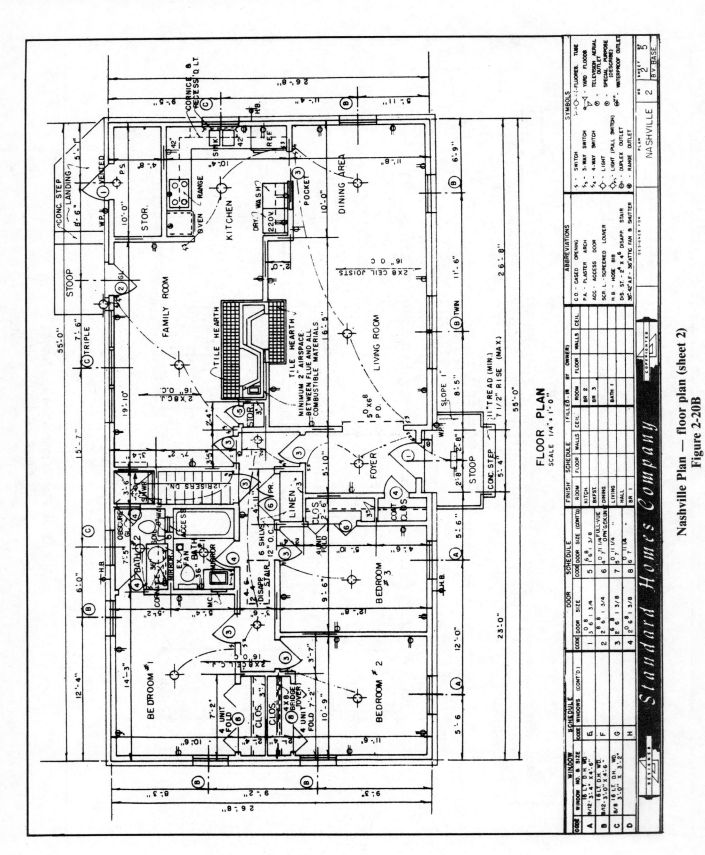

FLOOR PLAN
SCALE 1/4" = 1'-0"

Nashville Plan — floor plan (sheet 2)
Figure 2-20B

KITCH. CAB. ELEVATIONS

SCALE 3/8"=1'-0"
FOR SECTION THRU CABINETS
SEE DETAIL #21 ON COVER
SHEET.

Nashville Plan — floor plan (sheet 2)
Figure 2-20B (continued)

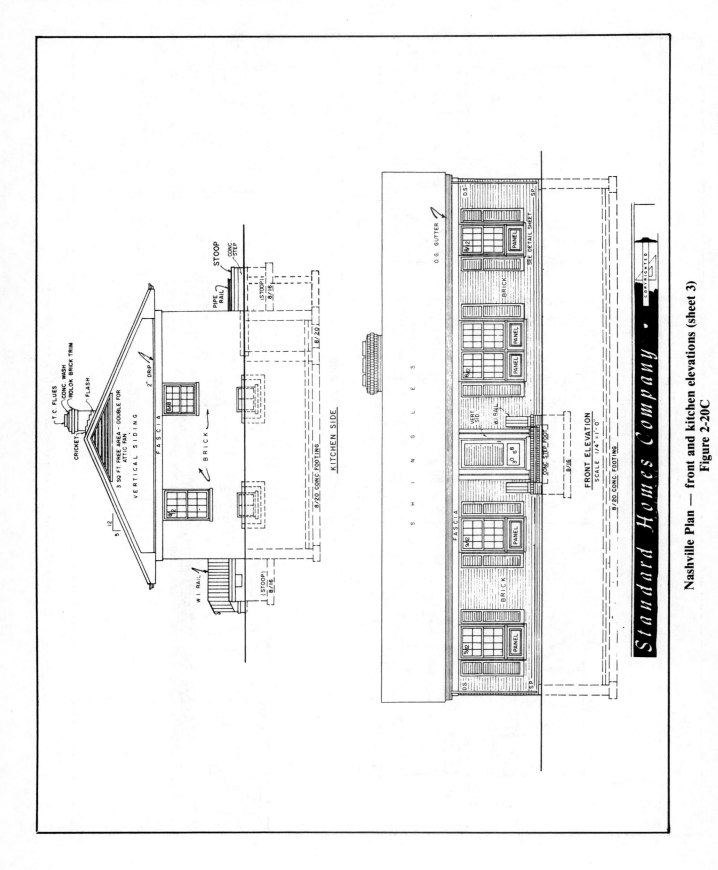

Nashville Plan — front and kitchen elevations (sheet 3)
Figure 2-20C

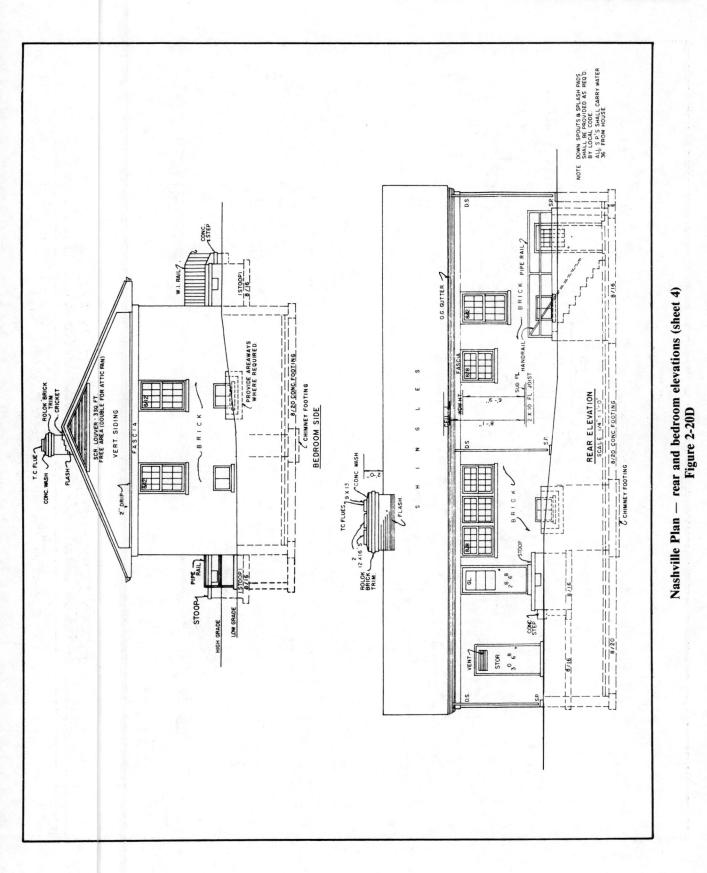

Nashville Plan — rear and bedroom elevations (sheet 4)
Figure 2-20D

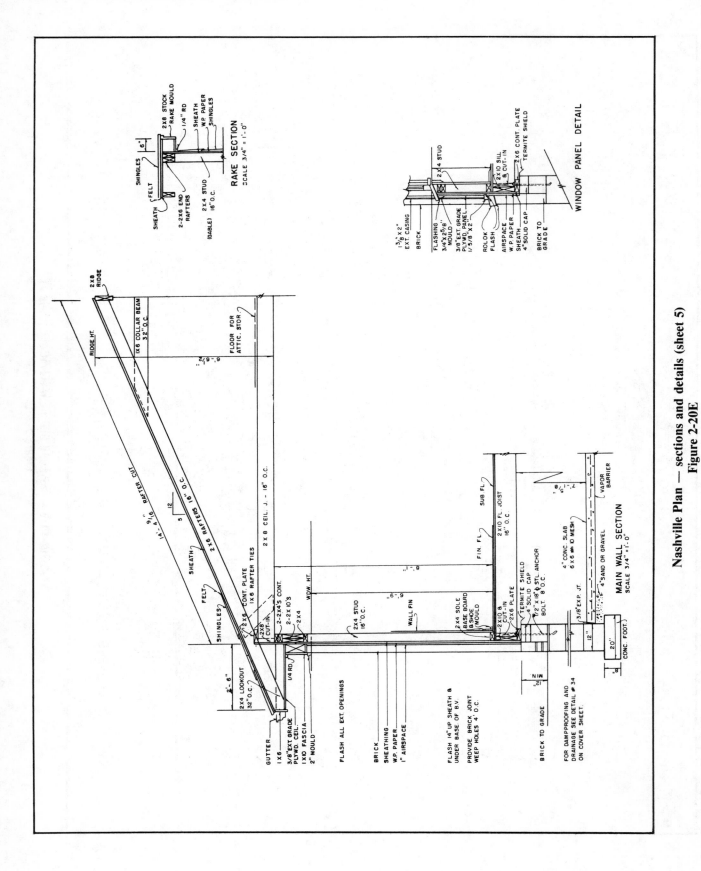

Nashville Plan — sections and details (sheet 5)
Figure 2-20E

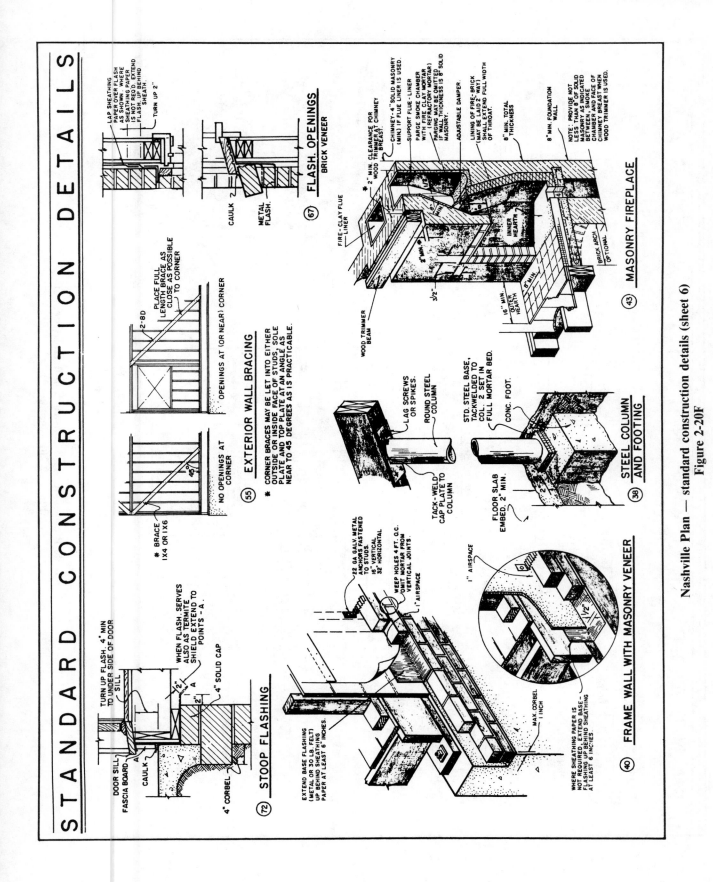

STANDARD CONSTRUCTION DETAILS

① FLASH. OPENINGS
BRICK VENEER

LAP SHEATHING PAPER OVER FLASH AS SHOWN WHERE SHEATHING PAPER IS NOT REQD. EXTEND FLASH. UP BEHIND SHEATH.
TURN UP 2"
CAULK
METAL FLASH.

⑤⑤ EXTERIOR WALL BRACING
PLACE FULL LENGTH BRACE AS CLOSE AS POSSIBLE TO CORNER
2-8D
OPENINGS AT (OR NEAR) CORNER
45°
NO OPENINGS AT CORNER
* BRACE 1X4 OR 1X6
* CORNER BRACES MAY BE LET INTO EITHER OUTSIDE OR INSIDE FACE OF STUDS, SOLE PLATE AND TOP PLATE AT AN ANGLE AS NEAR TO 45 DEGREES AS IS PRACTICABLE.

⑦② STOOP FLASHING
TURN UP FLASH. 4" MIN TO UNDER SIDE OF DOOR SILL
DOOR SILL
FASCIA BOARD
CAULK
WHEN FLASH. SERVES ALSO AS TERMITE SHIELD EXTEND TO POINTS - A.
4" SOLID CAP
2"
2"
4" CORBEL

④③ MASONRY FIREPLACE

2" MIN. CLEARANCE FOR WOOD TRIMMER AT CHIMNEY BREAST.
CHIMNEY - 4" SOLID MASONRY (MIN) IF FLUE LINER IS USED.
SUPPORT FLUE - LINER
PARGE SMOKE CHAMBER WITH FIRE CLAY MORTAR (REFRACTORY MORTAR) PARGING MAY BE OMITTED IF WALL THICKNESS IS 8" SOLID MASONRY.
ADJUSTABLE DAMPER.
LINING OF FIRE-BRICK (MAY BE LAID 2" WAY) SHALL EXTEND FULL WIDTH OF THROAT.
8" MIN. TOTAL THICKNESS
8" MIN. FOUNDATION WALL
NOTE: PROVIDE NOT LESS THAN 8" OF SOLID MASONRY AS INDICATED BETWEEN, SMOKE CHAMBER AND FACE OF CHIMNEY BREAST WHEN WOOD TRIMMER IS USED.
FIRE-CLAY FLUE LINER
2" MIN
INNER HEARTH
BRICK ARCH OPTIONAL
3/2
8" MIN.
16" OUTER HEARTH
OUTER HEARTH
WOOD TRIMMER BEAM

LAG SCREWS OR SPIKES.
ROUND STEEL COLUMN
TACK-WELD CAP PLATE TO COLUMN
STD. STEEL BASE, TACKWELDED TO COL. 2 SET IN FULL MORTAR BED.
CONC. FOOT.
FLOOR SLAB EMBED. 2" MIN.

③⑧ STEEL COLUMN AND FOOTING

④⓪ FRAME WALL WITH MASONRY VENEER
22 GA GALV. METAL ANCHORS FASTENED TO STUDS.
WEEP HOLES 4 FT. O.C. OMIT MORTAR FROM VERTICAL JOINTS.
16" VERTICAL 32" HORIZONTAL
1" AIRSPACE
1" AIRSPACE
1" AIRSPACE
MAX. CORBEL 1 INCH
1/2
WHERE SHEATHING PAPER IS NOT REQUIRED, EXTEND BASE-FLASHING UP BEHIND SHEATHING AT LEAST 6 INCHES.
EXTEND BASE FLASHING (METAL OR 30 LB. FELT) UP BEHIND SHEATHING PAPER AT LEAST 6 INCHES.

Nashville Plan — standard construction details (sheet 6)
Figure 2-20F

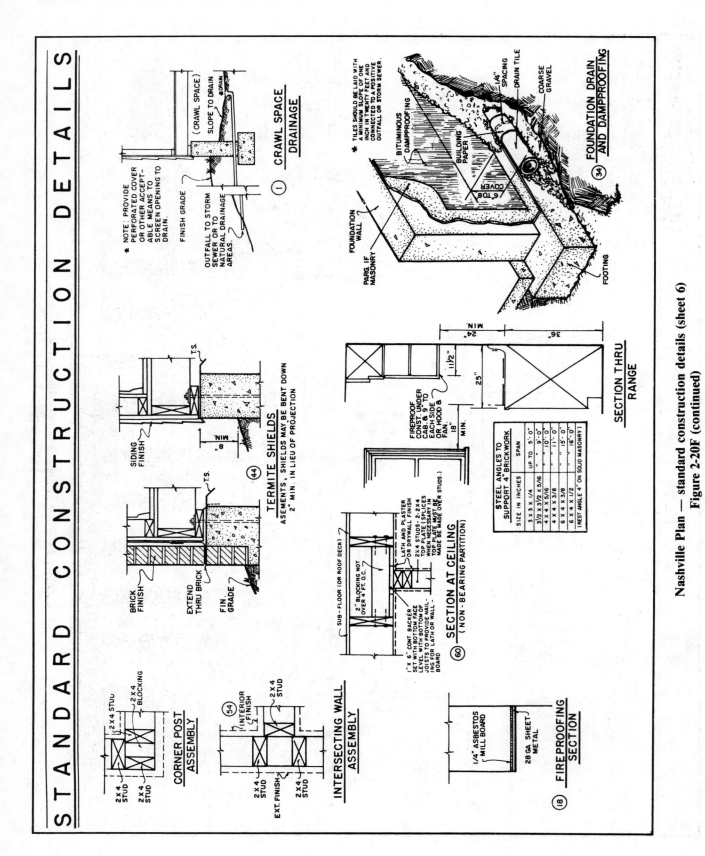

Nashville Plan — standard construction details (sheet 6)
Figure 2-20F (continued)

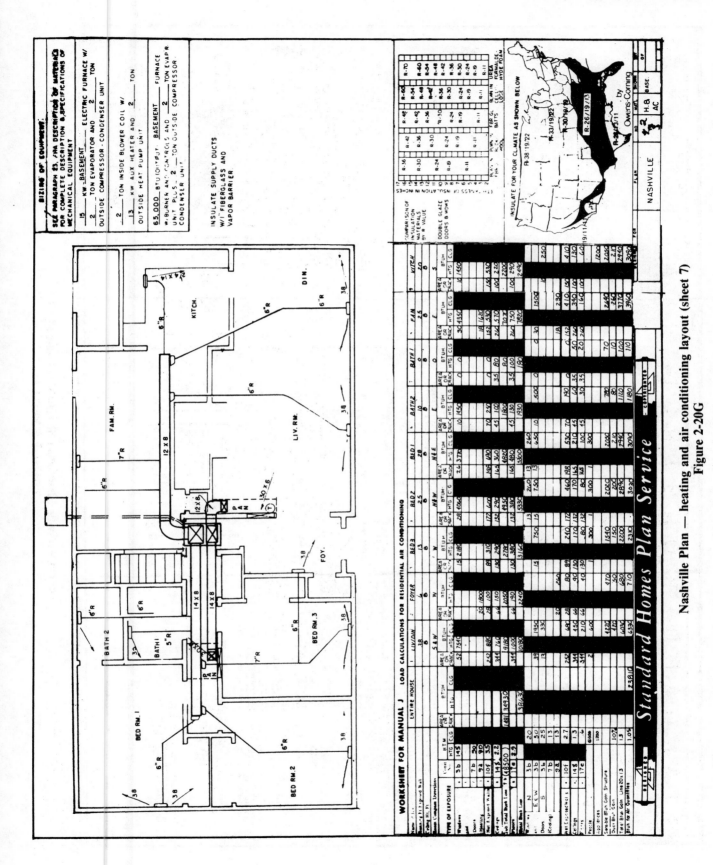

Nashville Plan — heating and air conditioning layout (sheet 7)
Figure 2-20G

NASHVILLE	PLAN # 2	WITH BASEMENT	B-V	SECTION 1
Concrete footings	10	yds	1-3-5 mix	
Stoop slabs, etc	2	"	1-2-4 mix	
Basement slab, steps	17½	"	1-2-4 mix	
Under-slab fill	22	tons	gravel	
Foundation walls	650		12x8x16 concrete blk	
(For full basement	875		8x8x16 " "	
height & 50% below grade)	150		8x4x16 solid cap blk	
Veneer	14,000		face brick	
	152	sacks	mortar mix	
	18	yds	sand	
Splash pads	5		12x4x30 concrete	
Chimney & fireplaces	1½	yds	1-3-5 mix concrete	
	4,000		common brick	
	2,000		face brick	
	150		fire brick	
	20	sq ft	4"x4" hearth tile	
	36	sacks	mortar mix	
	2	"	fire clay	
	5	yds	sand	
	28	1/f	12x16 T C flue liner	
	20	"	9x13 " " " "	
	1		T C thimble	
	2		C I dampers	
Steel angle lintels	2	pcs	3x3x½x4'0"	
Flashing	34	1/f	12" 16-oz copper	
Rough hardware	1700	sq ft	6x6 #10 wire mesh	
	2000		metal wall ties	
	34		½"x18" anchor bolts	
Foundation vents	3		C I screened 8x16	
	3		20" vent bars	
Steel angle lintels	2	pcs	3x3x¼x4'0"	
	11	"	" " "x3'8"	
	1	"	" " "x2'8"	
Steel pipe columns	5	"	3"x8' w/cap & base	
Steel pipe rail	26	1/f	2"	
	12	"	W I rail	
Termite shield	214	"	26 Ga G I	
Gutter	116	"	5" O G G I	
Right ends	2		5" " " " "	
Left ends	2		5" " " " "	
Downspout	70	1/f	4" rect G I	
Elbows	10		4" " " "	
Shoes	5		4" " " "	
Gutter hangers	40		adjustable	
Leader straps	18		G I	
Nails	250#		16d common	
	250#		8d "	
	50#		1" roofing	
	40#		8d cut flooring	
	40#		1-1/8" blued lath	
	100#		assorted	
Roofing, felt, etc	22	sqs	235# T B asphalt shingles	
	2	rls	30# felt	
	13	"	15# felt	
Expansion jnt filler	200	1/f	½"x4" asphalt	
Vapor barrier	1500	sq ft	.004 polyethlene	
Ceiling insulation	1400	" "	4" rockwool batts	
Flashing @ openings	280	" "	3-oz copper-coated paper	
Framing Lumber			Size Length Board Ft	
Sills & girders	6	pcs	2x10 16' 160	
	12	"	" 14 280	
	18	"	" 12 360	
	2	"	2x6 16 32	
	6	"	" 14 84	
	6	"	" 12 72	
Floor joists	88	"	2x10 14 2053	
	4	"	" 12 80	
	18	"	" 10 300	
Ceiling joists, etc	26	"	2x8 16 555	
	50	"	" 14 933	
	36	"	" 12 576	
Rafters, etc	90	"	2x6 18 1620	
	10	"	" 12 120	
Ridge rafter	2	"	2x8 16 43	
	2	"	" 14 37	
Bathwall studs, etc	6	"	2x8 8 64	
Headers	16	"	2x10 12 320	

NASHVILLE	PLAN # 2	WITH BASEMENT	B-V	SECTION 2
Framing Lumber			Size Length Board Ft	
Studs, etc	480	pcs	2x4 8' 2560	
Plates, etc	1500	1/f	2x4 1000	
Ledgers	120	"	2x2 40	
Bridging	550	"	1x4 183	
Roof & subflr shtg	4400	b/f	1x6-8 S4S	
Sidewall "	1800	"	1x6-8 S4S	
Vertical siding	190	"	12" V-joint	
Finish flooring	1100	"	#1 com oak 25/32x2¼	
Under flooring @ kit & family room	400	"	#2 pine T & G	
	470	1/f	4" moulded baseboard	
	470	"	oak shoe mould	
Closet shelving	100	"	1x12	
Hookstrip	16	"	1x4	
Shelf cleat	30	"	1x2	
Closet rod	28	"	1½"	
Disappearing stair	1		2-4x4-6	
Kitchen cabinets	14-1/3	1/f	base cabinets	
	18	"	wall cabinets	
Plaster (See alternate material below for dry wall finish)	1480	1/f	7/8" plaster ground	
	4352	sq ft	3/8" gypsum lath	
	240	" "	metal lath	
	1120	1/f	corner rite	
	110	"	corner bead	
	44	sacks	plaster	
	8	"	gauging plaster	
	30	"	finish lime	
	5	yds	sand	
Dry wall finish	4480	sq ft	3/8" sheetrock	
	9	pkgs	250' perf-a-tape	
Stairs	3	pcs	2x12 14'	
	2	"	2x2 14'	
	6	"	" 10'	
	11	"	10" tread 36"	
	13	"	8" riser 36"	
Outside door frames	2		3-0x6-8x5/4	
	2		2-8 " " "	
Inside door jambs	2		6-0x8-0x5/4	
	2		4-0 " " "	
	6		2-6x6-8x5/4	
	3		2-0 " " "	
	1		1-6 " " "	
Pocket door jamb	1		2-6x6-8	
Door trim	520	1/f	moulded casing	
	240	"	1-3/8" stop	
Front door, flush	1		3-0x6-8x1-3/4....3 lite	
Rear doors, flush	2		2-8 " " " " half glass	
Storage door, flush	1		3-0 " " " " exterior, vented	
Inside doors, flush	7		2-6x6-8x1-3/8	
	3		2-0 " " " "	
	1		1-6 " " " "	
Bi-fold doors	2	units	6-0x8-0 w/trim & hdw	
	1	"	4-0 " " " " " "	
Double hung doors	1	pr	4-0 " " " " " "	
Shower door	1		2-0x6-0 w/obscure glass, complete	
Door hardware	6	prs	4x4 loose-pin butts	
	10	"	3½x3½ " " "	
	1		front door lock	
	2		rear " "	
	1		storage " "	
	2		bathroom door locks	
	3		bed " " "	
	5		passage sets	
Pocket track & hdw	1	set	2-6	
	14		metal base knobs	
Front screen door	1		3-0x6-8x5/4 louvered	
Rear " "	1		2-8 " " "	
	2	sets	screen door hardware	

NASHVILLE	PLAN # 2	WITH BASEMENT	B-V	SECTION 3
D-H wood windows pre-fitted with stops, balances, weatherstrip, hardware & screens	1		triple unit 3-0x3-2....16 lite	
	1		twin " 3-0x4-6 16 "	
	2		single " 3-4x4-6 16 "	
	5		" " 3-0x4-6 16 "	
	2		" " 3-0x3-2 16 "	
Window trim	170	1/f	moulded casing	
	12	"	" mullion casing	
	50	"	" apron	
	50	"	" stool	
Ornamental shutters	4	prs	1-6x6-8 w/hardware	
Panels below windows	64	sq ft	3/8" exterior plywd	
	20	1/f	1-5/8"x2" sill stock	
	60	"	3/4"x2-5/8" O G mould	
	26	"	exterior casing to match above units	
Steel basement windows--complete	10		units 2-8x1-10....2 lite	
Eaves & rakes	32	sq ft	3/4" exterior plywd	
	288	" "	3/8" " "	
	180	1/f	1x10 S4S facia	
	170	"	1x6 S4S facia, etc	
	80	"	2x8 rake mould	
	60	"	2" drip mould	
	190	"	3/4" quarter round mould	
Screened louvers	2		triangular, 10'0" base, 5/12 pitch	
CONVERSION TABLE: BRICK VENEER & WOOD SIDING TO WOOD SIDING ONLY				
OMIT:	1	yd	1-3-5 mix concrete	
	625		12x8x16 concrete blk	
	14,000		face brick	
	80	sacks	mortar mix	
	9	yds	sand	
	2000		metal wall ties	
Steel angle lintels	1	pc	3x3x¼x4'0"	
	10	"	" " "x3'8"	
	2	rls	30# felt	
Vertical siding	190	b/f	12" V-joint	
ADD:	725		8x8x16 concrete blk	
	10		8x4x16 solid cap blk	
	1550	b/f	8" drop siding	
	6	sets	OS corner boards 8'	
	2		IS " " 8'	
	30#		8d casing nails	

MATERIAL LIST, NASHVILLE PLAN #2, WITH BASEMENT BRICK VENEER WITH CONVERSION TABLE FOR WOOD SIDING EXTERIOR WALLS

Nashville Plan — list of materials (sheet 8)
Figure 2-20H

The floor of the storage room is concrete slab on grade.

The plan shows that the stairs will have 12 risers coming up from the basement. There's a 3-way wall switch at the head (top) and foot (bottom) of the stairs. The switch controls a light located in the basement. The plan shows the location of the overhead basement lights and tells us the lights are P.S. (pull-switch). The owner will decide where to put the basement electrical outlets.

There are dotted lines showing the location of the furnace and pipe runs. The flue pipe joins a 9" x 13" terra-cotta-lined flue in the chimney.

The foundation plan also tells us there's a 66-gallon hot water heater in the basement. The height of the basement (from basement floor to finish floor overhead) is 8'⅝".

We've learned some important details from the foundation plan. Now let's take a look at the floor plan, Sheet 2.

Floor Plan (Sheet 2)

The view shown on Sheet 2 is a first-floor plan. It looks directly down on the layout of the first floor, including all doors and windows.

It gives us all the information we need to know about the location and size of wall openings, the direction doors open, room sizes, wall thicknesses, location of lights and switches and location of wall receptacles. In the kitchen, we see the placement of cabinets and appliances.

The plan shows the placement of bathroom fixtures. Study this part of the plan carefully. There's a lot of information packed into a small area.

Notice the 8" wall between the two baths. Did you find the plumbing access panel and soil stack location? Bath 1 shows an M.C. (medicine cabinet) with a mirror. Bath 2 shows a mirror over the lavatory. Next to the bathrooms are the basement stairs. The floor plan shows 12 risers going *down*. (The foundation plan showed 12 risers going *up*).

Look at the closet in bedroom 2 and the notation *4 x 8 bridge (over)*. This is a header which supports the ceiling joists and is built in line with the hall wall. Some builders would prefer to change the size of the closets slightly so a load-bearing wall would fit between the closets. A load-bearing wall would require 4" studs instead of the 3" partition shown on the plan.

The plan specifies 2 x 8 ceiling joists spaced 16" o.c. An arrow indicates the direction of the joist span or run.

The opening between the living room and foyer is a *5'0" x 6'8" P.O.* (plaster opening).

The floor plan shows all the measurements you'll need for laying out and framing the walls and partitions. Interior walls (parallel to the floor joists) must fall directly over the double joists shown on the foundation plan.

Front and Kitchen Elevations (Sheet 3)

Elevations show lines and proportions on vertical surfaces. This is important information for every spec builder. "Curb appeal" makes it easier to sell a home. A prospective buyer's first impression will be based on the front elevation.

The dotted lines show what's below ground level. We see the size of the concrete footing and can scale off the thickness of basement walls. Notice that areaways will be required for the basement windows.

The front elevation shows window panels, brick exterior finish and shingles on the roof. Brick extends to the top of the windows and is finished with a fascia board. Vertical siding is used to finish the entrance at the front door.

The figure *9/12* on a window tells us that the lite is 9" x 12". The house has open gutters (O.G.) and a downspout (D.S.) at each corner. There's a splash pad (S.P.) under each downspout.

The kitchen elevation shows the roof pitch (5/12). The gable-end finish is vertical siding. A 2" drip cap is required. If an attic fan is used, 3 SF of free area ventilation won't be enough. Double it. Chimney details are also shown in the kitchen elevation.

The elevations show a wrought iron (W.I.) rail for the front stairs and a pipe rail for the outside basement stairs. Notice the foundation vent in the front stoop.

Rear and Bedroom Elevations (Sheet 4)

These elevations show us the basement entrance, floor level, ceiling height and window height. The dotted lines show what's below ground level.

We can also see chimney flue sizes, chimney height above the roof, and exterior door sizes.

Sections and Details (Sheet 5)

The main wall section shows us the inside of the wall, from footing to ridge. It shows the details of a header that's commonly used for window and door openings in a load-bearing wall.

Studs are spaced 16'' o.c. Studs rest on a 2 x 4
sole plate and have a double 2 x 4 top plate. The
top of the window is 6'9'' above the finish floor.
The basement ceiling height is 7'1⅛''. The main-
floor ceiling height is 8'1''.

There's a 2 x 10 header joist, a 2 x 10 floor joist
and cut-in blocks between the floor joists. Ceiling
joists are 2 x 8's spaced 16'' o.c. They rest on the
top plate and have 2 x 8 cut-ins.

The rafters rest on a continuous 2 x 6 plate *on
top of the ceiling joists.* In most plans, the rafters
are sawed to fit directly on top of the top plate.

This plan allows for a horizontal soffit and a
2'6'' overhang at the eaves. The rafter cut is
14'4⁹⁄₁₆''. The roof pitch is 5/12. The distance from
the bottom of the ceiling joist to the top of the
ridge is 6'6½''.

Notice that the plan calls for 1 x 6 rafter ties at
each rafter to tie the roof system to the wall. The
rafter tie-in at the wall has two advantages. First, it
allows room for thicker ceiling insulation without
blocking air flow between the rafters. Second, it
provides more storage space and greater headroom
in the attic.

The window panel detail drawing shows construction details for the panels under the front windows. The 2 x 10 sill is a joist header.

The rake section shows construction details for
the roof at the gable end. The framing of the gable
must allow for the difference in thickness between
the brick veneer and the gable finish. The Nashville
Plan doesn't show this detail. But one method is to
use 2 x 6's (instead of 2 x 4's) for the gable studs.
(See Figure 2-21.)

Standard Construction Details (Sheet 6)
Stock plans usually have a sheet of standard construction details. Look at Figure 2-20F. This sheet
shows a series of detail drawings that apply to most
houses. But don't assume that all of the drawings
on the sheet will apply to the house *you're*
building. Some don't apply to the Nashville Plan at
all. For example, Detail 1 shows a procedure for
providing crawl space drainage. In the Nashville
Plan, no crawl space drainage is needed because
there is no crawl space. The house has a full basement.

Note also that it's common for local codes to
conflict with these standard details. If your code
covers the subject, follow the code, not the plan!

Heating & Air Conditioning (Sheet 7)
Many stock plans include a heating and air-

conditioning layout even though most HVAC contractors prefer to do their own layouts. Keep in
mind that the climate and type of fuel available
determine the size and type of equipment required.

Figure 2-20G lists the proposed heating and air-
conditioning equipment and shows the location of
duct runs on the Nashville Plan. It shows the size
of pipe, location of registers, and air direction.

The layout includes an insulation comparison
chart and a map with recommended R-values by
region. At the bottom of the sheet are load calculations if the home will have air conditioning. Some
communities require that new homes be built
within an "energy budget" if they are to be air con-
ditioned.

List of Materials (Sheet 8)
Figure 2-20H shows materials needed to build the
Nashville Plan. Even though you'll be making your
own detailed materials take-off for every project,
keep the list of materials that comes with the plan.
It may clarify parts of the plan and will help you
spot errors in your calculations.

Standard Material, Contract and Estimate Forms
Your stock plan may include some additional
documents: a description of materials form, an
owner-contractor agreement form, and a contractor's estimate form. The contractor's estimate
form is described in Figure 1-3 in Chapter 1. Let's
look at the material description form and owner-
contractor agreement form.

Material Description Form— Figure 2-22 picks up
where the blueprints leave off, giving additional information about all materials to be used in construction. There's no way to crowd all this information into the plans themselves. The Material
Description Form provides a single convenient
place to list information about quality, size, type,
style and manufacturer of materials.

Study this form carefully. I use this or a similar
form on all my jobs. It provides an accurate,
detailed, permanent record of all materials used,
and becomes a part of the construction contract.
This form becomes invaluable if you get into a con-
troversy over materials.

Owner-contractor agreement form— The form
shown in Figure 2-23 is a standard short form
builders often use. Some builders prefer to have
their own lawyer prepare a more detailed contract.

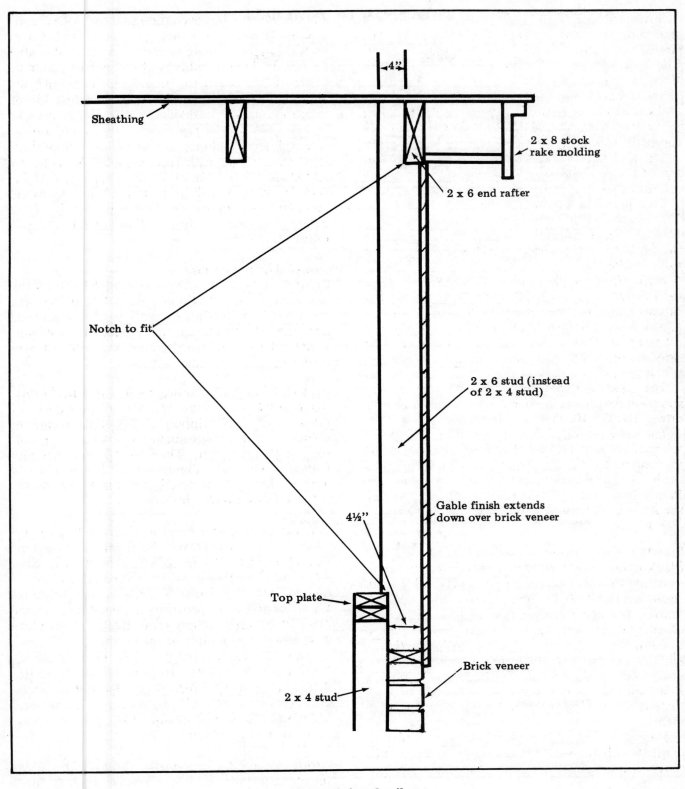

4"

Sheathing

2 x 8 stock
rake molding

2 x 6 end rafter

Notch to fit

2 x 6 stud (instead
of 2 x 4 stud)

Gable finish extends
down over brick veneer

4½"

Top plate

Brick veneer

2 x 4 stud

Rake section detail
Figure 2-21

☐ Proposed Construction
☐ Under Construction

DESCRIPTION OF MATERIALS   No. _____
(To be inserted by FHA, VA or FmHA)

Property address _____ City _____ State _____

Mortgagor or Sponsor _____
                                          (Name)   (Address)

Contractor or Builder _____
                                          (Name)   (Address)

INSTRUCTIONS

1. For additional information on how this form is to be submitted, number of copies, etc., see the instructions applicable to the FHA Application for Mortgage Insurance, VA Request for Determination of Reasonable Value, or FmHA Property Information and Appraisal Report, as the case may be.
2. Describe all materials and equipment to be used, whether or not shown on the drawings, by marking an X in each appropriate check-box and entering the information called for in each space. If space is inadequate, enter "See misc." and describe under item 27 or on an attached sheet. THE USE OF PAINT CONTAINING MORE THAN THE PERCENTAGE OF LEAD BY WEIGHT PERMITTED BY LAW IS PROHIBITED.
3. Work not specifically described or shown will not be considered unless

required, then the minimum acceptable will be assumed. Work exceeding minimum requirements cannot be considered unless specifically described.
4. Include no alternates, "or equal" phrases, or contradictory items. (Consideration of a request for acceptance of substitute materials or equipment is not thereby precluded.)
5. Include signatures required at the end of this form.
6. The construction shall be completed in compliance with the related drawings and specifications, as amended during processing. The specifications include this Description of Materials and the applicable Minimum Property Standards.

1. EXCAVATION:
Bearing soil, type _____

2. FOUNDATIONS:
Footings: concrete mix _____; strength psi _____   Reinforcing _____
Foundation wall: material _____   Reinforcing _____
Interior foundation wall: material _____   Party foundation wall _____
Columns: material and sizes _____   Piers: material and reinforcing _____
Girders: material and sizes _____   Sills: material _____
Basement entrance areaway _____   Window areaways _____
Waterproofing _____   Footing drains _____
Termite protection _____
Basementless space: ground cover _____; insulation _____; foundation vents _____
Special foundations _____
Additional information: _____

3. CHIMNEYS:
Material _____   Prefabricated *(make and size)* _____
Flue lining: material _____   Heater flue size _____   Fireplace flue size _____
Vents *(material and size)*: gas or oil heater _____; water heater _____
Additional information: _____

4. FIREPLACES:
Type: ☐ solid fuel; ☐ gas-burning; ☐ circulator *(make and size)* _____   Ash dump and clean-out _____
Fireplace: facing _____; lining _____; hearth _____; mantel _____
Additional information: _____

5. EXTERIOR WALLS:
Wood frame: wood grade, and species _____   ☐ Corner bracing. Building paper or felt _____
  Sheathing _____; thickness _____; width _____; ☐ solid; ☐ spaced _____" o. c.; ☐ diagonal; _____
  Siding _____; grade _____; type _____; size _____; exposure _____"; fastening _____
  Shingles _____; grade _____; type _____; size _____; exposure _____"; fastening _____
  Stucco _____; thickness _____"; Lath _____; weight _____ lb.
  Masonry veneer _____ Sills _____ Lintels _____ Base flashing _____
Masonry: ☐ solid ☐ faced ☐ stuccoed; total wall thickness _____"; facing thickness _____"; facing material _____
  Backup material _____; thickness _____"; bonding _____
  Door sills _____ Window sills _____ Lintels _____ Base flashing _____
Interior surfaces: dampproofing, _____ coats of _____; furring _____
Additional information: _____
Exterior painting: material _____; number of coats _____
Gable wall construction: ☐ same as main walls; ☐ other construction _____

6. FLOOR FRAMING:
Joists: wood, grade, and species _____; other _____; bridging _____; anchors _____
Concrete slab: ☐ basement floor; ☐ first floor; ☐ ground supported; ☐ self-supporting; mix _____; thickness _____";
  reinforcing _____; insulation _____; membrane _____
Fill under slab: material _____; thickness _____". Additional information: _____

7. SUBFLOORING: *(Describe underflooring for special floors under item 21.)*
Material: grade and species _____; size _____; type _____
Laid: ☐ first floor; ☐ second floor; ☐ attic _____ sq. ft.; ☐ diagonal; ☐ right angles. Additional information: _____

8. FINISH FLOORING: *(Wood only. Describe other finish flooring under item 21.)*

LOCATION	ROOMS	GRADE	SPECIES	THICKNESS	WIDTH	BLDG. PAPER	FINISH
First floor							
Second floor							
Attic floor _____ sq. ft.							
Additional information:							

DESCRIPTION OF MATERIALS

Description of materials form
Figure 2-22

DESCRIPTION OF MATERIALS

9. PARTITION FRAMING:
Studs: wood, grade, and species _____ size and spacing _____ Other _____
Additional information: _____

10. CEILING FRAMING:
Joists: wood, grade, and species _____ Other _____ Bridging _____
Additional information: _____

11. ROOF FRAMING:
Rafters: wood, grade, and species _____ Roof trusses (see detail): grade and species _____
Additional information: _____

12. ROOFING:
Sheathing: wood, grade, and species _____ ; □ solid; □ spaced _____ " o.c.
Roofing _____ ; grade _____ ; size _____ ; type _____
Underlay _____ ; weight or thickness _____ ; size _____ ; fastening _____
Built-up roofing _____ ; number of plies _____ ; surfacing material _____
Flashing: material _____ ; gage or weight _____ ; □ gravel stops; □ snow guards
Additional information: _____

13. GUTTERS AND DOWNSPOUTS:
Gutters: material _____ ; gage or weight _____ ; size _____ ; shape _____
Downspouts: material _____ ; gage or weight _____ ; size _____ ; shape _____ ; number _____
Downspouts connected to: □ Storm sewer; □ sanitary sewer; □ dry-well. □ Splash blocks: material and size _____
Additional information: _____

14. LATH AND PLASTER
Lath □ walls, □ ceilings: material _____ ; weight or thickness _____ Plaster: coats _____ ; finish _____
Dry-wall □ walls, □ ceilings: material _____ ; thickness _____ ; finish _____ ;
Joint treatment _____

15. DECORATING: *(Paint, wallpaper, etc.)*

ROOMS	WALL FINISH MATERIAL AND APPLICATION	CEILING FINISH MATERIAL AND APPLICATION
Kitchen _____		
Bath _____		
Other _____		

Additional information: _____

16. INTERIOR DOORS AND TRIM:
Doors: type _____ ; material _____ ; thickness _____
Door trim: type _____ ; material _____ Base: type _____ ; material _____ ; size _____
Finish: doors _____ ; trim _____
Other trim *(item, type and location)* _____
Additional information: _____

17. WINDOWS:
Windows: type _____ ; make _____ ; material _____ ; sash thickness _____
Glass: grade _____ ; □ sash weights; □ balances, type _____ ; head flashing _____
Trim: type _____ ; material _____ Paint _____ ; number coats _____
Weatherstripping: type _____ ; material _____ Storm sash, number _____
Screens: □ full; □ half; type _____ ; number _____ ; screen cloth material _____
Basement windows: type _____ ; material _____ ; screens, number _____ ; Storm sash, number _____
Special windows _____
Additional information: _____

18. ENTRANCES AND EXTERIOR DETAIL:
Main entrance door: material _____ ; width _____ ; thickness _____ ". Frame: material _____ ; thickness _____ "
Other entrance doors: material _____ ; width _____ ; thickness _____ ". Frame: material _____ ; thickness _____ "
Head flashing _____ Weatherstripping: type _____ ; saddles _____
Screen doors: thickness _____ "; number _____ ; screen cloth material _____ Storm doors: thickness _____ "; number _____
Combination storm and screen doors: thickness _____ "; number _____ ; screen cloth material _____
Shutters: □ hinged; □ fixed. Railings _____ , Attic louvers _____
Exterior millwork: grade and species _____ Paint _____ ; number coats _____
Additional information: _____

19. CABINETS AND INTERIOR DETAIL:
Kitchen cabinets, wall units: material _____ ; lineal feet of shelves _____ ; shelf width _____
 Base units: material _____ ; counter top _____ ; edging _____
 Back and end splash _____ Finish of cabinets _____ ; number coats _____
Medicine cabinets: make _____ ; model _____
Other cabinets and built-in furniture _____
Additional information: _____

20. STAIRS:

STAIR	TREADS		RISERS		STRINGS		HANDRAIL		BALUSTERS	
	Material	Thickness	Material	Thickness	Material	Size	Material	Size	Material	Size
Basement _____										
Main _____										
Attic _____										

Disappearing: make and model number _____
Additional information: _____

2

Description of materials form
Figure 2-22 (continued)

21. SPECIAL FLOORS AND WAINSCOT: *(Describe Carpet as listed in Certified Products Directory)*

	LOCATION	MATERIAL, COLOR, BORDER, SIZES, GAGE, ETC.	THRESHOLD MATERIAL	WALL BASE MATERIAL	UNDERFLOOR MATERIAL
FLOORS	Kitchen				
	Bath				

	LOCATION	MATERIAL, COLOR, BORDER, CAP. SIZES, GAGE, ETC.	HEIGHT	HEIGHT OVER TUB	HEIGHT IN SHOWERS (FROM FLOOR)
WAINSCOT	Bath				

Bathroom accessories: ☐ Recessed; material _____ ; number _____ ; ☐ Attached; material _____ ; number _____

Additional information: _____

22. PLUMBING:

FIXTURE	NUMBER	LOCATION	MAKE	MFR'S FIXTURE IDENTIFICATION NO.	SIZE	COLOR
Sink						
Lavatory						
Water closet						
Bathtub						
Shower over tub △						
Stall shower △						
Laundry trays						

△ ☐ Curtain rod   △ ☐ Door   ☐ Shower pan: material _____

Water supply: ☐ public; ☐ community system; ☐ individual (private) system. ★

Sewage disposal: ☐ public; ☐ community system; ☐ individual (private) system. ★

★*Show and describe individual system in complete detail in separate drawings and specifications according to requirements.*

House drain (inside): ☐ cast iron; ☐ tile; ☐ other _____   House sewer (outside): ☐ cast iron; ☐ tile; ☐ other _____

Water piping: ☐ galvanized steel; ☐ copper tubing; ☐ other _____   Sill cocks, number _____

Domestic water heater: type _____ ; make and model _____ ; heating capacity _____

_____ gph. 100° rise. Storage tank: material _____ ; capacity _____ gallons.

Gas service: ☐ utility company; ☐ liq. pet. gas; ☐ other _____   Gas piping: ☐ cooking; ☐ house heating.

Footing drains connected to: ☐ storm sewer; ☐ sanitary sewer; ☐ dry well.   Sump pump; make and model _____

_____ ; capacity _____ ; discharges into _____

23. HEATING:

☐ Hot water. ☐ Steam. ☐ Vapor. ☐ One-pipe system. ☐ Two-pipe system.

☐ Radiators. ☐ Convectors. ☐ Baseboard radiation. Make and model _____

Radiant panel: ☐ floor; ☐ wall; ☐ ceiling. Panel coil: material _____

☐ Circulator. ☐ Return pump. Make and model _____ ; capacity _____ gpm.

Boiler: make and model _____   Output _____ Btuh.; net rating _____ Btuh.

Additional information: _____

Warm air: ☐ Gravity. ☐ Forced. Type of system _____

Duct material: supply _____ ; return _____   Insulation _____ , thickness _____   ☐ Outside air intake.

Furnace: make and model _____   Input _____ Btuh.; output _____ Btuh.

Additional information: _____

☐ Space heater; ☐ floor furnace; ☐ wall heater. Input _____ Btuh.; output _____ Btuh.; number units _____

Make, model _____   Additional information: _____

Controls: make and types _____

Additional information: _____

Fuel: ☐ Coal; ☐ oil; ☐ gas; ☐ liq. pet. gas; ☐ electric; ☐ other _____ ; storage capacity _____

Additional information: _____

Firing equipment furnished separately: ☐ Gas burner, conversion type. ☐ Stoker: hopper feed ☐; bin feed ☐

Oil burner: ☐ pressure atomizing; ☐ vaporizing _____

Make and model _____   Control _____

Additional information: _____

Electric heating system: type _____   Input _____ watts; @ _____ volts; output _____ Btuh.

Additional information: _____

Ventilating equipment: attic fan, make and model _____ ; capacity _____ cfm.

kitchen exhaust fan, make and model _____

Other heating, ventilating, or cooling equipment _____

24. ELECTRIC WIRING:

Service: ☐ overhead; ☐ underground. Panel: ☐ fuse box; ☐ circuit-breaker; make _____ AMP's _____ No. circuits _____

Wiring: ☐ conduit; ☐ armored cable; ☐ nonmetallic cable; ☐ knob and tube; ☐ other _____

Special outlets: ☐ range; ☐ water heater; ☐ other _____

☐ Doorbell. ☐ Chimes. Push-button locations _____   Additional information: _____

25. LIGHTING FIXTURES:

Total number of fixtures _____   Total allowance for fixtures, typical installation, $ _____

Nontypical installation _____

Additional information: _____

3   ──────── DESCRIPTION OF MATERIALS ────────

Description of materials form
Figure 2-22 (continued)

DESCRIPTION OF MATERIALS

26. INSULATION:

Location	Thickness	Material, Type, and Method of Installation	Vapor Barrier
Roof			
Ceiling			
Wall			
Floor			

27. MISCELLANEOUS: (Describe any main dwelling materials, equipment, or construction items not shown elsewhere; or use to provide additional information where the space provided was inadequate. Always reference by item number to correspond to numbering used on this form.) _____

HARDWARE: (make, material, and finish.) _____

SPECIAL EQUIPMENT: (State material or make, model and quantity. Include only equipment and appliances which are acceptable by local law, custom and applicable FHA standards. Do not include items which, by established custom, are supplied by occupant and removed when he vacates premises or chattles prohibited by law from becoming realty.) _____

PORCHES:

TERRACES:

GARAGES:

WALKS AND DRIVEWAYS:
Driveway: width _____ ; base material _____ ; thickness _____ "; surfacing material _____ ; thickness _____ "
Front walk: width _____ ; material _____ ; thickness _____ ". Service walk: width _____ ; material _____ ; thickness _____ "
Steps: material _____ ; treads _____ "; risers _____ ". Cheek walls _____

OTHER ONSITE IMPROVEMENTS:
(Specify all exterior onsite improvements not described elsewhere, including items such as unusual grading, drainage structures, retaining walls, fence, railings, and accessory structures.)

LANDSCAPING, PLANTING, AND FINISH GRADING:
Topsoil _____ " thick: ☐ front yard; ☐ side yards; ☐ rear yard to _____ feet behind main building.
Lawns (seeded, sodded, or sprigged): ☐ front yard _____ ; ☐ side yards _____ ; ☐ rear yard _____
Planting: ☐ as specified and shown on drawings; ☐ as follows:
_____ Shade trees, deciduous, _____ " caliper.   _____ Evergreen trees. _____ ' to _____ ', B & B.
_____ Low flowering trees, deciduous, _____ ' to _____ '   _____ Evergreen shrubs. _____ ' to _____ ', B & B.
_____ High-growing shrubs, deciduous, _____ ' to _____ '   _____ Vines, 2-year _____
_____ Medium-growing shrubs, deciduous, _____ ' to _____ '
_____ Low-growing shrubs, deciduous, _____ ' to _____ '

IDENTIFICATION.—This exhibit shall be identified by the signature of the builder, or sponsor, and/or the proposed mortgagor if the latter is known at the time of application.

Date _____   Signature _____

Signature _____

Description of materials form
Figure 2-22 (continued)

━ AGREEMENT ━

This agreement made this ＿＿＿ day of ＿＿＿＿＿＿＿＿＿, 19 ＿, by and between

＿＿＿＿＿＿＿＿＿＿＿＿＿, herein known as the Owner, and ＿＿＿＿＿＿＿＿＿＿,
herein known as the Contractor.

WITNESSETH: That the Contractor and the Owner agree as follows:

1. That the Contractor shall furnish all labor and materials needed for constructing a house on ＿＿＿＿＿＿＿＿ Sts., ＿＿＿＿＿＿＿＿＿＿ according to the Standard Homes Plan Service Plans and Specifications attached hereto, and signed by the Owner and the Contractor.

2. That the Owner may make separate contracts to secure extra items, service or equipment, not included in the General Specifications or may award particular parts outlined therein directly to one or more Sub-Contractors provided the exact scope of their work shall be specifically shown and the contracts therefor hereto attached.

3. That no changes from the original Plans and Specifications shall be made, required, or collected for, unless both parties agree thereto in writing, as to the extent of the changes and the amount to be paid or deducted therefor, before work thereon shall have begun.

4. That Owner shall furnish an adequate survey of the property, and Contractor shall obtain and pay for all permits necessary to the prosecution of the work. He shall comply with all laws or regulations bearing on the conduct of the work, and shall notify Owner if the Drawings or Specifications are at variance therewith, or if unforeseen conditions arise. In all instances the Contractor shall bring building up to any local, state or mortgage loan standards.

5. The Contractor shall allow Owner or other interested parties to inspect all work and materials at all times and shall, after receiving notice in writing, remove all defective materials, whether completed or not. Materials rendered unfit for reworking are to be replaced at Contractor's own expense.

6. INSURANCE: Contractor shall carry such insurance as required by Workmen's Compensation Commission or other laws relating to compensation for personal injury to workmen. The Owner may carry additional insurance to protect himself from contingent liability and may object to and cause to be removed any condition dangerous to workmen or public. Owner shall carry Fire Insurance on building and materials as work progresses, either in binder form or in the usual manner so long as Owner and Contractor are protected.

7. Contractor shall complete all work included in this Agreement by ＿＿＿, ＿＿＿, 19 ＿, or within ＿＿＿ days from receipt of Loan Commitment, unless delayed by conditions beyond his control, in which case time shall be extended the actual number of days so delayed.

8. It is mutually agreed between the parties that the sum to be paid by Owner to Contractor for the erection of this building shall be:

$＿,＿,＿Thousand ＿ Hundred and ＿＿＿ Dollars.

1st Payment $＿.00, ＿＿＿ Dollars, Foundation Complete.

2nd Payment $＿.00, ＿＿＿ Dollars, house framed Roof complete.

3rd Payment $＿.00, ＿＿＿ Dollars, all outside work complete.

Where payments are to be made on labor or materials not actually paid for, Owner may make checks payable to Contractor and interested parties jointly or for their respective interests.

4th Payment $＿.00, ＿＿＿ Dollars, on completion, all specifications complied with and house ready to move in.

9. Final payment shall not be due until Contractor has delivered to Owner complete releases for all claims arising from this contract, or receipts in full covering all labor and material for which liens could be filed, or bond be furnished Owner against any such lien.

10. Any disagreement arising out of this Contract or the application of any provisions thereof shall be submitted to an Arbitrator or Arbitrators not interested in the finances of the contract. The parties hereto may agree on one Arbitrator, or may select one each and these two shall select a third. And it is mutually agreed that any such arbitration award shall be binding and have the same weight as a legal decision on any difference herein arising.

This AGREEMENT shall be binding on the parties hereto, their assigns, successors, representatives or administrators.

WITNESSES

＿＿＿＿＿＿＿＿＿＿＿＿ ＿＿＿＿＿＿＿＿＿＿ OWNER

＿＿＿＿＿＿＿＿＿＿＿＿ ＿＿＿＿＿＿＿＿＿＿ CONTRACTOR

Owner-contractor agreement form
Figure 2-23

In this chapter we've reviewed most of what you'll find on residential plans. Of course, every designer has his own technique and style. But most produce plans that you will have little trouble understanding if you have followed my analysis of the Nashville Plan.

Beware of those blueprints prepared by amateurs. If your client's plans were drawn by a relative or friend of his, they're probably filled with problems, contradictions and profit-shattering errors. It takes a great deal more than drafting ability to draw a workable house plan. You need a detailed understanding of construction, design, and minimum space requirements. Work only from plans drawn by a professional. That will cut down on errors, increase your profits, and protect your reputation as a builder.

Chapter **3**

Speculative Building

Many successful spec builders find a good basic house plan and then stick to it. When the first house is complete, you know the exact cost of labor and materials. You've found any faults or major problems in the plan and corrected them. When you build your next house with this same plan, you can build with "one eye and one hand." Your costs will be lower and your profit will be higher.

Sticking to one basic house plan doesn't mean that all houses have to look like carbon copies of each other. Look at the three models of Design 2161 in Figure 3-1.

Each house has its own character. The three models can be built side by side without looking too much alike. Yet they all have the same floor plan, exterior finish materials and are built slab-on-grade. They're different enough to make them attractive and marketable as a group. They're enough alike to be a profitable building project.

If you're developing a 100-house subdivision, you'll want to use more than one basic house plan. You may use three or four. But use as few as possible. Many builders have gone bankrupt from using too many different plans. Each new plan is a new learning experience. Each new experience has unexpected costs. Those unexpected costs come out of your pocket. Stick with a good basic house plan to reduce the number of surprises.

In this chapter, we'll take a detailed look at the

plans for a basic spec house. We'll cover the foundation plan, floor plan, plot plan, interior and exterior elevations, sections, and list of materials. Then we'll look at a baseline estimate and labor breakdown for this spec house.

Basic Spec House Plans
Design 2161 is a good layout for a spec house. Note that this plan has been reduced to fit in this book and is no longer in scale. Let's look at the important details shown on each sheet of the plans:

Sheet 1: Foundation Plan, Floor Plan and Interior Elevations
Figure 3-2 shows the foundation plan, first-floor plan and interior elevations for Design 2161.

Foundation plan— Here's what the foundation plan tells us:

The footing is 8'' wide. The 4'' slab is poured on unexcavated ground and over 4'' of gravel fill. In the house floor area, there's a plastic vapor barrier under the slab. In the terrace and garage areas, the slab is reinforced with 6'' x 6'' No. 8 welded wire mesh.

If you're using plant mix for the footing and slab concrete, here's a simple rule of thumb for figuring the number of cubic yards required:

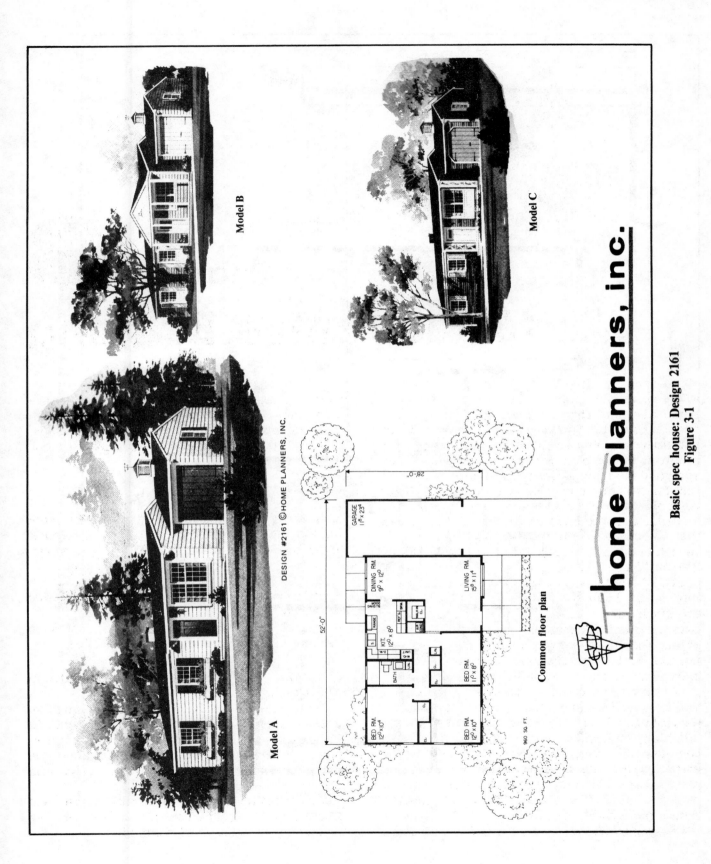

Model B

Model C

Model A

DESIGN #2161 © HOME PLANNERS, INC.

Common floor plan

home planners, inc.

Basic spec house: Design 2161
Figure 3-1

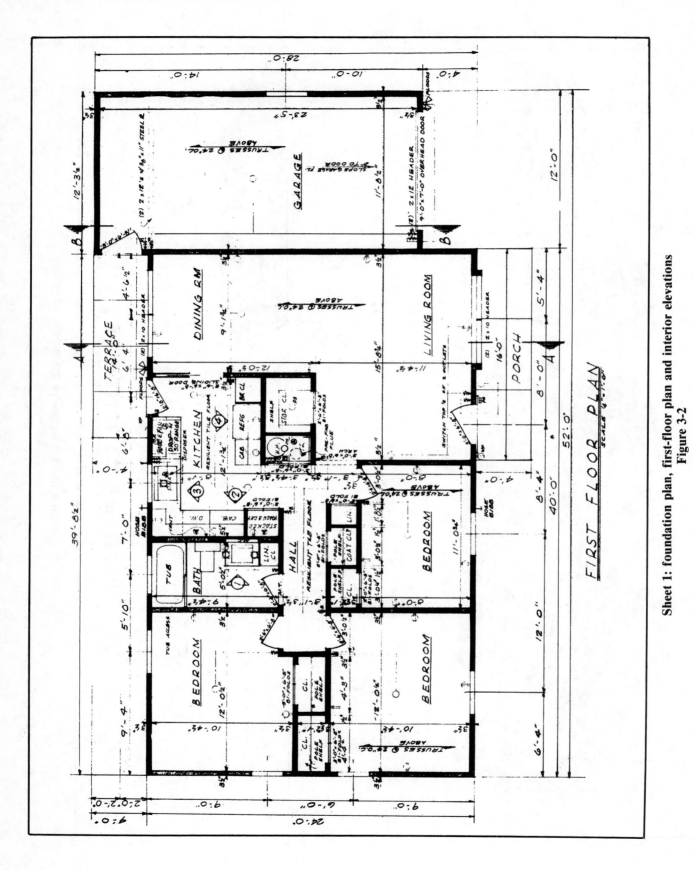

FIRST FLOOR PLAN
SCALE 1/4"=1'-0"

Sheet 1: foundation plan, first-floor plan and interior elevations
Figure 3-2

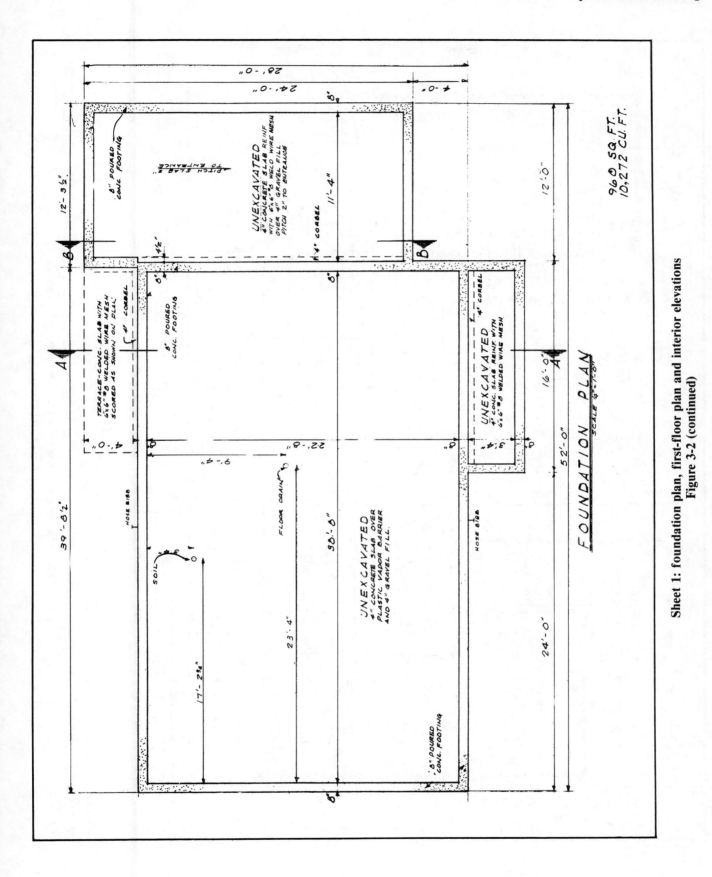

Sheet 1: foundation plan, first-floor plan and interior elevations
Figure 3-2 (continued)

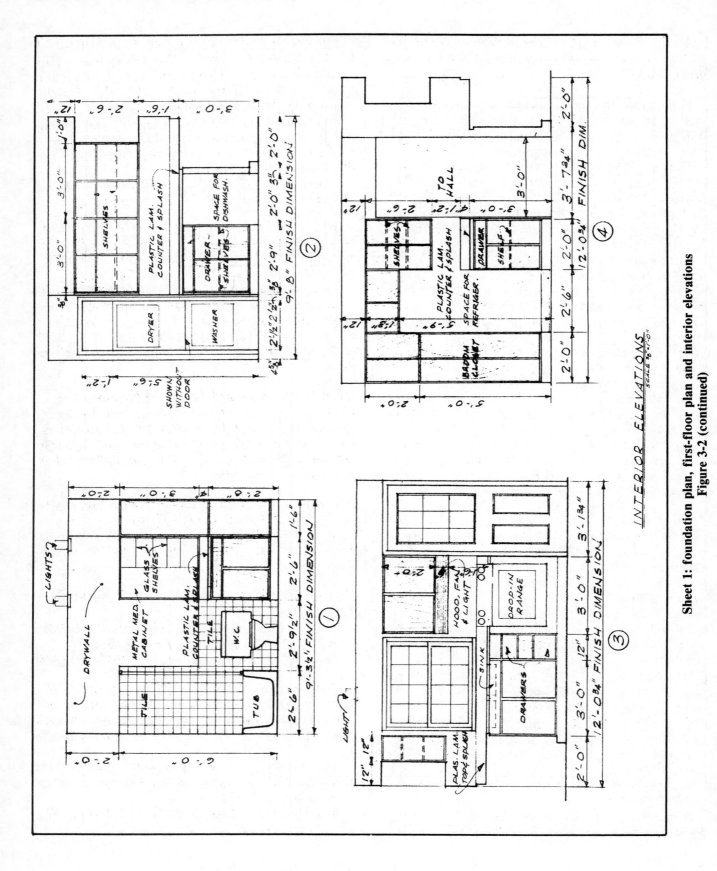

INTERIOR ELEVATIONS

**Sheet 1: foundation plan, first-floor plan and interior elevations
Figure 3-2 (continued)**

Multiply the length of the area (in feet) by the width (in feet) by the thickness (in inches) and divide by 314.

Notice that the footing width is often shown in inches, rather than feet. Be sure to convert this number to feet before calculating the number of cubic yards of concrete required.

Here's a sample calculation: Suppose the footing is 208' long, 8" (0.66667') wide and 12" thick. Multiply the length (208') by the width (0.66667') by the thickness (12") to get 1,664.01. Now divide by 314 to get 5.3 cubic yards of concrete required for the footing.

Here's another example: Let's say you're pouring a slab 39'8" long, 22'8" wide and 4" thick. How many cubic yards of concrete will you need? Multiply the length (39.66667') by the width (22.66667') by the thickness (4") to get 3,596.45. Now divide by 314. That gives you 11.45 cubic yards of concrete required for the slab. Round off to 11.5 cubic yards.

If you're planning to mix footing and slab concrete on the job site, you'll find the mixing proportions on the standard list of materials that comes with the plans.

Sheet 1 also shows us the locations of the soil stack and floor drain. Notice section lines A-A and B-B. (Sections A-A and B-B will be shown in detail in Figure 3-5.)

Floor plan— The first-floor plan for Design 2161 offers good traffic flow, kitchen privacy, plenty of window area in the dining room, and adequate closet and storage space. The plan shows three bedrooms. There are 960 square feet of living space in this floor plan.

The frame wall between the bathroom and kitchen is 5½" thick. The vent pipe is located in this wall. The access to the tub plumbing is in the opposite wall.

The floor plan shows all door sizes and the direction of swing. Note the sliding (pocket) door between the kitchen and dining room and a 3'0" x 6'8" arch between the living room and hall.

Section lines A-A and B-B on the floor plan coincide with section lines A-A and B-B on the foundation plan.

Notice the dimension lines drawn between window centerlines.

The heater (H) and water heater (W.H.) are located in the same closet. There's a prefab flue for the heater. A 3'0" x 6'8" bifold door is specified for this equipment closet. The closet has a pull switch (P.S.) light. There's an electric circuit for direct tie-in (electric drop) for the heater, dishwasher (D.W.) and garbage disposal.

Two 2 x 10 headers are shown for the double window openings in the front and rear walls. Two 2 x 12's with a 3/8" x 11" steel plate are shown for the carport ceiling beam.

Trusses are spaced 24" o.c. The arrow shows the direction of run or span.

Interior elevations— There are four interior elevations shown on Sheet 1. These elevations give the dimensions for fixtures and cabinets in the kitchen and bathroom.

Pay special attention to the style of kitchens and bathrooms. Color and design are particularly important in these two rooms. A little extra touch can increase the marketability of the house. A skylight in the kitchen, for example, can be a strong selling point.

Sheet 2: Plot Plan and Exterior Elevations

Figure 3-3 shows the plot plan and the following exterior elevations: front elevation for model A, right-side elevation, rear elevation and left-side elevation. Let's look first at the plot plan and then at the exterior elevations.

Plot plan— A plot plan defines the lot and shows the location of the buildings on the lot. The plot plan shown in Figure 3-3 was not drawn to scale. This allows you to fill in the appropriate dimensions.

Exterior elevations— Here's what these exterior elevations tell us:

The top of the house slab is 8" above the finish grade. This is the minimum for most codes.

The depth of the foundation is specified by local code.

The living room double window projects 6". Each window is double hung (D.H.) and is 36" wide. The top sash is 24" high and the bottom sash is 36" high. Bedroom windows are flashed at the top.

The siding is 8" exposed bevel siding. This means that 8" is exposed to the weather. Conduit downspouts are used.

The elevations show shingles on the roof. The roof pitch is 4/12.

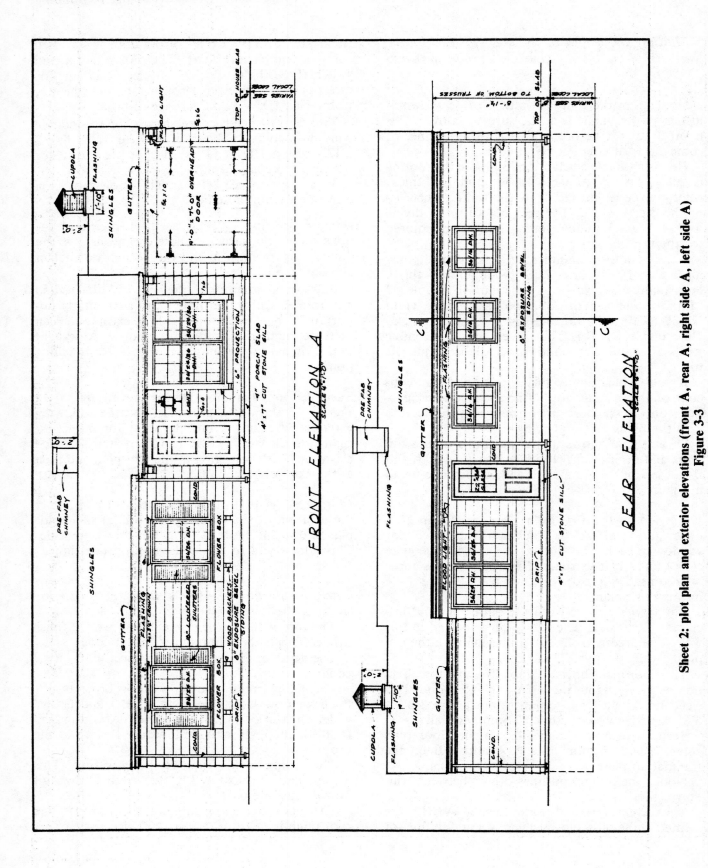

Sheet 2: plot plan and exterior elevations (front A, rear A, right side A, left side A)
Figure 3-3

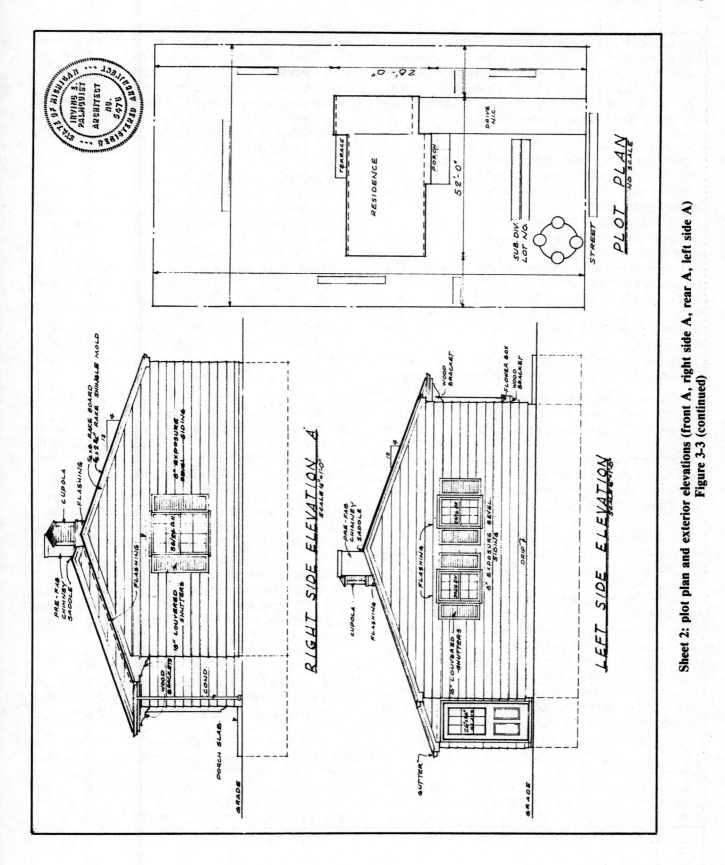

Sheet 2: plot plan and exterior elevations (front A, right side A, rear A, left side A)
Figure 3-3 (continued)

Sheet 3: Exterior Elevations and Sections

Figure 3-4 shows the following exterior elevations: front elevation for model B, right-side elevation for model B, front elevation for model C and right-side elevation for model C. This sheet also shows sections D-D and E-E. Let's look first at the elevations:

Exterior elevations— Here's what the exterior elevations for models B and C tell us:

For model B, the porch roof is 6/12 pitch, steeper than the house roof, which is 4/12 pitch. The porch is 16'0'' x 4'0''.

Notice the 1/8'' cement asbestos window panels in both models B and C.

The side elevations show the rake details. The cupola is essential for attic ventilation since there are no gable vents specified. Screened vents are located in the soffit under the eaves. Air enters the attic through these vents and passes out through the cupola vent.

Notice section line D-D shown on front elevation B, and section line E-E shown on front elevation C.

Sections— Section D-D shows construction details for the porch. The rafters and ceiling joists are 2 x 4's spaced 16'' o.c. The porch ceiling is 8'8'' above the porch slab.

Section E-E shows a cross-section of front elevation C. There's a double 2 x 6 door header. The elevation also shows eave treatment, foundation and floor slab details.

Sheet 4: Sections

Figure 3-5 shows sections A-A, B-B and C-C. (Section lines A-A, B-B and C-C were shown in Figures 3-2 and 3-3.)

Notice the ceiling height (from top of slab to bottom of truss) in the main house and in the garage. These sections also show the window height from top of slab to top of window.

There's an 8'' x 8'' x 16'' masonry solid "L" block for foundation and floor tie-in.

In many parts of the country, the foundation area and the ground underneath the slab must be chemically treated for termites. Be sure to check the code before you build.

List of Materials

Figure 3-6 shows the standard list of materials for Design 2161. Even though you'll be making your own detailed materials take-off sheet for every project, be sure to keep the list of materials that comes with the plans. It may help clarify difficult parts of the plans. And you can use it as a checklist against your own take-off sheet to make sure you haven't left anything out.

Now that we've looked at each sheet of the plans for Design 2161, let's compute a baseline estimate for the cost of construction. We'll also look at a labor breakdown for the project.

Baseline Estimate for Design 2161

We learned how to calculate a baseline cost estimate in Chapter 1. The baseline estimate includes all labor and materials, plus the lot. Let's review the formula:

Multiply the carpenter's hourly union wage by 3. Then multiply this new wage by the total area of the house (in square feet).

Design 2161 shows a main house with an attached garage, front porch and terrace. Let's do a baseline cost estimate for each of these components:

1) Main house: Suppose the *union* hourly wage is $20.00 for a carpenter in your area. Multiply the union wage ($20.00) by 3 to get a formula wage of $60.00. Then multiply $60.00 by the total area of the main house (960 SF). This gives you a total cost of $57,600.00, or $60.00 per square foot.

2) Garage: An attached garage with no interior finish will run about half the cost of the main house. For the garage, your formula wage will be $30.00 instead of $60.00. The total area of the garage is 295 square feet. Here's our baseline estimate for the cost of the garage:

Multiply the formula wage ($30.00) by the total area of the garage (295 SF). This gives you a garage cost estimate of $8,850.00. This includes a proportionate share of the lot cost.

A garage with a finished interior will cost more. The formula wage is $40.00, or twice the union wage. Multiply the formula wage ($40.00) by the total area of the garage (295 SF). A garage with a finished interior will cost about $11,800.00.

3) Front porch: The porch floor is concrete. The porch roof is gabled, has a ceiling, and ties into the main roof. The porch is 16'0'' x 4'0'' for a total area of 64 square feet. The formula wage for this type of porch is $30.00. Multiply the formula wage ($30.00) by the porch area (64 SF). The porch will cost $1,920.00.

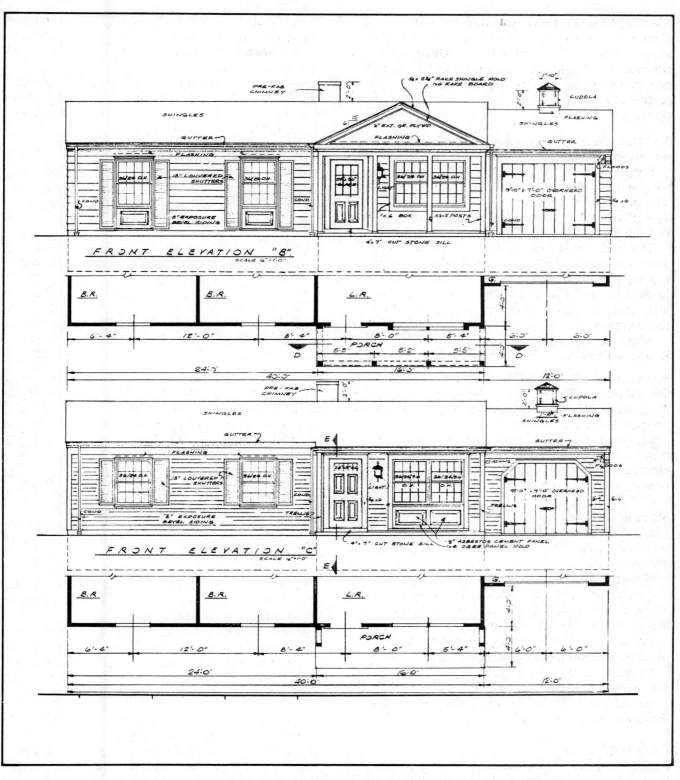

**Sheet 3: exterior elevations (front B, front C, right side B, right side C)
and sections (D-D and E-E)
Figure 3-4**

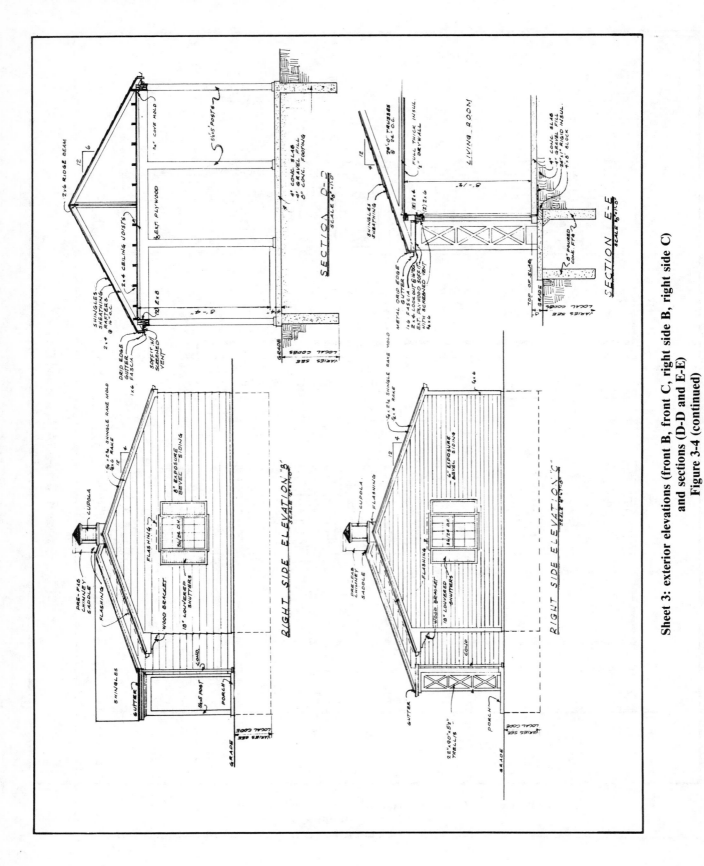

Sheet 3: exterior elevations (front B, front C, right side B, right side C) and sections (D-D and E-E)
Figure 3-4 (continued)

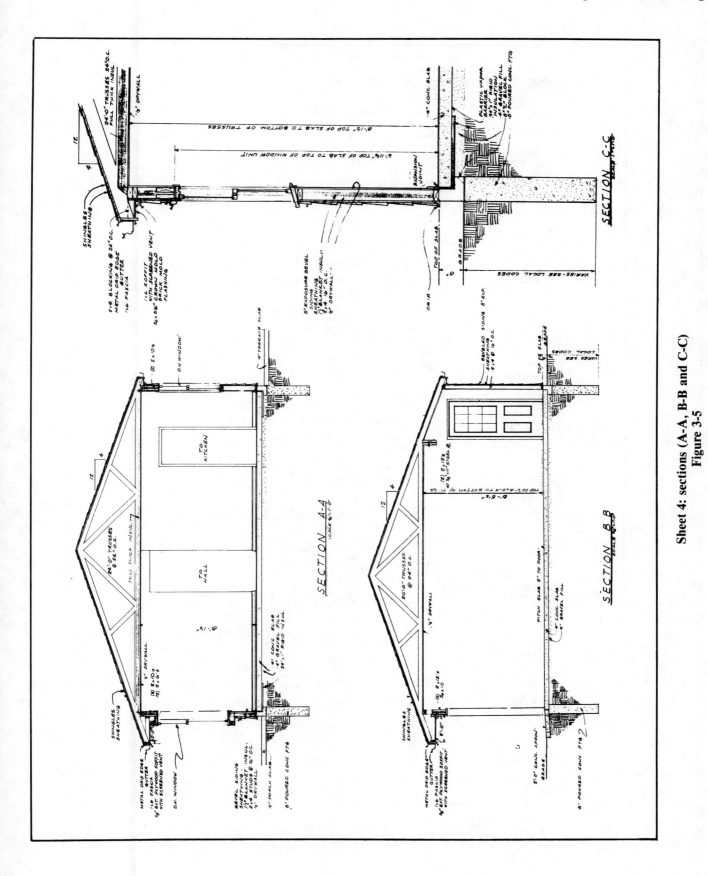

Sheet 4: sections (A-A, B-B and C-C)
Figure 3-5

LINE NO.	ITEM COLUMN NO. 1	QUANTITY & UNIT MEAS.	MATERIAL (TYPE and/or SIZE)	UNIT COST	TOTAL COST
1	DESIGN NO. HP 2161 WITH OPTIONAL ELEVATIONS A, B, & C				
2	MASONRY				
3	Footings	21 Cu.Yds.	60/40 Gravel		
4	208 Lin.Ft. 8x42"	94 Sacks	Cement		
5		280 Sq.Ft.	1" Fiberglass Perimeter Insulation		
6					
7	Masonry Block Walls	132 Pcs.	8 x 8 x 16" Masonry Solid "L" Blocks		
8		1 Cu.Yd.	50/50 Mason Sand		
9		7 Sacks	Mortar		
10	Prefabricated				
11	Chimneys	1 Unit	Prefabricated Furnace Flue		
12		1 Unit	" Roof Housings & Flashings		
13	Concrete Slabs				
14	Exterior Slabs	(Sidewalks and Driveway Not Included)			
15		420 Sq.Ft.	6"x6" #8/8 Welded Wire Reinforcing Mesh		
16		6 Cu.Yds.	Fill Gravel		
17		6 Cu.Yds.	60/40 "		
18		30 Sacks	Cement		
19		4 Gals.	Liquid Cement Hardener		
20	Interior Slabs				
21		14 Cu.Yds.	Fill Gravel		
22		14 Cu.Yds.	60/40 "		
23		70 Sacks	Cement		
24		1000 Sq.Ft.	Plastic Membrane Vapor Barrier		
25					
26	FRAMING LUMBER	(Lumber Sizes Based on Design Stress, f=1450)			
27	Exterior Walls	122 Pcs.	2"x 4"x 8'-0" Studs (ADD 8 PCS. FOR ELEVATIONS B & C)		
28		27 Pcs.	2"x 4"x 10'-0" "		
29		656 Lin.Ft.	2" x 4" Plates		
30		2 Pcs.	2"x 12"x 14'-0" Headers reinforced with ONE		
31			3/8"x 11"x 12'-0" Steel Plate - Garage Header		
32		2 Pcs.	2"x 12"x 10'-0" Headers		
33		2 Pcs.	2"x 10"x 14'-0" "		
34		3 Pcs.	2"x 8"x 16'-0" " (ELEVATION B ONLY)		
35		80 Lin.Ft.	2" x 6" Headers		
36		1300 Sq.Ft.	Exterior Wall Sheathing		
37		1500 Sq.Ft.	15# Building Felt (Optional)		
38	Interior Partitions				
39		126 Pcs.	2"x 4"x 8'-0" Studs		
40		9 Pcs.	2"x 6"x 8'-0" "		
41		648 Lin.Ft.	2" x 4" Plates		
42		30 Lin.Ft.	2" x 6" "		
43		84 Lin.Ft.	2" x 6" Headers		
44		1 Unit	2'-6"x 6'-8"x 3-5/8" Sliding Door Wall Pocket		
45	Ceiling Framing				
46		7 Pcs.	2"x 4"x 10'-0" Ceiling Joists (ELEVATION B ONLY)		
47	Roof Framing				
48		25 Units	24'-0" - 4/12 Pitch 2" x 4" Wood Trusses		
49		16 Pcs.	2"x 4"x 10'-0" Rafters (ELEVATION B ONLY)		
50		12 Lin.Ft.	2" x 6" Ridge Member (ELEVATION B ONLY)		
51		1500 Sq.Ft.	Roof Sheathing (ADD 118 SQ.FT. FOR ELEVATION B)		
52		1700 Sq.Ft.	15# Building Felt (ADD 100 SQ.FT. FOR ELEVATION B)		
53	ROOFING & SHEET METAL				
54		1 Square	Asphalt Ridge Shingles		
55		28 Lin.Ft.	" Valley Roll Roofing (ELEVATION B ONLY)		
56		16 Squares	" Self-Sealing Shingles (ADD 1 SQ. FOR ELEV. B)		
57		188 Lin.Ft.	Metal Drip Edge (ADD 24 LIN.FT. FOR ELEVATION B)		
58		32 Pcs.	5" x 7" Metal Step Flashings		
59		17 Lin.Ft.	Wall to Roof Flashing (ELEVATION B)		
60		48 Lin.Ft.	Window & Door Head Flashing		
61		50 Lbs.	Roofing Nails (ADD 3 LBS. FOR ELEVATION B)		
62		116 Lin.Ft.	12" Girth Sheet Metal Gutter		
63		55 Lin.Ft.	2" x 3" Sheet Metal Conductor Pipe & Fittings (ADD		
64			8 LIN.FT. FOR ELEVATION B)		
65	EXTERIOR FINISH MATERIALS				
66	Windows		ALL WINDOWS WITH BRICK MOLD CASING		
67		2 Single	24"x16/16" Double Hung Cut 3-Wide 4-High		
68		3 Single	36"x16/16" " " " " 4-Wide 4-High		
69		3 Single	36"x24/24" " " " " 4-Wide 4-High		
70		1 Double	36"x24/24" " " " " 4-Wide 4-High		
71		1 Double	36"x24/36" " " " " 4-Wide 5-High (ELEV. A ONLY)		
72		1 Double	36"x24/36" " " " " 4-Wide 4-High (ELEVS. B & C)		
73	Door Frames				
74		1 Front	3'-0"x 6'-8" Rabbeted 1-3/4" - 5/4" x 6" Casing		
75		1 Rear	3'-0"x 6'-8" " 1-3/4" - Brick Mold Casing		
76		1 Rear	2'-8"x 6'-8" " 1-3/4" " " " "		
77		1 Garage	9'-0" x 7'-0" 2"x6" Overhead Frame - Loose Casing		
78	Field Applied Door				
79	& Window Trim	10 Lin.Ft.	5/4"x 10" Garage Door Trim (ELEVATIONS A & B)		
80		16 Lin.Ft.	5/4"x 6" " (" A & B)		
81		56 Lin.Ft.	1" x 6" Projected Box Window Trim (ELEVATIONS A & B)		
82		24 Lin.Ft.	5/4"x 6" Window Casing (ELEVATION C ONLY)		
83		14 Sq.Ft.	1/8" Cement Asbestos Window Panel (ELEVATIONS B & C)		
84		20 Lin.Ft.	1" x 4" Ogee Panel Mold (ELEVATIONS B & C)		
85		24 Lin.Ft.	5/4" x 4" Garage Door Trim (ELEVATION C)		
86					
87					
88					
89					
90					

Sheet 5: list of materials for Design 2161
Figure 3-6

LINE NO.	ITEM COLUMN NO. 2	QUANTITY & UNIT MEAS.	MATERIAL (TYPE and/or SIZE)	UNIT COST	TOTAL COST
1	DESIGN NO. HP 2161 WITH OPTIONAL ELEVATIONS A, B & C				
2	EXTERIOR FINISH MATERIALS (Continued)				
3	Cornice	120 Lin.Ft.	1" x 6" Fascia Board		
4		80 Lin.Ft.	1" x 6" Soffit " Vented & Screened (ELEVS. A & C)		
5		104 Lin.Ft.	1" x 6" " " " (ELEV. B)		
6		26 Sq.Ft.	3/8" Ext. Grade Vented & Screened Soffit (ELEV.B)		
7		64 Sq.Ft.	3/8" Ext. Grade Plywood Ceiling (ELEV. B)		
8		60 Sq.Ft.	3/8" Ext. Grade Vented & Screened Soffit (ELEV. A & C)		
9		108 Lin.Ft.	3/4" x 3¼" Crown Mold (ELEVATIONS A & C)		
10		116 Lin.Ft.	3/4" x 3¼" " " B)		
11		40 Lin.Ft.	3/4" x 3/4" Cove Mold (" B)		
12		24 Lin.Ft.	1" x 4" Beam Soffit (" B)		
13		40 Lin.Ft.	1" x 8" " Facing (" B)		
14		78 Lin.Ft.	5/4" x 6" Rake Board (" A, B & C)		
15		20 Lin.Ft.	1" x 6" " (" B)		
16		80 Lin.Ft.	5/4" x 2 3/4" Rake Shingle Mold (ADD 20 LIN.FT. FOR		
17			ELEVATION B)		
18		3 Pcs.	18"x 10"x 6" Ornamental Wood Cornice Brackets (ELEV. A)		
19		1 Pc.	18"x 10"x 6" " " " (ELEV. A)		
20			(ELEVATIONS B & C)		
21	Porch & Balcony				
22	Trim	10 Pcs.	1"x 6"x 8'-0" Column Facing (ELEVATION B)		
23		10 Pcs.	1"x 4"x 8'-0" " " " B)		
24	Siding				
25		1600 Lin.Ft.	5/8" x 8¼" Bevel Siding (ELEVATIONS A & B)		
26		2400 Lin.Ft.	1/2" x 6¼" Bevel Siding (" C)		
27		92 Lin.Ft.	5/4" x 6" Corner Boards (" C)		
28	Miscellaneous				
29	Millwork	2 Pairs	16" x 42" Louvered Shutters		
30		3 Pairs	20" x 54" " " (ELEVATIONS A & C)		
31		1 Pair	20" x 54" " " (" B)		
32		2 Pairs	20" x 80" " " (" B)		
33		2 Lin.Ft.	22"x 5¼"x 9'-6" Trellis		
34		1 Unit	24"x 24"x 36" Louvered Cupola Complete with		
35			Sheet Metal Roof		
36	INTERIOR FINISH				
37	Exterior Doors	1 Front	3'-0"x 6'-8"x 1-3/4" 6 Panel (ELEVATION A)		
38		1 Front	3'-0"x 6'-8"x 1-3/4" 2 " Glazed 26" x 36"		
39			(ELEVATION B)		
40		1 Front	3'-0"x 6'-8"x 1-3/4" 4 Panel Glazed 26" x 8" (ELEV. C)		
41		1 Rear	2'-8"x 6'-8"x 1-3/4" 2 " 22"x36"		
42		1 Rear	3'-0"x 6'-8"x 1-3/4" 2 " 26"x36"		
43		1 Garage	9'-0"x 7'-0"x 1-3/4" Overhead Door, Track & Hardware		
44	Insulation				
45		960 Sq.Ft.	Full Thick Floor & Ceiling Insulation		
46		900 Sq.Ft.	1½" Blanket Sidewall Insulation		
47	Drywall Wallboard				
48		4336 Sq.Ft.	1/2" Gypsum Drywall Wallboard		
49		1445 Lin.Ft.	Perforated Joint Tape		
50		44 Lbs.	Drywall Joint Compound		
51		38 Lbs.	Holdtite Drywall Nails		
52		256 Lin.Ft.	Metal Drywall Cornerbeads		
53	Tilework				
54		1 Pc.	Marble Threshold		
55		26 Sq.Ft.	Unglazed Ceramic Tile Floor		
56		86 Sq.Ft.	Glazed " " Walls		
57		1 Pc.	Paper Holder		
58		1 Pc.	Soap & Grab Bar		
59		2 Pcs.	Towel Bars		
60	Finish Flooring				
61		150 Sq.Ft.	Resilient Floor Tile		
62	Interior Doors				
63		1 Interior	2'-4"x 6'-8"x 1-3/8" Flush		
64		4 Interior	2'-6"x 6'-8"x 1-3/8" "		
65		4 Interior	2'-0"x 6'-8"x 1-3/8" Bifold Door		
66		3 Interior	3'-0"x 6'-8"x 1-3/8" " "		
67	Door Frames				
68		1 Set	2'-4"x 6'-8"x 4-5/8" Jambs With Stops		
69		4 Sets	2'-6"x 6'-8"x 4-5/8" " " "		
70	Door Trim				
71		2 Sides	2'-4" x 6'-8" Casing		
72		8 Sides	2'-6" x 6'-8" "		
73		1 Side	2'-8" x 6'-8" "		
74		2 Side	3'-0" x 6'-8" "		
75	Window Trim				
76		180 Lin.Ft.	1/2" x 1-3/4" Casing & Apron		
77	Running Trim				
78		264 Lin.Ft.	1/2" x 2-1/2" Base		
79		278 Lin.Ft.	1/2" x 3/4" Shoe		
80		16 Lin.Ft.	1" x 12" Shelving		
81		36 Lin.Ft.	1" x 2" Shelf Cleat		
82		24 Lin.Ft.	1" x 3" Hook Strip		
83		1 Pc.	4'x 8'x 1/2" Plywood Shelving		

Sheet 5: list of materials for Design 2161
Figure 3-6 (continued)

LINE NO.	ITEM COLUMN NO. 3	QUANTITY & UNIT MEAS.	MATERIAL (TYPE and/or SIZE)	UNIT COST	TOTAL COST
1	DESIGN NO. HP 2161 WITH OPTIONAL ELEVATIONS A, B & C				
2	INTERIOR FINISH (Continued)				
3	Cabinets				
4	Baths	1 Base	18" x 72" x 21" Linen Storage Cabinet		
5		1 Base	30" x 32" x 21" Lavatory Cabinet		
6		1 Pc.	30" x 21" Plastic Laminate Counter & Splash		
7		1 Pc.	16" x 36" Medicine Cabinet		
8		1 Unit	60" x 60" Aluminum & Glass Tub Enclosure		
9	Kitchen & Laundry				
10		1 Base	36" x 36" x 24" Sink and Storage Cupboard		
11		1 Base	36" x 36" x 24" Drop-In Range & " "		
12		1 Base	12" x 36" x 24" 4-Drawer " "		
13		1 Base	24" x 36" x 24" " "		
14		1 Base	33" x 36" x 24" " "		
15		1 Base	24" x 60" x 24" Broom " "		
16		2 Upper	36" x 30" x 12" 1 shelf " "		
17		1 Upper	12" x 30" x 12" 1 shelf " "		
18		1 Upper	36" x 24" x 12" 2 shelf " "		
19		1 Upper	24" x 24" x 12" 2 shelf " "		
20		1 Upper	30" x 15" x 12" 1 shelf " "		
21		1 Upper	24" x 30" x 12" 1 shelf " "		
22		1 Pc.	L-shaped 84" x24" + 48" x24" Plastic Laminate Counter		
23			& Splash		
24		1 Pc.	24" x 24" Plastic Laminate Counter & Splash		
25		1 Unit	36" Range Hood, Exhaust Fan & Duct		
26	FINISH HARDWARE				
27		1 Front	Door Lock Set		
28		2 Rear	" " Sets		
29		1 Set	Safety Chain		
30		1 Set	Mail Slot Hardware		
31		1 Set	Mortise Dead Bolts		
32		3 Pairs	4" x 4" Butts		
33		1 Interior	Privacy Lock Set		
34		3 Interior	Passage Latch Sets		
35		1 Interior	Sliding Door Flush Pull & Latch Combination		
36		4 Pairs	3½" x 3½" Butts		
37		20 Lin.Ft.	Bifold Sliding Door Top Guide Track		
38		7 Pcs.	" " " " " Pins		
39		7 Pairs	" " " " " and Bottom Pivots		
40		7 Pairs	3" x 3" Butts		
41		7 Pcs.	Door Pulls		
42		4 Pcs.	Metal Adjustable Closet Poles		
43		12 Sets	Double Hung Window Sash Locks		
44		24 Sets	" " " " Lifts		
45		1 Pint	Glue		
46		1 Ream	Sand Paper		
47					
48	ROUGH HARDWARE & MISCELLANEOUS STEEL CONNECTOR PARTS				
49		30 Lbs.	8d Masonry Nails		
50		100 Lbs.	16d Common "		
51		150 Lbs.	8d " "		
52		25 Lbs.	Roofing "		
53		25 Lbs.	8d Casing "		
54		10 Lbs.	6d " "		
55		35 Lbs.	8d Aluminum Sinker Head Siding Nails		
56		20 Lbs.	8d Finish Nails		
57		5 Lbs.	6d " "		
58		1 Lb.	4d " "		
59					
60					
61					
62					
63					
64					
65					
66					
67					
68					
69					
70					
71					
72					
73					
74					
75					
76					
77					
78					
79					
80					
81					
82					
83					
84					
85					
86					
87					
88					
89					
90					

Sheet 5: list of materials for Design 2161
Figure 3-6 (continued)

4) Terrace: The terrace is 14'0'' x 4'0'' for a total area of 56 square feet. The terrace is an uncovered slab and can normally be built for 1/10 of the cost of the main house. Therefore the formula wage for the terrace is $6.00. Multiply the formula wage ($6.00) by the terrace area (56 SF). The terrace will cost about $336.00.

The complete baseline estimate for the main house, garage (with no interior finish), and terrace looks like this:

Main house	$57,600.00
Garage	8,850.00
Porch	1,920.00
Terrace	336.00
Total	$68,706.00

Remember: A baseline estimate is only a guideline. It's part of your cost estimating procedure. Don't sign any contracts until you've completed a *detailed, itemized, material and manhour checklist* for each phase of construction.

Labor breakdown— In Chapter 1, we also learned a guideline for computing the number of manhours required to build a basementless home:

Allow 1 manhour per square foot of floor space (2/3 skilled labor and 1/3 semi-skilled labor).

Our baseline estimate already includes labor and materials. But you'll need a labor breakdown for the main house, garage and porch. Let's apply our manhour formula to Design 2161:

1) Main house: The main house has 960 SF of floor area. Allowing 1 manhour per square foot of floor space, the main house will require 960 manhours (640 skilled manhours and 320 semi-skilled manhours).

2) Garage: The attached garage with no interior finish requires only 0.5 manhours per square foot of floor area. The area of the garage is 295 SF. Multiply the area (295 SF) by the manhour factor (0.5) to get a total of 147.5 manhours required to do the job. Round this up to 148. This comes to 99 skilled manhours and 49 semi-skilled manhours.

If you're building a garage with a finished interior, apply the normal manhour guideline: allow 1 manhour per square foot of floor space. You'd need 295 manhours (197 skilled manhours and 98 semi-skilled manhours) to complete a garage with a finished interior.

3) Front porch: For our porch with gable roof, ceiling and main-roof tie-in, use the same manhour factor as for the garage with no interior finish. Multiply the porch area (64 SF) by the manhour factor (0.5) to get a total of 32 manhours required to do the job. This comes to 21 skilled manhours and 11 semi-skilled manhours.

The complete labor breakdown looks like this:

	Total manhours	Skilled manhours	Semiskilled manhours
Main house	960	640	320
Garage	148	99	49
Porch	32	21	11
Total	1,140	760	380

Let's use our manhour breakdown to figure the labor costs for Design 2161. Multiply the number of skilled manhours by the hourly union wage in your area. Multiply the number of semi-skilled manhours by the hourly semi-skilled wage. Let's say the union wage in your area is $20.00, and the wage for semi-skilled labor is $8.00. Here are your labor costs for Design 2161:

760 manhours @$20.00	=	$15,200.00
380 manhours @$ 8.00	=	3,040.00
Subtotal		$18,240.00
Taxes and insurance @20%	=	3,648.00
Total labor cost		$21,888.00

Add another 15 to 25% for employee taxes and insurance. If you allow 20% for taxes and insurance, this comes to $3,648.00. Your total labor cost for Design 2161 is $21,888.00.

Your baseline construction estimate and labor breakdown can be adjusted to fit your own construction company and crew, as well as any work that you decide to sub out.

The most accurate material and labor estimates will be based on your own experience. Keep detailed records of your own actual costs to streamline estimating and increase your profits.

Every estimate you make has to include overhead, fire and liability insurance, taxes and interest. Then add on a percentage for profit.

Profit is the money you earn *in addition to* the wages you pay yourself for the work you do on the job. If your total costs for Design 2161 come to $69,000.00, a 10% profit would be $6,900.00. You might earn more. You might earn less. It depends on the market and on your selling cost.

A spec builder who builds one or two houses at a time can spend a lifetime building just three or four basic plans. The idea is to select a good plan — one that will appeal to the largest number of buyers — and stick with it until the market changes.

Remember that big houses don't necessarily mean big profits. Sometimes they just mean big headaches. Accurate estimates, tight cost control, quality construction, and knowing what your buyers want, are what lead to consistent profits.

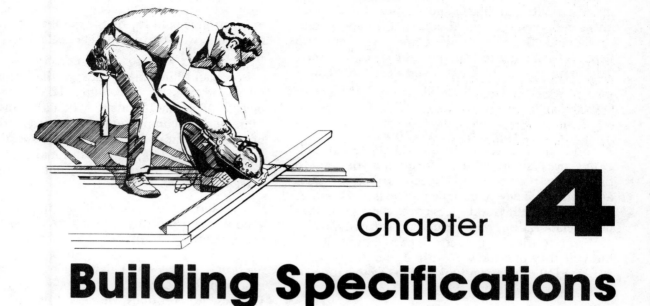

Chapter **4**

Building Specifications

The last two chapters covered the most basic construction document, the blueprints or plans. In Chapter 1 we touched on another important construction document, the contract between the owner and the builder. In this chapter we'll examine a third document that establishes the obligation of the builder — the construction specifications, usually called the "specs."

Construction specifications describe (sometimes in great detail) the grade of each material to be used, and may explain how each material must be installed. Specs also establish the required color and finish, and may identify the manufacturer, trade name and model number. On a small job the specs may be just a few notes stapled to the plans. Specifications for a larger project may be as thick as a phone book.

Specs supplement the working drawings. The drawings will show the outlines of the building, all the walls, partitions, windows, and doors. They'll show the arrangement and connection of all materials. But drawings for a hospital or high-rise office building couldn't possibly include everything the builder needs to know. The specs fill in the gaps with detailed descriptions of materials, workmanship and installation instructions.

The specs are just as much a part of the contract as the plans. Never sign a construction contract without reading and understanding the specs. If the specifications are not 100% clear, get answers either from the architect or the owner. Just skimming over the specs isn't good enough. Leave any detail unclear and you've planted the seed for a dispute over some detail later. Disputes and unresolved problems erode profit on any job.

In this chapter we'll take a detailed look at building specifications. Good specs are divided into sections. Generally, the work of each trade is listed in a separate section. We'll look at important information that should be included in all specs. We'll note some points that will help you identify well-written specs and suggest ways to deal with poorly-written specs. Like good contracts, good specs are written in language any builder can understand. If a set of specs seems designed to confuse or obscure the point rather than clarify it, you may be bidding the wrong job.

Finally, we'll walk through a sample set of building specifications. Questions in this chapter will test your understanding of the specs we've included here. If you can answer the questions, you've learned how to read and *understand* the specs.

Remember this point throughout the chapter. Specs are written to protect the owner. They're the owner's insurance against haphazard performance. The owner can require anything as long as he's willing to pay for it. Never assume that the owner doesn't really intend what the specs require. You may be right. The owner didn't intend and probably never even thought about what the specs seem to demand. But if a dispute comes up, there's no defense for failing to live up to the specs precisely. Too many contractors have paid dearly for ignoring what seemed like an unimportant part of some specification.

Proper Specifications

A good set of specs can be a builder's best friend. Good specs lay out all the information you need to build the structure. Good specifications are clear and concise. They're designed to be understood as well as enforced. Good specs cover every point that needs elaboration, and include no surprises.

When you look at a set of specs, how will you know if they provide you with all the information you need? Let's take a look at the organization of specs.

Spec Organization

The specs should be divided into sections, with one section for each trade (or related trades). For example, all sitework should be listed in one section. The excavation contractor should find every spec that relates to his work in the sitework section. That promotes easier, more accurate estimating.

The Construction Specification Institute (C.S.I.) has developed a spec system called the Uniform Construction Index (U.C.I.). This system divides specifications into 16 sections. Each section has several subsections. Here's the breakdown of sections and subsections and some important information the specification writer should include in each section:

Section 1, General Requirements: This section describes the owner-contractor relationship. Be sure it includes provisions for insurance, arbitration of disputes, certificates of payment, correction of work improperly done, labor and materials furnished at stated intervals, and a bond for completion of the contract.

Section 2, Sitework: This includes clearing, demolition, earthwork, site utilities, paving, fencing, surfacing, and landscaping. For excavation, the specs should explain the extent of excavation required for basement, footings, piers, plumbing lines, wells, septic tanks and drains. After reading the specs, you should have no question about the depth below finish grade for each item and which contractor will handle each item. On many jobs, excavation is paid for by the number of units moved. How will these units be calculated and by whom?

For backfilling and grading, the specs should reveal the extent of backfilling and the materials to be used. They should identify quality standards for fill materials, compaction requirements and testing procedure, if any. The specs should specify the extent of all grading, any additional materials required, who will supply them, and the quality of workmanship.

For landscaping, there will usually be some provision for protecting existing trees, shrubs and other natural features.

Section 3, Concrete: This includes reinforcement, cast-in-place concrete, precast concrete, and finishes. For footings, there will usually be a section that describes the method of construction, materials required and the size for each class of footing (under walls, piers and chimneys) if this isn't obvious in a plan detail.

For foundation walls and piers, the materials and methods of construction will be spelled out. If a portion of the building has no basement, some section should state the distance from the ground level to the first-floor joists, and describe the method of ventilation for this area.

For basement floors, there will be a description of the subbase, finish, thickness, reinforcement, and drainage.

Section 4, Masonry: This section includes brick, block, stone, and reinforcing for masonry. If there's a chimney on the job, the specs will include a list of materials and descriptions: flues, caps, fireplaces and hearths. The specs will state what thimbles and cleanouts are to be used. If the job is a wood-frame house, there should be some description of how the frame is formed around the masonry of the chimney.

Section 5, Metals: It will include structural steel framing, metal fabrications, steel decking and ornamental metals. The specs should indicate where steel lintels, beams, columns and anchor bolts are required. A section should describe each item and the method of installation. If ornamental railings or grilles are required, the specs will include descriptions, methods of installation and possibly a recommended supplier.

Section 6, Wood and Plastics: This covers all rough carpentry, finish carpentry and rough hardware. For wood framing, the specs will usually identify species, grades and sizes of lumber for sills, plates, girders, joists, studs, corner posts, bracing, bridging, fire stops, rafters, sheathing, and subflooring. Spacing, method of erection, and style of framing for floors, walls and partitions will be identified. It's common to require that the lumber used be grade-marked so it's clear that each piece complies with the specs.

Finish carpentry includes items such as interior and exterior trim, baseboards, moldings and cornices. Watch out for custom trim items here. Some specs require custom trim rather than stock items available at most lumber yards. Custom trim costs several times more than stock items in small quantities. If stock items are to be used, style numbers should be listed. Mantels or custom stairways should be described in detail here. The specs should identify the quality and type for each item.

The specs will give details on rough hardware such as anchor bolts, rough screws, nails, hangers, and strapping.

Section 7, Thermal and Moisture Protection: This section covers waterproofing, insulation, shingles, roofing tile, membrane roofing, flashing vents, and caulking. The specs will describe the type, weight and installation method for flashing on chimneys, dormers, windows, doors, valleys and roof-to-wall intersections, and will state where counter-flashing is required, the type of counter-flashing, and method of installation. Any gutters and downspouts in the job will be covered here. Look for descriptions of materials and details on how they are to be installed.

Section 8, Doors and Windows: You'll find all doors, windows, glazing and hardware listed here. Descriptions of doors, frames and windows will include information on style, material, quality, thickness, and size. There will also be something on how windows are controlled — double hung, sliding, casement, etc. If any patented devices are required, the manufacturer's name, style, and the product number will be included.

If there are screens or storm sash, the specs will describe what screens or storm sash are required, the quality needed, and where they are to be installed.

For glazing, the specs will describe the quality and thickness of glass and method of installation.

The section on finish hardware will usually state a dollar amount that's allowed under the contract. The idea is that the owner selects the hardware from a catalog or at a building material dealer when it's time to install it. The contractor pays for the hardware. Any excess over the allowance must be reimbursed to the contractor by the owner.

If there isn't an allowance in the specifications, manufacturer's name and catalog number will usually be listed for each item. Finish hardware includes butts, hinges, locks, door knobs, window fasteners and drawer pulls.

Section 9, Finishes: This includes a lot — lath and plaster, metal studs, gypsum wallboard, tile, terrazzo, suspended ceilings, flooring, carpet, and painting.

The section on interior wall and ceiling finishes should include a detailed description of the types of wall and ceiling finishes to be used. Information on finish floors will probably include the grade and brand name of the material required and a reference to the manufacturer's installation instructions.

If the job includes lath and plaster, the specs will specify how it must be mixed, how it must be applied, the type of finish, and the rooms to be plastered. The lath will be identified as either metal or plasterboard and will include something on the method of installation. The thickness of grounds to be used should be listed.

In the paint section, look for the manufacturer's product number and color if ready-mixed paints are used. The specs will state the number of coats required in each location.

Section 10, Specialties: This part of the specifications may be quite short or even non-existent. It covers items such as compartments and cubicles, vents, flagpoles, partitions, and toilet accessories. Nearly everything here will be specified by model number and manufacturer.

Section 11, Equipment: This may be a large section in the specifications for some types of buildings. It covers equipment for banks, such as vaults, counters, automatic tellers; for churches, such as pulpits, pews, altars; for schools, such as desks, tables, video and audio devices; food preparation equipment for restaurants; gymnasium and service station equipment, and the like. For a single family home, you won't find much listed here. If there is anything in this section, it will include a manufacturer's name and catalog number. Note carefully who has to install each item.

Section 12, Furnishings: This covers primarily cabinets (kitchen cabinets, bookcases, medicine cabinets), seating, drapes, rugs and mats. There will be a detailed description of each item including the manufacturer's name and catalog number. Again, be sure of who will install each item.

Note carefully the interior fittings in all cabinets. Sometimes the hardware that's required in the cabinet has to be ordered separately. Look for the type, size, quality of materials and a detailed drawing for custom items. Stock items will be listed by catalog number and manufacturer.

Section 13, Special Construction: This section includes miscellaneous items that don't fit anywhere else: air-supported structures, incinerators, observatories, swimming pools. You'll get a list by type, size, quality and manufacturer. There will be detail drawings for custom items. Be sure you know who has to install each item.

Section 14, Conveying Systems: The most important item here is elevators. But you'll also find pneumatic tube systems, dumbwaiters, conveyor belts and baggage handlers in this section.

Section 15, Mechanical: This section, and the electrical section that follows, may comprise more than half of the specifications in a large commercial or industrial project. Mechanical specifications cover piping, valves, tanks, plumbing fixtures, heating and cooling systems, air distribution systems, ductwork and controls for heating and cooling systems.

Notice that both Section 2, Sitework, and Section 15, Mechanical, include piping. This can be confusing. Piping in Section 2 will be sewer pipe, site drainage pipe, and water or gas mains away from the building itself. Piping in and around the building will usually be listed in Section 15. But whoever wrote the specifications may not follow this distinction. Look for pipe and plumbing in both Sections 2 and 15.

The section on heating and cooling equipment will include a system description and probably give the name of the manufacturer and catalog number for all major components. There may also be a list of radiators or registers by size and location.

Plumbing specifications will include the type and size of material required for all soil lines, vents, drains, hot and cold water lines, and gas lines. Fittings will be identified and the method of installation will be noted. A common problem is the question of who does the cutting for the plumber and who patches up after he's finished. The specifications should address this issue.

The specs should also identify who is to make the connection with the public water supply and sewage systems. Fixtures will be listed by manufacturer's name and catalog number.

Section 16, Electrical: The last section includes electrical conduit and duct, wire and cable, supports, switches and receptacles, service entrance switchgear, transformers, generators and telephone systems.

Wiring and lighting specs will include information on the type of wire and gauge, the number of circuits to be installed, and the number of outlets per circuit. Outlets and switches will be shown on the plans. But the quality or manufacturer's catalog number will probably be listed in the specs.

If ornamental lighting fixtures are included, an allowance for these fixtures will be listed in the specs so the owner can pick the fixtures out later.

That pretty well summarizes what you'll find in a set of specifications. Of course, a small residential job may have very brief specifications and will omit several of the 16 sections we've just described. And not every set of specifications will follow the 16 section U.C.I. system exactly. Many government jobs, and especially jobs for the U.S. Army or Navy, use a slightly different order. But the similarities will be much greater than the differences on nearly all jobs.

Spec Language
Specifications must give you a clear, accurate understanding of the exact construction required. A good set of specs will have short sentences, exact terminology, exact punctuation, no blanket clauses, and accurate descriptions. Poorly-written specs are often filled with unclear descriptions, unreasonable requirements, blanket clauses and errors.

Let's look at the keys to well-written specs. Then we'll cover some of the common problems found in substandard specs and how to handle them.

Well-written specs— Here's what to look for:

1) Short sentences written in simple language. Spec descriptions should be easily understood by contractors, subcontractors, superintendents and tradesmen.

Specs aren't just a construction aid. They're also a legal document. But this doesn't mean they have to be complicated and confusing. They should be easily understood by all the people who have to use them to construct the building.

2) Exact terminology. Each word in a spec description must have one meaning and only one meaning in the context in which it appears. If a word or description has more than one meaning, get it clarified *in writing* before signing the contract.

3) Exact punctuation. Be sure you read each description carefully. Read the punctuation as well as the words. Commas left out or misplaced can change how the building is constructed. When the building requirements change, so does your estimate, materials order, cost . . . and profit. Make sure you understand the meaning of each description.

4) Steer clear of "blanket clauses." Examples of blanket clauses are: "to the satisfaction of the architect," "as directed," "if required," or "except where otherwise instructed."

The trouble with a blanket clause is that it puts you at the mercy of someone's arbitrary discretion. No contractor would bid a job if the only specification read, "Build this home to the satisfaction of the owner." That's a very clear and objective statement. But who can predict what will satisfy the owner? Building under a specification like that is like writing a blank check that can be filled in at any time and for any amount — by someone else.

You can't draw up an accurate bid if you don't know exactly what's required. Accurate estimates and tight cost controls are the keys to profit. Don't sign away your profits. Get blanket clauses clarified or eliminated *before you sign.*

5) Accurate, detailed descriptions. Have the drawings in front of you when you read the specs. And ask yourself, "Do these specs fit with these drawings? Do these specs make sense? Can this building be built this way? Are these materials available locally? Do the specs give me the details necessary to make an accurate estimate and construct the right building?"

Good specs can help you become a profitable builder with a good reputation. And when you consider that several thousand separate items go into the average house, you can see how important it is that you understand the exact meaning of each spec description.

Substandard specs— Many contractors either refuse to bid on jobs that have substandard specs or increase the allowance for contingency so there's a cushion to fall back on. If you feel you must bid on a job that has specs that aren't up to standard, here are some of the common problems and how to handle them:

1) Unclear descriptions. Here's an example of a spec description that's unclear. It's taken from a paragraph on framing:

Such lumber as is required shall be of a suitable character for the purpose intended, of straight grain and free from all defects.

The first part of this spec says we can use any lumber suitable for framing. But the next part of the spec tells us the lumber must be straight grain and free from all defects. You won't find much lumber like that at your lumber yard. Probably not one stud in 100 is "free from all defects."

Be cautious. If you can't get the spec description changed, bid a price that will assure you of a reasonable profit even if you have to send a lot of lumber back to the lumber yard. Don't assume that you'll be able to use the same kind of lumber you use on most jobs.

2) Unreasonable requirements. Here's an example of an unreasonable spec. It's taken from a clause on concrete:

Nowhere on the job shall the concrete be dropped into place through a greater distance than six inches.

This requirement is unreasonable because it isn't practical. Somewhere on every job the chute pops up a little and drops the mud more than 6 inches. It may be possible to keep the drop from the chute to less than 6 inches on nearly all of the job. But doing that will slow production considerably and produce no advantage to anyone. A better specification would require that the drop from the chute be reduced to the minimum amount practical.

Here's another example of an unreasonable or unfair specification:

Whenever the bottom of the trench is not sound, at no cost to the owner, the contractor must expand the size of the trench until it reaches soil which (in the opinion of the architect) is suitable for supporting a concrete foundation.

The architect is just trying to protect his client. But this spec is unfair because it's turned the contractor into an insurance company, covering the risk that the soil isn't sound. Sure, you can't pour a footing in a bog. But the owner bought the lot, not the contractor. If the footing has to be expanded, the owner can pay for the expansion by the cubic yard of soil excavated and cubic yard of extra concrete poured.

3) Broad exceptions. Clauses that say, "except where otherwise noted" create a lot of problems. This means that everyone using the specs has to search through them and try to locate each "otherwise." Specification writers should state where and what the exceptions are. Here's an example of a clause that will give you a problem:

All floors throughout the residence shall be grade 15 continuous filament level loop nylon carpet with a 100-ounce rebond pad, except where otherwise noted.

Here's what the clause should say:

All floors except the kitchen and bathrooms shall be grade 15 continuous filament level loop nylon carpet with a 100-ounce rebond pad.

4) Incorrect specs. Sometimes you'll find that your specs contradict your blueprints. Here's an example of a spec description that doesn't fit with the plans.

Lay a 4" cement floor over the entire basement, to be composed of one layer 3½" thick concrete consisting of 1 part portland cement, 3 parts clean, dry sand, and 5 parts crushed stone. Finish with a 1/2" top dressing composed of 1 part cement to 2 parts sand.

When we look at the plans, we find that the floor is only 3½" thick, not 4" thick.

What should you do when the specs and blueprints don't agree? Look at the first page of your specs. The first page often tells you how to handle this problem. Here's an example:

Discrepancies: The drawings herein referred to are intended to cooperate with this specification and any work shown on one and not mentioned in the other, or vice versa, is to be performed without extra charge. All figured dimensions are to take precedence to scaled dimensions and larger drawings to those of smaller scale, and should any error be discovered, it must be referred to the architect for correction.

For our sample problem with the 3½" floor, we would refer the error to the architect. Don't do whatever seems easier for you and figure that once it's done, no one's going to care. Here's another example of instructions for handling discrepancies between specs and plans. You'll find these instructions in the "general requirements" section of the specifications:

The documents forming the contract are mutually accessory and complementary, and shall be given precedence in the following order, provided they are in existence at the time of the closing of the contract:
  (1) General requirements of the contract. (2) Specifications. (3) Full-size detail drawings. (4) Large scale drawings. (5) General drawings.

These instructions tell us:

• When your specs and blueprints don't agree, do what the specs tell you to do. Specs have priority over blueprints.

• When your detail drawings or large scale drawings don't agree with your general drawings, do what the detail drawings or large scale drawings tell you to do. Detail drawings and large scale drawings have priority over general drawings.

We've looked at the organization and language in a good set of specs. And we've seen how to handle some of the common problems found in substandard specs. Now we're ready to walk through a complete set of building specifications. Let's study the plans and look at the specs for the Alfred B. Stone residence.

The Alfred B. Stone Residence

Assume you're negotiating with Mr. Stone to build the house shown in the perspective drawing in Figure 4-1. Figure 4-2 shows the floor plan.

The owner has provided you with the plans. You take them back to your office to study. Here are some of the major features:

1) Part of the exterior is rock faced. The balance is wood siding. The chimney is built of ledge rock with a concrete cap on top. The roof has a 4/12 pitch.

2) The garage is a part of the house and can be entered from the terrace. The garage door is an overhead door and slides on tracks suspended below the ceiling of the garage.

3) There's a basement under the main house. The basement plan shows the foundation walls. The concrete footings (under the foundation walls) are 1'4'' wide. The footings are wider than the walls they support. This allows the load (carried by the footings) to be spread over a larger ground area.

4) The basement has metal windows.

5) The walls are constructed of 2 x 4 studs. On the outside of the studs there's insulated sheathing, building paper and wood siding. On the inside there's lath and plaster. The entire wall thickness for these walls is about 6''. The 4'' studs actually measure 3½''. The studs between the closets are set

flat to gain more space.

6) There's a fireplace in the living room. Detailed drawings of the fireplace and chimney are included in the plan.

7) The plan calls for a flagstone terrace and gives the dimensions.

8) Screens and storm windows and doors are required.

9) Gutters and downspouts are to be installed.

10) A radiant heating system is specified as the home's heating plant.

Now that we've covered some of the important details shown in the plans, let's look at the specifications for the Stone residence. If this were a real job, you would have your plans in hand when you read the specs. Just as you're always checking the floor plans, elevations, sections and detail drawings against each other, you'll also be checking your plans against your specs.

Specs for A. B. Stone Residence

Here are the specifications for the Stone residence. Be sure to read each spec requirement carefully. Always check specs against the plans, and make sure you *understand* each spec requirement. *Your success as a builder depends on it!* At the end of these specifications, you'll find a series of questions you can use to test your understanding of the specs for this job.

You'll need to refer to the specs during each phase of construction. Be sure you read the specs carefully and understand them before you begin construction. And don't hesitate to refer back to them as often as necessary.

Courtesy: Home Planners, Inc.

Perspective drawing of Alfred B. Stone residence
Figure 4-1

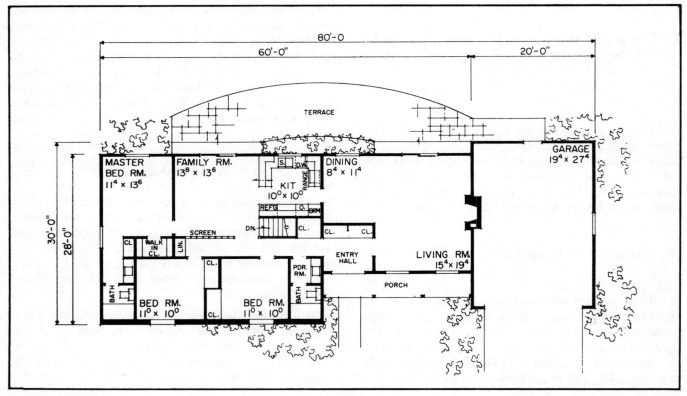

Courtesy: Home Planners, Inc.

Floor plan of Stone residence
Figure 4-2

Section I - General Conditions of the Contract

1) **General note:** The contract documents consist of the drawings and specifications, the agreement and the general conditions of the contract.

2) **Foreman:** The contractor, if not on the work personally, shall keep a competent foreman on the job. The foreman shall represent the contractor in the latter's absence and all directions given to him shall be binding as if given to the contractor. On request, all such directions will be confirmed in writing to the contractor.

3) **Changes in the work:** No changes or extra work shall be allowed or paid for unless agreed upon in writing by the owner before the same is executed.

4) **Permits:** The building laws of this county and city relating to the subject matter of this specification, as now in force, are to form a part of this specification, and all of their requirements are to be fulfilled by each contractor. Contractors shall obtain all required construction permits for the completion of their parts of the work and pay for the cost of the same.

5) **Property surveys:** The owner will establish the lot lines. All other lines and levels necessary to the location and erection of the building shall be established by a competent surveyor who shall be employed by the masonry contractor. The contractor is to check carefully all levels given on the drawings with existing levels and is to call the owner's attention to discrepancies before proceeding with the work.

6) **Materials and protection:** The contractor shall provide and pay for all materials, labor, water, tools, equipment, etc. necessary, unless otherwise noted. All materials and workmanship shall be of the best quality. The contractor shall provide and maintain complete protection of all his work and protect the owner's and adjacent properties from damage, injury, or loss.

7) **Discrepancies:** This specification and the drawings are intended to describe and provide a finished piece of work. They are intended to be cooperative, and what is called for by either shall be as binding as if called for by both. The contractor will understand that the work herein described shall be complete in every detail, notwithstanding every item necessarily involved is not particularly mentioned, and the contractor will be held to provide all labor and materials necessary for the entire completion of the work intended to be described, and shall not avail himself of any manifestly unintentional error or omission should such exist. Should any error or inconsistency appear in the drawings or specifications, the contractor, before proceeding with the work, shall make mention of the same to the architect for proper adjustment, and in no case shall he proceed with the work in uncertainty.

8) **Patching and replacing damaged work:** Each contractor will be held responsible for all damage that is caused by his work or workmen. Patching and replacing of damaged work shall be done by the contractor who installed the work, as directed by the architect, but the cost of same shall be paid by the contractor who is responsible for the damage.

9) **Liability insurance:** The contractor shall secure and protect the owner from any liability or damage whatsoever for injury (including death) to any persons, or property; including any liability or damage which may arise by virtue of any statute or law now in force or which may hereafter be enacted. All insurance policies are to be issued by companies authorized to do business under the laws of this state, and certificates of such insurance shall be filed with the owner or architect as requested.

10) **Social Security insurance, etc.:** Each contractor shall pay the contributions measured by the wages of his employees required by the Social Security Act and by the public laws of the state and assumes exclusive liability for said contributions. The contractor shall further agree to hold harmless the owner of account of any contributions measured by the wages of his employees as above stated, assessed against the owner under the authority of said act and the public laws of the state. The contractor shall also be responsible for the payment of any payroll taxes or contributions for unemployment insurance or annuities which are measured by the wages, salaries, or other remunerations paid to the employees of the contractor.

11) **Fire insurance:** The contractor will secure a fire insurance policy in a company duly licensed, covering the building under construction and (while on the premises thereof) equipment, tools, materials, supplies and temporary structures. Such policy shall include the perils of fire and lightning, those of extended coverage and vandalism and malicious mischief, as well as the standard form of unearned premium endorsements. Such policy shall be issued on a "Completed Value Form" and include the contractor and owner as insureds. The cost of this insurance shall be borne by the owner.

12) **Guarantee:** The general contractor shall execute and deliver to the owner, before final certificates will be issued, a written guarantee covering all work under the contract. Guarantee to be for a period of one year (unless a different period of time is specified under the several trade sections). Should any defect develop during said period or periods due to improper materials, workmanship or arrangements, the same shall be made good by contractor without expense to owner. Any other work affected in making good such imperfections must also be made good.

13) **Completion of work:** Should the contractor become bankrupt, or in case of any unusual or unnecessary delay on the part of the contractor in providing the necessary material and performing the necessary labor at the time the same is required, the owner, within three days after notifying the contractor of his intention to do so, shall have the right to enter upon the premises and procure such necessary materials or labor as the case may require and complete the work in such a manner as shall be proper and right under this contract, charging the cost to the contractor and deducting such charges from the amount of the contract price. The expense incurred by the owner as provided herein either for furnishing materials or for finishing the work, and any damage incurred through such default, shall be audited and certified by the owner whose certificate thereof shall be conclusive upon the parties.

14) **Certificates of payment:** Certificates for payment are to be issued by the architect in accordance with the terms of the contract. In case of delivery of materials or prepared work which cannot be conveniently fixed in place, the owner may, if the architect approves what has been done, grant a certificate of payment on account of the same, which payment shall be reckoned one of the payments on account of the contract.

All materials necessary to the construction of the building, delivered upon the premises, shall be held to be the property of the owner, and no materials shall be removed from the premises without the written consent of the owner.

Before the issuing of each certificate for payment, the contractor shall, if required by the owner, furnish him with a sworn statement of the amount due at the time of the application for the certificate in question, to any subcontractors or parties supplying materials. The contractor shall also, when requested by the owner, furnish waivers of lien for subcontractors and material suppliers.

15) **Cleanup:** The contractor shall keep the premises clean at all times and will remove all surplus rubbish at completion of the work. The building is to be left clean.

16) **Subcontractor:** Subcontractors will not be recognized as parties to this contract.

17) **Bids:** The owner reserves the right to reject any or all bids.

Section II - Excavating and Grading

1) **General:** The contractor performing the work herein specified shall be responsible for and governed by all the requirements of the "General Conditions." He shall do all excavating for foundation walls, footings and floors as required on the drawings. The bottoms of all footing excavations shall be exactly level on solid undisturbed earth. Excavation shall include space for installation of forms and an extra space of 1'0" around the entire outside, to permit cementing and waterproofing the outside of all walls below grade. Excavations are to be kept free of standing water.

2) **Top soil removal:** Remove all top soil over the building area and for a distance of ten (10) feet beyond the outside walls. Place same in a neat pile on the lot as directed.

3) **Backfilling and grading:** The contractor shall bring the finish grades to the lines shown on the drawings. If additional dirt is needed or excess dirt is to be hauled away, the owner will pay for same. When the walls are built to grade line, the contractor shall backfill against them with approved material, free from debris. Backfill is to be well puddled and tamped.

Excavating and backfilling for underground plumbing, etc. will be done by those performing the mechanical trade contract.

Section III - Masonry and Concrete

1) **General:** The "General Conditions" preceding these specifications shall be considered as a part of these specifications. The contractor is to furnish all materials and labor pertaining to the masonry and concrete work as shown on the drawings and specified herein. All walls are to be level and true.

2) **Concrete work:** All concrete work used in construction for floors, walls, or drives shall be mixed in the proportions necessary to develop a compressive strength of 2,000 pounds per square inch in 28 days. The proportions of fine and coarse aggregates used may be varied slightly as directed by the architect. The cement shall be a first-class approved brand of portland cement. Fine aggregate shall be clean, sharp sand. Coarse aggregate shall be graded, washed gravel or crushed hard stone. Mixing shall be done in a batch-machine mixer until the cement is thoroughly distributed and the mass uniform in color. For pavements, expansion joint material shall be held down 1/2" from finish surface. At completion of job, fill the top portion of all joints in pavement with hot asphalt. Construction and contraction joints in pavements shall be provided as recommended by the Portland Cement Association for concrete pavements.

3) **Concrete blocks:** Hollow blocks, where shown, shall be first quality, smooth face, well seasoned, square edge, and sound, with a 8" x 16" face. Same to be laid true and plumb with all courses level. Provide half joint, corner and bond blocks as required.

4) **Concrete floors:** Concrete floors in basement shall be two coat work 4" thick, troweled to a hard smooth surface. The finish coat shall not be less than 1" thick and of a mixture consisting of one part cement to one and one-half parts clean, coarse sand.

5) **Zonolite concrete:** Where called for on drawings, provide a 6:1 mix Zonolite insulating concrete slab. Trowel to a reasonably smooth finish.

6) **Walks, driveways, garage floors:** Walks, driveway, and garage floor shall be 4" one-coat concrete slabs reinforced with a 6" x 6" wire mesh and troweled smooth.

7) **Drain tile:** Furnish and install as shown on drawings 4" agricultural drain tile around the basement wall, with 1/4" joints covered two-thirds with tar paper. Lay to a pitch of 1/4" to the foot. Cover same with 12" of gravel, and connect with a tight-joint glazed tile to building drainage system.

8) **Foundation walls:** To be built of approved type cement blocks completely bedded in mortar. Provide a 4" solid slab at top of wall for joist bearing. Waterproofing to consist of a 1/2" coat of 2:1 mix waterproof cement and sand, troweled to a smooth finish and a coat of asphalt waterproofing applied as recommended by the manufacturer.

9) **Mortar:** All mortar for brick, concrete block, and any other masonry units shall be composed of one part portland cement, one part lime putty or hydrated lime and six parts sand. Same to be laid with 3/8" joints, tooled smooth for a weathered joint. Lime shall be U.S. Gypsum Company's Red Top hydrated lime or freshly burned lump lime slaked on the job for 48 hours, allowed to cool and kept moist until used.

10) **Ledge rock:** To be natural strata-type stone similar to Tennessee hard sand stone. Random coursing or rubble to be laid with no more than three vertical joints in a line.

11) **Chimney:** Build the chimney with common, hard-burned brick and in no case allow less than 2" between any woodwork and smoke flues. Exposed parts of chimney are to be faced with ledge rock properly bonded. Smoke flues shall be lined with approved type fire-clay lining. Provide and set necessary thimble and cleanout doors.

12) **Fireplace:** Build as detailed, with sides, back and floor lined with fireclay brick laid flat with thoroughly filled joints. Equip same with chain control cast iron damper. Provide ash dump where shown. Flue size as shown, same to be approved flue lining with smooth cement joints. Hearth to be built as shown. This contractor is to furnish complete and built-up mantel as shown on plans and details. Provide separate clay flue lining in fireplace masonry for heating plant.

13) **Flagstone terrace:** Lay random patterned flagstone where shown on drawings. Same to be laid on a level bed of sand.

14) **Lannon stone:** Furnish and lay rock-faced Lannon stone as shown on the drawings. The stones shall be 4" thick and properly tied to the framing with metal wall ties. These stones shall be laid in a nonstaining cement mortar composed of one part of nonstaining cement to three parts of sand with the addition of one-fifth part of hydrated lime.

15) **Miscellaneous iron and steel:** Furnish and install miscellaneous items such as grilles, package receivers, fireplace dampers, cleanout doors, ash dumps, etc. Paint all steel and iron with one coat of alkyd enamel in addition to the shop coat before installation.

Furnish and erect all metal windows shown for the basement and as hereinafter specified. All sections of frames, sash and muntins shall be solid hot rolled steel of size, thickness and shape as manufactured by Detroit Steel Products Company, Ceco Steel Products Company or approved equal. All sash shall be stock as specified, divided into the number of units shown on the drawings. All sash shall be arranged for inside glazing by means of wire clips and putty. Glazing and clips will be furnished and installed by the painting and glazing contractor. All metal windows will receive one coat of rust-resisting primer before shipment.

Section IV - Carpentry and Millwork

1) **General:** The work performed under this heading is subject to the requirements of the "General Conditions" preceding these specifications. This contractor shall furnish and install all necessary materials and labor under this heading as specified herein and indicated on drawings. All framing lumber shall be of Number 1 yellow pine or fir. Lumber shall be free from any defects that will weaken the structural strength.

2) **Framing:** This contractor shall do all cutting and framing required by the plumbing, heating and electrical contractors or any other mechanics to install their work. No woodwork shall be fastened to the chimney. All floor joists are to be 2 x 10's and framed on the ends with material of the same size as the joists. Cutting of the floor joists to a depth of one-sixth of the joist depth shall be permitted except in the middle third of the span. No stud shall be cut more than one-half its depth to receive piping and ducts.

Trimmers and headers of double joists shall be put around all chimneys, stairways, fireplace openings, etc. Allow no less than 2" clearance around chimney masonry. Place double joists under all partitions running parallel to same.

Bridging to be 1 x 3 white pine. All rows of joists 12' and under to have one row of bridging. Spans of joists over 12' to have two rows of bridging.

Studding shall be 2 x 4's spaced 24" on center and doubled at sides of all windows and door openings. All outside and inside corners to be formed with three pieces of 2 x 4 spiked together. All partitions to have 2 x 4 plates doubled at the top and so installed as to tie into intersecting portions. Door and window headers shall be framed with double framing set on edge, and are to be of the following sizes:

Openings 3 feet and under: 2 x 4
Openings 3 feet to 5 feet: 2 x 6
Openings 5 feet to 7 feet: 2 x 8
Openings over 7 feet: 2 x 12
In bearing walls, use trusses where necessary.

3) **Sheathing:** Cover all outside walls with 25/32 insulating sheathing. Provide 1 x 4 diagonal corner braces let into the face of the studs and extending from sill to plate where possible. All exterior sheathing to be covered with one layer of building paper before finished material is put on.

4) **Rough floor:** Rough floor shall be sheathing plywood of C-D grade, unsanded 5-ply, 5/8" thick.

5) **Stairs:** Basement stairs to be 2 x 10 plank closed stairs with 7/8"-thick treads. A stair railing of proper design shall be provided. The stairs will have 6'8" of continuous clear headroom measured vertically from the front edge of tread to a line parallel to the stair run.

6) **Roof framing:** Cover roof with Number 2 sheathing (1 x 6) laid tight and double nailed at each rafter. All rafters of the roof shall be 2 x 6.

7) **Roofing:** Furnish and install strictly in accordance with manufacturer's directions three-tab asphalt shingles laid in strips 12" wide and 36" long. Colors will be selected by owner. Provide a layer of saturated 15-pound-per-square roofing felt between the shingles and sheathing. All asphalt shingles shall bear the Underwriters' class "C" fire label.

8) **Exterior walls:** All outside sheathing to be covered with one layer of building paper manufactured by Johns-Manville Company before finished material is applied. Exterior siding shall be Number 1 redwood of thickness and width called for on plans.

9) **Outside trim:** All outside trim shall be Number 1 white pine and according to detail drawings.

10) **Exterior door frames:** The material for outside door frames shall be the same as specified for outside trim. Rabbet 1⅛" jambs to receive doors and screens as scheduled on plans.

11) **Exterior doors:** All outside doors to be of sizes and designs shown on drawings, 1¾" thick. Material to be oak veneered, sandpapered, scraped, and hand smoothed.

12) **Windows and glazing:** All double hung windows shall be of ponderosa pine, factory built, nonsticking type with factory-fitted sash. All glass to be glazed with double strength "A" quality glass except where otherwise noted on drawings. Provide all special window frames and glass as called for on drawings.

13) **Interior trim and millwork:** All millwork stock shall be thoroughly seasoned, kiln dried, and free from imperfections. All interior trim with exception of bathroom and kitchen shall be red oak, common sawed; all other trim shall be selected birch. Included in this contract is all the detailed work shown such as: (1) wardrobe closets, (2) linen closets, (3) cupboards and cases, (4) medicine cabinets, (5) bookshelves, (6) fireplace mantel, (7) miscellaneous shelving, (8) paneling, (9) door and window trim, (10) cornices and special moldings.

14) **Interior doors:** All interior doors except as shown on drawings to be 1⅜" thick, flush type with built-up white pine cores, cross-banded with 1/8" veneers, sanded and faced vertically on both sides with 1/20" veneer. All closets as shown on drawing to have approved sliding doors.

15) **Finish flooring:** Furnish and install Number 1 oak flooring 3/4" x 2¼" for all rooms except kitchen and bathrooms. Kitchen and bathroom to have Armstrong 12 mil gauge Solarian sheet vinyl placed on a layer of Armstrong S200 adhesive. Pattern to be selected by owner. Between all finished wood floors and wood subfloors provide a single layer of deafening felt as supplied by Johns-Manville Company. All wood flooring to be well drawn together, joints broken and sanded to receive finish.

16) **Insulation:** Furnish and install in accordance with manufacturer's directions ceiling insulation of a 4" standard batt type. Pack all door and window frame spaces with insulation.

17) **Rough and finish hardware:** Rough hardware of every nature such as nails, screws, bolts, etc., shall be furnished and installed as required to execute and properly complete the work. Finish hardware

will be selected by the owner. Contractor shall include in his bid an allowance of $450.00 for this item. Finish hardware shall be installed by this contractor.

18) **Wall tile:** Furnish and install polystyrene plastic wall tile on the walls of the bathroom. Tiles to be 4½" x 4½" marbleized, and are applied on the plaster. Formica will be placed on sink and countertops with a rubber base tile on the return. All material shall be cemented according to the manufacturer's directions and guaranteed against moving and bulging.

19) **Caulking:** All exterior window frames set in masonry and all other trim intersections of wood and masonry shall be caulked with the best grade nonhardening, nonstaining elastic compound and applied as soon after installation of surrounding work as possible.

20) **Weatherstripping:** Furnish and install spring bronze weatherstrips complete with interlocking sills for all exterior doors. Nothing in these specifications shall prevent the installation of prehung doors meeting the requirements listed in these specifications.

21) **Screens and storm sash windows and doors:** Furnish and fit all exterior double hung windows, unless otherwise noted, with half screens, using 14 mesh best quality copper screening. Furnish and fit for all windows and exterior doors, unless otherwise noted, 1⅛"-thick storm windows, glazed with D.S.A. glass. Furnish and fit for all exterior doors, combination copper screen and storm doors.

22) **Medicine cabinets and mirrors:** Provide a recess 24" square for a medicine cabinet, and face it with a 1/4" hinged plate glass mirror, flanked by fluorescent tubes. This will be selected by the owner but set by this contractor.

23) **Shades and rods:** Shades and rods will be furnished by the owner, and an allowance will be provided in the contract. They will be installed by this contractor.

Section V - Lathing and Plastering

1) **General:** The contractor performing the work under this heading is governed by all the requirements of the "General Conditions" which preceded these specifications.

2) **Work included:** The work included under this trade heading consists of the furnishing of all labor, materials, equipment and services required for all plastering work as hereinafter called for and described in the various subheadings of this specification and on the contract drawings. Also, all incidental items to effect a finished and complete job, even though each incidental item in connection with the execution and completion of the plastering work is not particularly mentioned herein.

3) **Lathing:** Gypsum lath shall be applied in sheets 16" x 48", and applied in strict accordance with manufacturer's directions, with all joints broken on studding and well nailed.
Metal lath shall be expanded metal, weighing not less than 3.4 pounds per square yard and used on all interior corners from floor to ceiling and on ceiling of first floor. Use expanded cornerites at junction of side walls and ceiling of rooms. Furnish and install metal corner beads at all exposed plaster corners, plaster openings, archways, etc. Install metal lath strips 18 inches long diagonally across all window and door openings at heads and sills.

4) **Plastering:** Plastering shall be three coat work on all walls and ceilings throughout except in the basement. The first or scratch coat shall be an approved brand of fibered plaster mixed and applied in accordance with manufacturer's directions, scratched on surface to provide bond. Basement ceiling shall be two coat work on gypsum lath. The brown or second coat shall be an approved brand of plaster, mixed and applied in accordance with manufacturer's directions. This coat shall be rodded true with all angles straight, plumb and true. The finish coat shall consist of lime putty and gauging plaster mixed in the proper proportions. This coat shall be applied when brown coat is dry and shall be troweled, smoothed and worked to a perfect surface, free from winds, irregularities and all other defects. Garage is to receive 1/2" cement plaster on metal lath on side walls and ceiling.

5) **Patching and cleaning:** Do all necessary patching and repairing of the plastering after other workmen have finished their work. Thoroughly clean up after the work is completed, and remove all rubbish from the premises. Plastering with cracks, blisters, pits, checks or discoloration will not be accepted. In all cases, the plastering throughout shall be delivered clean and perfect in every respect.

6) **Guarantee:** All plastering shall be guaranteed for a period of 18 months from date of completion and acceptance of building, including guarantee against popping, peeling or other defects.

Section VI - Sheet Metal Work

1) **General:** The work performed under this heading is subject to the requirements of the "General Conditions" preceding these specifications, and the contractor is responsible for and governed by all the requirements specified therein. This contractor shall furnish and install all labor and materials necessary to make watertight all flashings, chimney saddles, ridges, conductors, etc.

2) **Materials:** All sheet metal shall be of Number 26 standard galvanized iron. All copper used shall be 16-ounce hard copper. Solder shall be the best grade, composed of one-half piglead and one-half block tin conforming to A.S.T.M. (American Society for Testing Materials) Standard Specifications. Wherever possible all work shall be prefabricated in the shop and installed with as few field soldering joints as possible. Paint all galvanized iron work with galvanized metal primer after installation.

3) **Gutters and downspouts:** Gutters shall be of galvanized iron, half-round with closed ends, and shall be installed with lap and soldered joints. Gutters shall have 6" long leaders extending into downspouts and all outlets shall be equipped with heavy copper wire ball type strainer. Gutter hangers shall be tinned malleable iron spaced not more than 4' apart. Downspouts shall be corrugated round of the size necessary for the building but not less than 3". Downspouts shall have elbows at the bottom and be cemented into tile pipe at the bottom; also be held in place at 10'0" intervals and with no less than two fastenings per downspout of the same material as the gutter fastenings.

4) **Miscellaneous:** Furnish and install all other items such as kitchen vent fan, ducts, etc. Provide carpenter with all metal flashings required.

5) **Guarantee:** Work under this specification must carry a three-year guarantee against defects of every description.

Section VII - Plumbing

The installation of the plumbing is not a part of this contract.

Section VIII - Electrical

1) **General:** The "General Conditions" for the work attached hereto are hereby made a part of this specification. All electrical work shall be installed to comply with all laws applying to electrical installations in effect in the community. If no codes exist the work shall conform to the regulations of the National Electrical Code and electric utility company regulations. All materials shall conform to the standards of the Underwriters' Laboratories Inc. and at the completion of the work evidence shall be furnished showing compliance with laws and regulations in effect.

2) **Service connections:** This contractor shall provide and install the conduit and entrance fitting from the point at which the utility company shall bring their service to the building and the main service switch. Cabinets shall be of code gauge sheet steel with baked enamel finish. From the outside service connection to the distribution panel the wires shall be Number 2/0 or larger.

3) **Wires and cables:** All wire and cable shall be N.E.C. Standard. No wire smaller than Number 12 will be permitted. Wire Number 8 and larger shall be stranded with a double-braid outer covering. All wires shall be new and of type "R.U." construction, except as otherwise noted. Clamp type connectors of approved manufacture shall be used throughout.

4) **Circuits:** Branch circuits shall be wired with Number 12 gauge wire or larger if required. Appliance circuits shall be of the following sizes: (1) Range - Number 6, 3-wire; (2) Heater - Number 10, 2-wire with ground; (3) Oil Burner - Number 12, 2-wire with ground, (4) Small motors - Number 12, 2-wire with ground. All other circuits shall have Number 12, 2-wire with ground.

5) **Switches, receptacles and plates:** Provide switches and double wall receptacles as shown on plans. All switches shall be 125/250 volts, single or double pole as required, tumbler type, rated 10 amperes, except as noted, and shall be either Bryant or Hubbell manufacture. Flush switches shall have bakelite plates, and surface mounted switches in conduit shall have steel plates. When switches are grouped, furnish gang plates. Duplex convenience outlets shall be 15 ampere, 125 volt, except as noted, with bakelite plates for flush mounting and steel plates for surface mounting. Receptacles shall be General Electric Hubbell flush type. Mounting height of receptacles to be 1'0" above floor, except as otherwise noted.

6) **Radio and television plugs:** Locate as directed on the plans. A 3/4" nonmetallic raceway for antenna lead-in shall be run in unfinished portion of attic and be thoroughly grounded.

7) **Electric clocks and telephone:** Where called for, this contractor shall furnish and install outlets for electric clocks to operate on 120 volt, 60 Hz, single phase service. He shall furnish wall mounting box of adequate size. Such conduit as may be necessary for the telephone company to install service shall be furnished and installed by this contractor. This includes standard boxes and telephone plates for the main telephone and any branch telephones.

8) **Temperature controls:** This contractor to furnish and install the necessary conduit, wiring and control for all thermostats, ventilating fans and radiant heating circulating pumps as required.

9) **Bells and buzzers:** Provide and install where indicated all bells or buzzers, transformers, wiring and push buttons. Exterior bells or buzzer stations to be ornamental push buttons to match finish hardware. Interior bell or buzzer stations shall be standard metal boxes and single polarity plug with plates, and shall be equipped with 10' of approved cord per station having one end terminating in a push button and the other end equipped with plug.

10) **Fixtures:** Fluorescent and incandescent lighting fixtures will be furnished by the owner but installed by this contractor. All circuit interruption equipment necessary for the various circuits is to be provided and installed by this contractor.

Section IX - Radiant Heating

1) **General:** The "General Conditions" for the work preceding this specification are hereby made a part of this specification. The heating contractor will furnish all materials and labor under the scope of the contract unless otherwise noted, and shall install the complete radiant hot water heating system. Anything accepted as a standard trade practice reasonably incidental to the completion of the system shall be furnished under this specification without additional cost to the owner.

2) **Work by others:** The following work and items are to be furnished by others and are not included in the heating contract:
  (a) Openings in walls, floors or ceiling together with any framing necessary to finish the openings.
  (b) Water service to boiler room.
  (c) Furnace chimney with smoke thimble placed at proper length for size boiler specified.
  (d) Electric wiring of all controls and the furnishing of a fused line switch for the entire system.
  (e) Cleaning and decorating of all exposed piping.

3) **Boiler:** Furnish and install one hot water boiler equipped for oil heat and capable of delivering the required number of square feet standing radiation based on 150 Btu per square foot, equipped with pressure and relief valve, also combination altitude, pressure, and temperature gauge. Furnish and install two oil tanks each of 275-gallon capacity in basement.

4) **Pump:** Furnish and install one Bell and Gossett, or approved equal, H-2" booster pump with bronze impeller and cast iron body capable of delivering the necessary water at not more than a 6' head.

5) **Control:** Furnish and mount only one low-voltage thermostat and relay to operate pump and one line-voltage direct-acting aquastat to maintain boiler temperature. A Crane Company or approved equal three-way bimetal water mixing valve is to be furnished and installed as shown on plans. A flow valve shall be located just above the water heater connection.

6) **Piping and grids:** Furnish 3/8" I.D. or 1/2" O.D. copper grids for ceiling of first floor; these grids to be limited to 200' in length. Each grid to originate from the main in basement and terminate in return in basement. Each grid shall have a gate valve located at take-off from main for balancing the system. An air vent connection to each grid shall be located at the highest point with 1/4" tube connection. Grids will be fastened to metal lath with a metal strap on each joist, and care must be exercised to prevent traps in the grids. Grids shall be arranged so the Btu output will not exceed 65 Btu per square foot.

For the basement, 3/4" O.D. copper grids for the floor with output not to exceed 40 Btu per square foot; grids to be mounted on templates not to exceed 1½" from surface of floor, and care must be exercised to prevent trapping. In no case may a single grid be more than 400' in length. All joints must be sweat type with solder containing 95% of tin and 5% antimony, and all ends to be opened to prevent any restrictions at joints due to burrs from cutting.

When grids are completed, all piping must be tested with a water pressure of 200 pounds per square inch before plastering or pouring of floors, as a guarantee against leaks. A gate valve is to be placed to each grid in floor from main to balance the system. Both concrete and plaster must be thoroughly dry before any heat is applied to the system; at least two weeks with good ventilation should be allowed. The concrete slab shall be poured directly over the heating pipes. In no instance shall the thickness of concrete above the top of piping be less than 1½".

This contractor shall be responsible for maintaining all radiant heating pipes in their proper position during the pouring of the concrete floors. This shall include all necessary instruction, supervision and inspection.

7) **Expansion tank:** Furnish and install one 30-gallon expansion tank; with wheel handle stop and waste on inlet and 1/2" boiler drain on outlet; also protection against freezing must be provided by furnishing a good permanent type of antifreeze.

Section X - Painting

1) **General:** The work performed under this heading is subject to the requirements of the "General Conditions" preceding these specifications. This contractor shall properly protect finished work, hardware, etc., from damage and paint spots and he shall make good all damage done to such items through the neglect or carelessness of his employees or from his failure to properly protect the same.

All finished surfaces to be smooth, even, and free from any defects. Brushwork shall show even coatings free from brush marks.

Prepare all surfaces before painting, removing dust and marks from the surface. Sandpaper all woodwork before starting work. All check and nail holes are to be puttied after the priming coat is applied and shall be flush with adjoining surfaces. Use tinted putty to match woodwork on stained and varnished surfaces. All knots, sap or pitch streaks shall be shellacked before priming. All millwork shall receive primer coat before or after installation. Exterior work shall be primed at once after erection, but no painting shall be done in freezing or wet weather.

Furnish and arrange for inside glazing by means of wire clips and putty for all metal windows in basement.

2) **Colors:** All colors not covered by the specifications shall be as selected by the owner and samples shall be submitted for his approval.

3) **Materials:** All materials shall be the best of their respective kinds, manufactured by DuPont & Company, Pratt & Lambert, Sherwin-Williams or approved equal. Linseed oil and turpentine shall conform to A.S.T.M. specifications. All materials shall be delivered to the job in the original containers of the manufacturer with labels intact and seals unbroken.

4) **Exterior woodwork:** The painting of all exterior trim shall consist of two coats of white exterior latex paint in addition to the primer coat. The exterior stained woodwork shall receive two coats of stain and penetrating oil.

5) **Sheet metal work:** Paint all sheet metal work except copper, lead, and zinc with two coats of exterior paint after erection. These coats are in addition to the shop coat required.

6) **Interior painting:** All wood and trim shall be stained and varnished. It shall be filled with paste filler, mixed with oil stain and well rubbed in. Clean lightly with sandpaper and apply one coat of white shellac. Finish with two coats of varnish sanded between coats. Varnish to have finish as selected.

Now that you've read through the specs for the Stone residence, let's test your knowledge with the following questions.

Quiz on A. B. Stone Residence
Here are some questions that might come to mind when you're building the Stone residence. I've provided the answers immediately following the questions, but you should be able to find all the answers in the specifications.

1) Can we use door stops for exterior doors?

2) Is sheet insulation used throughout the building?

3) What size are the rafters?

4) How many coats of paint are required for the exterior trim?

5) How long is the guarantee for the sheet metal work? The plastering? The concrete work?

6) Can the owner come on the job with his own labor or materials and proceed with the work?

7) Who pays for the cost of fire insurance?

8) How far outside the building line does the excavation extend?

9) What size bridging material is required and what type of wood?

10) Assume that the general contractor on this job subs out the electrical wiring. After the building is completed, the owner finds that the wiring isn't any good. Who is responsible?

11) The section on chimneys tells us that in no case shall any woodwork be less than 2'' from the smoke flues. Let's assume the contractor follows this specification. But when the building inspector comes to inspect the work, he tells the contractor that the distance between the woodwork and flues must be at least 5''. Who's responsible for this error?

Answers

1) No. The door jambs will be rabbeted 1¼'' for the door and screen.

2) No. Standard 4'' batt type ceiling insulation must be installed.

3) You'll use 2'' x 6'' rafters.

4) Exterior trim will have three coats of paint.

5) Sheet metal: 3 years. Plastering: 18 months. Concrete: 1 year.

6) The owner can do this if there's any unusual or unnecessary delay.

7) *The owner.*

8) *It extends 1'0".*

9) *Use 1" x 3" white pine.*

10) *The "General Conditions" section tells us that a subcontractor isn't recognized as a party to this contract. This means that the owner will hold the general contractor responsible for the faulty wiring. As the general contractor, you can hold the subcontractor responsible. But you'll still have to make sure that the work is redone to the satisfaction of the owner.*

11) *This question may cause a dispute and perhaps even a lawsuit — though probably one the contractor will lose. The specifications require 2" between the woodwork and smoke flues. But the specs also state that the contractor must comply with the building laws of the city and county and that the contractor is wholly liable and responsible. This means that the contractor is held responsible even though the architect made the error.* Be alert. Look for this type of error, and report it to the architect before construction begins.

Here's another series of questions you can use to test your understanding of the specs for the A. B. Stone residence. If you don't remember the answers, look again at the specs. You can find all of the answers there.

1) What will you do with the rubbish?

2) Is any grading required? What will you do with the dirt?

3) What material will you use in the foundation wall?

4) How will you mix the concrete?

5) What is the thickness of the cement floor? What mixture will you use for the base and top of floor?

6) What kind of flue lining will you use?

7) What material will you use for first-floor joists?

8) Which partitions have double plates on top?

9) When will you use two rows of bridging?

10) Is the contractor paid extra for work not mentioned in the specifications but shown on the drawings?

11) If there is any question about the meaning of part of the specifications or drawings, whose decision is final?

12) Who's responsible for social security payments? Payroll taxes?

13) What kind of shingles will you use? What type of roofing felt?

14) What kind of lath and plaster are required? What is the finish material for the interior walls of the garage?

15) What must you do to knots in woodwork before putting on the priming coat?

16) What kind of floor will you put in the bathroom? How are the walls finished?

17) The specifications for the Stone residence are divided into sections. In which section will you find the following items?

- Type of brick

- Size and spacing of studs

- Grade of interior finish wood

- Type of gutters required

- Type of screens and storm windows

Now that you've taken a detailed look at building specifications, you can see how a bad set of specs can eat up your profits and undermine your reputation as a quality builder. Read your specs carefully. Make sure you understand every spec requirement. And get clarified *in writing* any unclear requirements before you start to build. Your specs should give you a clear, accurate understanding of the exact construction required. And when they do, you're on the road to success.

Chapter 5

Lumber

As a builder of houses and light commercial structures, lumber will probably be your greatest material cost. To get the most for your lumber dollar, you should know how to select the right species, grade and dimension for every part of every project. In this chapter, I'll show you how.

We'll look at the 17 most common species of construction lumber. We'll learn how lumber is seasoned, graded and measured for sale. We'll take a detailed look at both Southern pine lumber and Canadian lumber.

Lumber Species

Construction lumber comes from two classes of trees: hardwood and softwood. Hardwood trees include oak, maple and poplar. These trees have broad leaves, and they drop their leaves every autumn. Softwood trees include fir, pine and spruce. These are the needle-leaf conifers that keep their leaves all year.

There's no rule that says how hard or soft the wood must be to be classified as "hardwood" or "softwood." In fact, many hardwoods are actually softer than the average softwood. Nearly all construction lumber is softwood.

In the center of a tree is a soft, cream-colored core called the *pith*. Around the pith, wood grows in concentric rings. These rings are called *annual* rings, because a new one is added every growing season. Each annual ring is made up of one band of light "springwood" and one band of dark "summerwood." To figure out the age of the tree, you can count the number of annual rings.

The outer rings of the tree are called the *sapwood*. The sapwood carries water and food from the roots of the tree up to the leaves. The inner rings are called the *heartwood*. Heartwood is darker and more durable than the sapwood. That's why heartwood is preferred for construction uses.

Now let's look at the 17 species of wood most commonly used for construction lumber. Each type of wood has advantages and disadvantages. For your beams, you'll want a wood that can support heavy loads. For exterior siding, you'll want a weather-resistant wood. For interior surfaces, you'll want a wood that takes a good finish. Anyone who buys and uses thousands of dollars worth of lumber a year should be able to select the right species and grade for each task.

Douglas Fir

Western Oregon and Washington are called the Douglas fir regions of the United States. Douglas fir has more uses than any other kind of wood. It's nearly all heartwood and doesn't warp or twist. It's free from blue stain, a fungus-caused sap stain which penetrates deep into the wood and can't be removed by surfacing. The large amount of heartwood makes Douglas fir as decay-resistant as the white oak and longleaf pine. The color of the wood runs from yellow to red-brown, depending on growth conditions.

You can use Douglas fir for framing, sheathing, floors, doors, siding, studs, joists, lath and finish. It meets government requirements for framing lumber: high stiffness, good bending strength, good nail-holding power, hardness, and durability. Douglas fir is sometimes called just "fir" or "DF." If you hear a carpenter or builder using those terms, you can bet the reference is to Douglas fir.

West Coast Hemlock

This species grows only in the Pacific Northwest, from Oregon to Alaska. It has an attractive straight, even grain formed of long, tough fibers. It's free from pitch and gum, lightweight when dry, light in color, takes a finish and holds nails well.

West Coast hemlock is good for general construction because of its stiffness and strength-to-weight ratio. Machined West Coast hemlock is light in color. The wood gets harder with age, but it's one of the few woods that doesn't turn dark as it gets old. No other softwood is so much like the hardwoods in color, texture, aging, and finishing qualities.

Sitka Spruce

Sitka spruce is the largest of the spruces. It grows along the coast from northern California to Alaska. The wood is lightweight and moderately hard. Its long fibers give it a high strength-to-weight ratio. Scaffolding and ladders are commonly made from Sitka spruce.

The key quality of Sitka spruce is resilience. It can take shock, bend and recover after carrying a load. It makes good finish material for houses, even though it isn't as attractive as fir, cedar or Western hemlock. It makes good siding, and is easy to paint. It's specified for rough drainboards and large doors for houses, garages, warehouses and freight sheds.

Sitka spruce has small, tight knots and no pitch or odor. It takes nails without splitting. It's commonly used for built-in fixtures, large crating boxes and food containers.

Western Red Cedar

This is the largest and finest of the cedars. This tree grows in the moist regions of Washington and Oregon. Its key quality is decay-resistance. But it's also known for its light weight and even grain. The uniform structure and light cell walls provide air spaces that make the wood a good insulator.

Besides being the best wood for shingles, Western red cedar is also excellent for bevel siding. It doesn't warp, buckle, shrink, swell or check. It shows an attractive grain and texture under natural finishes, but it also takes paint and enamel well. Western red cedar is used for clear and knotty paneling, interior and exterior trim, window sash, doors and frames.

Western red cedar is especially good for mitered corners and tight joints. Its uniform fibers allow precision machining. The wood cuts to sharp and true edges, and holds its position and shape. It's ideal for roof planks and other structures with high humidity, such as trellises, greenhouses and posts.

Incense Cedar

This is a durable and decay-resistant native American wood. It comes from a tree found in California, southern Oregon and western Nevada. The wood has very little resin.

The sapwood is white or cream in color. Heartwood is light brown or light red-brown. The texture is fine and uniform with small, evenly-arranged cells. The wood has a pleasant, spicy smell that's common in all cedars. This wood is used to line closets, either in board or plywood paneling.

Incense cedar doesn't conduct heat or cold well. This makes it a good insulator. Use it for sheathing, siding, floor and roof decking. It's lightweight and easy to work with. This cuts down on handling and construction costs.

Incense cedar takes paint or stain easily and can be machined to a smooth, silken surface. The rich brown-red color, small, sound knots and flowing grain make it distinctive paneling material. In its clear grades, incense cedar is ideal for stained interior trim.

Redwood

California is the main source of redwood. These trees grow to heights of 275 or 300 feet, with diameters of 6 to 10 feet or more. Redwood is famous for durability, low shrinkage and high paint retention. Redwood heart has natural preservative that makes it as decay resistant as other lumber that has been treated with chemical preservatives.

Foundation grade redwood is used when lumber will be in contact with the ground or concrete. It's also widely used for siding and exterior trim. Industrial uses for redwood include tanks, vats, and mill roofs.

Red Cypress

This tree grows in the swamps of the coastal plains of the Southeastern states and along the Gulf of

Mexico. The wood is known for its fine texture and beautiful grain. It looks good with a natural finish.

Like redwood, red cypress has its own preservative. The natural oils in red cypress prevent the growth of plant life that cause decay. These oils also resist termites and help preserve paint. The wood is resistant to many of the chemical solutions used in industrial plants. Red cypress is used for tanks, vats, pipes, troughs and conveyers. It's also been used in breweries, corn refineries and wineries, because it doesn't add any color, taste or odor to food products that come in contact with it.

Idaho White Pine

This tree grows in northern Idaho, Washington and Montana. The wood is light in color, straight-grained and soft. Although it's soft and lightweight, Idaho white pine lumber doesn't split easily. You can nail it right up to the end of the piece.

Use Idaho white pine for flush sidings, paneling, shelving, cornice lumber and exterior trim. Just seal the knots with aluminum house paint or shellac before applying the finish coat. You can also use Idaho white pine for sheathing and subfloors.

The clear grades of Idaho white pine are nearly identical to the Northern white pine found in New England. Many homes were built of Northern white pine in the 1600's and some of them are still standing! All the white pines are weather-resistant, because they have a smooth, even texture. Idaho white pine also takes paint well.

Ponderosa Pine

This tree grows throughout the twelve Western states, at altitudes of 2,000 to 6,000 feet. It has a close, even texture, is tough, and lightweight. This makes it the most popular material for grain chutes, heavy-duty flooring and truck beds.

Ponderosa pine is used for doors, sash, window frames, cabinets, shelving and other stock millwork items. Most softwood doors are made of Ponderosa pine. But it's also used for the cores of hardwood-veneer doors, as it's lightweight, doesn't warp or twist, and takes glue well. Interior trim and moldings are often made of Ponderosa pine — especially small moldings that might be hard to nail without splitting.

Use lower grades of Ponderosa pine for rough carpentry items such as sheathing and subflooring. The knotted material that goes into the upper board grades makes fine shelving and knotty pine paneling. High-speed planing surfaces the knotty boards

to a smooth, polished finish. These boards can also be machined into a variety of patterns.

Sugar Pine

This is the largest of all the pines. It can grow to a diameter of 12 feet and a height of 250 feet. These trees are so tall that it might be 75 feet from the ground to the first limb. Many sugar pines are 6 or 7 feet thick when cut. These trees grow in the Sierra Nevada of California and parts of southern Oregon.

Like Idaho white pine and Northern white pine, sugar pine is a true white pine. The wood of the young tree is creamy white. As it ages, it darkens to a pale brown. It's lightweight and has a soft, even texture, with strong fibers. The wood is firm, and you can cut it smoothly in any direction. Pattern makers and wood carvers like sugar pine because it's easy to do precision work with hand tools in the corky wood. Sugar pine is used for piano keys, organ pipes, foundry patterns and sash.

The sugar pine tree is perfect for the production of wide, thick, clear pine lumber. Architects often request it for special doors, built-ins and enameled interior trim on expensive jobs. Sugar pine is also used for window frames, millwork, drain boards, exterior trim and siding. In southern California, where there are high temperatures and low humidity, sugar pine is used in most doors and sash.

Southern or Yellow Pine

This tree grows mainly in the twelve Southern states. Four types of Southern pine are used in construction: longleaf, shortleaf, loblolly and slash.

Southern pine makes excellent framing material. Builders throughout the South and Midwest use it for siding, flooring, sheathing, wall panels, sash and doors, rails, newel posts and all other dressed or turned exterior and interior finishes. It's also used in beams, girders, posts, columns, joists, rafters, roof trusses, factory flooring, and bridges. Key characteristics of Southern pine are strength, durability, stiffness, nail- and screw-holding power, and attractive texture and grain. It takes stain, varnish and paint if no resin is present.

Arkansas Soft Pine

This tree is also a shortleaf pine, but it's a little different from the Southern pine. Arkansas soft pine has a soft texture, close fibers, is lightweight, brightly colored and has no heavy pitch. These trees grow in the mountains of southwestern Arkansas. The soft texture is tough, resilient and resists splitting from nails and screws.

This wood makes tight, firm joints in framework and mirror-smooth surfaces on paneling. Arkansas soft pine products include interior trim, "Trim Pak" finish moldings, siding, jambs and flooring.

Oak
Oak is the hardwood you're most likely to use for interior finish and floors. The two most common types of oak are red oak and white oak. White oak is heavier and stronger and takes a better finish than red oak. Red oak has an open grain, is more red and is commonly used in floors. Both red and white oak are used for interior finish.

Oak is a porous wood and will shrink. It's kiln-dried, not air-seasoned, and is expensive. Most oak furniture, paneling and doors that you buy today have only a thin veneer of oak over a core of some other less expensive wood.

Maple
Northern hard maple is strong and porous. The tiny, evenly-scattered pores give the wood a fine, uniform texture. Maple doesn't shell, splinter or disintegrate in ordinary use. It wears evenly, even under abrasion. The color of the heartwood is brown. The sapwood is cream colored. Make sure the maple you use has been kiln-dried.

Maple takes a high polish. A heavy-duty finish seals the pores, keeps out dirt, and resists marring and scratching. A hard maple floor with a natural finish has a golden glow, and it becomes more attractive with age. To get a darker colored floor, use an acid stain to prepare the wood for a transparent or permanent color finish. These finishes will show off the grain, bird's-eyes and burls.

Mahogany
Mahogany is the most common tropical hardwood used in this country. The color ranges from light red to nearly brown and the grain is striking. It's a good choice for interior trim. You can stain the wood or leave it natural. Philippine mahogany is common in doors, fireplace mantels, staircases, china closets and paneling.

Birch
Birch is another good interior trim material. It's cut from yellow and sweet birch trees grown in the Michigan-Wisconsin area. The red heartwood is called red birch. The white sapwood is called white birch. Unselected (mixed-color) birch may contain both red heartwood and white sapwood in the same piece of wood.

Use unselected birch when you want a contrasting effect under light stains or under a natural finish. Unselected birch is excellent for veneer work on paneling and doors.

When you use unselected birch under a dark stain, the heartwood and sapwood will blend together. You can also use unselected birch under all paints, enamels and lacquers.

Appalachian Yellow Poplar
Appalachian yellow poplar is a strong, lightweight hardwood. It's been used for fine painted or enameled woodwork since early colonial days.

This wood is moderately soft with a straight, fine grain and uniform texture. The heartwood is yellow-green and is a good material for painted exterior woodwork. The sapwood is white.

That covers the 17 most important lumber species used in construction. Most builders will seldom have contact with species other than the 17 I've discussed to this point.

Now let's see how lumber is seasoned, measured and graded for sale.

Seasoning Lumber
After the tree is cut and sawed into lumber at the mill, the next step is drying or seasoning. Freshly cut timber contains a lot of sap and water and is commonly called "green." This moisture must be removed before you can use the lumber for construction. There are two ways to dry lumber: air seasoning and kiln drying. Let's look at both methods. We'll also take a look at lumber shrinkage and swelling.

Air Seasoning
Air-seasoned lumber has been stacked in piles in the storage yard. Each layer of boards is separated by strips so that air can circulate between the layers. In a few months, the wood will be dry enough to use for framing work. But don't use air-seasoned wood for interior finish. Air-seasoned wood is usually too green for interior trim, sash, doors, frames, rails, cornices — everything except framing. Green lumber continues to shrink as it dries, causing splits and cracks that mar the surface.

Be especially careful to check your lumber deliveries during "boom" construction periods. There probably won't be enough well-seasoned lumber to go around, which means you're likely to get some lumber in many deliveries that's too green to use. Make sure green lumber doesn't find its way into the finish work on your construction projects.

Kiln Drying

For interior finish work, sashes, doors, frames, porches, cornices and all nonframing construction, use kiln-dried lumber. Here's how the kiln-drying process works.

The lumber is placed in a large chamber called a kiln. Air temperature and humidity are scientifically controlled so moisture content of the lumber reaches the desired level after three or four days.

Orders for kiln-dried lumber can be filled on short notice because there's less time required between manufacture and shipping. Most important, it can be dried regardless of the season or weather. Also, kiln drying tends to reduce losses due to splitting and cracking. Another advantage is that the drying process kills the insects and organisms that cause decay.

Your carpenters will have fewer construction problems with kiln-dried lumber. This means your jobs will go smoother, there will be less waste and a better chance of finishing the job on schedule and within the estimate.

The disadvantage, of course, is that kiln-dried lumber costs more. It isn't economical to kiln dry lumber for framing in most cases.

Shrinkage and Swelling

Suppose you've just received a shipment of well-seasoned lumber and you plan to frame the walls even though rain is expected. Will the lumber stay the same size? Not necessarily. Dry lumber swells when exposed to moisture just as green lumber shrinks when used in dry weather.

When the moisture content (MC) of a piece of lumber is between 0 and 30%, the moisture is held within the walls of the wood cells. At about 30% MC, the cell walls reach their limit. This limit is called the *fiber saturation point*. Any additional water must be held in the cell cavities. This is why the lumber swells when the moisture increases beyond the saturation point. Water is filling the space *between* the cells, causing a slight separation of the wood cells.

When the moisture content drops below the fiber saturation point, the lumber begins to give up moisture to the air and starts to shrink. The more moisture lost, the more shrinkage. When commercial softwood dimension lumber is dried from its fiber saturation point (about 30% MC) to an average of 19% MC, the lumber shrinks 2.35% in thickness and 2.8% in width.

Nominal size (inches)	Surfaced green net size (inches)	Surfaced dry net size (inches)
2 × 2	1 9/16 × 1 9/16	1 1/2 × 1 1/2
2 × 3	1 9/16 × 2 9/16	1 1/2 × 2 1/2
2 × 4	1 9/16 × 3 9/16	1 1/2 × 3 1/2
2 × 6	1 9/16 × 5 5/8	1 1/2 × 5 1/2
2 × 8	1 9/16 × 7 1/2	1 1/2 × 7 1/4
2 × 10	1 9/16 × 9 1/2	1 1/2 × 9 1/4
2 × 12	1 9/16 × 11 1/2	1 1/2 × 11 1/4
3 × 3	2 9/16 × 2 9/16	2 1/2 × 2 1/2
3 × 4	2 9/16 × 3 9/16	2 1/2 × 3 1/2
3 × 6	2 9/16 × 5 5/8	2 1/2 × 5 1/2
3 × 8	2 9/16 × 7 1/2	2 1/2 × 7 1/4
3 × 10	2 9/16 × 9 1/2	2 1/2 × 9 1/4
3 × 12	2 9/16 × 11 1/2	2 1/2 × 11 1/4
4 × 4	3 9/16 × 3 9/16	3 1/2 × 3 1/2
4 × 6	3 9/16 × 5 5/8	3 1/2 × 5 1/2
4 × 8	3 9/16 × 7 1/2	3 1/2 × 7 1/4
4 × 10	3 9/16 × 9 1/2	3 1/2 × 9 1/4
4 × 12	3 9/16 × 11 1/2	3 1/2 × 11 1/4

**Size difference between unseasoned and dry lumber
Figure 5-1**

Green lumber is slightly wider and thicker than the same nominal dimension of dry lumber. A piece of green lumber will shrink to approximately the standard dry size as it dries down to about 15% MC. For example, a nominal 2 x 4 measures 1½ by 3½ when dry and 1 9/16 by 3 9/16 when green. Many types of lumber are dried before surfacing and only dry sizes for these products are given in the standard.

The American Lumber Standard definition of "dry" is a moisture content of 19% or less. Most dimension lumber is used in locations where its moisture content will continue to drop until it reaches about 15%. Figure 5-1 shows the difference in size between unseasoned and dry dimension lumber.

Measuring the Lumber

Lumber is measured and sold in *board feet*. A board foot is the amount of lumber needed to fill a space 12" wide, 12" long and 1" thick. Thus, a board 12" wide, 12" long and 2" thick has 2 board feet of lumber. A 2 x 6 that's one foot long has one board foot. A 2 x 4 has 0.6666 board feet per foot of length.

Here's how to find the number of board feet in a piece of lumber:

Multiply the thickness (in inches) by the width (in inches). Divide by 12. Multiply by the length of the lumber (in feet).

Here's an example. Suppose you want to find the number of board feet in a 2 x 8 that's 10 feet long. Multiply the thickness (2'') by the width (8'') to get a total of 16. Divide the total by 12. This gives you 1.333. Multiply 1.333 by the length (10') to get a total of 13.333 board feet.

Here's an exception to remember. If you buy 5/8'' or 3/4'' thick siding by the board foot, each square foot of siding is considered to be one board foot. That's because any board less than 1'' thick is considered to be 1'' thick when using the board foot system. And be sure to use the *nominal* (not actual) dimensions of the lumber. Nominal means the named size, like *2 x 4*, not the actual size, which is closer to 1½'' by 3½''.

The actual size depends on how the wood is surfaced (dressed), and whether it's green or dry. Rough lumber has marks left by the saw blade. It's usually larger in both thickness and width than the standard size. A rough-sawn surface is common in post and timber products. Because of surface roughness, grading of rough lumber can be difficult.

Surfaced lumber has been planed or sanded on one side (S1S), two sides (S2S), one edge (S1E), two edges (S2E), or combinations of sides and edges (S1S1E, S2S1E, S1S2E, or S4S).

Figure 5-2 shows nominal and actual sizes for most common pattern siding. Figure 5-3 shows nominal and actual sizes, and board foot measures for most common lumber.

Grading Lumber

Every piece of lumber you receive should have a *grade mark* stamped on it. This is your guarantee that the lumber has been properly inspected and classified. The grade mark will show: grade, moisture content, species group, grading agency and mill identification number. Figure 5-4 shows a sample grade mark.

In the early 1970's the U.S. Department of Commerce sponsored the American Softwood Lumber Standard (PS 20-70) to provide a uniform set of rules for grading framing lumber. Before PS 20-70 there were as many ways to grade lumber as there were lumber grading agencies. That was no problem if you always bought lumber from the same mill. You just learned to use their standards. But the first time you bought lumber from a different area, everything was changed. You can imagine how confusing this was for architects and specification writers who prepared contract documents for jobs in different parts of the country. PS 20-70 changed

all that. Now there is one uniform standard and nearly all mills follow it.

PS 20-70 provides a national grading rule that shows grade strength ratios and grade descriptions for dimension lumber. Take a look at Figure 5-5. The national grading rule separates dimension lumber into *width* categories. This figure shows the width categories for framing lumber. Notice that the grading rule only deals with *dimension lumber*, lumber from 2'' up to (but not including) 5'' in nominal thickness.

PS 20-70 covers nearly all construction lumber but is not applied to crossarms, factory and shop lumber, finish (select) lumber, foundation lumber, industrial clears, ladder stock, laminating stock, railroad stock, scaffold plank, ship decking, plank stock, stadium seats and worked lumber.

Under PS 20-70, all construction lumber is placed in one of three categories: stress-graded, nonstress-graded, and appearance lumber. Stress-graded lumber is mostly framing lumber. Nonstress-graded lumber is mostly 1'' boards. Appearance lumber is used mostly for siding and flooring where appearance is more important than strength. Let's look closely at those three categories.

Stress-Graded Lumber

Almost all softwood lumber from 2'' to 4'' thick is stress graded under PS 20-70. Stress grading doesn't require that each piece actually be tested for strength. Mechanical stress grading is used on key structural members. But most stress-graded lumber is visually graded. Someone inspects each piece and classifies it by the number and type of defects, such as knots and other imperfections.

For visually-graded lumber a single set of grade names and descriptions is used throughout the United States and Canada (Figure 5-5). Other stress-graded products include posts, stringers, beams, decking, and some boards.

Dimension lumber is the most common stress-graded item in lumber yards. Nearly all dimension lumber is 2'' or wider. Dimension visually graded lumber is divided into five categories: Light Framing, Structural Framing, Studs, Structural Joists and Planks, and Appearance Framing.

Most of the lumber you use is *Light Framing*. The grades are either Construction, Standard or Utility. These are the lower strength materials and are available up to 4'' thick. The best Light Framing grade is Construction. Utility grade has the lowest strength.

Name of Product	Nominal Size		Actual Size	
	Column No. 1		Column No. 2	
	Thickness inches	Width inches	Thickness inches	Width inches
Bevel siding	1 1 1	4 5 6	1/2 by 3/16 5/8 by 3/16 --	3-1/2 4-1/2 5-1/2
V'd-rustic siding Round edge drop siding	1 1 1 1	4 5 6 8	9/16 3/4 -- --	3-1/8 4-1/8 5-1/8 7-1/8
Flooring S-2-S	1 1 1 1 1¼ 1½	2 3 4 5 6 --	5/16 7/16 9/16 25/32 1-1/16 1-5/16	1-1/2 2-3/8 3-1/4 4-1/4 5-1/4 --
D. & M. ceiling Beaded-1-S S-2-S	-- -- -- --	3 4 5 6	5/16 7/16 9/16 11/16	2-3/8 3-1/4 4-1/4 5-1/4
D. & M. beaded partition S-2-S		3 4 5 6	3/4	2-3/8 3-1/4 4-1/4 5-1/4
Shiplap	1	4 6 8 10 12	25/32	3-1/8 5-1/8 7-1/8 9-1/8 11-1/8
D. M. & V'd partition S-2-S	1 1¼ 1½ 2 --	4 6 8 10 12	25/32 1-1/16 1-5/16 1-5/8 --	3-1/4 5-1/4 7-1/4 9-1/4 11-1/4

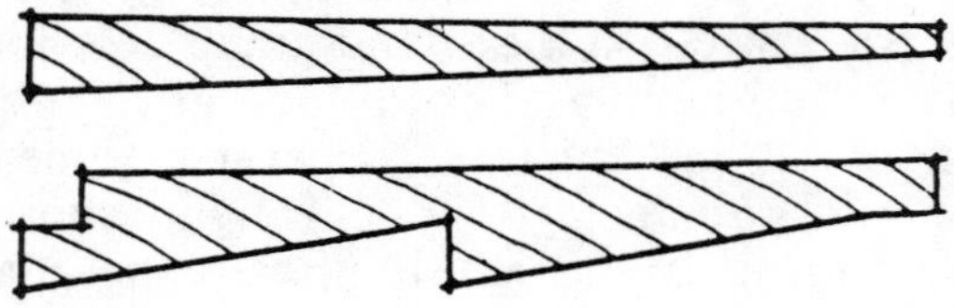

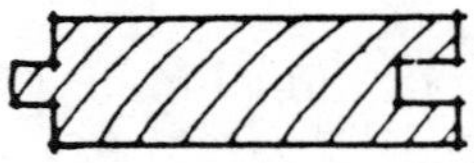

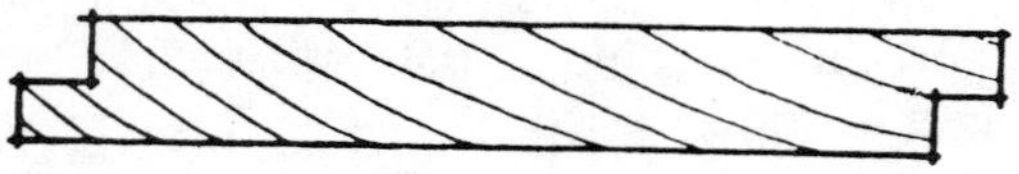

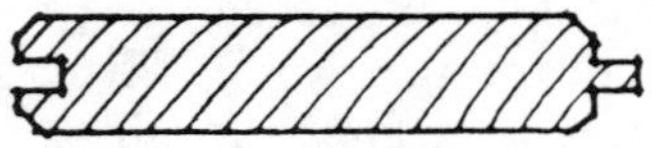

S-2-S means "smooth 2 sides."
D. & M. means "dressed & matched."

Nominal and actual sizes of common pattern siding
Figure 5-2

Nominal size inches	Actual size inches	Board measure per linear foot
2 x 2	1-1/2 x 1-1/2	.33
3	2-1/2	.50
4	3-1/2	.67
2 x 5	1-1/2 x 4-1/2	.83
6	5-1/2	1.00
8	7-1/4	1.33
10	9-1/4	1.67
12	11-1/4	2.00
14	13-1/4	2.33
3 x 3	2-1/2 x 2-1/2	.75
4	3-1/2	1.00
6	5-1/2	1.50
8	7-1/4	2.00
10	9-1/4	2.50
12	11-1/4	3.00
14	13-1/4	3.50
4 x 4	3-1/2 x 3-1/2	1.33
6	5-1/2	2.00
8	7-1/4	2.67
10	9-1/4	3.33
12	11-1/4	4.00
14	13-1/4	4.67
6 x 6	5-1/2 x 5-1/2	3
8	7-1/2	4
10	9-1/2	5
12	11-1/2	6
14	13-1/2	7
8 x 8	7-1/2 x 7-1/2	5.33
10	9-1/2	6.67
12	11-1/2	8
14	13-1/2	9.33
10 x 10	9-1/2 x 9-1/2	8.33
12	11-1/2	10
14	13-1/2	11.67
12 x 12	11-1/2 x 11-1/2	12
14	13-1/2	14
14 x 14	13-1/2 x 13-1/2	16.33

Nominal and actual sizes of common lumber and number of board feet
Figure 5-3

A grade mark "STD&BTR" means that the lumber was originally graded both Standard and Construction. That alone doesn't explain how much was Standard and how much was Construction

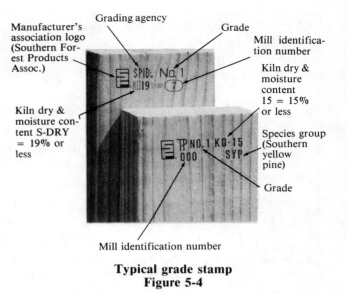

Typical grade stamp
Figure 5-4

grade. The mill's agreement with the original buyer probably makes that clear, but you have no way of telling the percentages specified. Even if you did know that at least 75% was Construction grade, the stack may have been broken and resorted before it reached your yard.

Structural Framing includes thicker material such as 2" x 6", 2" x 8", and 2" x 10". The grades are Select Structural, No. 1, No. 2, and No. 3. These grades are used where bending strength is critical. Rafters and trusses are common applications. Wider material would be used for posts, beams and stringers. Most structural lumber is graded in combinations such as No. 2&BTR or possibly No. 3&BTR.

Stud grade may be 2" x 2" to 4" x 4" and up to 10 feet long. Naturally, 2" x 4" is the most common dimension. Stud grade is usually identified as "PET" which means it has been precision end trimmed to save end cutting waste.

Structural Joists and Planks and *Appearance Framing* are used less than other dimension lumber. Grades for Structural Joists and Planks are the same as for Structural Framing. Appearance Framing is available only as Appearance grade.

Strength of lumber grades— Figure 5-6 gives the bending strength ratio for the five grades of visually graded dimension lumber. The ratio shown at the right is the percentage of strength compared to the average for a clear, straight-grained piece of the same species. Notice that Stud grade has less bending strength than Construction grade under the

Structural Light Framing
(2" to 4" thick, 2" to 4" wide)

Select structural and No. 1	Intended primarily for use where high strength, stiffness and good appearance are desired.
No. 2	Popular for most general construction uses.
No. 3	Appropriate for use in general construction where appearance is not a factor.

Light Framing
(2" to 4" thick, 2" to 4" wide)

Construction and standard	Widely used for general framing purposes. Pieces are of good appearance but graded primarily for strength and serviceability.
Utility	Widely used where a combination of good strength and economical construction is desired for such purposes as studding, blocking, plates, bracing and rafters.

Studs
(2" to 4" thick, 2" to 6" wide, 10' and shorter)

Stud	Special purpose grade intended for all stud uses including load-bearing walls.

Structural Joists and Planks
(2" to 4" thick, 5" and wider)

Select structural and No. 1	Intended primarily for use where high strength, stiffness and good appearance are desired.
No. 2	Popular for most general construction uses.
No. 3	Appropriate for use in general construction where appearance is not a factor.

Appearance Framing
(2" to 4" thick, 2" and wider)

Appearance	For use in general housing light construction where lumber permitting knots but of high strength and fine appearance is desired.

Sample grade classifications for framing lumber
Figure 5-5

Lumber classification	Bending strength ratio
Light framing (2" to 4" thick, 4" wide)[1]	Percent
Construction	34
Standard	19
Utility	9
Structural light framing (2" to 4" thick, 2" to 4" wide)	
Select structural	67
1	55
2	45
3	26
Studs (2" to 4" thick, 2" to 4" wide)	
Stud	26
Structural joists and planks (2" to 4" thick, 2" to 4" or wider)	
Select structural	65
1	55
2	45
3	26
Appearance framing (2" to 4" thick, 2" to 4" wide)	
Appearance	55

[1]Widths narrower than 4" may vary in strength

Strength ratio table
Figure 5-6

Light Framing category, but more than Standard or Utility. Notice also that the two Structural classes have very similar bending strength ratios.

PS 20-70 has established these classes, grade names and the minimum bending strength ratios. This strength ratio is your index of relative quality. Actual strength for any grade depends on the type of tree the lumber was cut from. That's why bending strength is given in a ratio. For example, Construction Douglas fir will have a higher bending strength than redwood of the same grade. But the strength ratio of both will be the same because grades compare the average of clear wood of the same species

Descriptions of what visual graded lumber classes look like are in the grading rule books published by the grading associations. Grades of lumber will have about the same appearance regardless of species. So Stud grade Southern yellow pine will have about the same number of knots, checks and other imperfections as Stud grade fir. But the strength will be different because fir and pine have different bending strengths.

The National Grading Rule also establishes some limits on sizes of edge knots and other defects for lumber that is graded by a combination of mechanical and visual methods. Here are the common defects and blemishes recognized by the American Lumber Standards committee:

Defect: Any irregularity in or on the wood that may lessen its strength, durability or utility value.

Blemish: Anything (not classified as a defect) marring the appearance of the wood.

Bark pocket: Patch of bark partially or wholly enclosed in the wood.

Bird's-eye: Small central spot with the wood fibers arranged around it in the form of an ellipse. This gives the appearance of an eye. The bird's-eye should not be considered a defect unless it's hollow or unsound.

Check: Lengthwise separation of the wood. A check usually occurs across the rings of annual growth. See Figure 5-7.

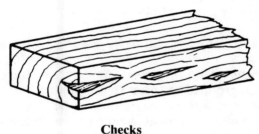

**Checks
Figure 5-7**

Cross-break: Separation of the wood cells across the grain. This can be caused by unequal shrinkage or mechanical stress.

Decay: Disintegration caused by fungi. The words *dote* and *rot* mean the same as decay. Worm holes are also decay. See Figure 5-8.

Gum spot or *streak:*   Accumulation of a gumlike substance in the form of a small patch or streak.

Imperfect manufacture: All defects or blemishes produced during manufacture. These include chipped grain, loosened grain, raised grain, torn grain, skips in dressing, hit and miss, variation in sawing,

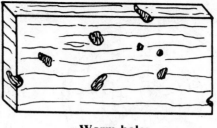

**Worm holes
Figure 5-8**

miscut lumber, machine burn, machine gouge, mismatching, and insufficient tongue or groove.

Knot: A branch or limb may be embedded in a tree. When this tree is cut into lumber, the embedded limb will show up as a knot. Knots are classified by size, form, quality and occurrence. Figure 5-9 shows a pointed knot and a round knot.

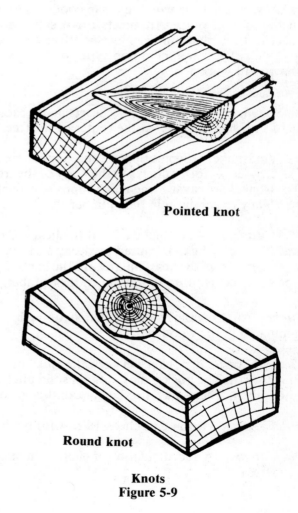

Pointed knot

Round knot

**Knots
Figure 5-9**

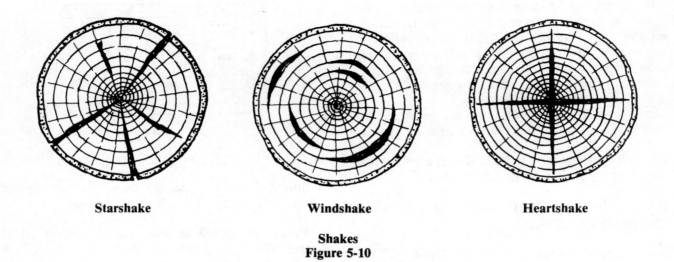

| Starshake | Windshake | Heartshake |

Shakes
Figure 5-10

Shake: Lengthwise separation of the wood, usually between the heartwood and sapwood or parallel to the annual rings. Architects often state in the specifications that lumber must be "free from sap and shakes." Three common types of shakes are shown in Figure 5-10:

• Starshake— Similar to a heartshake, except there's no sign of decay at the center of the tree.

• Windshake— Separation of the annual rings. The separation forms cracks in the body of the tree. Windshakes are caused when high winds wrench the tree. They're common in pine lumber.

• Heartshake— A small cavity at the heart of the tree. The heartshake is caused by decay and cracks that extend from the heart out toward the bark. This type of defect is often found in hemlock timber.

Pitch: An irregularly-shaped patch of resin accumulated in the wood cells.

Pitch pocket: A pocket between the rings of annual growth. The pocket may contain solid pitch or liquid pitch. There may also be bark in the pocket.

Pitch seam: A shake or check filled with pitch.

Pitch streak: An accumulation of pitch in the shape of a streak.

Split: A lengthwise separation of the wood due to the wood cells tearing apart.

Stain: Any discoloration that causes the lumber to be other than its natural color.

Wane: Bark or tapering on the edge or corner of a piece of lumber.

Warp: Any variation from a true or flat surface.

Grouping of species— Some species have always been grouped together so the lumber from them is considered the same. This is done for species that have about the same bending strength and are very similar in appearance. Properties assumed for the group may not apply to a particular species in that group. Each group of lumber species is known by a unique name approved by the American Lumber Standards Committee.

The association that grades the lumber has very precise definitions of the identities, properties, and characteristics of individual species for each group. The major U.S. grading associations are shown in Figure 5-11. Each publishes reference materials about the lumber produced by association members and will sell you the grading rule book currently in use. Get a copy of the rule book that covers the lumber you use most. The book costs only a few dollars and can settle a lot of arguments.

Most lumber grades and sizes serve more than one purpose in construction. Some species or species

Name and address	Species covered by grading rules
Northeastern Lumber Manufacturers Association, Inc. 13 South Street Glen Falls, New York 12801	Balsam fir, Eastern white pine, red pine, Eastern hemlock, black spruce, white spruce, red spruce, pitch pine, tamarack, jack pine, Northern white cedar
Northern Hardwood and Pine Manufacturers Association 305 E. Walnut Street Green Bay, Wisconsin 54301	Bigtooth aspen, quaking aspen, Eastern white pine, red pine, jack pine, black spruce, white spruce, red spruce, balsam fir, Eastern hemlock, tamarack
Redwood Inspection Service 617 Montgomery Street San Francisco, California 94111	Redwood
Southern Pine Inspection Bureau Box 846 Pensacola, Florida 32502	Longleaf pine, slash pine, shortleaf pine, loblolly pine, Virginia pine, pond pine, pitch pine
West Coast Lumber Inspection Bureau Box 25406 1750 SW. Skyline Boulevard Portland, Oregon 97225	Douglas fir, Western hemlock, Western red cedar, incense cedar, Port Orford cedar, Alaska cedar, Western true firs, mountain hemlock, Sitka spruce
Western Wood Products Association 700 Yeon Building Portland, Oregon 97204	Ponderosa pine, Western white pine, Douglas fir, sugar pine, Western true firs, Western larch, Englemann spruce, incense cedar, Western hemlock, lodgepole pine, Western red cedar, mountain hemlock, red alder

Principal grading associations
Figure 5-11

groups are available at the retail level only in grade groups. For example, joist and plank material is often sold as No. 2 and Better (2&BTR). If you need No. 1 material for a special application, you may have to sort No. 2 and better lumber or make a special purchase.

Not all grades, sizes, and species described by the grade rules are produced and not all those produced will be available in your area. Regional interest, building code requirements and transportation costs all affect the way lumber is distributed. Small retail yards stock only a limited number of species and grades. Large yards usually cater to particular construction needs and carry more dry dimension grades along with clears, finish, and decking.

Joist and rafter spans— Floor joists, ceiling joists and rafters must be adequate to carry the intended load. That's clear to every carpenter and builder.

But how do you know which size, grade and species will do the job? A structural engineer could calculate exactly what's needed. But if you aren't an engineer, use a span table for the type of lumber that's available. Your lumber yard will probably have span tables for the types of lumber they sell.

Span tables show the minimum size and grade required for the *horizontal* projection of a member. For floor joists and ceiling joists, the horizontal projection is the same as the actual length of the joists. To determine the span of sloped members, such as roof rafters, you'll have to convert the sloping distance to a horizontal distance. Figure 5-12 shows how to do this.

Let's say your sloping distance is 24' and your roof slope is 8 in 12. Look at Figure 5-12. Locate the point that marks the 24' sloping distance. Now follow along the arc from this point until it intersects with a line drawn from the 8/12 slope marker. From

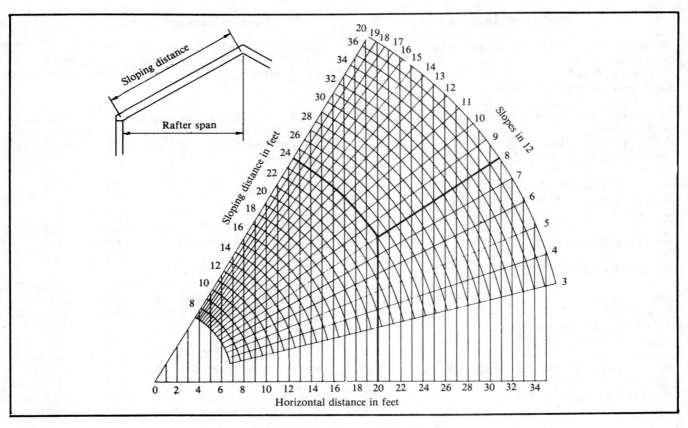

Conversion diagram for rafters
Figure 5-12

this intersection, draw a vertical line straight down to the horizontal distance marker. Your horizontal distance or span will be 20'.

Joists are usually spaced 12", 16" or 24" o.c. But you can also space them 19.2" and 13.7" o.c. When you use 19.2" spacing, you can divide 8' lengths of sheet material into 5 equal sections. The 13.7" spacing divides 8' panels into 7 equal sections. This gives the designer more flexibility. And these new spacings can cut down waste and doubling of members.

The National Forest Products Association publishes the span tables for joists and rafters. These tables show the allowable spans for specific grades of lumber.

Figure 5-13 shows a sample span table for Southern pine floor joists. It assumes a 30 pound per square foot live load. That's the assumed weight of the people and furnishings in the bedroom. Your code will specify what the assumed live load has to be. For example, Figure 5-13 shows that if you're using No. 1 select structural 2 x 10's at 16" o.c., the maximum span between supporting walls is 18'5".

Figure 5-14 compares the relative strength of common species when used for floor joists. Notice that Southern pine and Douglas fir are among the strongest species listed.

Nonstress-Graded Lumber
For most of this century most lumber intended for general building purposes was not stress graded. This category of lumber has been referred to as *yard lumber*. Today, many former yard items have been given defined properties. So "yard" lumber doesn't mean what it used to.

Boards are the most common nonstress-graded construction lumber. Boards are usually sold as nominal 1" thickness. The actual thickness is usually less. Standard nominal widths are 2", 3", 4", 6", 8", 10", and 12". Grades generally available in retail yards are No. 1, No. 2, and No. 3 (or Construction, Standard, and Utility). These will often be combined in grade groups. Boards are solid square edged, dressed and matched (tongued and

Size and spacing inches	Grade inches o.c.	Dense Sel Str KD and No. 1 Dense KD	Dense Sel Str, Sel Str KD, No. 1 Dense and No. 1 KD	Sel Str, No. 1 and No. 2 Dense KD	No. 2 Dense, No. 2 KD and No. 2	No. 3 Dense KD	No. 3 Dense	No. 3 KD	No. 3
2 x 5	12.0	10-3	10-0	9-10	9-8	**9-3**	**8-11**	**8-6**	**8-3**
	13.7	9-9	9-7	9-5	9-3	**8-7**	**8-4**	**8-0**	**7-9**
	16.0	9-3	9-1	8-11	8-9	**8-0**	**7-9**	**7-5**	**7-2**
	19.2	8-9	8-7	8-5	8-3	**7-3**	**7-1**	**6-9**	**6-6**
	24.0	8-1	8-0	7-10	7-8	**6-6**	**6-4**	**6-0**	**5-10**
2 x 6	12.0	12.6	12-3	12-0	11-10	**11-3**	**10-11**	**10-5**	**10-1**
	13.7	11-11	11-9	11-6	11-3	**10-6**	**10-3**	**9-9**	**9-5**
	16.0	11-4	11-2	10-11	10-9	**9-9**	**9-6**	**9-0**	**8-9**
	19.2	10-8	10-6	10-4	10-1	**8-11**	**8-8**	**8-3**	**8-0**
	24.0	9-11	9-9	9-7	9-4	**8-0**	**7-9**	**7-4**	**7-1**
2 x 8	12.0	16-6	16-2	15-10	15-7	**14-10**	**14-5**	**13-9**	**13.3**
	13.7	15-9	15-6	15-2	14-11	**13-11**	**13-6**	**12-10**	**12-5**
	16.0	15-0	14-8	14-5	14-2	**12-10**	**12-6**	**11-11**	**11-6**
	19.2	14-1	13-10	13-7	13-4	**11-9**	**11-5**	**10-10**	**10-6**
	24-0	13-1	12-10	12-7	12-4	**10-6**	**10-2**	**9-9**	**9-5**
2 x 10	12.0	21-0	20-8	20-3	19-10	**18-11**	**18-5**	**17-6**	**16-11**
	13.7	20-1	19-9	19-4	19-0	**17-9**	**17-2**	**16-5**	**15-10**
	16.0	19-1	18-9	18-5	18.0	**16-5**	**15-11**	**15-2**	**14-8**
	19.2	18-0	17-8	17-4	17-0	**15-0**	**14-6**	**13-10**	**13-5**
	24.0	16-8	16-5	16-1	15-9	**13-5**	**13-0**	**12-5**	**12-0**
2 x 12	12.0	25-7	25-1	24-8	24-2	**23-0**	**22-4**	**21-4**	**20-7**
	13.7	24-5	24-0	23-7	23-1	**21-7**	**20-11**	**19-11**	**19-3**
	16.0	23-3	22-10	22-5	21-11	**19-11**	**19-4**	**18-6**	**17-10**
	19.2	21-10	21-6	21-1	20-8	**18-3**	**17-8**	**16-10**	**16-3**
	24.0	20-3	19-11	19-7	19-2	**16-3**	**15-10**	**15-1**	**14-7**

Floor Joists—30 psf live load. Sleeping rooms and attic floors. (Spans shown in light face type are based on a deflection limitation of L/360. Spans shown in bold face type are limited by the recommended extreme fiber stress in bending value of the grade and includes a 10 psf dead load.)

**Southern pine floor joists
Sample span table
Figure 5-13**

grooved) or with a shiplapped joint. Boards formed by end-jointing shorter sections are common.

Boards are usually identified either as *selects* or *commons*. Selects are graded for appearance. Common grades are used in sheathing and for utility purposes. They are separated into three to five different grades, depending on the species and grading rule used. Typical grades are by number (No. 1, No. 2) or by descriptive terms (Construction, Standard).

Unfortunately, there is no uniform system for identifying various species and grades of boards. First-grade boards are usually graded primarily for strength, but appearance is also considered. This grade is used for siding, cornice, shelving, and paneling. Knots and knotholes will be larger and more frequent in the lower grades. Second- and third-grade boards are often used together for subfloors, roof and wall sheathing, and rough concrete work. Fourth-grade boards have only adequate strength. They are used for roof and wall sheathing, subfloor, and concrete forms.

Grading of nonstress-graded lumber varies by species, product, and grading association. Lath, for example, is available generally in two grades, No. 1 and No. 2. The appropriate lumber grading rule book has complete descriptions of these products.

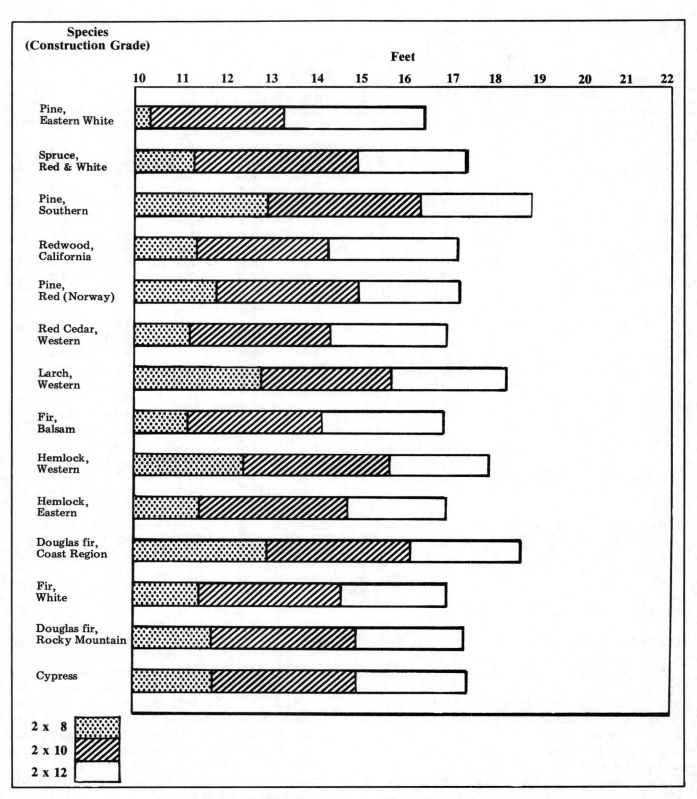

Floor joist span 16'' o.c. 40 pound live load
Figure 5-14

Appearance Lumber

Appearance lumber is nonstress-graded but is in a separate category because looks are important in the grading process. This category includes most lumber worked to a pattern. There is a trend toward prefinishing many of these items. The appearance category of lumber includes trim, siding, flooring, ceiling, paneling, casing, base, stepping, and finish boards. Finish boards are commonly used for shelving and built-in cabinet work.

Most appearance lumber grades are described by letters and combinations of letters (B&BTR, C&BTR, D) and are also known as select grades. Descriptive terms such as "prime" and "clear" are applied to some species. The letters FG (flat grain), VG (vertical grain), or MG (mixed grain) are options for some appearance lumber products.

In cedar and redwood there is a clear difference in color between heartwood and sapwood. Heartwood has high natural resistance to decay and will be identified as *heart*.

In some species, two or three appearance grades are available for some products. A typical example is casing and base in the grades of C&BTR and D. Other species are sold as B&BTR, C, C&BTR, and D. Although several grades may be described in grade rules, only one or two will be offered at your lumber yard.

Grade B&BTR allows a few small imperfections, mainly in the form of minor skips in finishing ("hit and miss"), small checks or stains due to seasoning. Depending on the species, small pitch areas and pin knots may be present. Since appearance grades are based on the quality of one face, the reverse side may be lower quality.

Grade C&BTR is the combination most common in construction. It's used for high-quality interior and exterior trim, paneling and cabinet work, especially where a clear finish will be used. It's also used for flooring in homes, offices, and public buildings.

The number and size of flaws in the wood increases as the grades drop from B&BTR to D and E. Your only key to what the letters mean is the grade rule used by the producing mill. C is used for many of the same purposes as B&BTR. Grade D will have larger and more surface blemishes than C grade. But this may not detract from the appearance when painted. Grade D is used in finish construction for many of the same uses as C.

Redwood and cedar have different grade designations. Grades such as Clear Heart, A, or B are used in cedar; Clear All Heart, Clear, and Select are typical redwood grades. Finish boards are usually a nominal 1" thick. When dressed on two sides they measure about 3/4".

Siding— Beveled siding is ordinarily stocked only in white pine, Ponderosa pine, Western red cedar, cypress, or redwood. Drop siding, also known as rustic siding or barn siding, is usually stocked in the same species as beveled siding. Siding may be stocked as B&BTR except in cedar where Clear, A, and B may be available. In redwood, Clear All Heart and Clear are the usual designations. Vertical grain (VG) is sometimes a part of the grade designation. Drop siding sometimes is stocked also in sound knotted C and D grades of Southern pine, Douglas fir, and hemlock. Drop siding may be dressed, matched, or shiplapped.

Flooring— Oak, maple and harder softwood species such as Douglas fir, Western larch, and Southern pine are used for flooring. Your lumber yard probably has at least one softwood and one hardwood flooring in stock. Flooring is usually nominal 1" thick dressed to 25/32" and 3" or 4" nominal width. Thicker flooring is available for heavy-duty floors both in hardwoods and softwoods. Thinner flooring is available in hardwoods, especially for recovering old floors.

Vertical- and flat-grain (also called quarter-sawed and plain-sawed respectively) flooring is manufactured from both softwoods and hardwoods. Vertical-grained flooring shrinks and swells less than flat-grained flooring and resists opening of joints better.

The chief grades of maple are Clear No. 1 and No. 2. Quarter-sawed oak comes in Clear and Select. Plain-sawed is available in Clear, Select and Number 1 Common.

Shingles and shakes— Shingles are sawn from Western red cedar, Northern white cedar, and redwood. The shingle grades are: Western red cedar, No. 1, No. 2, No. 3; Northern white cedar, Extra Clear, 2nd Clear, Clear Wall, Utility; redwood, No. 1, No. 2 VG, and No. 2 MG.

Shingles that are all heartwood resist decay better than shingles that contain sapwood. Edge-grained shingles are less likely to warp than flat-grained shingles. Thick-butted shingles and narrow shingles are also less likely to warp. The thickness of shingles may be described as 4/2 or 5/2¼, for example. This means that four shingles have 2'' of butt thickness or five shingles have 2¼'' of butt thickness. Lengths may be 16'', 18'', or 24''.

Shingles are usually packed four bundles to the square. A square of shingles will cover 100 square feet of roof area when applied at standard weather exposures.

Shakes are handsplit or handsplit and resawn from Western red cedar. There is only one grade: 100 percent clear, graded from the split face in the case of handsplit and resawn material. Handsplit shakes are graded from the best face. Shakes must be 100 percent heartwood free of bark and sapwood. The standard thickness of shakes ranges from 3/8'' to 1¼''. Lengths are 18'' and 24'', and a 15-inch ''starter-finish course'' length.

That finishes our look at seasoning, measuring and grading lumber. Now let's talk about lumber availability and take a specific look at a commonly used wood.

The Species You Use

Not all species of lumber are available in all parts of the country. For example, very little Southern pine is sold in the Pacific Northwest. In Oregon and Washington there's plenty of Douglas fir available locally. Likewise, it's hard to find Douglas fir framing lumber in Virginia lumberyards. There's no reason to haul lumber thousands of miles when other species with similar characteristics are available locally. For that reason, the lumber species you use probably depends on the area of the country where you build.

But over the last 20 years there has been a broad shift in transportation patterns for framing lumber. Southern pine has always been used in the South. But in the 1970's it began to find wider markets in the Midwest, East and even in the West. Prices for Southern pine were highly competitive in many areas where Douglas fir had been the lumber of choice for many years. Whether this trend continues is a matter of speculation. But the abundant Southern pine forests and reasonable transportation costs may make Southern pine the first choice for more and more residential and light commercial builders.

Let's take a detailed look at Southern pine lumber, and use it as an example for what we learned above. We'll look at the grading rules for Southern pine and the grade descriptions. We'll also cover recommended preservatives, seasoning requirements and lumber sizes.

Grade Descriptions

Figure 5-15 shows grade descriptions for Southern pine lumber. This figure is a good example of the types of grades and grading method that we covered earlier. Here are some important points to remember:

• The stress-rated grades have ''SR'' in the grade name. For example, stress-rated 5'' x 5'' timbers are labeled No. 1 SR, No. 1 Dense SR, No. 2 SR and No. 2 Dense SR. You can cut these graded members into shorter lengths without changing the stress rating. The shorter pieces have the same stress rating as the original length. And the stress ratings apply to members used either flat or on edge.

• Unseasoned clear wood has a strength ratio of 100%. Now compare the strength of unseasoned clear wood to that of green lumber with defects and blemishes. The strength ratio of the green lumber is also shown as a percentage. But the percentage drops because of the defects. Common strength ratios for green lumber are: 86%, 72%, 65%, 55% and 45%. These percentages appear next to some of the grade names shown in Figure 5-15.

• Grade restrictions apply to the entire length of each piece. This means you can use each piece in continuous spans, over double spans or under concentrated loads without allowing for special shear stress requirements.

Figure 5-16 shows the Southern pine grades recommended for different parts of your construction projects.

Figure 5-17 shows the preservatives, retentions and applicable American Wood Preservers Association standards for Southern pine lumber.

Seasoning Requirements

When you see a grade mark on lumber, you know it's been properly seasoned. Southern pine grading rules restrict moisture content to a maximum of 19% for lumber that is 2'' thick or less. If your specs call for lumber that is ''KD 15'' (kiln-dried) or ''MC 15,'' the maximum moisture content is 15%. The grade mark MC 15 can be used for air-dried lumber or kiln-dried lumber.

PRODUCT	GRADE	CHARACTER OF GRADE AND TYPICAL USES
FINISH	***B&Btr**	Highest recognized grade of finish. Generally clear although a limited number of pin knots permitted. Finest quality for natural or stain finish.
	C	Excellent for painted or natural finish where requirements are less exacting. Reasonably clear but permits limited number of surface checks. and small tight knots.
	C&Btr	Combination of "B&Btr" and "C" grades, satisfies requirements for high quality finish.
	D	Economical. serviceable grade for natural or painted finish.
*PANELING INCLUDING FILLETS	**B&Btr** **C** **C&Btr**	Similar to above grades with additional restrictions on stain and wane.
	D	Top quality knotty pine paneling for natural or stained finish. Knots are smooth and even with surrounding surface.
	No. 1	Not contained in current SPIB Grading Rules; however. if specified. will be designated and graded as "D" grade.
	No. 2	Knotty pine grade somewhat less exacting than "D" but suitable for natural or stained finish. Tight-knotted. with knots generally smooth across surface. Minor surface pits and cavities permitted. Wane not permitted on face.
	No. 3	More manufacturing imperfections allowed than in No. 2 but suitable for economical use.
*FLOORING CEILING PARTITION DROP SIDING BEVEL SIDING	**B&Btr** **C** **C&Btr** **D**	See Finish grades.
	No. 1	No. 1 Flooring not provided under SPIB Grading Rules as separate grade. but if specified. will be designated and graded as "D". No. 1 drop siding is graded as No. 1 boards.
	No. 2	Slightly better than No. 2 boards. High utility value where appearance is not factor.
	No. 3	More manufacturing imperfections allowed than in No. 2 but suitable for economical use.
BOARDS S4S SHIPLAP S2S&CM	**No. 1**	High quality with good appearance characteristics. Generally sound and tight-knotted. Largest hole permitted is 1/16". A superior product suitable for wide range of uses including shelving. form and crating lumber.
	No. 2	High quality sheathing material. characterized by tight knots. Generally free of holes.
	No. 3	Good. serviceable sheathing. usable for many applications without waste.
	No. 4	Admit pieces below No. 3 which can be used without waste or contain usable portions at least 24" in length.
STRUCTURAL LUMBER	***Dense Str. 86** ***Dense Str. 72** ***Dense Str. 65**	Number at end of grade names indicates the percentage stress of clear wood value. (All grades identified as "Structural" contain only sound wood free from any form of decay.)

Southern pine grade descriptions
Figure 5-15

PRODUCT	GRADE	CHARACTER OF GRADE AND TYPICAL USES
DIMENSION Structural Light Framing 2″ to 4″ thick 2″ to 4″ wide	*Select Structural *Dense Select Structural	High quality, relatively free of characteristics which impair strength or stiffness. Recommended for uses where high strength, stiffness and good appearance are required.
	No. 1 No. 1 Dense	Provide high strength, recommended for general utility and construction purposes. Good appearance, especially suitable where exposed because of the knot limitations.
	No. 2 No. 2 Dense	Although less restricted than No. 1, suitable for all types of construction. Tight knots.
	No. 3 No. 3 Dense	Assigned design values meet wide range of design requirements. Recommended for general construction purposes where appearance is not a controlling factor. Many pieces included in this grade would qualify as No. 2 except for single limiting characteristic. Provides high quality and low cost construction.
STUDS 2″ to 4″ thick 2″ to 6″ wide 10′ and Shorter	Stud	Stringent requirements as to straightness, strength and stiffness adapt this grade to all stud uses, including load-bearing walls. Crook restricted in 2″ x 4″ — 8′ to ¼″, with wane restricted to 1/3 of thickness.
Structural Joists & Planks 2″ to 4″ thick 5″ and wider	*Select Structural Dense Select *Structural	High quality, relatively free of characteristics which impair strength or stiffness. Recommended for uses where high strength, stiffness and good appearance are required.
	No. 1 No. 1 Dense	Provide high strength, recommended for general utility and construction purposes. Good appearance, especially suitable where exposed because of the knot limitations.
	No. 2 No. 2 Dense	Although less restricted than No. 1, suitable for all types of construction. Tight knots.
	No. 3 No. 3 Dense	Assigned stress values meet wide range of design requirements. Recommended for general construction purposes where appearance is not a controlling factor. Many pieces included in this grade would qualify as No. 2 except for single limiting characteristic. Provides high quality and low cost construction.
***Light Framing 2″ to 4″ thick 2″ to 4″ wide**	*Construction	Recommended for general framing purposes. Good appearance, strong and serviceable.
	*Standard	Recommended for same uses as Construction grade, but allows larger defects.
	*Utility	Recommended where combination of strength and economy is desired. Excellent for blocking, plates and bracing.
	*Economy	Usable lengths suitable for bracing, blocking, bulkheading and other utility purposes where strength and appearance not controlling factors.
Appearance Framing 2″ to 4″ thick 2″ and wider	*Appearance	Designed for uses such as exposed-beam roof systems. Combines strength characteristics of No. 1 with appearance of "C&Btr."

Southern pine grade descriptions
Figure 5-15 (continued)

PRODUCT	GRADE	CHARACTER OF GRADE AND TYPICAL USES
TIMBERS 5'' x 5'' & larger	**No. 1 SR** **No. 1** **Dense SR** **No. 2 SR** **No. 2** **Dense SR**	No. 1 and No. 2 are similar in appearance to corresponding grades of 2'' dimension. Recommended for general construction uses. SR in grade name STRESS RATED.
	Square Edge and Sound **No. 1, No. 2, No. 3**	Not stress-rated but economical for general construction purposes.
INDUSTRIAL LUMBER	***Dense Industrial 86** ***Industrial 86** ***Dense Industrial 72** ***Industrial 72** ***Dense Industrial 65** ***Industrial 65**	These classifications cover a variety of industrial grades where resistance to abrasive action, mechanical wear, or ability to absorb shock is desirable on specific use conditions.
*FACTORY FLOORING AND DECKING	**Dense** **Standard**	High quality product, suitable for plank floor where face serves as finish floor. Has a better appearance than No. 1 Dense because of additional restrictions on pitch, knots, pith and wane.
	Select **Dense Select**	Slightly less restrictive than Dense Standard but more restrictive than No. 1 dimension. Sound, solid appearance.
	Commercial **Dense Commercial**	Same requirements as corresponding grades of No. 2 dimension.
SCAFFOLD PLANK	***Dense Industrial 72 Scaffold Plank**	Extra high quality. Available in dimensions 2'' and thicker and all widths.
	Dense Industrial 65 Scaffold Plank	High quality. Available in dimensions 2'' and thicker and all widths.
STADIUM SEATS	***No. 1 Dense Stadium Grade**	Superior material with one face free of pitch and otherwise complying with No. 1 Dense dimension.
	*No. 1 Stadium Grade	Similar to No. 1 Dense Stadium Grade, except density not required.

*Caution! Most mills do not manufacture all products and make all grade separations. Those products and grades manufactured by relatively few mills are noted with an asterisk.

Southern pine grade descriptions
Figure 5-15 (continued)

Use-Item	MINIMUM GRADES RECOMMENDED
FRAMING	
Sills on Foundation Walls or Slab on Ground*.	Utility
Sills on Piers*—Built-up.	No. 2
Joists, Rafters, Headers.	No. 3
Plates, Caps, Bucks.	Utility
Studs.	Stud Grade
Ribbon Boards, Bracing, Ridge Boards (1″ nominal thickness).	No. 2
Collar Beams.	No. 2
Furring Grounds 1″ nominal thickness.	No. 3
Subflooring.	No. 3
Wall Sheathing.	No. 3
Roof Sheathing, Pitched.	No. 3
Roof Decking, Flat 1″ thick.	No. 2 KD
2″ thick.	No. 2 KD
Exposed Decking—where appearance is of prime concern 3″ & 4″ thick.	Dense Standard DT&G Deck
Industrial—appearance not prime concern 3″ & 4″ thick.	Commercial DT&G
Stair Stringers or Carriages.	No. 1
Cellar and Attic Stair Treads and Risers.	No. 1 Dense
Roof Truss Members 2″ to 4″ thick Upper and Lower Chords.	No. 2
Other.	No. 3
5″ & thicker.	No. 2 SR
Heavy Timber Construction Beams Built-up—2″ to 4″ thick.	No. 2
Solid—over 5″ thick.	No. 2 SR
Posts and Columns 2″ to 4″ thick.	No. 2
over 5″ thick.	No. 2 SR
SIDING, PANELING, FINISH AND MILLWORK	
Siding Bevel, Drop, Rough Sawn For rustic applications.	No. 2
For appearance applications	C&Btr
Exterior Trim Cornice.	C Finish or C Ceiling
Mouldings, Drip Cap, Water Table . . .	C Mouldings
Trim, Facia, Corner Boards, Soffits. . .	No. 1
Window and Door Frames, Sash, Shutters, Screens. . . .	C
Doors, Garage and Warehouse.	No. 1

Use-Item	MINIMUM GRADES RECOMMENDED
SIDING, PANELING, FINISH AND MILLWORK (cont'd)	
Porch Ceiling.	No. 2 Ceiling
Flooring**††.	No. 2
Stair Treads.	No. 1 Dense
Stair Stringers or Carriages & Risers††.	No. 1
Columns, Built-up††	No. 2
Newel Posts, Railings, Balustrades††.	No. 1
Finished or Top Flooring**†† Uncovered Floors, Natural, Stained.	C Flooring
Covered Floors.	No. 2 Flooring
Industrial or Workroom Floors.	No. 2 Flooring or End Grain Block Flooring
Interior Finish and Trim Stair Treads or Stepping.	C
Trim.	C
Mouldings.	C
Ceiling.	C
Partition.	C
Closet Lining.	No. 2 Ceiling
Shelving.	No. 1
Paneling For rustic application.	No. 2 KD
For appearance application.	C
FENCING AND ACCESSORIES††	
Fencing Framing, Posts, Boards.	No. 3
Pickets.	No. 2
Gates, 1-inch thick.	No. 2
Gates, 2-inch thick.	No. 3
BALCONY, DECKS, PATIOS AND BOARDWALKS††	
Posts and Caps 2″ to 4″ thick.	No. 1 Dense
5″ & thicker.	No. 1 SR
Sills 2″ to 4″ thick.	No. 1
5″ & thicker.	No. 1 SR
Beams, Stringers 2″ to 4″ thick.	No. 1
5″ & thicker.	No. 1 SR
Railings, Rail Posts.	No. 1
Steps and Ramps.	No. 2
Decking.	No. 2 (Specify "Bark side up")
Decking, Laminated, on Edge.	No. 2

Recommended grades of Southern pine
Figure 5-16

Use-Item	MINIMUM GRADES RECOMMENDED
HEAVY FALSEWORK, HEAVY FORMS AND CAISSONS	
Sills, Mud Sills, Posts and Caps......	No. 1 SR
Stringers.....................	No. 1 SR
Truss Members	
Compression and Tension Members	
2" to 4" thick..............	No. 2
5" and thicker..............	No. 2 SR
Centering, Lagging and Wedges......	No. 1 SR
Bracing.....................	No. 1
SCAFFOLDING	
Uprights and Bracing	
2" to 4" thick (19% MC).........	No. 1
2½" and thicker (over 19%MC)	Dense Select Structural
Planking......................	Dense Industrial 65 Scaffold Plank
CONCRETE FORMS	
Shoring and Plates...............	No. 1
Joists and Beam Forms............	No. 2
Bracing.......................	No. 2
Boarding......................	No. 2
STADIUM SEATS††	
Seats.........................	No. 1 Stadium Grade
(Specify "Bark side up")	
HIGHWAY STRUCTURES/BRIDGES ††	
Sills, Posts, Caps.................	No. 1 SR
Bracing, Sway...................	No. 1 SR
Truss Members	
Compression and Tension........	No. 1 SR
Floor Beams, Stringers...........	No. 1 SR
Nailing Strips...................	No. 1
Sub Decking, Plank or Laminated Decking, Top	
2" to 4" thick..................	No. 1 Dense Select
5" and thicker.................	No. 1 Dense SR
Bulkhead and Plank	
2" to 4" thick..................	No. 1
5" and thicker.................	No. 1 SR
Sidewalk Plank..................	No. 1 Dense
Cleats and Scupper Blocks	
2" to 4" thick..................	No. 2
5" and thicker.................	No. 2 SR
Railings and Rail Posts	
2" to 4" thick..................	No. 1 Dense
5" and thicker.................	No. 1 SR
Wheel and Fellow Guards.........	No. 1 Dense
Fire Stops.....................	No. 2

Use-Item	MINIMUM GRADES RECOMMENDED
CULVERTS AND DRAINS††	
2" to 4" thick...................	No. 1
5" and thicker..................	No. 1 SR
RIGHT OF WAY FENCING††	
Posts.........................	No. 2
Framing.......................	No. 2
Boards........................	No. 2
Gates	
1" thick......................	No. 2
2" to 4" thick.................	No. 3
Stakes........................	Utility
GUARD RAIL††	
Railing & Rail Posts	
2" to 4" thick.................	No. 1 Dense
5" and thicker.................	No. 1 SR
SIGNPOSTS††	
Posts	
2" to 4" thick.................	No. 2
5" and thicker.................	No. 2 SR
PIERS AND WHARVES (OPEN CONSTRUCTION)††	
Timber Sheet Piling	
2" to 4" thick.................	No. 1
5" and thicker.................	No. 1 SR
Timber in Cribs.................	No. 1 SR
Caps..........................	No. 1 SR
Stringers......................	No. 1 SR
Bracing	
2" to 4" thick.................	No. 1
5" and thicker.................	No. 1 SR
Decking	No. 1 Dense
2" to 4" thick.................	Select
5" and thicker.................	No. 1 Dense SR
Decking, Laminated, on Edge......	No. 1
Guard Timbers..................	No. 1 Dense SR
Mooring Posts..................	No. 1 Dense SR
Fenders and Wales..............	No. 1 Dense SR

Note: Grade selection should be based on engineering requirements. All above grades are minimum grades for particular use.

*For slab on ground where sill on foundation wall or piers is within 18" of ground on inside or 12" on outside, use pressure preservative treated sills. See SFPA Technical Bulletin No. 6 regarding preservative treatment.

**Flooring may be specified "end-matched" the grade being the same as if plain-end.

††See Technical Bulletin No. 6 regarding preservative treatments and Bulletin No. 14 regarding deck, patio and fence applications.

Recommended grades of Southern pine
Figure 5-16 (continued)

PRESERVATIVE TYPES, RETENTIONS AND APPLICABLE AWPA STANDARDS[1],[2]

Waterborne Preservatives [3]. See Note [4] for trade names.

MATERIAL AND USAGE	Creosote [5]	Creosote-Coal Tar [5]	Creosote-Petroleum	Pentachlorophenol [3]	Acid Copper Chromate (ACC)	Ammoniacal Copper Arsenate (ACA)	Chromated Copper Arsenate (CCA)	Chromated Zinc Chloride (CZC)	Fluor Chrome Arsenate Phenol (FCAP)	AWPA Standards [6]
LUMBER AND TIMBER					Retention Assay of Treated Wood — lbs./cu. ft.					
Above ground	8[9]	8[9]	8[9]	0.40	0.25	0.25	0.25	0.45	0.25	C2
Ground contact Nonstructural	10[9]	10[9]	10[9]	0.50	0.50	0.40	0.40	NR[7]	NR	C2
Ground contact Structural	12[9]	12[9]	12[9]	0.60	NR	0.60	0.60	NR	NR	C14
In salt water	25	25	NR	NR	NR	2.5	2.5	NR	NR	C14
PILES										
Land or fresh water use and foundations	12	12	12	0.60	NR	0.80	0.80	NR	NR	C3
Salt Water										
Prevalent Marine Organism										
Teredo only	20	20	NR	NR	NR	2.5[8] and 1.5	2.5[8] and 1.5	NR	NR	C18
Pholads only	20	20	NR	NR	NR	NR	NR	NR	NR	C18
Limnoria tripunctata only	NR	NR	NR	NR	NR	2.5[8] and 1.5	2.5[8] and 1.5	NR	NR	C18
For both pholads and limnoria tripunctata use a dual treatment										
First treatment	—	—	—	—	—	1.0	1.0	—	—	C18
Second treatment	20	20	—	—	—	—	—	—	—	C18
POLES										
Utility										
Normal	7.5	7.5	7.5	0.38	NR	0.60	0.60	NR	NR	C4
Severe service conditions (high incidence of decay and termite attack)	9.0	9.0	9.0	0.45	NR	0.60	0.60	NR	NR	C4
Building poles—structural	9.0[9]	NR	NR	0.45	NR	0.60	0.60	NR	NR	C3
POSTS										
Fence, guide, and sight										
Round, half-round, and quarter-round	8	8	8	0.40	0.50	0.40	0.40	NR	NR	C14
Sawn four sides	10	10	10	0.50	0.62	0.50	0.50	NR	NR	C14
Guardrail and sign (incl. spacer blocks)										
Round	10	10	10	0.50	NR	0.50	0.50	NR	NR	C14
Sawn four sides	12	12	12	0.60	NR	0.60	0.60	NR	NR	C14

Footnotes

1. Southern Pine protected from weather or exposed in a manner not to permit water to stand for any appreciate length of time does not require preservative treatment. Building codes generally require wood floors closer than 18 inches, or wood girders closer than 12 inches, to exposed ground be pressure treated (preservative):

2. AWPA Standards detail plant operating procedures for pressure treatment of wood. These standards include minimum vacuum, pressure, penetration requirements, maximum steaming and temperature allowances. AWPA also details retention and assay zone requirements for each commodity, preservative and wood species. AWPA standards make it unnecessary for specifications to include detailed requirements on penetration and allowable processes. Generally, it is desirable to specify the preservative desired, the intended application, necessary retention and to reference appropriate AWPA standards. AWPA standard C-1 applies to each of the treating processes and all types of material.

3. Pentachlorophenol in suitable solvents or waterborne preservatives can provide a clean, paintable, odorless, dry surface. When one or more of these features is required, the processor should be so advised when the order is placed.

4. Trade names of waterborne preservatives: Acid Copper Chromate (ACC) (Celcure*). Ammoniacal Copper Arsenate (ACA) (Chemonite*), Chromated Copper Arsenate, Type A (CCA Type A) (Greensalt), Chromated Copper Arsenate, Type B (CCA Type B) (Boliden* CCA) (Koppers CCA-B) (Osmose K-33*), Chromated Copper Arsenate, Type C (CCA Type C) (Chrome-Ar-Cu-CAC*) (Langwood*) (Wolman* CCA), Chromated Zinc Chloride (CZC), Fluor Chrome Arsenate Phenol (FCAP) (Osmosalts*—Osmosar*) (Tanalith) (Wolman* Salts FCAP) (Wolman* Salts FM P).

5. When these preservatives are specified for material to be used in salt water the creosote-coal tar shall conform to Standard P2 or P12, and the creosote shall conform to Standard P1 or P13.

6. These retentions are taken from among those given in various standards of the American Wood Preservers Association (AWPA) and have been selected by SFPA on the basis of structural importance.

7. NR — Not recommended.

8. The assay retentions are based on two assay zones—0 to 0.50 inch and 0.50 to 2.0 inches.

9. Not recommended where cleanliness and freedom from odor are necessary.

*Reg. U.S. Pat. Off.

For information on pressure treating standards, quality control programs, and other technical assistance write or contact any of the following:

American Wood Preservers Association
7735 Old Georgetown Rd.
Bethesda, MD 20014
(301) 652-3109

American Wood Preservers Institute
1651 Old Meadow Road
McLean, VA 22101
(703) 893-4005

Society of American Wood Preservers, Inc.
1401 Wilson Blvd./Suite 205
Arlington, VA 22209
(703) 841-1500

Southern Pressure Treaters Association
2920 Knight Street, Room 121
Shreveport, LA 71105
(318) 861-2479

Preservatives for Southern pine
Figure 5-17

ITEMS (Nominal)	MOISTURE CONTENT LIMIT	
	MAXIMUM (DRY)	KILN-DRIED (KD 15 or MC 15)
D&Btr Grades 1'' & 1¼''	15%	12% on 90% of pieces 15% on remainder
1½'', 1¾'' & 2''	18%	15%
Over 2'' not over 4''	19%	15%
Over 4''	20%	18%
Paneling[1] 1''	---	12%
Boards 2'' and less & Dimension 2'' to 4''	19%	15%
Decking[2] 2'' thick 3'' and 4'' thick	19%	15% 15% on 90% of pieces 18% on remainder
Heavy Dimension[3] Over 2'' not over 4''	19%	15%
Timbers[3] 5'' and thicker	23%	20%

1. Required to be kiln-dried to 12% maximum moisture content.
2. All thicknesses of Decking should be specified at 15% maximum moisture content.
3. Moisture content provisions must be specified since seasoning is not mandatory in these sizes.

Southern pine seasoning requirements
Figure 5-18

Moisture content restrictions apply at the time of lumber shipment and at the time of delivery. They also apply at the time of dressing, if dressed lumber is used. Figure 5-18 shows seasoning requirements for various dimensions of Southern pine.

Lumber Sizes
Look again at Figure 5-3. This figure shows the nominal and actual dimensions of the most commonly used lumber sizes. Figure 5-19 shows the standard sizes for Southern pine lumber.

No matter what type of lumber you use, shop around for the best buy. Prices vary from yard to yard. Select an economical lumber that can do the job. Don't use No. 2 grade when No. 3 is acceptable. Doing quality work that is affordable is important to your reputation as a builder.

Canadian Lumber
Southern pine isn't the only species of lumber that's winning new markets. More and more Canadian lumber is being shipped into the U.S. As long as transportation costs remain reasonable and the value of the U.S. dollar is high in relation to the Canadian dollar, builders in states close to Canadian mills will find their lumberyards stocked with a good selection of Canadian lumber selling at competitive prices.

If you live in an area where Canadian lumber is sold, you should be familiar with the Standard Grading Rules for Canadian Lumber, published by the National Lumber Grades Authority. Canadian dimension lumber grades are identical to American grades and meet all the requirements of the American Softwood Lumber Standard PS 20-70. Species combinations, assigned stress values and spans all conform to standards used by U.S. mills.

Species Combinations
Certain species of Canadian lumber are combined and marketed together. The 15 Canadian commercial species combinations are shown in Figure 5-20. Here are the key characteristics of these species.

Douglas fir-larch (North): Includes Douglas fir and Western larch. The two woods have similar strength and weight, are hard, decay-resistant, and have good nail-holding, gluing and painting qualities. Colors range from red-brown to yellow-white.

Hem-fir (North): Includes Western hemlock and Amabilis fir. These are light woods with moderate strength. They're easy to work, take paint and glue and hold nails well. Colors range from pale yellow-brown to white.

Eastern hemlock-tamarack (North): includes Eastern hemlock and tamarack. Both woods are moderately strong and are good for general construction. They're fairly hard and durable. Colors range from yellow-brown to white.

Spruce-pine-fir: This is the broadest class of combination species. This group includes white spruce, red spruce, black spruce, Engelmann spruce, lodgepole pine, jack pine, alpine fir, and balsam fir. These are moderate-strength woods, easy to work with. They take paint well and have good nail-holding qualities. Colors range from white to pale yellow.

Western hemlock (North): includes only Western hemlock. It's a moderately light, hard, strong wood. Colors range from white to pale brown, with little difference between the appearance of the sapwood and heartwood.

Rough structural lumber

THICKNESS[1] (Inches)			WIDTH (Inches)		
	Rough			Rough	
Nominal	Dry	Green	Nominal	Dry	Green
2	1-5/8[4]		2	1-5/8	
3	2-5/8[4]	2-11/16	3	2-5/8	2-11/16[2]
4	3-5/8[4]	3-11/16	4	3-5/8	3-11/16[2]
			5	4-5/8	4-3/4[2]
			6	5-5/8	5-3/4[2]
			8	7-3/8[2]	7-5/8
			10	9-3/8[2]	9-5/8
			12	11-3/8[2]	11-5/8
			14	13-3/8[2]	13-5/8
			16	15-3/8[2]	15-5/8
			18	17-3/8[2]	17-5/8
			20	19-3/8[2]	19-5/8
5" & thicker	3/8" off nominal[4]	3/8" off nominal	5" & wider	3/8" off nominal[3]	3/8" off nominal[3]

[1]Thicknesses apply to their corresponding widths as squares and wider, except 2" green thickness of 1-11/16 applies to widths of 14" and over.
[2]These widths apply only to thicknesses of less than 5".
[3]These widths apply only to thicknesses of 5" and over.
[4]These minimum thicknesses apply to 80% of the pieces of an item or shipment and the remainder (20%) may be 1/32" thinner.

Dimension and structural lumber, dressed

THICKNESS[1] (Inches)			WIDTH (Inches)		
	Standard ALS Minimum Dressed			Standard ALS Minimum Dressed	
Nominal	DRY	GREEN	Nominal	DRY	GREEN
2	1-1/2		2	1-1/2	
2-1/2	2[3]	2-1/16	3	2-1/2	2-9/16[2]
3	2-1/2[3]	2-9/16	4	3-1/2	3-9/16[2]
3-1/2	3[3]	3-1/16	5	4-1/2	4-5/8[2]
4	3-1/2[3]	3-9/16	6	5-1/2	5-5/8[2]
			8	7-1/4	7-1/2
			10	9-1/4	9-1/2
			12	11-1/4	11-1/2
			14	13-1/4	13-1/2
			16	15-1/4	15-1/2
			18	17-1/4	17-1/2
			20	19-1/4	19-1/2
5" & thicker	1/2" off nominal[4]	1/2" off nominal	5" & wider	1/2" off nominal	1/2" off nominal

[1]2"dressed green thickness of 1-9/16 applies to widths of 14" and over.
[2]These green widths apply to thicknesses of 3" and 4" only, except as provided in Footnote (1).
[3]Not required to be dry unless specified.

Board dimension

	THICKNESS (Inches)		WIDTH (Inches)	
	Nominal	Dressed	Nominal	Dressed
Finish	3/8	5/16	2	1-1/2
	1/2	7/16	3	2-1/2
	5/8	9/16	4	3-1/2
	3/4	5/8	5	4-1/2
	1	3/4	6	5-1/2
	1-1/4	1	7	6-1/2
	1-1/2	1-1/4	8	7-1/4
	1-3/4	1-3/8	9	8-1/4
	2	1-1/2	10	9-1/4
	2-1/2	2	11	10-1/4
	3	2-1/2	12	11-1/4
	3-1/2	3	14	13-1/4
	4	3-1/2	16	15-1/4
Boards	1	3/4[1]	2	1-1/2
	1-1/4	1	3	2-1/2
	1-1/2	1-1/4	4	3-1/2
			5	4-1/2
			6	5-1/2
			7	6-1/2
			8	7-1/4
			9	8-1/4
			10	9-1/4
			11	10-1/4
			12	11-1/4
			over 12	off 3/4

[1]Boards less than the minimum dressed thickness for 1" nominal but which are 5/8" or greater thickness dry may be regarded as American Standard Lumber, but such boards shall be marked to show the size and condition of seasoning at the time of dressing. They shall also be distinguished from 1" boards on invoices and certificates.

Finish dimension

	THICKNESS (Inches)		WIDTH (Inches)		
	Nominal	Worked	Nominal	Face	Over-all
Bevel Siding	1/2	3/16x7/16	4	3-1/2	3-1/2
	5/8	3/16x9/16	5	4-1/2	4-1/2
	3/4	3/16x11/16	6	5-1/2	5-1/2
	1	3/16x3/4	8	7-1/4	7-1/4
Drop Siding Rustic and Drop Siding (dressed and matched)	5/8	9/16	4	3-1/8	3-3/8•
	1	23/32	5	4-1/8	4-3/8•
			6	5-1/8	5-3/8•
			8	6-7/8	7-1/8•
			10	8-7/8	9-1/8•
Rustic and Drop Siding (shiplapped)	5/8	9/16	4	3	3-3/8•
	1	23/32	5	4	4-3/8•
			6	5	5-3/8•
			8	6-5/8	7-1/8•
			10	8-5/8	9-1/8•
			12	10-5/8	11-1/8•
Flooring	3/8	5/16	2	1-1/8	1-3/8••
	1/2	7/16	3	2-1/8	2-3/8••
	5/8	9/16	4	3-1/8	3-3/8••
	1	3/4	5	4-1/8	4-3/8••
	1-1/4	1	6	5-1/8	5-3/8••
	1-1/2	1-1/4			
Ceiling	3/8	5/16	3	2-1/8	2-3/8••
	1/2	7/16	4	3-1/8	3-3/8••
	5/8	9/16	5	4-1/8	4-3/8••
	3/4	11/16	6	5-1/8	5-3/8••
Partition	1	23/32	3	2-1/8	2-3/8
			4	3-1/8	3-3/8
			5	4-1/8	4-3/8
			6	5-1/8	5-3/8
Paneling	1	23/32	3	2-1/8	2-3/8
			4	3-1/8	3-3/8
			5	4-1/8	4-3/8
			6	5-1/8	5-3/8
			8	6-7/8	7-1/8
			10	8-7/8	9-1/8
			12	10-7/8	11-1/8
Shiplap	1	3/4	4	3-1/8	3-1/2
			6	5-1/8	5-1/2
			8	6-7/8	7-1/4
			10	8-7/8	9-1/4
			12	10-7/8	11-1/4
Dressed and Matched	1	3/4	4	3-1/8	3-3/8
	1-1/4	1	5	4-1/8	4-3/8
	1-1/2	1-1/4	6	5-1/8	5-3/8
			8	6-7/8	7-1/8
			10	8-7/8	9-1/8
			12	10-7/8	11-1/8

•Over-all widths for 5/8" thickness are 1/16" less.
••Over-all widths for 3/8", 1/2" and 5/8" thicknesses are 1/16" less.

Factory flooring and decking Heavy roofing and shiplap

THICKNESS (Inches)		WIDTH (Inches)			
			Dressed		
Nominal	Dressed	Nominal	D & M	Ship-lapped	For Splines
2	1-1/2	4	3	3	3-1/2
2-1/2	2	6	5	5	5-1/2
3	2-1/2	8	6-3/4	6-3/4	7-1/4
4	3-1/2	10	8-3/4	8-3/4	9-1/4
5	4-1/2	12	10-3/4	10-3/4	11-1/4

Standard sizes for Southern pine products
Figure 5-19

Commercial species combinations designations	Species in combination
Douglas fir-larch (north)	Douglas fir western larch
Hem-fir (north)	western hemlock amabilis fir
Eastern hemlock-tamarack (north)	eastern hemlock tamarack
Spruce-pine-fir	white spruce red spruce black spruce Engelmann spruce lodgepole pine jack pine alpine fir balsam fir
Western hemlock (north)	western hemlock
Coast sitka spruce	coast sitka spruce
Ponderosa pine	ponderosa pine
Western cedars (north)	western red cedar Pacific coast yellow cedar
Western white pine	western white pine
Red pine (north)	red pine
Eastern white pine (north)	eastern white pine
Northern aspen	quaking aspen bigtooth aspen balsam poplar
Black cottonwood	black cottonwood
Northern species	Any species included in this table except those in the northern aspen and black cottonwood species combinations
Coast species	Douglas fir western larch western hemlock amabilis fir coast sitka spruce

Canadian commercial species combinations
Figure 5-20

Coast Sitka spruce: contains only coast Sitka spruce. It's a light, resilient, moderately-strong wood that's easy to work with, takes paint and holds nails well. Colors range from creamy white to light pink, with a large proportion of clear wood.

Ponderosa pine: contains only Ponderosa pine. It's moderately strong and is easily worked to a smooth, uniform finish. It takes paint, stain and varnish well. The wood seasons readily and holds nails well. Sapwood color is pale yellow. Heartwood color ranges from deep yellow to red-brown.

Western cedars (North): This group includes Western red cedar and Pacific Coast yellow cedar. The two woods aren't equal in strength, but both are decay-resistant and have a nice appearance. Both are easy to work with and take finish well. Red cedar has red-brown heartwood and light sapwood. Yellow cedar has a uniform yellow color.

Red pine (North): contains only red pine. It's fairly strong and easy to work with, takes a good finish and holds nails and screws well. Moderately durable, it seasons with little checking or cupping. The sapwood is thick and pale yellow in color. Heartwood color ranges from pale brown to red.

Eastern white pine (North): contains only Eastern white pine. It's the softest of the Canadian pines. Eastern white pine isn't as strong as most pines, but it doesn't split or splinter. It holds nails well and shrinks less than any other Canadian species except the cedars. The wood takes stain, paint and varnish well. The color of the sapwood is almost white. The heartwood color ranges from creamy white to light straw-brown.

Western white pine: contains only Western white pine. Characteristics are similar to Eastern white pine.

Northern aspen: contains quaking aspen, bigtooth aspen and balsam poplar. These are lightweight woods of relatively low strength. Colors range from almost white to gray-white.

Black cottonwood: includes only black cottonwood, a hardwood. Characteristics are similar to those of the Northern aspen group, but it's lower in strength and stiffness.

Northern species: Any Canadian softwood species included in NLGA Standard Grading Rules for Canadian Lumber can be classed as "Northern Species."

Coast species: includes Douglas fir, Western larch, Western hemlock, amabilis fir and coast Sitka spruce. The characteristics of each of these species are listed above.

Not all of these species combinations are available in all grades and all sizes. Figure 5-21 shows the availability of Canadian dimension lumber. Grades and sizes marked with a black dot will be available at many yards that stock Canadian lumber.

Figure 5-22 shows maximum spans for floor joists when using Canadian lumber. Your lumber dealer will have more complete span tables. Figure 5-23 is sample Canadian grade stamps.

Species	Grade	2×4	2×6	2×8	2×10	2×12
Douglas fir-larch (North)	Select structural	■	■	■	■	■
	No. 1	●	●	●	●	●
	No. 2	●	●	●	●	●
	No. 3	●	●	●	●	●
	Construction	●	□	□	□	□
	Standard	●	□	□	□	□
	Utility	●	□	□	□	□
Hem-fir (North)	Select structural	■	■	■	■	■
	No. 1	●	●	●	●	●
	No. 2	●	●	●	●	●
	No. 3	●	●	●	●	●
	Construction	●	□	□	□	□
	Standard	●	□	□	□	□
	Utility	●	□	□	□	□
Eastern hemlock-tamarack (North)	Select structural	○	○	○	○	○
	No. 1	●	■	■	■	■
	No. 2	●	●	●	●	●
	No. 3	●	●	●	●	●
	Construction	●	□	□	□	□
	Standard	●	□	□	□	□
	Utility	●	□	□	□	□
Spruce-pine-fir	Select structural	■	■	■	■	■
	No. 1	●	●	●	●	●
	No. 2	●	●	●	●	●
	No. 3	●	●	●	●	●
	Construction	●	□	□	□	□
	Standard	●	□	□	□	□
	Utility	●	□	□	□	□
Western hemlock (North)	Select structural	■	■	■	■	■
	No. 1	■	■	■	■	■
	No. 2	■	■	■	■	■
	No. 3	■	■	■	■	■
	Construction	■	□	□	□	□
	Standard	■	□	□	□	□
	Utility	■	□	□	□	□
Coast sitka spruce	Select structural	■	■	■	■	■
	No. 1	■	■	■	■	■
	No. 2	■	■	■	■	■
	No. 3	■	■	■	■	■
	Construction	■	□	□	□	□
	Standard	■	□	□	□	□
	Utility	■	□	□	□	□
Ponderosa pine	Select structural	■	■	■	■	■
	No. 1	■	■	■	■	■
	No. 2	■	■	■	■	■
	No. 3	■	■	■	■	■
	Construction	■	□	□	□	□
	Standard	■	□	□	□	□
	Utility	■	□	□	□	□
Western cedars (North)	Select structural	■	■	■	■	■
	No. 1	●	●	●	●	■
	No. 2	●	●	●	●	■
	No. 3	●	●	●	●	■
	Construction	●	□	□	□	□
	Standard	●	□	□	□	□
	Utility	●	□	□	□	□
Western white pine	Select structural	■	■	■	■	■
	No. 1	■	■	■	■	■
	No. 2	■	■	■	■	■
	No. 3	■	■	■	■	■
	Construction	■	□	□	□	□
	Standard	■	□	□	□	□
	Utility	■	□	□	□	□

Availability of Canadian dimension lumber
Figure 5-21

Species	Grade	2×4	2×6	2×8	2×10	2×12
Red pine (North)	Select structural	■	■	■	■	■
	No. 1	■	■	■	■	■
	No. 2	●	●	●	■	■
	No. 3	●	●	●	■	■
	Construction	●	□	□	□	□
	Standard	●	□	□	□	□
	Utility	●	□	□	□	□
Eastern white pine (North)	Select structural	○	○	○	○	○
	No. 1	■	■	■	■	■
	No. 2	●	●	●	●	●
	No. 3	●	●	●	●	●
	Construction	●	□	□	□	□
	Standard	●	□	□	□	□
	Utility	●	□	□	□	□
Northern aspen	Select structural	○	○	○	○	○
	No. 1	■	■	○	○	○
	No. 2	■	■	○	○	○
	No. 3	■	■	○	○	○
	Construction	■	□	□	□	□
	Standard	■	□	□	□	□
	Utility	■	□	□	□	□
Black cottonwood	Select structural	■	■	○	○	○
	No. 1	●	●	○	○	○
	No. 2	●	●	○	○	○
	No. 3	●	●	○	○	○
	Construction	●	□	□	□	□
	Standard	●	□	□	□	□
	Utility	●	□	□	□	□
Northern species	Select structural	■	■	■	■	■
	No. 1	■	■	■	■	■
	No. 2	■	■	■	■	■
	No. 3	■	■	■	■	■
	Construction	■	□	□	□	□
	Standard	■	□	□	□	□
	Utility	■	□	□	□	□
Coast species	Select structural	■	■	■	■	■
	No. 1	■	■	■	■	■
	No. 2	■	■	■	■	■
	No. 3	■	■	■	■	■
	Construction	■	□	□	□	□
	Standard	■	□	□	□	□
	Utility	■	□	□	□	□
MSR Lumber	1200f-1.2E	○	○	○	○	○
	1450f-1.3E	●	●	○	○	○
	1500f-1.4E	○	○	○	○	○
	1650f-1.5E	●	●	○	○	○
	1800f-1.6E	●	●	○	○	○
	1950f-1.7E	○	○	○	○	○
	2100f-1.8E	●	●	○	○	○
	2250f-1.9E	○	○	○	○	○
	2400f-2.0E	●	●	○	○	○
	2550f-2.1E	○	○	○	○	○
	2700f-2.2E	○	○	○	○	○
	3000f-2.4E	○	○	○	○	○
	900f-1.0E	○	○	○	○	○
	900f-1.2E	○	○	○	○	○
	1200f-1.5E	○	○	○	○	○
	1350f-1.8E	○	○	○	○	○
	1800f-2.1E	○	○	○	○	○

Notes:

● Denotes lumber that is readily available from Canada.

■ Denotes lumber that is available in limited supply or on special order, depending partly on regional differences; check availability before ordering.

○ Indicates that item is not marketed in that size or grade in Canada at present time.

□ Size is not applicable for that grade.

1. Appearance Grade is not available in any size or species from Canada.

2. Stud Grade is readily available in 2 x 3 and 2 x 4 sizes, and available in limited supply in 2 x 6 size for all species.

Availability of Canadian dimension lumber
Figure 5-21 (continued)

Floor Joists (30 psf Live Load) (10 psf Dead Load) - Sleeping Rooms and Attic Floors
Maximum Allowable Span (ft.-in.)

Species	Grade	2×6			2×8			2×10			2×12		
		Joist spacing (inches)											
		12	16	24	12	16	24	12	16	24	12	16	24
Douglas fir-larch (North)	Select structural	12-3	11-2	9-9	16-2	14-8	12-10	20-8	18-9	16-5	25-1	22-10	19-11
	No. 1/appearance	12-3	11-2	9-9	16-2	14-8	12-10	20-8	18-9	16-5	25-1	22-10	19-11
	No. 2	12-0	10-11	9-7	15-10	14-5	12-7	20-3	18-5	16-1	24-8	22-5	19-7
	No. 3	10-4	9-0	7-4	13-8	11-10	9-8	17-5	15-1	12-4	21-2	18-4	15-0
Hem-fir (North)	Select structural	11-7	10-6	9-2	15-3	13-10	12-1	19-5	17-8	15-5	23-7	21-6	18-9
	No. 1/appearance	11-7	10-6	9-2	15-3	13-10	12-1	19-5	17-8	15-5	23-7	21-6	18-9
	No. 2	11-3	10-2	8-4	14-11	13-5	11-0	19-0	17-2	14-0	23-1	20-10	17-0
	No. 3	9-1	7-10	6-5	11-11	10-4	8-5	15-3	13-2	10-9	18-6	16-0	13-1
Eastern hemlock-tamarack (North)	Select structural	11-0	10-0	8-9	14-6	13-2	11-6	18-6	16-10	14-8	22-6	20-6	17-11
	No. 1/appearance	11-0	10-0	8-9	14-6	13-2	11-6	18-6	16-10	14-8	22-6	20-6	17-11
	No. 2	10-5	9-6	8-3	13-9	12-6	10-11	17-6	15-11	13-11	21-4	19-4	16-11
	No. 3	9-7	8-3	6-9	12-7	10-11	8-11	16-1	13-11	11-4	19-7	16-11	13-10
Spruce-pine-fir	Select structural	11-7	10-6	9-2	15-3	13-10	12-1	19-5	17-8	15-5	23-7	21-6	18-9
	No. 1/appearance	11-7	10-6	8-8	15-3	13-10	11-6	19-5	17-8	14-8	23-7	21-6	17-9
	No. 2	11-0	9-9	7-11	14-6	12-10	10-6	18-6	16-4	13-4	22-6	19-11	16-3
	No. 3	8-6	7-4	6-0	11-3	9-9	7-11	14-4	12-5	10-1	17-5	15-1	12-4
Western hemlock (North)	Select structural	11-10	10-9	9-4	15-7	14-2	12-4	19-10	18-0	15-9	24-2	21-11	19-2
	No. 1/appearance	11-10	10-9	9-4	15-7	14-2	12-4	19-10	18-0	15-9	24-2	21-11	19-2
	No. 2	11-3	10-3	8-11	14-11	13-6	11-8	19-0	17-3	14-11	23-1	21-0	18-2
	No. 3	9-9	8-5	6-11	12-10	11-1	9-1	16-4	14-2	11-7	19-11	17-3	14-1
Coast sitka spruce	Select structural	12-0	10-11	9-7	15-10	14-5	12-7	20-3	18-5	16-1	24-8	22-5	19-7
	No. 1/appearance	12-0	10-10	8-11	15-10	14-4	11-8	20-3	18-3	14-11	24-8	22-3	18-2
	No. 2	11-6	10-0	8-2	15-2	13-2	10-9	19-4	16-9	13-8	23-6	20-5	16-8
	No. 3	8-8	7-6	6-2	11-6	9-11	8-1	14-8	12-8	10-4	17-9	15-5	12-7
Ponderosa pine	Select structural	10-9	9-9	8-6	14-2	12-10	11-3	18-0	16-5	14-4	21-11	19-11	17-5
	No. 1/appearance	10-9	9-9	8-6	14-2	12-10	11-3	18-0	16-5	14-4	21-11	19-11	17-5
	No. 2	10-5	9-6	7-10	13-9	12-6	10-4	17-6	15-11	13-2	21-4	19-4	16-0
	No. 3	8-6	7-4	6-0	11-3	9-9	7-11	14-4	12-5	10-1	17-5	15-1	12-4
Western cedars (North)	Select structural	10-5	9-6	8-3	13-9	12-6	10-11	17-6	15-11	13-11	21-4	19-4	16-11
	No. 1/appearance	10-5	9-6	8-3	13-9	12-6	10-11	17-6	15-11	13-11	21-4	19-4	16-11
	No. 2	10-1	9-2	7-11	13-4	12-1	10-6	17-0	15-5	13-4	20-8	18-9	16-3
	No. 3	8-8	7-6	6-2	11-6	9-11	8-1	14-8	12-8	10-4	17-9	15-5	12-7
Western white pine	Select structural	11-3	10-3	8-11	14-11	13-6	11-10	19-0	17-3	15-1	23-1	21-0	18-4
	No. 1/appearance	11-3	10-3	8-6	14-11	13-6	11-3	19-0	17-3	14-4	23-1	21-0	17-5
	No. 2	10-10	9-4	7-8	14-3	12-4	10-1	18-2	15-9	12-10	22-1	19-2	15-7
	No. 3	8-4	7-3	5-11	11-0	9-6	7-9	14-0	12-2	9-11	17-0	14-9	12-1
Red pine (North)	Select structural	11-0	10-0	8-9	14-6	13-2	11-6	18-6	16-10	14-8	22-6	20-6	17-11
	No. 1/appearance	11-0	10-0	8-6	14-6	13-2	11-3	18-6	16-10	14-4	22-6	20-6	17-5
	No. 2	10-9	9-6	7-9	14-2	12-6	10-2	18-0	15-11	13-0	21-11	19-5	15-10
	No. 3	8-4	7-3	5-11	11-0	9-6	7-9	14-0	12-2	9-11	17-0	14-9	12-1

Floor joist spans for Canadian lumber
Figure 5-22

Floor Joists (30 psf Live Load) (10 psf Dead Load) - Sleeping Rooms and Attic Floors (continued)
Maximum Allowable Span (ft.-in.)

Species	Grade	2 × 6			2 × 8			2 × 10			2 × 12		
		Joist spacing (inches)											
		12	16	24	12	16	24	12	16	24	12	16	24
Eastern white pine (North)	Select structural	10-9	9-9	8-6	14-2	12-10	11-3	18-0	16-5	14-4	21-11	19-11	17-5
	No. 1/appearance	10-9	9-9	8-6	14-2	12-10	11-3	18-0	16-5	14-4	21-11	19-11	17-5
	No. 2	10-5	9-6	7-9	13-9	12-6	10-2	17-6	15-11	13-0	21-4	19-4	15-10
	No. 3	8-4	7-3	5-11	11-0	9-6	7-9	14-0	12-2	9-11	17-0	14-9	12-1
Northern aspen	Select structural	11-3	10-3	8-11	14-11	13-6	11-8	19-0	17-3	14-11	23-1	21-0	18-2
	No. 1/appearance	11-3	10-2	8-4	14-11	13-5	11-0	19-0	17-2	14-0	23-1	20-10	17-0
	No. 2	10-8	9-3	7-6	14-0	12-2	9-11	17-11	15-6	12-8	21-9	18-10	15-5
	No. 3	8-2	7-1	5-9	10-9	9-3	7-7	13-8	11-10	9-8	16-8	14-5	11-9
Black cottonwood	Select structural	10-9	9-9	7-11	14-2	12-10	10-6	18-0	16-4	13-4	21-11	19-11	16-3
	No. 1/appearance	10-6	9-1	7-5	13-10	12-0	9-9	17-8	15-4	12-6	21-6	18-7	15-2
	No. 2	9-5	8-2	6-8	12-5	10-9	8-9	15-10	13-8	11-2	19-3	16-8	13-7
	No. 3	7-4	6-4	5-2	9-8	8-4	6-10	12-4	10-8	8-8	15-0	13-0	10-7
Northern species	Select structural	10-5	9-6	8-3	13-9	12-6	10-11	17-6	15-11	13-11	21-4	19-4	16-11
	No. 1/appearance	10-5	9-6	8-3	13-9	12-6	10-11	17-6	15-11	13-11	21-4	19-4	16-11
	No. 2	10-1	9-2	7-8	13-4	12-1	10-1	17-0	15-5	12-10	20-8	18-9	15-7
	No. 3	8-4	7-3	5-11	11-0	9-6	7-9	14-0	12-2	9-11	17-0	14-9	12-1
Coast species	Select structural	11-7	10-6	9-2	15-3	13-10	12-1	19-5	17-8	15-5	23-7	21-6	18-9
	No. 1/appearance	11-7	10-6	8-11	15-3	13-10	11-8	19-5	17-8	14-11	23-7	21-6	18-2
	No. 2	11-3	10-0	8-2	14-11	13-2	10-9	19-0	16-9	13-8	23-1	20-5	16-8
	No. 3	8-8	7-6	6-2	11-6	9-11	8-1	14-8	12-8	10-4	17-9	15-5	12-7
MSR Lumber f-E Classification	1200f-1.2E	10-9	9-9	8-6	14-2	12-10	11-3	18-0	16-5	14-4	21-11	19-11	17-5
	1450f-1.3E	11-0	10-0	8-9	14-6	13-2	11-6	18-6	16-10	14-8	22-6	20-6	17-11
	1500f-1.4E	11-3	10-3	8-11	14-11	13-6	11-10	19-0	17-3	15-1	23-1	21-0	18-4
	1650f-1.5E	11-7	10-6	9-2	15-3	13-10	12-1	19-5	17-8	15-5	23-7	21-6	18-9
	1800f-1.6E	11-10	10-9	9-4	15-7	14-2	12-4	19-10	18-0	15-9	24-2	21-11	19-2
	1950f-1.7E	12-0	10-11	9-7	15-10	14-5	12-7	20-3	18-5	16-1	24-8	22-5	19-7
	2100f-1.8E	12-3	11-2	9-9	16-2	14-8	12-10	20-8	18-9	16-5	25-1	22-10	19-11
	2250f-1.9E	12-6	11-4	9-11	16-6	15-0	13-1	21-0	19-1	16-8	25-7	23-3	20-3
	2400f-2.0E	12-9	11-7	10-1	16-9	15-3	13-4	21-5	19-5	17-0	26-0	23-7	20-8
	2550f-2.1E	12-11	11-9	10-3	17-0	15-6	13-6	21-9	19-9	17-3	26-5	24-0	21-0
	2700f-2.2E	13-1	11-11	10-5	17-4	15-9	13-9	22-1	20-1	17-6	26-10	24-5	21-4
	3000f-2.4E	13-6	12-3	10-9	17-10	16-2	14-2	22-9	20-8	18-0	27-8	25-1	21-11
	900f-1.0E	10-1	9-2	8-0	13-4	12-1	10-7	17-0	15-5	13-6	20-8	18-9	16-5
	900f-1.2E	10-9	9-9	8-2	14-2	12-10	10-9	18-0	16-5	13-8	21-11	19-11	16-8
	1200f-1.5E	11-7	10-6	9-2	15-3	13-10	12-1	19-5	17-8	15-5	23-7	21-6	18-9
	1350f-1.8E	12-3	11-2	9-9	16-2	14-8	12-10	20-8	18-9	16-5	25-1	22-10	19-11
	1800f-2.1E	12-11	11-9	10-3	17-0	15-6	13-6	21-9	19-9	17-3	26-5	24-0	21-0

Floor joist spans for Canadian lumber
Figure 5-22 (continued)

A.F.P.A.® 00
S—P—F
S-DRY STAND

Alberta Forest Products Association
204–11710 Kingsway Avenue
Edmonton, Alberta T5G 0X5

C L A
S·P·F
100
No. 2
S-GRN.

Canadian Lumbermen's Association
27 Goulburn Avenue
Ottawa, Ontario K1N 8C7

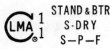

STAND & BTR
S-DRY
S—P—F

Cariboo Lumber Manufacturers Association
301–197 Second Avenue North
Williams Lake, British Columbia V2G 1Z5

CFPA® 00
S-P-F S-DRY
CONST

Central Forest Products Association
14G–1975 Corydon Avenue
Winnipeg, Manitoba R3P 0R1

Also using this stamp is:

Saskatchewan Forest Products Corporation
101 First Avenue East
Prince Albert, Saskatchewan S6V 2A5

S-P-F
S-DRY
100 No 2

Council of Forest Industries
of British Columbia
1500–1055 West Hastings Street
Vancouver, British Columbia V6E 2H1

Also using this stamp is:

Council of Forest Industries
of British Columbia
Northern Interior Lumber Sector
803–299 Victoria Street
Prince George, British Columbia V2L 2J5

ILMA S-DRY 1
00 S—P—F

Interior Lumber Manufacturers Association
295–333 Martin Street
Penticton, British Columbia V2A 5K7

M L B SPRUCE PINE FIR **STAND** S-GRN

MILL 11 — 466

Maritime Lumber Bureau
P.O. Box 459
Amherst, Nova Scotia B4H 4A1

O.L.M.A.® 01-1
CONST. S-DRY
SPRUCE · PINE · FIR

Ontario Lumber Manufacturers Association
159 Bay Street, Suite 414
Toronto, Ontario M5J 1J7

Quebec Lumber Manufacturers Association
3555 Boulevard Hamel West, Suite 200
Quebec, Quebec G2E 2G6

No 1
S-DRY
D FIR (N)

MacDonald Inspection
125 East 4th Avenue
Vancouver, British Columbia V5T 1G4

NLGA RULE
No 1
S-GRN
HEM-FIR-N

Pacific Lumber Inspection Bureau
1460–1055 West Hastings Street
Vancouver, British Columbia V6E 2G8

Sample Canadian grade stamps
Figure 5-23

Selecting the Right Lumber
By now you should feel comfortable with lumber grades and what the names mean. Of course, there's more to learn. The lumber grading books have far more detail than the overview provided here. I recommend that you get the grading rule for the species you use most. It only costs a few dollars, and it will make you an expert.

Chapter 6

Framing the Floor and Walls

The term *rough carpentry* describes the framing of a building. This doesn't mean that the quality of the work is "rough" or poor. It refers to the work of constructing the *skeleton*, or structural part of a building. Framing includes sills, girders, floor joists, wall studs, headers, plates, ceiling joists, rafters, posts, and columns. All interior and exterior finish material is attached to the framing.

In this chapter, we'll look at two important parts of rough carpentry: floor framing and wall framing. (The third part, roof framing, will be covered in the next chapter.)

We'll begin this chapter with a look at basic framing principles. We'll explore four common framing methods. We'll also look at modular coordination. Then we'll take a detailed look at the procedures for installing quality, cost-efficient floor and wall framing.

Basic Framing Principles

The lumber in a building will undergo many changes of temperature and humidity. Even well-seasoned lumber loses moisture during the winter when the building is heated. The lumber will shrink as the moisture content lessens. The amount of this shrinkage depends on three factors:

- The moisture content (MC) of the lumber at the time it's set in place

- The temperature and humidity of the building

- The compression caused by the load on the lumber

Most shrinkage is across the width of the board. There isn't much shrinkage along the length of the board. This means that *the more horizontal weight-bearing members there are in a house frame, the more the frame will shrink*. Remember this when you select your framing method and lumber.

To brace the side and end walls of a frame you'll either: (1) use diagonal or knee braces at the corner posts of the building, (2) run the sheathing diagonally to the studs, or (3) use plywood sheets at corners. Be sure to use enough joist ties to secure the walls together. You may need to use bridging to lock the joists to each other. Also, make sure the headers of openings in walls and floors are strong enough to carry the dead load plus live loads.

Methods of Framing

The four basic types of framing are: braced frame, platform frame, balloon frame, and plank-and-beam frame. You may find only one or two types in your area. Each region of the United States — the Pacific Coast, the Midwest, the South and the Atlantic Coast — uses a different type of framing.

But the same basic principles of framing and bracing are used throughout the country.

Braced Frame

The two kinds of braced frame construction are: early braced frame and modern braced frame. You're not likely to find early braced framing in modern buildings. But I'll describe it because you need to know the advantages of both methods.

Early braced frame— In early braced frame construction, most of the bearing members are vertical. This means that the frame of the building won't shrink very much.

Each post and beam is mortised and tenoned together. Diagonal braces keep the angles (formed by the posts and beams) in place. Wooden pins or dowels through the joints secure the braces.

Early braced frame construction doesn't rely on the sheathing for rigidity. The sheathing can even run vertically up the sides of the building. Barns were once built this way, and many of them are still in use.

Modern braced frame— Figure 6-1 shows the modern braced frame. The studs of the side walls and center partitions are all exactly the same length. The side wall studs are attached to the sill plate, and the partition studs are attached to the girder and plate. If you use a steel girder, you'll have the same amount of shrinkage in the outside and inside walls.

To make the frame rigid, let diagonal braces into the studs at the corners of the building. Or install plywood, structural fiberboard, or steel bands at the corners. Your choice of bracing material will depend on the locations of the openings in the side or end walls.

Lap and spike the joists to the side wall studs. You can make the joists continuous between the two side walls. Or you can side lap and spike them at the center partition. In either case, these joists tie the two side walls together. The joists are supported by the full width of the sill and girder. Bridge the joists where necessary, and put fire stops in at the ends of all joists.

If you use plank subflooring, lay it diagonally to the floor joists. Also, lay it in opposite directions on the upper and lower floors. This makes the whole building, as well as the floors, more rigid. If you use plywood subflooring, install it with the long dimension across the supporting joists.

Platform (or Western) Frame

Figure 6-2 shows the platform, or Western, frame. Each story of the building is built as a separate unit. You lay the subfloor before you raise the walls. The ends of the joists support the side and center wall studs. This causes equal shrinkage at the side and center walls. Otherwise, the features of platform framing are similar to modern braced framing.

Platform framing has become more popular since kiln-dried lumber and precut studs have come into use. It's probably the fastest and safest method of frame construction. You can use shorter materials for studding since the studs go up through only one story. The second-floor joists rest on a two-piece plate, or cap, set on top of the first-floor studding. No wood fire stops are needed, since the rough flooring extends to the outside edges of the framing.

Run a continuous header around the entire building. This header should be the same size as the joists. At the first floor, the bottom edge of the header rests on the sill. Make the outside face of the header flush with the outside edge of the sill. This forms what is known as a "box sill." Instead of using a continuous header at the second floor, you can extend the joists flush to the outside of the plate and put blocking between the joists. Take another look at Figure 6-2.

Build and support partitions just as you would in any other framing system. Use double joists, sole pieces, and plates for support. Figure 6-45, later in this chapter, shows the framing of double joists under a partition.

Frame the outside walls the same as the interior partitions. This won't eliminate shrinkage, but it will make the shrinkage equal in both outside walls and interior partitions. This means that your floors and ceiling will remain level no matter how much the wood shrinks.

If you use steel I-beams instead of wood girders in your load-bearing inside walls, you'll need to use a wood member on top of the steel beam. The inset in Figure 6-2 shows this method. Make the wood member the same cross-section size as your sill members. This will insure uniform shrinkage in the sill and girder.

Platform framing can reduce your lumber and labor costs. You'll need shorter lengths of lumber, which you can cut to size before construction begins. You can assemble the large units, such as trusses, headers, and window frames, before the

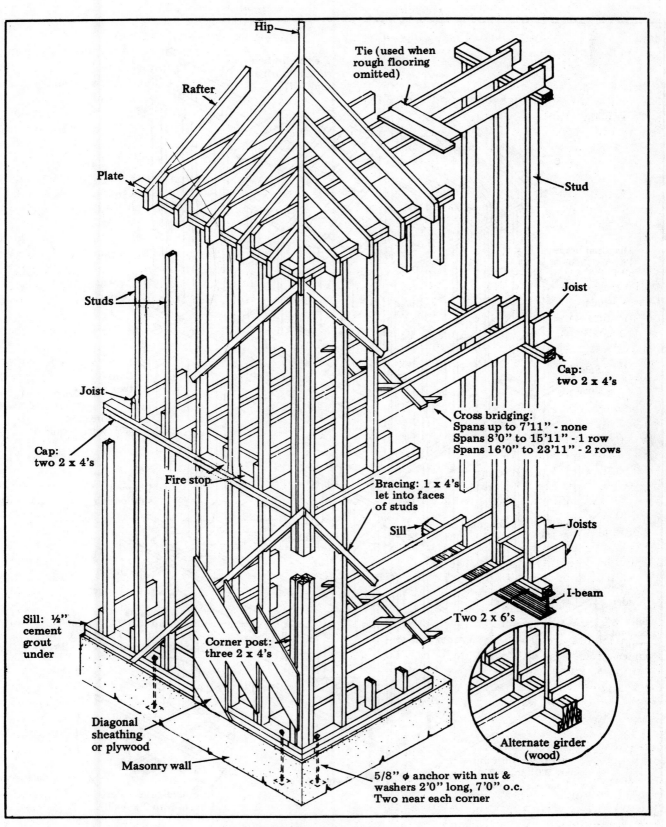

Hip

Tie (used when rough flooring omitted)

Rafter

Plate

Stud

Studs

Joist

Joist

Cap: two 2 x 4's

Cap: two 2 x 4's

Cross bridging:
Spans up to 7'11" - none
Spans 8'0" to 15'11" - 1 row
Spans 16'0" to 23'11" - 2 rows

Fire stop

Bracing: 1 x 4's let into faces of studs

Sill

Joists

I-beam

Two 2 x 6's

Sill: ½" cement grout under

Corner post: three 2 x 4's

Alternate girder (wood)

Diagonal sheathing or plywood

Masonry wall

5/8" ø anchor with nut & washers 2'0" long, 7'0" o.c. Two near each corner

Modern braced framing
Figure 6-1

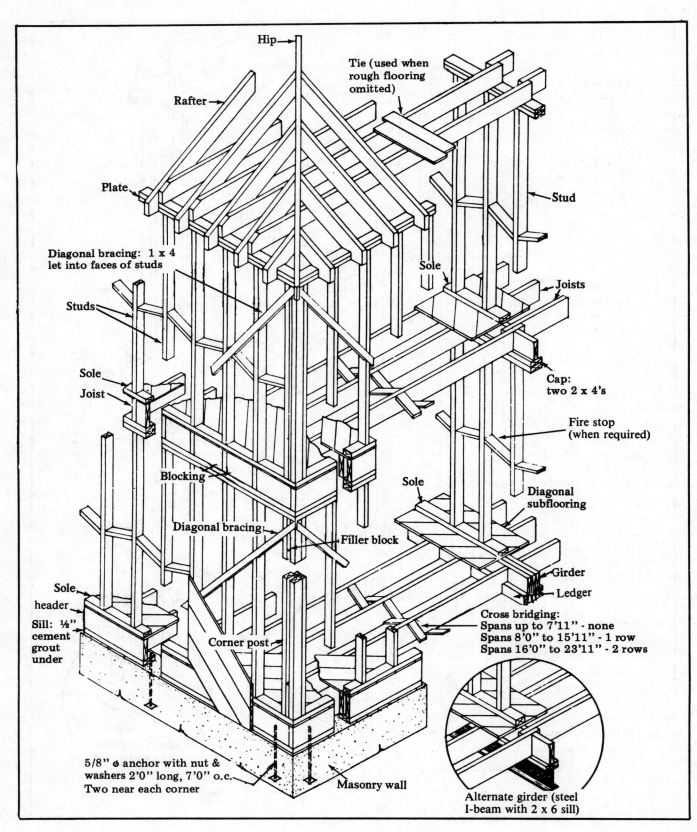

Hip

Tie (used when rough flooring omitted)

Rafter

Plate

Stud

Diagonal bracing: 1 x 4 let into faces of studs

Sole

Joists

Studs

Sole
Joist

Cap:
two 2 x 4's

Fire stop
(when required)

Blocking

Sole

Diagonal
subflooring

Diagonal bracing

Filler block

Girder
Ledger

Sole,
header

Cross bridging:
Spans up to 7'11" - none
Spans 8'0" to 15'11" - 1 row
Spans 16'0" to 23'11" - 2 rows

Sill: ½"
cement
grout
under

Corner post

5/8" ø anchor with nut &
washers 2'0" long, 7'0" o.c.
Two near each corner

Masonry wall

Alternate girder (steel
I-beam with 2 x 6 sill)

Platform (Western) framing
Figure 6-2

studs go up. Then you just raise the large units into place in one single operation.

Balloon Frame

Figure 6-3 shows the balloon frame. This type of framing replaced the early braced frame until modern braced framing and platform framing were developed.

The balloon frame has many features to recommend it. You can put the outside studs up quickly because they extend the full two stories, from sill to roof plate. You can put up the load-bearing partitions the same way. At the second floor, you rest the joists on a ledger (or ribbon) and nail them against the studs. The attic or ceiling joists rest on the doubled top plate.

The openings between the long studs in the balloon frame cause a strong flue effect. You *must* install fire stops between the studs. Without fire stops, flames from the basement can sweep up to the top plate and into the attic.

Once the sheathing is in place, a balloon frame structure is as strong and rigid as a building with braced framing. But until you put that sheathing on, the frame won't be as sturdy. Keep this in mind when you're prancing on the second floor joists!

Balloon framing often costs less than other methods, because you can put it up faster. You can save money on materials if you leave out the 1 x 4 bracing. Instead, you'd run the sheathing diagonally from the corner post to the second floor sill, or use plywood (or structural fiberboard). Balloon framing shrinks less because there's less cross-section lumber in each wall. To be sure of uniform shrinkage in the outside wall and any partitions, use a steel girder with a wood top member the same size as your sill members.

Plank-and-Beam Frame

The plank-and-beam system is used for floors and roofs in modern one-story structures with large glass areas, modular coordination, open-space planning, and "natural finish" materials.

The plank-and-beam system concentrates structural loads on fewer and larger members. You can make planks continuous over two or more spans. And site assembly goes quickly, needing fewer manhours. This gives you more cost savings.

You can use plank-and-beam framing with ordinary wood stud or masonry walls. You can also use it as the skeleton frame in curtain wall construction. Figure 6-4 shows how basic this system

is. Notice how this method is similar to structural steel framing.

If you're going to use this system, you need an engineered design, high quality materials, and careful workmanship. All joints must fit exactly, since there are fewer contacts between members.

Plank-and-beam structures often need a roof overhang to keep the summer sun out of the large glass areas. Constructing the overhang is easy. You don't need rafter tails, soffits, fascia, moldings or eave vents. Insulation, roofing and gravel all stop at a simple wood member which is covered with metal to provide a drip. The overhang is nearly completed when the carpenters finish sheathing the roof.

If you can space the beams to get the most efficient use of such materials as planks and drywall, you'll shorten time spent cutting and fitting materials. You might decide on 8' spacing for beams and columns. That way, 16' planks will be continuous over two spans. With 8' beam spacing, you can fit 4' x 8' sheets of drywall and plywood on the walls without any cutting.

You'll find further cost savings in the planking. The 2 x 6 or 2 x 8 tongue-and-groove (T&G) planks can be finished with "natural finishes" before you put them up. You don't need interior scaffolding. Also, you won't have the cost of ceiling joists, bridging, lath and plaster, or wallboard. When you nail the planks into place, the ceiling is complete.

Usually, you won't put ceiling material on the underside of the planking. But some codes may require it.

When you need to put ceiling tiles or sheet material on the planking, make sure you install a lower grade of plank. Since the interior planking won't show, you don't need to worry about how the plank looks.

If you use the plank-and-beam system for a floor, you can use 2" planking as both subfloor and finish flooring (if local code permits).

We've seen the advantages of a plank-and-beam framing system. Let's look at an example. Figures 6-5 and 6-6 show the differences between the plank-and-beam frame system and conventional framing. Both figures show a typical 40' wall section. They both have the same wall openings, the same floor-to-ceiling height, and the same height of the rough floor over the outside grade. The studs in Figure 6-5 are on 24" centers, and those in Figure 6-6 are on 16" centers.

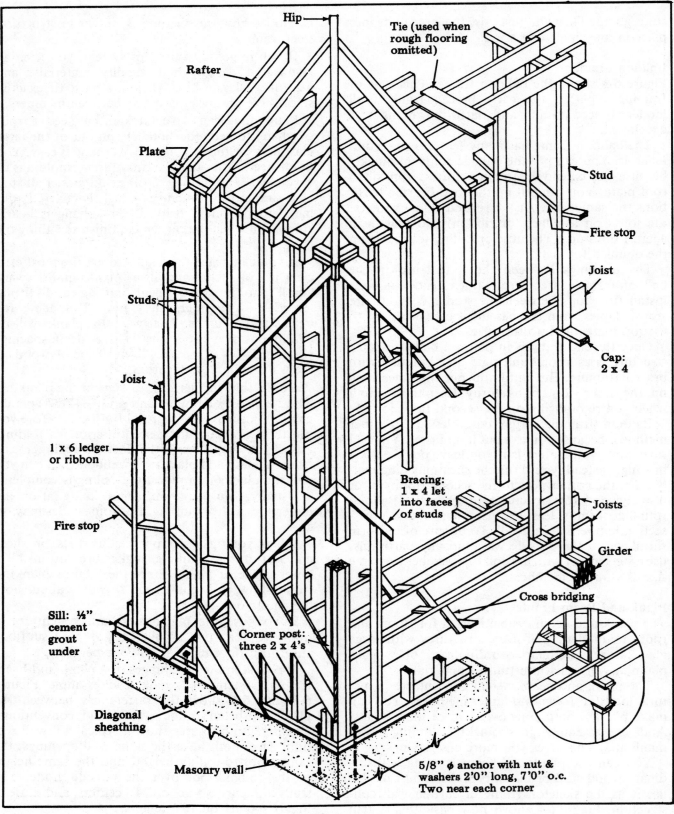

Balloon framing
Figure 6-3

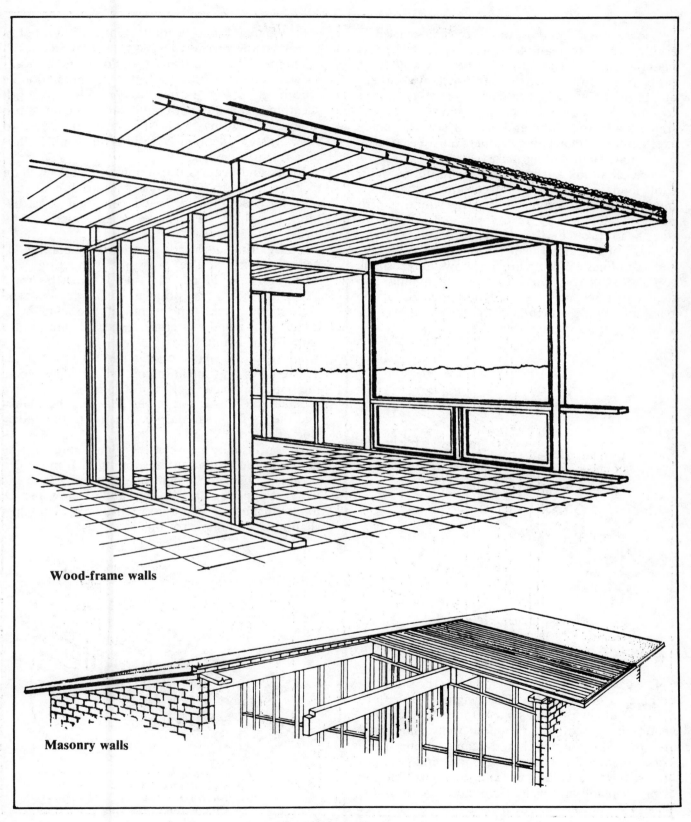

Wood-frame walls

Masonry walls

Plank-and-beam framing
Figure 6-4

Look at the number of pieces of lumber used in each system. The plank-and-beam system uses less material. This means your carpenters will have fewer pieces to handle, saw, and nail. Also, there are fewer points where members must be nailed together. Since the members must support concentrated loads, the plank-and-beam system demands excellent craftsmanship. Members must be cut true and square, and nailing must be exact.

Next let's look at the windows and doors. With roof loads concentrated on columns spaced as much as 10' apart, you can use large windows with no heavy lintels. Figure 6-5 shows window and door heads brought directly to the top plate. There are no lintel beams (headers) or cripple studs, as shown in Figure 6-6. Figure 6-5 shows only a single top plate. But you'll need to provide a nailing strip (equal to a second plate) on the underside of the planking. You'll attach interior finish and exterior sheathing and siding to the nailing strip.

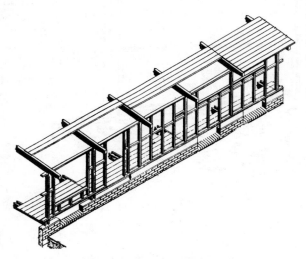

**Plank-and-beam framing
Figure 6-5**

The basement window openings are the same size in Figures 6-5 and 6-6. The height of the floor above grade is also the same. Notice that, in conventional construction, window wells and lintels are used. In the plank-and-beam system, these same windows are above ground and without lintels. If you decide to use window wells, you can bring the plank-and-beam floor closer to the grade. This way, you can cut down on the cost of entrance steps and rails.

In the plank-and-beam system, you probably won't need to provide double studs around all wall openings. There isn't any roof load on the walls between columns. But don't just cut out all double studding. If you use low-racking-strength sheathing on a wall with many large openings, too much bending may cause windows and doors to stick. Install adequate sheathing or double studding on such walls.

Plank-and-beam construction also works well on pier foundations. You can use piers to directly support the floor beams. In conventional construction, the floor joists have to be supported by girders set in piers.

Considering all the advantages to the plank-and-beam system, you're probably wondering why every house in the country isn't framed this way. Plank-and-beam is popular for vacation-type homes. But there are technical problems with design limits and materials, especially in certain parts of the country.

The exposed plank ceiling has no attic space. This creates a problem in cold-winter areas where the buildings need thick roof insulation. You have to place the insulation either between the planks and the roofing material, or fasten it to the underside of the planking.

If you insulate between the planks and roofing material, the insulation must be rigid. It must also be able to bear the weight of the roofing (and the workers applying the roofing), and of snow and ice loads. It must remain a good insulator even when slightly wet. It can't rot or decompose.

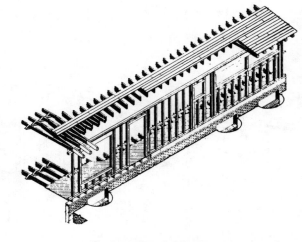

**Conventional framing
Figure 6-6**

If you fasten the insulation to the underside of the planking, you have to think about appearance. The insulation must be attractive as well as functional. And you'll find that many insulation materials can't be nailed to planking. They must be laid in mastic on roof decks that are limited to a maximum pitch of 3 in 12.

Rigid insulation for a plank-and-beam roof often costs more than insulation for a conventional frame roof. Normally, the other cost savings of the plank-and-beam system make up for the higher cost of insulation. When insulating for extreme climates, however, the plank-and-beam system may not be economical. Be sure to look at your total costs before you decide on a framing method.

Modular Coordination

Most builders have their favorite story about the board feet of lumber that were wasted because plan dimensions weren't coordinated with material dimensions. And it isn't just lumber that's wasted due to lack of modular coordination. Vinyl flooring comes in rolls that are 8' wide, roofing comes in rolls 3' wide, and wallboard comes in 4' x 8' sheets. Most construction materials that are delivered to the site in rigid form are sold in a limited number of standard sizes.

Take advantage of this standard sizing. Plan your houses to make full use of standard-size materials. You'll cut down on the amount of material. And you'll spend less time cutting and fitting small pieces or adding extra sections. That means more house for your buyer's dollar and more profit in your pocket.

Let's look at the five important parts of a modular plan: the planning grid, horizontal and vertical planes, wall modules, roof modules, and floor modules.

Planning grid— Every modular plan uses three dimensions: length, width and height. The modular length and width are the two most important dimensions. Your modular plan should be based on a planning grid like the one shown in Figure 6-7.

A modular plan divides the horizontal plane of the house into units ("modules") of 4", 16", 24" or 48". You'll usually want overall dimensions to be in multiples of 4.

The 16" unit gives you flexibility in the spacing of windows and doors. Increments of 24" and 48" are used for the exterior dimensions of the house. You can easily match floor, ceiling, and roof construction to these dimensions.

The planning grid gives you an easy, accurate way to lay out a modular plan. It doesn't matter whether you're building a small home or a mansion. Modular planning works for both.

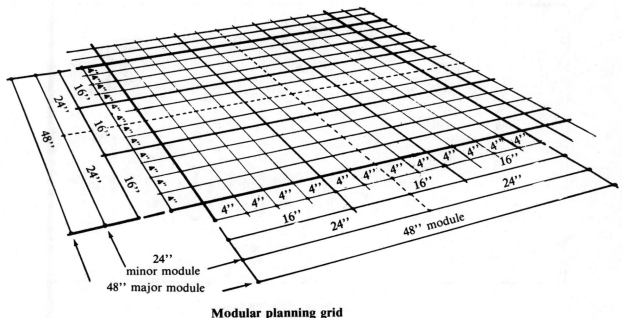

Modular planning grid
Figure 6-7

Horizontal and vertical planes— Once the house is laid out on a modular grid, the entire house is divided into horizontal and vertical planes, as shown in Figure 6-8 A. These planes don't show any thickness. The allowance for wall thickness and tolerance is based on fixed module lines at the outside face of the studs. Take a look at Figure 6-8 B.

Exterior walls and partitions can have many thickness variables. Floor and roof construction may also vary in thickness, depending upon structural requirements and the type of framing and finishing.

Wall modules— Your next step is to separate exterior wall elements at natural division points. In Figure 6-9, overall house dimensions are based on 48'' and 32'' modules. The 16'' module gives you

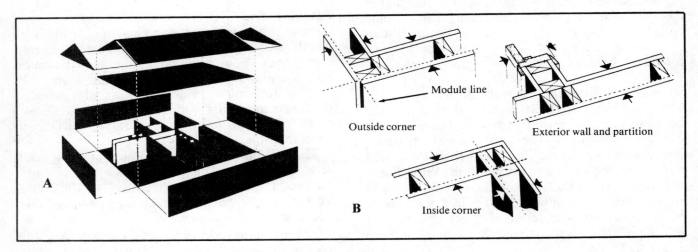

Horizontal and vertical plane sections
Figure 6-8

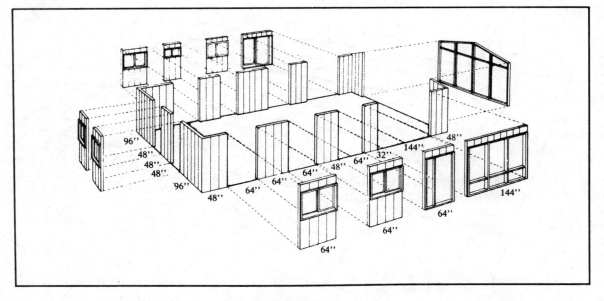

Exterior wall, modules
Figure 6-9

the most flexibility in placing door and window openings. You can put wall openings exactly on the 16″ module. Then you don't have to worry about the extra wall framing that nonmodular house construction often needs.

Roof modules— Increments for house depths are in 24″ multiples. Eight 48″ module depths (32′) and ten 24″ module depths (20′) take care of most roof span requirements. Figure 6-10 shows the major and minor module roof increments.

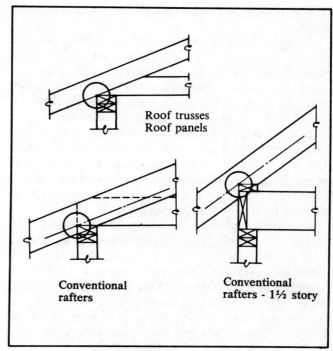

Pivotal point for modular roof planning
Figure 6-11

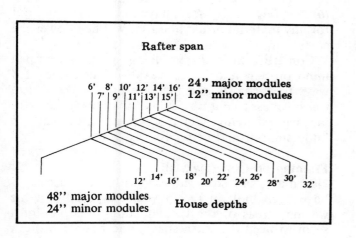

Roof modules
Figure 6-10

The "pivotal point" shown in Figure 6-11 is the fixed point of reference in the modular line of the exterior wall. Modular roof design and construction dimensions are determined from this point.

Floor modules— You can cut down on flooring costs by using the 48″ module. It lets you use full 4′ x 8′ plywood panels for subflooring. There's less cutting and less waste.

Floor joists are spaced 12″, 16″ or 24″ o.c. for conventional construction. Joists can be "lapped" or "in-line". You can also space joists 13.7″ and 19.2″ o.c. The 13.7″ spacing gives you seven equal sections from an 8′ length of plywood subflooring. The 19.2″ spacing gives you five equal sections.

Framing the Floor
Now, let's start from the bottom and work our way up, as we take a detailed look at floor framing. We'll learn about basement and crawl-space con-

struction, framing on slabs, sills, posts and girders, joist lengths, spans, grades, and sizes. Then we'll look at bridging, subflooring and field-glued floor systems.

Basement and Crawl-space Construction
You'll normally install floor framing on top of a basement or a crawl-space. Let's see what's supporting that floor frame you're putting together so carefully as we take a look at basements and crawl-spaces.

Treated wood basement construction— Treated wood basement construction has been accepted by the U.S. Department of Housing and Urban Development (HUD) for FHA-insured loans.

The preservative treatment for wood foundations must comply with the American Wood Preservers Bureau Standard AWPB-FDN. Each piece of lumber or plywood should have the AWPB stamp, or that of another approved inspection agency. Lumber and plywood treated under this standard will last a long time. Treated wood basement walls, like basement walls made of other materials, are affected by soil conditions and backfill heights. Be sure to check code or engineering requirements for stud size for fill over 48" high.

In treated wood basement construction, you can install the wiring, insulation, and finish walls the same way you do in stud-wall construction. It's ideal for prefab construction assembly. You can build sections in the shop, then assemble them on the site. This cuts down on your labor costs.

Excavate the basement normally. Install the plumbing lines below the basement floor. Cover the basement floor with at least 4" of crushed stone or gravel. Make sure the gravel extends at least 6" beyond the footing line, so it can be a base for the treated wood footing.

Put plywood wall panels over the studs and fasten the panels in place. Seal the joints with a waterproof caulk. Use pressure-treated 1/2"-thick standard C-D (exterior glue) plywood, with the grain going *across* studs. You don't need to block at horizontal plywood joints if the joints are at least 4' above the bottom plate. (These specifications are based on a soil condition with 30 pounds per cubic foot equivalent fluid weight.)

Apply 6-mil polyethylene in a continuous sheet over the exterior side of the wall below grade level. Cover the gravel bed with 6-mil polyethylene as well. You can pour a standard concrete floor over this base. This basement setup is shown in Figure 6-12.

To resist the thrust of the backfill pressure, be sure the first-story floor is securely fastened to the top of the wood basement walls. Use solid blocking on the end walls so that the foundation wall loads will be evenly shared with the solid floor frame. Don't do any backfilling until the basement floor and first-story floor are in place.

Treated wood crawl-space construction— Figure 6-13 shows the basics of wood crawl-space construction. You build the pressure-treated wall panels with continuous, buried footing plates. Use only lumber and plywood treated as required by the AWPB-FDN.

You'd build a crawl-space foundation about the same as you'd build a treated wood basement, using treated studs, plates and plywood. The difference, of course, is that crawl-spaces don't need floors. Anchor studs to a treated footing on top of a gravel base at the frost line. Then run the studs from the gravel base up to the sill plate. You can space the treated 2 x 4 studs 24" o.c. for single-story construction. Two-story construction needs a stud spacing of 12" o.c.

Remember that a crawl-space doesn't need more than 24" of headroom. This means that you don't need to extend the 1/2"-thick plywood facing all the way down to the footing. The plywood siding covers only from the sill plate down to the level of the ground, the crawl space floor.

Some builders are reluctant to try treated wood foundations. You've probably been building with concrete foundations for years and like the results. Why try something new when the old way works fine? A lot of us looked the other way when wallboard first became popular. But to be successful in the construction business, and to *continue to be successful*, we have to be on the lookout for any material or method that will help lower the costs of quality construction.

Construction is always changing. Keep an open mind on new and better ways to build. Look for ways to cut costs and increase profits. You're in this business to put up quality buildings while making a good living. Let me suggest some ways to build smarter and better for less money.

Framing on Concrete Slabs

Not many one-story houses with full basements have been built in recent years, especially in the warmer areas of the United States. People don't seem to need the extra space, and houses without basements cost less to build.

In the past, the main use of a basement was as a place to put the coal or wood central heating plant, and for storing the fuel and ashes. The basement also had room for laundry and utilities. Now that gas and electric heating systems are used in most houses, the basement is no longer needed for storage. And it's easy enough to make space on the ground floor for the heating equipment, laundry, and utilities.

When you build without a basement, you can anchor your rough carpentry to one of three common types of concrete foundations: slab-on-ground, thickened-edge slab, or the independent slab and wall. Here's a look at each of these concrete support systems:

Slab-on-ground— The on-ground concrete slab is a popular construction method. But don't use this kind of construction on sloping ground or in low-lying areas, unless you're building a split-level home. Structural and drainage problems can take all the profit out of the job. However, split-level houses may have a part of the foundation designed for an on-grade slab. In this case, the problem-

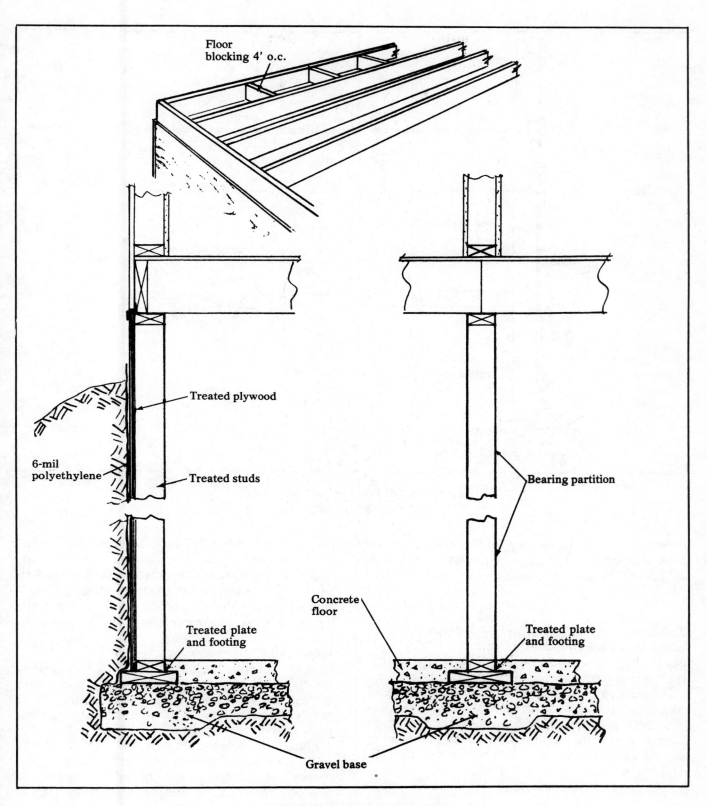

Floor
blocking 4' o.c.

Treated plywood

6-mil
polyethylene

Treated studs

Bearing partition

Concrete
floor

Treated plate
and footing

Treated plate
and footing

Gravel base

Treated wood basement construction
Figure 6-12

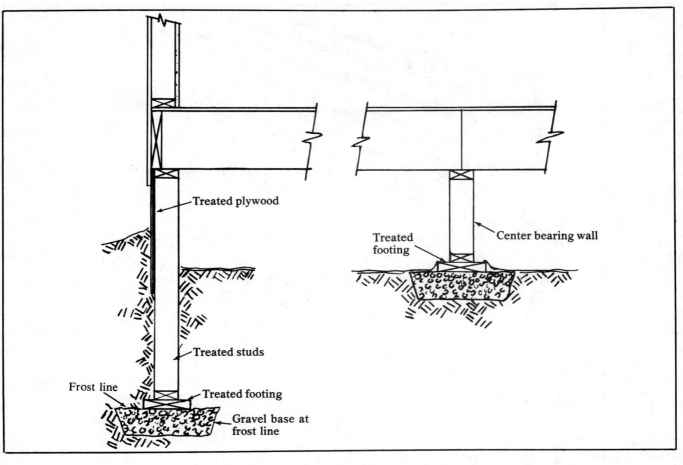

Treated wood crawl-space construction
Figure 6-13

causing features of a sloping lot become an advantage. You can use the slope of the lot to step up a split-level house with a minimum of excavation or fill.

Thickened-edge slab— Figure 6-14 shows the combined slab and foundation, called a "thickened-edge" slab. It's useful in warm climates where frost isn't a problem, and where soil conditions are suitable. You pour a shallow perimeter-reinforced footing, together with the slab, over a vapor barrier. You should have the bottom of the footing at least 1' below the natural grade line. The footing should be supported on solid, unfilled, and well-drained ground.

Independent slab and wall— In areas where the ground freezes during winter, you have to support the walls of the house with foundations or piers that go below the frost line to solid bearing on unfilled soil. In this case, the concrete slab and foundation wall are usually separate. Figures 6-15, 6-16 and 6-17 show three common methods of independent slab and wall construction. Be sure to use sills treated with wood preservative.

Wood Sill Construction
In a wood-frame building, the *sill plate* is usually the first member set in place on the foundation wall, unless you're installing termite shields. There are three common types of wood sill construction. Let's look at each method.

Platform construction usually uses the "box sill," as shown is Figure 6-18. The box sill provides support and fastening for the joists and for the header at the ends of the joists. A 2" or thicker wood plate is anchored to the foundation wall. When you use a sill sealer, the plate sits on top of the sealer.

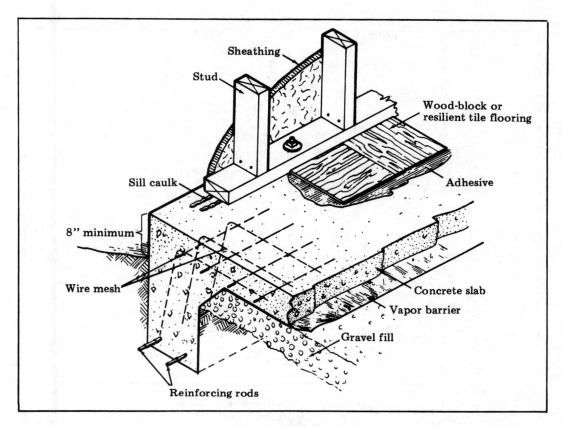

Thickened-edge slab
Figure 6-14

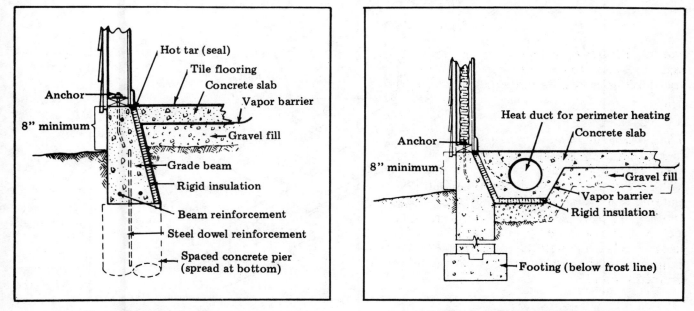

Reinforced grade beam for concrete slab
Figure 6-15

Full foundation wall for cold climates
Figure 6-16

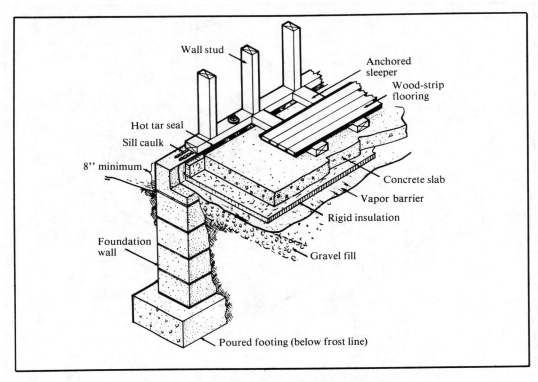

**Independent concrete floor slab and wall
Figure 6-17**

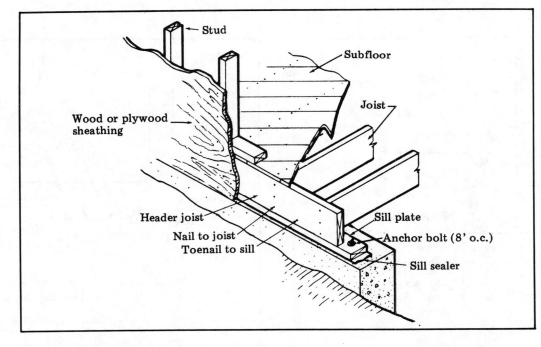

**Box sill
Figure 6-18**

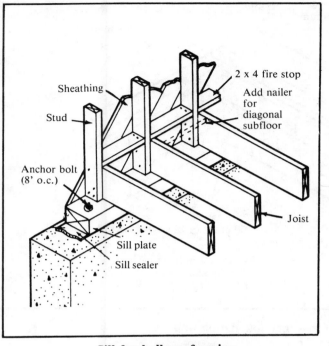

Sill for balloon framing
Figure 6-19

Balloon-frame construction also uses a nominal 2'' or thicker wood sill. As you can see in Figure 6-19, the joists rest on top of the sill. The studs also bear on the sill and are nailed to both the sill and the floor joists. You lay the subfloor either at right angles to the joists (for panels) or diagonally (for planks). If you use diagonal subflooring, you'll need a nailing member between joists and studs at the wall line. Install fire stops between the studs at the floor line.

If you use freestanding exterior piers instead of a solid foundation wall, use the sill construction method shown in Figure 6-20. You can either use a large timber for the sill beam, or you can build one up from nominal 2-by members, with a ledger to support the joists. When using a beam, make the bottom of the joists fit flush with the top of the beam.

Figure 6-21 shows how to anchor the sills to either masonry or concrete foundation walls.

Posts and Girders
When you build basements, you can use wood or steel posts to support the wood girders or steel beams. In crawl-space houses, masonry piers are often used.

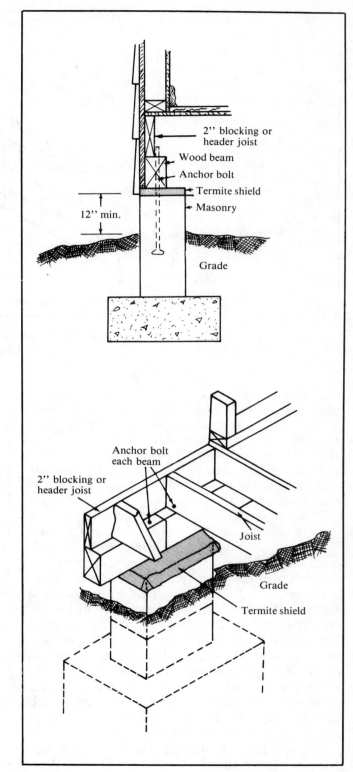

Freestanding exterior pier
Figure 6-20

133

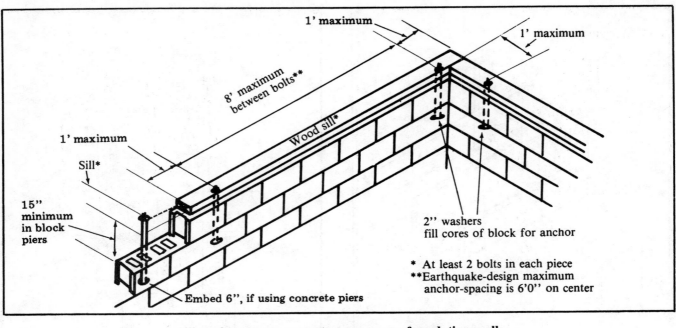

Sill anchorage on concrete or masonry foundation walls
Figure 6-21

A round steel post usually comes with a steel bearing plate at each end. Be sure to anchor the top of the post securely to the beam or girder. There are two ways you can attach the steel post to the floor. Either mount the base plate to the footing and use the floor as an anchor (as shown in Figure 6-22), or use the pedestal as a base and attach the plate to the top of it.

Wood posts that you use free-standing in a basement should be solid and not smaller than 6" x 6". When you use them with a frame wall, the wood posts can be 4" x 6", to match the depth of the studs. Square the wood posts at both ends.

Fasten the wood posts securely to the girder, as shown in Figure 6-23. Rest the bottom of the post on a masonry pedestal that is 2" to 3" above the finish floor. Pin the post to the pedestal. In moist or wet conditions, you should treat the bottom end of the post with a wood preservative.

Both wood girders and steel beams are common in modern construction. Be sure to compare the cost of each method. Steel is expensive, so it's usually more economical to use wood.

The standard I-beam and the wide-flange beam are the most common steel beam shapes. Wood girders come in two types: solid and built-up. Most builders prefer the built-up girders because they are

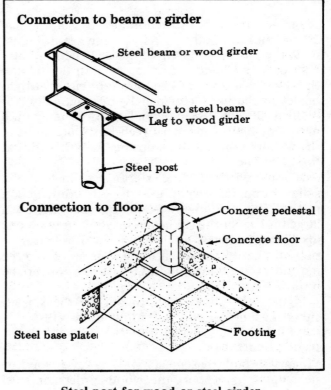

Steel post for wood or steel girder
Figure 6-22

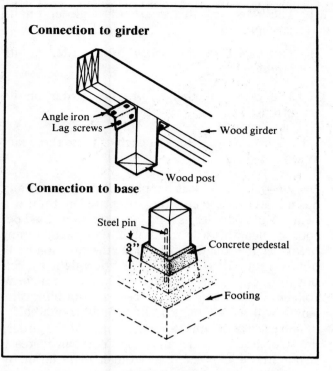

Wood post for wood girder
Figure 6-23

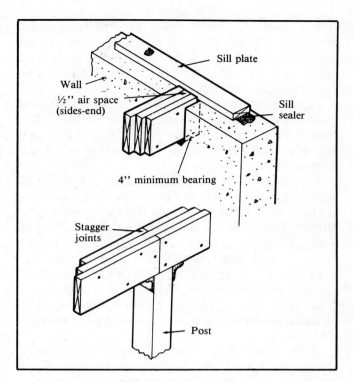

Built-up wood girder
Figure 6-24

normally made from drier dimension lumber and are more stable.

A built-up wood girder is shown in Figure 6-24. It's made of two or more pieces of 2'' dimension lumber spiked together. Join the ends of the lumber together over a supporting post.

You can nail a two-piece girder from one side with 10d nails. Use two nails at the end of each piece and drive the other nails 16'' apart, staggered. Nail a three-piece girder from each side with 20d nails. Use two nails near each end of each piece and stagger the other nails 32'' apart. Figure 6-25 shows the sizes of common wire nails.

Ends of wood girders should bear at least 4'' on the masonry walls or pilasters. If you're going to have joists resting on the girder, the top of the girder should be level with the top of the sill plates on the foundation walls. When wood is untreated, leave a 1/2'' air space at the ends and sides of the wood girders, as shown in Figure 6-24. In termite-infested areas, line these pockets with metal.

You may want to use commercially available glue-laminated beams for exposed beams in finished basement rooms. These beams come in long lengths and are made with a ''crown'' that allows

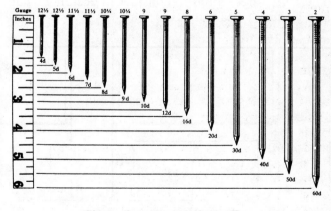

Sizes of common wire nails
Figure 6-25

for a certain amount of sag under weight. Use glue-lam beams when you want to use the fewest possible posts.

Girder Spacing

When you decide on the spacing for the girders, you have to consider (1) the length and depth of the joists, and (2) the location of the bearing partitions on the floor above the joists.

It's good practice to never have more than 16' between girders. For a span of 25', one girder should do if you put it half way between each of the other supports. If the span is 35', use two equally-spaced girders.

Figure 6-26 shows how the location of bearing partitions affects the spacing of girders. Put girders underneath the bearing partitions. You'll find that doing so will often change the spacing of girders between side wall supports.

Section A in Figure 6-26 shows the placement of a girder under joists that have no bearing partitions on the floor above. Section B shows a bearing partition placed midway between two side walls. The girder is directly underneath this partition. In section C, the bearing partition is 8' from one side wall and 16' from the other. Again, the girder is directly underneath the partition. If you have two bearing partitions, as shown in section D, you should also have two girders (even if the spacing is less than 16').

Use Southern yellow pine, Douglas fir, redwood, or cypress for sill and girder construction. These members are often treated with a preservative to protect against rot and insects.

Girder Size and Strength Criteria
There are three things to consider in deciding what size girder or beam to use:

1) The effect of the *length* of a girder on its strength

2) The effect of the *width* of a girder on its strength

3) The effect of the *depth* of a girder on its strength

Let's take a closer look at each of these three factors.

Length— If a plank is supported at each end and has a load that's even along its entire length, it will bend. A plank twice as long, with the same load per foot of length, will bend much more and may break. If the length is doubled, the safe load will not be reduced to one-half, as you might expect, but to one-quarter. However, if you have a *single concentrated load* at the center, doubling the length will reduce the safe load only to one-half.

The greater the unsupported length of the girder, the stronger the girder must be. You can increase girder strength by using a stronger material or by using a larger beam. Enlarge the beam by increasing the width or depth or both.

Width— Doubling the width of a girder doubles its strength. One double-width girder can carry the

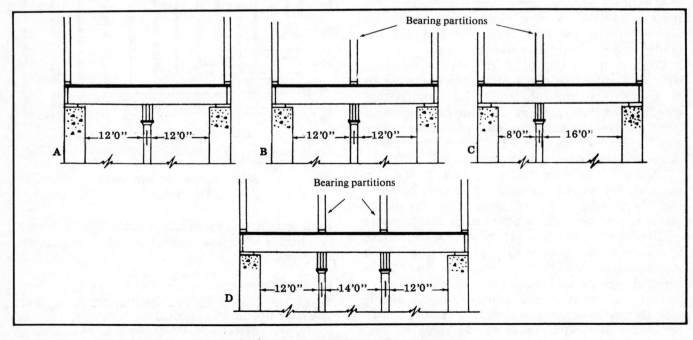

Placement of girders under partitions
Figure 6-26

same load as two single-width girders placed side by side.

Depth—Doubling the depth of a girder means it can carry four times the weight. Thus, a beam 3" wide and 12" deep can carry four times the load of a beam 3" wide and 6" deep. So if you want to increase the strength of a beam, it's more economical to increase the depth than the width.

But avoid making the girder much more than 10" in depth because a deeper girder will cut down on headroom in the basement. If you've increased the girder to its maximum depth, and you still need more strength, you can increase the width, add supports to reduce the span, or use a stronger material.

Girder-Joist Installation
The next step is to lay the wood joists across the girders and secure them there. The simplest method of floor-joist framing has the joists bearing directly on the wood girder or steel beam. In this method, the top of the beam must be flush with the top of the sill. You can frame this way if there will be enough headroom below the girder in the basement. But remember that shrinkage is usually greater at the girder than at the foundation in this type of girder-joist construction.

Joist supports— Let's go over some of the other ways to connect joists to girders. Joist hangers, stirrups, and supporting ledger strips all provide greater headroom. And since they cut down on the amount of horizontal wood subject to shrinkage, the amount of shrinkage at the inner beam and the outer walls will be more nearly even. Figure 6-27 shows how to use a stirrup.

Now take a look at Figure 6-28 A. Depending on the size of the joists and wood girders, the joists can be supported on a ledger strip in several different ways. Each method provides about equal shrinkage at the outer walls and at the center girder, since the 2" ledger is the same thickness as the outer wall sill.

Joists must always bear on the ledger. In Figure 6-28 B, the connecting scab at each pair of joists gives an unbroken horizontal tie, and also a nailing area for the subfloor. Figure 6-28 C shows a steel strap that ties the joists together. Use this method when the top of the beam is level with the top of the joists.

Be sure to leave a small space above the beam, for shrinking of the joists. You don't need scabs or strap ties when the subfloor provides ties across the joists on both sides of the girder. I prefer to use plywood subfloor panels to tie the joists together.

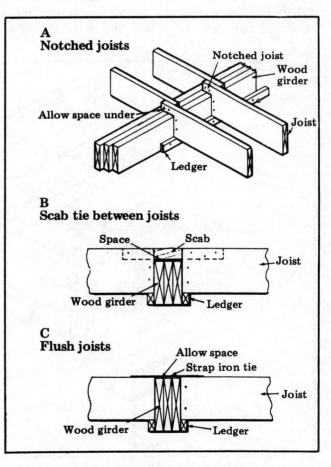

Ledger on center wood girder
Figure 6-28

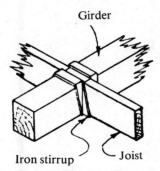

Joist supported by an iron stirrup
Figure 6-27

This eliminates the cost of the extra materials and labor needed by the other systems.

Spaced wood girders— When you need a space for heating ducts in a partition supported on the girder, use a spaced wood girder. Take a look at Figure 6-29. To make a spaced girder, use solid wood blocking between the two girder members. *Make sure that the blocking doesn't interfere with the duct runs.* You'll usually need a metal bolster on a single-post support for a spaced girder. Be sure that the bolster is large enough to support both members of the spaced girder.

Steel beams— When you're using a steel beam instead of a wood girder, you have three ways to arrange the joists. (1) The joists can rest directly on top of the beam. (2) The joists can rest on a wood ledger or steel angle iron bolted to the web, as shown in A, in Figure 6-30. Or, (3) the joists can bear directly on the flange of the beam, as shown in Figure 6-30 B. In this last method, you need wood blocking between the joists and near the beam flange to prevent overturning.

Figure 6-31 shows a steel I-beam girder with a 2 x 4 on top of it. The 2 x 4 is bolted to the I-beam, and the joists are nailed to the 2 x 4. In this system, the horizontal grain of the wood supporting the load is the same as the horizontal grain of the wood in the outside walls. This reduces shrinkage. More important, it eliminates the difference in shrinkage between girders and exterior walls.

Calculating Joist Lengths

Now let's talk more about these joists that stretch from side to side, or from side to girder. Remember back when we talked about modular planning? In many houses, 6% to 17% of the floor framing material gets wasted. Check the front-to-back dimension, or house depth, of your floor plan. (Depth is measured between the outside surfaces of the exterior wall studs.) If your house depth can't be evenly divided by 4, you're throwing away usable joist lumber. Take a look at Figure 6-32 to see what's meant by house depth.

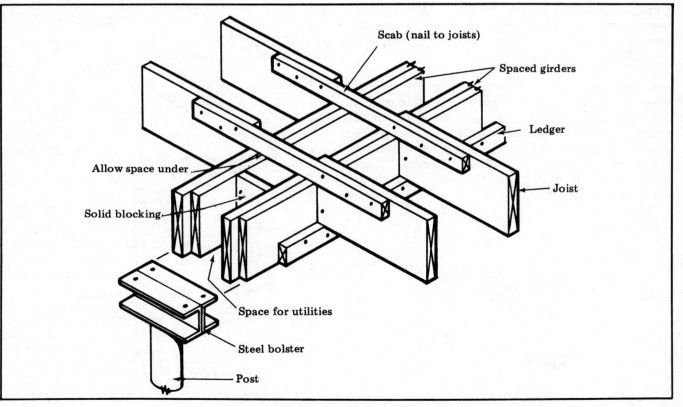

Spaced wood girder
Figure 6-29

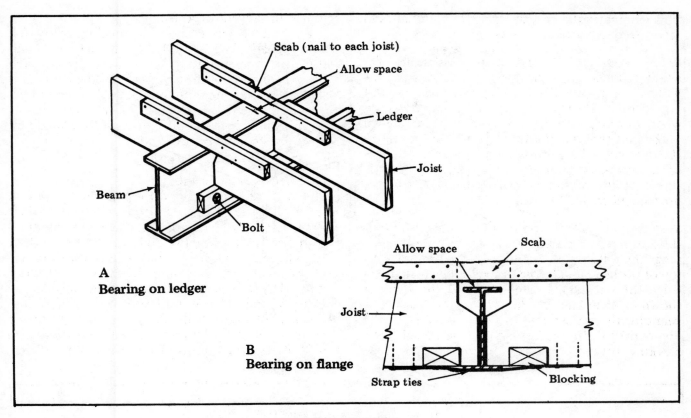

Steel beam and joists
Figure 6-30

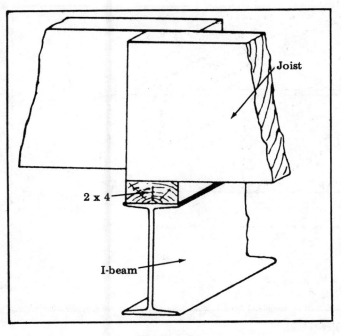

Steel girder
Figure 6-31

Wood joists are made in length increments of 2', with a tolerance of minus 0'' and plus 3''. Joists 12' and longer normally have a plus tolerance of 1/2'' or more. This allows for cutting into shorter standard lengths.

Here's a rule of thumb you can use to calculate the joist length:

In normal platform construction, with lapped joists bearing on top of a center girder, the length of the joist is one-half the house depth, minus the thickness of the band joist, plus the overlap of the joists at the center support.

Figure 6-33 illustrates this rule of thumb. Notice that if half of the lap length is a 1½'' minimum, then the *total* joist overlap must be at least 3''.

Joist lengths match standard lumber sizes when the house depth is on the 4' module of 24', 28', or 32'. Look at Figure 6-34. Notice that a 25' house takes just as many linear feet of joists to frame the floor as a house with a 28' depth. Compare the amount of joist material needed for 29' and 32'

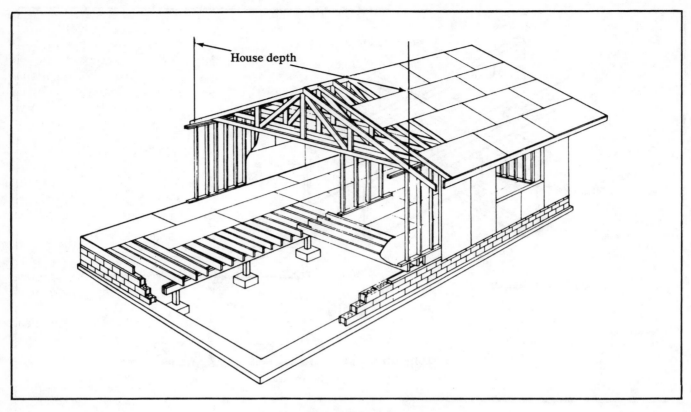

House depth measurement
Figure 6-32

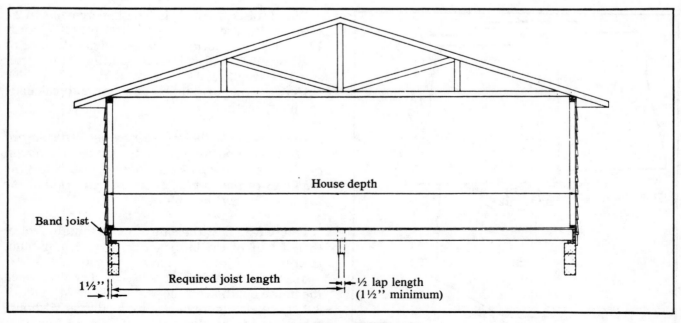

Required joist length
Figure 6-33

House depth (feet)	Joist required (feet)	Standard length (feet)
23	10½	12
22	11	12
23	11½	12
24	12	12
25	12½	14
26	13	14
27	13½	14
28	14	14
29	14½	16
30	15	16
31	15½	16
32	16	16

**Standard joist length
Figure 6-34**

house depths. You can buy joist material for the 32' house at the same cost as for the 29' house. This is true for in-line as well as for lapped joists.

In *nonmodular* house depths, when you use in-line joists you usually cut off and waste the extra length of joist. If you use lapped joists, the extra length is taken up in extra lapping over the center support. But this extra lap doesn't make the floor any stronger.

If changing to the 4' module increases the joist spans, recheck the allowable span for each species and grade being used. It will probably be workable in most cases. If changing to the 4' module reduces the joist spans, maybe you can use smaller size joists.

Allowable Joist Spans

The term "allowable span" for lumber joists refers to the clear span between supports, as shown in Figure 6-35.

The three things that determine what the clear span of joist should be are:

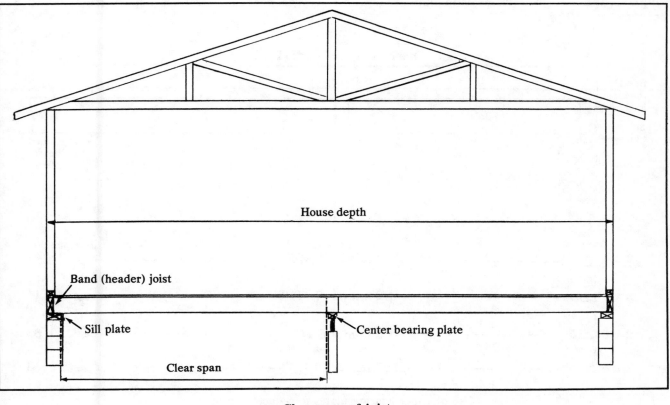

**Clear span of joist
Figure 6-35**

- The working stress for the joist grade and species

- The joist size and spacing

- The design load specified in the building code

Most standards limit joist deflection to 1/360 of the span under a 40-pound per square foot (or smaller) uniform live load. The joists must be strong enough to carry the live load (contents of the room) plus a dead load (the building itself) of 10 pounds per square foot.

Here's a rule of thumb you can use to find the clear span of joist:

In normal platform construction, where joists bear on a center girder, the clear span of each joist is one-half the house depth (measured between the outside surfaces of the exterior studs), minus the width of the sill plate, minus one-half the width of the bearing plate on the center support.

Look at the examples in Figure 6-36. Where 2 x 4 sill plates and a 2 x 4 center bearing plate are used,

the clear span of each joist is one-half the house depth minus 5¼''. If you try to figure this out yourself, you'll find it won't turn out the same unless you use *actual* dimensions. A 2 x 4 is actually closer to 1½'' x 3½''.

Figure 6-36 also shows that you can use wider sills, or center bearing plates, to reduce the clear span. By reducing the clear span you can save the expense of using larger joists or closer joist spacings. The cost of the wider sills and plates is low compared to the overall savings.

When 2 x 6 sill plates are used with a 2 x 4 center bearing plate, the clear span of each joist is one-half the house depth minus 7¼'', or 2'' less than if a 2 x 4 sill plate had been used. If 2 x 6's are used for both sill and center plates, the clear span is one-half the house depth minus 8¼'', or 3'' less than if 2 x 4 plates had been used. The clear span is reduced even more as you continue to increase the sill and center bearing plates, as shown in Figure 6-36.

Minimum Joist Grades and Sizes

Figure 6-37 shows the minimum grades and sizes of joists needed for house depths from 20' to 32'.

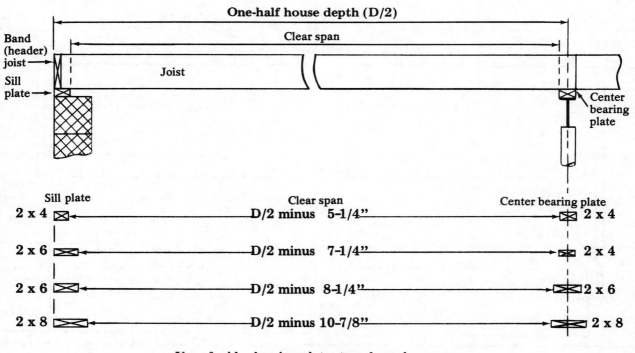

Use of wider bearing plates to reduce clear spans
Figure 6-36

Species	Joist spacing	Joist size	House depth (measured between outside surfaces of exterior studs), in feet													Grading rule agency*
			20	21	22	23	24	25	26	27	28	29	30	31	32	
Balsam fir	16"	2x8	No.2	No.2	No.2	No.2	No.2[c]	No.1[c]								NeLMA NH&PMA
		2x10	No.3	No.3	No.3	No.3[a]	No.2	No.2	No.2	No.2	No.2	No.2	No.2[c]	No.1[a]		
	24"	2x8	No.2[c]	No.1[a]												
		2x10	No.2	No.2	No.2	No.2	No.2	No.2[c]	No.1	No.1[c]						
California redwood (open grain)	16"	2x8	No.3[c]	No.2	No.2	No.2[a]	No.1[b]									RIS
		2x10	No.3	No.3	No.3	No.3	No.3	No.3[c]	No.2	No.2	No.2	No.2[a]	No.1[a]			
		2x12						No.3	No.3	No.3	No.3	No.3	No.3[b]	No.2	No.2	
	24"	2x8	No.2	No.1[a]												
		2x10	No.3	No.3[c]	No.2	No.2	No.2	No.2	No.2[c]	No.1[c]						
		2x12				No.3	No.3	No.3	No.3	No.3[c]	No.2	No.2	No.2	No.2[a]	No.1[b]	
Douglas-fir-larch	16"	2x8	No.3	No.3	No.3	No.2	No.2	No.2	No.2	No.2	No.1[b]					NLGA WCLIB WWPA
		2x10				No.3	No.3	No.3	No.3	No.3	No.3[a]	No.2	No.2	No.2	No.2	
		2x12									No.3	No.3	No.3	No.3	No.3	
	24"	2x8	No.2	No.2	No.2	No.2	No.2[c]	No.1[c]								
		2x10	No.3	No.3	No.3	No.3[a]	No.2		No.2	No.2	No.2	No.2	No.2[b]	No.1[a]	No.1[c] DENSE	
		2x12				No.3	No.3	No.3	No.3	No.3	No.3[b]		No.2	No.2	No.2	
Douglas-fir-south	16"	2x8	No.3	No.3	No.3[c]	No.2	No.2	No.2[a]	No.1[c]							WWPA
		2x10			No.3	No.3	No.3	No.3	No.3	No.2	No.2	No.2	No.2	No.2[c]		
		2x12									No.3	No.3	No.3	No.3	No.3	
	24"	2x8	No.2	No.2	No.2[a]	No.1[c]										
		2x10	No.3	No.3	No.3	No.3[c]	No.2	No.2	No.2	No.2	No.2[b]	No.1[c]				
		2x12					No.3	No.3	No.3	No.3	No.3[a]	No.2	No.2	No.2	No.2	
Eastern hemlock-tamarack	16"	2x8	No.3	No.3[c]	No.2	No.2	No.2[b]	No.1[a]								NeLMA NLGA
		2x10			No.3	No.3	No.3	No.3	No.3[a]	No.2	No.2	No.2	No.2[a]	No.1	No.1[c]	
	24"	2x8	No.2	No.2[a]	No.1[a]											
		2x10	No.3	No.3	No.3[c]	No.2	No.2	No.2	No.2	No.2[c]	No.1[b]					
Eastern spruce	16"	2x8	No.2	No.2	No.2	No.2	No.2[a]	No.1	No.1[c]							NeLMA NH&PMA
		2x10	No.3	No.3	No.3	No.3	No.3[b]	No.2	No.2	No.2	No.2	No.2	No.2[c]	No.1		
	24"	2x8	No.2[b]	No.1	No.1[a]											
		2x10	No.3[c]	No.2	No.2	No.2	No.2	No.2[a]	No.1	No.1	No.1[b]					
Englemann spruce-alpine fir or Englemann spruce-lodgepole pine	16"	2x8	No.2	No.2	No.2	No.2	No.2[c]	No.1[a]								WWPA
		2x10	No.3	No.3	No.3	No.3[a]	No.2	No.2	No.2	No.2	No.2		No.2[c]	No.1	No.1[c]	
		2x12						No.3	No.3	No.3	No.3	No.3	No.2	No.2	No.2	
	24"	2x8	No.2[c]	No.1[a]												
		2x10	No.2	No.2	No.2	No.2	No.2	No.2[c]	No.1	No.1[c]						
		2x12	No.3	No.3	No.3	No.3[c]			No.2	No.2	No.2	No.2	No.2	No.2[c]	No.1	No.1
Hem-fir	16"	2x8	No.3[a]	No.2	No.2	No.2	No.2	No.2	No.2[c]							NLGA WCLIB WWPA
		2x10		No.3	No.3	No.3	No.3	No.3[a]	No.2	No.2	No.2	No.2	No.2	No.2	No.2	
		2x12							No.3	No.3	No.3	No.3	No.3	No.3	No.3[c]	
	24"	2x8	No.2	No.2	No.1	No.1										
		2x10	No.3	No.3	No.2	No.2	No.2	No.2	No.2	No.2[c]	No.1	No.1[a]				
		2x12			No.3	No.3	No.3	No.3	No.3[a]		No.2	No.2	No.2	No.2	No.2	
Idaho white pine or Western white pine	16"	2x8	No.2	No.2	No.2	No.2[a]	No.1	No.1[a]								WWPA
		2x10	No.3	No.3	No.3	No.3[a]	No.2	No.2	No.2	No.2	No.2	No.2		No.1	No.1[b]	
		2x12					No.3	No.3	No.3	No.3	No.3	No.3[c]	No.2	No.2	No.2	
	24	2x8	No.1	No.1[b]												NLGA
		2x10	No.2	No.2	No.2	No.2	No.2[a]	No.1	No.1[b]							
		2x12	No.3	No.3	No.3	No.3[c]	No.2	No.2	No.2	No.2	No.2	No.2[a]	No.1	No.1	No.1[c]	

Based on construction with center girder, nominal 2-inch-thick band (header) joists and, except where footnoted, nominal 2 x 4 sill and 2 x 4 center bearing plates.

Minimum joist grades for different house depths 40 psf uniform live load
Figure 6-37

Species	Joist spacing	Joist size	House depth (measured between outside surfaces of exterior studs), in feet 20	21	22	23	24	25	26	27	28	29	30	31	32	Grading rule agency*
Lodgepole pine	16"	2x8	No.2	No.2	No.2	No.2	No.2	No.1a / No.2c	No.2	No.2	No.2	No.2	No.2	No.2a	No.1c	WWPA
		2x10	No.3	No.3	No.3	No.3	No.3		No.3	No.3	No.3	No.3	No.3a	No.2	No.2	
		2x12														
	24"	2x8	No.2	No.1	No.1a											
		2x10	No.3a	No.2	No.2	No.2	No.2	No.2	No.2c	No.1	No.1b					
		2x12									No.2	No.2	No.2	No.2b	No.1	
Northern pine	16"	2x8	No.3c	No.2	No.2	No.2	No.2	No.2a	No.1c							NeLMA NH&PMA
		2x10	No.3	No.3	No.3	No.3	No.3	No.3c	No.2	No.2	No.2	No.2	No.2	No.2	No.2c	
	24"	2x8	No.2	No.2b	No.1	No.1c										
		2x10	No.3	No.3c	No.2	No.2	No.2	No.2	No.2a	No.1	No.1	No.1c				
Ponderosa pine-sugar pine	16"	2x8	No.2	No.2	No.2	No.2	No.2b	No.1c								WWPA
		2x10	No.3	No.3	No.3	No.3	No.2	No.2	No.2	No.2	No.2	No.2	No.2a	No.1a		
		2x12						No.3	No.3	No.3	No.3	No.3a	No.2	No.2	No.2	
	24"	2x8	No.2c	No.1	No.1c											
		2x10	No.2	No.2	No.2	No.2	No.2	No.2c	No.1	No.1a						
		2x12	No.3	No.3	No.3	No.3a		No.2	No.2	No.2	No.2	No.2	No.2b	No.1	No.1	
Southern pine	16"	2x8	No.3	No.3	No.3a	No.2	No.2	No.2	No.2c	MG No.2b	No.1b					SPIB
		2x10				No.3	No.3	No.3	No.3	No.3	No.3c	No.2	No.2	No.2	No.2	
		2x12									No.3	No.3	No.3	No.3	No.3	
	24"	2x8	No.2	No.2	No.2c	No.2	MG No.2c	No.1c								
		2x10	No.3	No.3	No.3	No.3b	No.2	No.2	No.2	No.2	MG No.2	No.2	No.2b	No.1a	No.1c DENSE	
		2x12				No.3	No.3	No.3	No.3	No.3	No.3c	No.2	MG No.2	No.2	No.2	
Southern pine KD (15% mc)	16"	2x8	No.3	No.3	No.3	No.3b	No.2	No.2	No.2	MG No.2	No.1	No.1b DENSE				SPIB
		2x10					No.3	No.3	No.3	No.3	No.3	No.3b	No.2	No.2	No.2	
		2x12										No.3	No.3	No.3	No.3	
	24"	2x8	No.2	No.2	No.2	No.2a	MG No.2a	No.1c								
		2x10	No.3	No.3	No.3	No.3	No.3b	No.2	No.2	No.2	No.2	No.2	MG No.2c	No.1	No.1c	
		2x12				No.3	No.3	No.3	No.3	No.3	No.3	No.3a	No.2	No.2		
Spruce-pine fir Coast Sitka spruce or Sitka spruce	16"	2x8	No.2	No.2	No.2	No.2	No.2a	No.1	No.1a							NLGA / WCLIB
		2x10	No.3	No.3	No.3	No.3	No.2	No.2	No.2	No.2	No.2	No.2	No.2	No.2c	No.1	
		2x12							No.3	No.3	No.3	No.3	No.3a	No.2	No.2	
	24"	2x8	No.2b	No.1	No.1c											
		2x10	No.2	No.2	No.2	No.2	No.2	No.2a	No.1	No.1						
		2x12	No.3	No.3	No.3	No.3a		No.2	No.2	No.2	No.2	No.2	No.2a	No.1	No.1	
Western hemlock	16"	2x8	No.3	No.3a	No.2	No.2	No.2	No.2	No.2	No.1b						WWPA
		2x10			No.3	No.3	No.3	No.3	No.3	No.3c	No.2	No.2	No.2	No.2	No.2	
		2x12								No.3	No.3	No.3	No.3	No.3	No.3a	
	24"	2x8	No.2	No.2	No.2a	No.1	No.1c									
		2x10	No.3	No.3	No.3b	No.2	No.2	No.2	No.2	No.2b	No.1	No.1b				
		2x12			No.3	No.3	No.3	No.3	No.3		No.2	No.2	No.2	No.2	No.2	
White woods (western woods)	16"	2x8	No.2	No.2	No.2	No.2a	No.1b									WWPA
		2x10	No.3	No.3	No.3	No.3a	No.2	No.2	No.2	No.2	No.2	No.2a	No.1a			
		2x12						No.3	No.3	No.3	No.3	No.3c	No.2	No.2	No.2	
	24"	2x8	No.1	No.1b												
		2x10	No.2	No.2	No.2	No.2	No.2a	No.1	No.1a							
		2x12	No.3	No.3	No.3	No.3c	No.2	No.2	No.2	No.2	No.2	No.2a	No.1	No.1	No.1c	

[a]Nominal 2 x 6 sill plate and 2 x 4 center bearing plate; or width of sill plate plus one-half width of center bearing equal to 7¼" or more.
[b]Nominal 2 x 6 sill plate and 2 x 6 center bearing plate; or width of sill plate plus one-half width of center bearing equal to 8¼" or more.
[c]Nominal 2 x 8 sill plate and 2 x 8 center bearing plate; or width of sill plate plus one-half width of center bearing equal to 10¾" or more.
*NeLMA—Northeastern Lumber Manufacturers Association.
 NH&PMA—Northern Hardwood and Pine Manufacturers Association.
 RIS—Redwood Inspection Service.
 NLGA—National Lumber Grades Authority, a Canadian Agency.
 WCLIB—West Coast Lumber Inspection Bureau.
 WWPA—Western Wood Products Association.
 SPIB—Southern Pine Inspection Bureau.

Minimum joist grades for different house depths 40 psf uniform live load
Figure 6-37 (continued)

These requirements are based on 40 pounds per square foot uniform live load. The numbers assume that you're using platform construction with center girder, nominal 2''-thick band joists, and 2 x 4 sill and center bearing plates. In some cases, using a 2 x 6 or 2 x 8 sill or bearing plate allows you to use a smaller size and grade of joist for a larger house depth.

Here's how to use this chart. The plans may call for No. 2 Southern pine 2 x 10's spaced 16'' o.c. for a house depth of 30'. Now look at Figure 6-37. The same joist grade and size can be used for a house depth of 32'. By changing the floor plan to a major 4' module of 32', you stop the waste of joist length and still get the full span use of the grade.

This doesn't mean that you should run right out and change any set of house plans that doesn't fit the major 4' module. *Be sure to consider all parts of the construction before you change anything.* But remember that the idea is to *look* for ways to cut costs and make the most cost-efficient use of materials.

You'll also find times when a more economical joist grade or size can be used without changing the floor plans. Here's an example: To span a 28' house depth, you can reduce the cost by using No. 3 Douglas fir-larch 2 x 10's, spaced 16'' o.c., with 2 x 6 sill plates; instead of the more expensive 2 x 10's, No. 2 grade of the same species, spaced the same, with 2 x 4 sill plates. You could also save money by using a higher grade, but a smaller size. For example, try using No. 1 Douglas fir-larch 2 x 8's, 16'' o.c., with 2 x 6 sill and center bearing plates.

In some cases, you might get the best use of your joist grade and size by increasing your joist spacing. No. 2 Douglas fir-larch or Southern pine 2 x 10's can span a 28' house when spaced 24'' o.c., as well as when they're spaced 16'' o.c. But before you decide to do this, look at all the cost variables. The 24'' spacing must have a thicker plywood subfloor than the 16'' spacing. Compare the added cost of the thicker subfloor to the savings in material and labor you'd get from having to use fewer joists.

Field-glued Floor Systems

The maximum clear span is often limited by lack of stiffness. But you can increase the maximum span by field-gluing plywood subflooring or underlayment to the joists. The glued floor system will also reduce squeaks and nail pops.

Figure 6-38 shows the minimum requirements for field-glued floor systems for house depths from 20' to 32'. The table only shows the joist grades and sizes that can increase the maximum span through field-gluing. The joist and plywood requirements agree with the maximum clear spans for glued plywood floor systems accepted by most building codes.

Species	Joist spacing	Joist size	Plywood underlayment thickness	House depth (measured between outside surfaces of exterior studs), feet												
				20	21	22	23	24	25	26	27	28	29	30	31	32
Balsam fir	16"	2x8	1/2	—	—	—	—	—	No.1	No.1c	—	—	—	—	—	—
		2x10	1/2	—	—	—	—	—	—	—	—	—	—	—	No.1	No.1
California redwood (open grain)	16"	2x8	1/2	—	—	—	No.2	No.2	No.2a	No.1b	—	—	—	—	—	—
			19/32	—	—	—	—	—	—	—	No.1c	—	—	—	—	—
			23/32	—	—	—	—	—	—	—	No.1a	—	—	—	—	—
		2x10	1/2	—	—	—	—	—	—	—	—	—	—	No.2	No.2	No.1a
			19/32	—	—	—	—	—	—	—	—	—	—	—	—	No.2b
	24"	2x8	23/32	No.2	No.2	No.1	No.1b	—	—	—	—	—	—	—	—	—
		2x10	23/32	—	—	—	—	—	—	No.2a	No.1	No.1	No.1b	—	—	—
		2x12	23/32	—	—	—	—	—	—	—	—	—	—	—	No.2	No.2
Douglas fir-larch	16"	2x8	1/2	—	—	—	—	—	—	—	No.2	No.2a	No.1c	—	—	—
			19/32	—	—	—	—	—	—	—	—	—	No.1a	—	—	—
			23/32	—	—	—	—	—	—	—	—	—	—	No.1c	—	—
	24"	2x8	23/32	—	—	—	—	—	No.1	No.1c	—	—	—	—	No.1	No.1
		2x10	23/32	—	—	—	—	—	—	—	—	—	—	—	No.1	No.1

Joist grades having additional house depth spanning capability through field-gluing of plywood subfloor-underlayment to joists 40 psf live load

Figure 6-38

Species	Joist spacing	Joist size	Plywood underlayment thickness	20	21	22	23	24	25	26	27	28	29	30	31	32
Douglas fir-south	16"	2x8	1/2	—	—	—	—	—	No.2	No.2	No.2[a]	No.1[c]	—	—	—	—
			19/32	—	—	—	—	—	—	—	—	No.2[c]	—	—	—	—
			23/32	—	—	—	—	—	—	—	No.2[b]	No.1[c]	—	—	—	—
	24"	2x10	1/2	—	—	—	—	—	—	—	—	—	—	—	—	No.2
		2x8	23/32	—	—	No.2	No.2[b]	No.1	No.1[a]	—	—	—	—	—	—	—
		2x10	23/32	—	—	—	—	—	—	—	—	No.2	No.2[b]	No.1	No.1	No.1[b]
Eastern hemlock Tamarack	16"	2x8	1/2	—	—	—	—	No.2	No.2	No.2[b]	No.1[a]	—	—	—	—	—
			19/32	—	—	—	—	—	—	No.2[a]	—	No.1[c]	—	—	—	—
			23/32	—	—	—	—	—	—	—	—	No.1[a]	—	—	—	—
		2x10	1/2	—	—	—	—	—	—	—	—	—	—	No.2	No.2	No.2[a]
			19/32	—	—	—	—	—	—	—	—	—	—	—	—	No.2
	24"	2x8	23/32	—	No.2	No.2[c]	No.1	No.1[a]	—	—	—	—	—	—	—	—
		2x10	23/32	—	—	—	—	—	—	—	—	No.2	No.1	No.1	No.1	No.1[c]
Eastern spruce	16"	2x8	1/2	—	—	—	—	—	—	No.1	No.1[b]	—	—	—	—	—
Englemann spruce-Alpine fir	16"	2x8	1/2	—	—	—	—	—	No.1	No.1[c]	—	—	—	—	—	—
		2x10	1/2	—	—	—	—	—	—	—	—	—	—	—	—	No.1
Hem-fir	16"	2x8	1/2	—	—	—	—	—	—	—	No.1	No.1[a]	—	—	—	—
			19/32	—	—	—	—	—	—	—	—	No.1	—	—	—	—
	24"	2x10	23/32	—	—	—	—	—	—	—	—	—	No.1	No.1[c]	—	—
Lodge pine	16"	2x8	1/2	—	—	—	—	—	—	No.1	No.1[a]	—	—	—	—	—
		2x10	1/2	—	—	—	—	—	—	—	—	—	—	—	—	No.1
	24"	2x8	23/32	—	—	No.1	No.1[c]	—	—	—	—	—	—	—	—	—
		2x10	23/32	—	—	—	—	—	—	—	—	No.1	No.1[c]	—	—	—
Northern pine	16"	2x8	1/2	—	—	—	—	—	—	No.1	No.1	No.1[c]	—	—	—	—
			19/32	—	—	—	—	—	—	—	—	No.1[a]	—	—	—	—
			23/32	—	—	—	—	—	—	—	—	No.1	—	—	—	—
		2x10	1/2	—	—	—	—	—	—	—	—	—	—	—	—	No.2[b]
	24"	2x8	23/32	—	—	—	No.1	—	—	—	—	—	—	—	—	—
		2x10	23/32	—	—	—	—	—	—	—	—	—	No.1	No.1[c]	—	—
Ponderosa pine-Sugar pine	16"	2x8	1/2	—	—	—	—	—	No.1	No.1[a]	—	—	—	—	—	—
		2x10	1/2	—	—	—	—	—	—	—	—	—	—	—	No.1	No.1
	24"	2x10	23/32	—	—	—	—	—	—	—	No.1	—	—	—	—	—
Southern pine	16"	2x8	1/2	—	—	—	—	—	—	No.2[a]	MG No.2	MG No.2	MG No.2	No.1[c]	—	—
			19/32	—	—	—	—	—	—	—	—	—	—	No.1[a]	No.1[c]	—
	24"	2x8	23/32	—	—	—	—	—	No.1	No.1[b]	—	—	—	—	—	—
		2x10	23/32	—	—	—	—	—	—	—	—	—	—	—	No.1	No.1
Southern pine KD (15% mc)	16"	2x8	1/2	—	—	—	—	—	—	—	No.2	MG No.2	MG No.2[a]	No.1[a]	—	—
			19/32	—	—	—	—	—	—	—	—	—	—	No.2[c]	No.1[c]	—
			23/32	—	—	—	—	—	—	—	—	—	—	—	No.1[b]	—
	24"	2x8	23/32	—	—	—	—	MG No.2	MG No.2[c]	No.1	No.1[a]	—	—	—	—	—
		2x10	23/32	—	—	—	—	—	—	—	—	—	—	—	MG No.2[c]	No.1
Western hemlock	16"	2x8	1/2	—	—	—	—	—	—	No.2	No.2[b]	No.1	No.1[c]	—	—	—
			19/32	—	—	—	—	—	—	—	—	—	No.1[a]	—	—	—
			23/32	—	—	—	—	—	—	—	—	—	—	No.1[c]	—	—
	24"	2x8	23/32	—	—	—	—	No.1	No.1[c]	—	—	—	—	—	—	—
		2x10	23/32	—	—	—	—	—	—	—	—	—	—	No.1	No.1[b]	—
White woods (Western woods)	16"	2x8	1/2	—	—	—	—	No.1	No.1[a]	—	—	—	—	—	—	—
		2x10	1/2	—	—	—	—	—	—	—	—	—	—	No.2	No.1	No.1[c]

[a] Nominal 2x6 sill plate and 2x4 center bearing plate; or width of sill plate plus one-half width of center bearing equal to 7¼" or more.
[b] Nominal 2x6 sill plate and 2x6 center bearing plate; or width of sill plate plus one-half width of center bearing equal to 8¼" or more.
[c] Nominal 2x8 sill plate and 2x8 center bearing plate; or width of sill plate plus one-half width of center bearing equal to 10⅞" or more.

Joist grades having additional house depth spanning capability through field-gluing of plywood subfloor-underlayment to joists 40 psf live load
Figure 6-38 (continued)

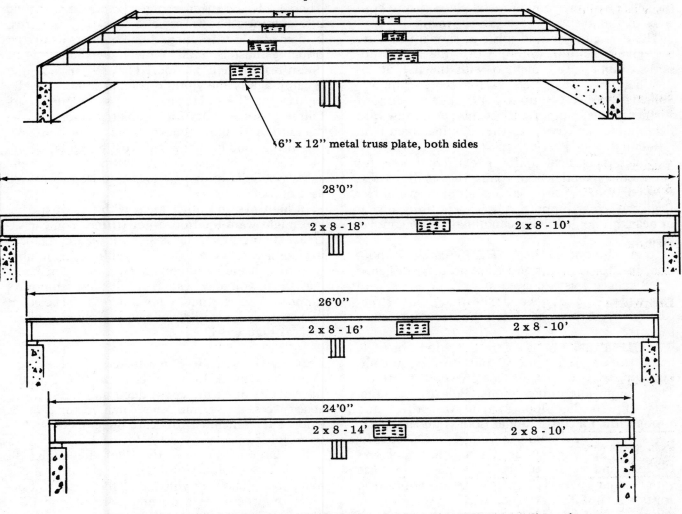

Splices alternated

6" x 12" metal truss plate, both sides

28'0"

2 x 8 - 18'   2 x 8 - 10'

26'0"

2 x 8 - 16'   2 x 8 - 10'

24'0"

2 x 8 - 14'   2 x 8 - 10'

Off-center spliced joist designs span further than simple joists of the same dimension
Figure 6-39

Field-gluing can also reduce the joist size or grade. This saves you more money. Look again at Figures 6-37 and 6-38. Compare the joist sizes shown for the same house depth for a house using field-gluing and one that doesn't. Here's an example that saves on sill size:

For a 28' house depth, without field-gluing you'll use No. 1 Douglas fir-larch or Southern pine 2 x 8's, 16" o.c., with 2 x 6 sill and center bearing plates. For a 28' house depth, with field-gluing you can use No. 2 joists of the same size and species, on the same spacing and with only 2 x 4 sill and center bearing plates.

Using a lower, easier-to-get grade, or using smaller-sized joists, often saves you more money when using a field-glued system. For example, nails are spaced 12" o.c. in field gluing, compared to 6" and 10" for conventional construction. Check this out on your next job. It may increase your profits.

Off-center Spliced Joists
Another way you can increase the joist span is to use off-center spliced joists. This practice may be unfamiliar to many builders. Let's see how to do it.

If two unequal joist lengths are spliced together so that the splice is somewhere other than in the center of the span, the joist span can be longer. Always alternate the splices for extra support, just as you would stagger subflooring. You can make the splice with plywood or metal plates tied to both sides of the joists.

The off-center spliced joists as shown in Figure 6-39 give structural continuity over the center support. Compare this method with the use of individual joists that meet at the center support. Splicing can give you up to a 40% increase in joist stiffness. This is important because, as we saw with field-gluing, stiffness is the limiting factor in residential floor joist design.

Under theoretical uniform load conditions for common residential floor spans, the bending stress on a continuous joist is zero at a point several feet from the center support. A spliced joint located at or near this point will have minimal bending movement.

Research done by the NAHB Research Foundation, Inc. for HUD, showed that properly-designed 2 x 8 spliced joists, spaced at 24'' o.c., and with glue-nailed 5/8'' plywood sheathing, are structurally adequate for a 28'-deep house with a center bearing.

NAHB actually built a demonstration house. It had a floor section built of spliced joists, in combination with a glue-nailed plywood subfloor. They found interesting results. Full-scale loading tests showed this floor system was stiffer and stronger than a conventional floor. Full length 28' joists were preassembled from No. 2 Hem-fir lumber by splicing two 2 x 8's together. One was 18' long and the other 10'. They made the splice with standard 6'' x 12'' truss plates on both sides. Joists were installed 24'' o.c., with the splices alternating on either side of the center support, as shown in Figure 6-39. NAHB also used a field-glued floor system. T&G 5/8''-thick plywood sheathing was glue-nailed to the joists with the T&G joint glued. The full-length spliced joists make good floors. Construction is easier and the cost lower.

For houses of up to 28' depth with center bearing, you can use a 2 x 8 off-center spliced-joist floor system. For other house depths, make spliced joists from standard lengths of the same quality 2 x 8 lumber, as shown in Figure 6-39. The off-center splicing system offers all the benefits of the in-line joist system, plus an increased span capability. Figure 6-40 shows the layout of the in-line floor joist systems over a center support for both preassembled and unassembled joists.

Bridging

There's disagreement about whether to use solid bridging (also known as solid blocking) and cross bridging in floor framing. Some builders insist that bridging must be used. Others say that bridging doesn't strengthen the floor. They say, in fact, that it can magnify the vibrations from someone walking on the floor. However, cross bridging is required in some codes for residential floors.

Figure 6-41 shows conventional floor construction using solid bridging and cross bridging. Bridging keeps the joist from twisting or buckling sideways. Solid bridging serves as a fire stop as well as a brace.

In hallways and other areas with lots of traffic, you might want the floor to be stiffer. You can increase stiffness by using straight or diagonal blocking (solid or cross bridging). If you're using lumber with a high moisture content and that you know may warp, you may need to give it some temporary support. You can nail a 1 x 3 across the joists at mid-span with two 8d nails at each joist. Take a look at Figure 6-42.

Framing Openings in Floor Joists

When you space floor joists at 24'' o.c., you'll have plenty of clearance for ducts and flues, crawl-space access doors, and other small openings in the floor. Just remember to locate these openings *between* joists.

For large openings in the floor, like stairwells and chimney holes, you'll have to cut at least one main joist. The way you frame the joists will depend on where these openings are on the floor plan.

You might need to do some cutting of openings for heating and plumbing fixtures. Maybe you won't need to cut the full depth of the joist, but you may have to notch the top or bottom. You should have the carpenter, who understands joist structure, do this job — not the heating or plumbing contractor.

Figures 6-43 and 6-44 show the layouts for framing around large openings in the floor joists. The framing members around these openings are generally the same depth as the joists.

Figure 6-43 shows a part of a conventional floor that has a stairwell and a chimney hole. The headers are short joists on two sides of any large floor opening. They support the ends of the joists that have been cut off and are usually doubled, just

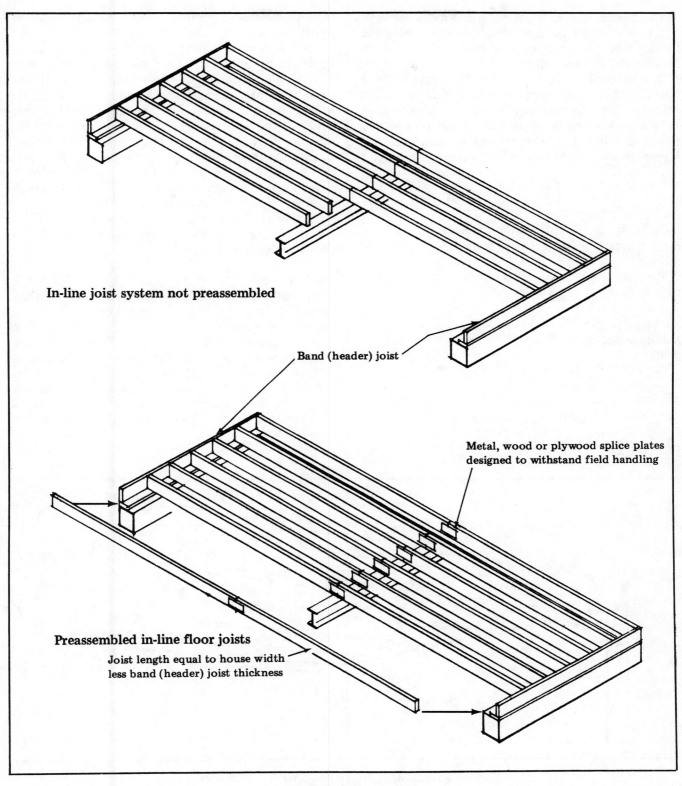

In-line joist system not preassembled

Band (header) joist

Metal, wood or plywood splice plates
designed to withstand field handling

Preassembled in-line floor joists

Joist length equal to house width
less band (header) joist thickness

In-line floor joist systems
Figure 6-40

as this figure shows. If these cut-off joists are very short, you don't need a double header. A single header will do. Take another look at Figure 6-43 around the chimney hole.

The trimmers are the joists that support the headers. They are either regular joists, or extra joists parallel to the regular joists, as shown in Figure 6-43. If stairwell partition studs support the trimmer joists, you don't need to reinforce the trimmers. But in the case of the chimney hole, you

should double the trimmer joists because they support the headers.

Mark the location of trimmer joists on the sill and the girder at the same time you space the other joists. The location of the trimmer joists should not affect the spacing of the regular joists. On the trimmers on each side of the wall, mark the location of the headers. Then spike the inside (single) header between the side trimmers. Mark the tail joists to match the regular joist spacing, and spike (end nail) the tail joists into the single header. If you need a double header, spike it to the single header after the tail joists are in place.

Spike the short trimmers between the headers. You can use metal joist hangers as shown in Figure 6-44 for attaching joists over 6' long to the header.

Subflooring

Now let's take a look at what goes over the floor joists — the flooring. Conventional systems use subflooring over the floor joists. Subflooring gives you a working platform and a base for finish flooring. You can use square-edge or T&G boards no wider than 8'' and at least 3/4'' thick. Or you can use 1/2''- to 3/4''-thick plywood for the subflooring. The thickness of the plywood depends on the species you use, the joist spacing, and the type of finish floor.

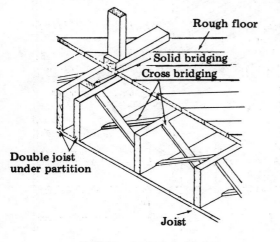

Solid and cross bridging
Figure 6-41

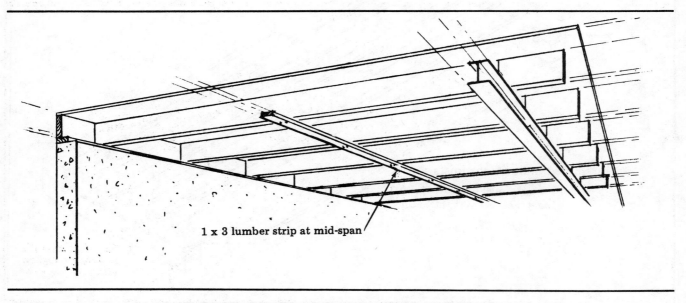

Where green joists have a known tendency to warp,
a 1 x 3 strip will provide restraint
Figure 6-42

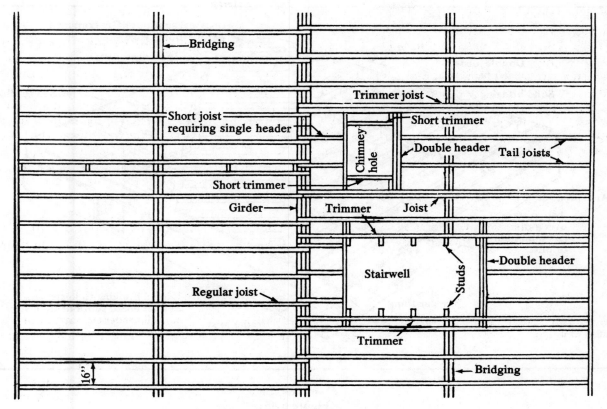

Floor plan for joist layout
Figure 6-43

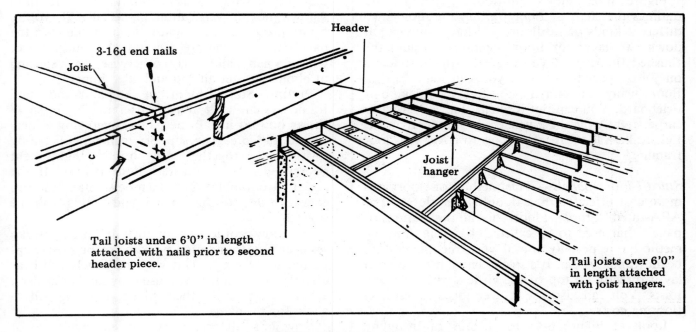

Joist hangers used in framing openings
Figure 6-44

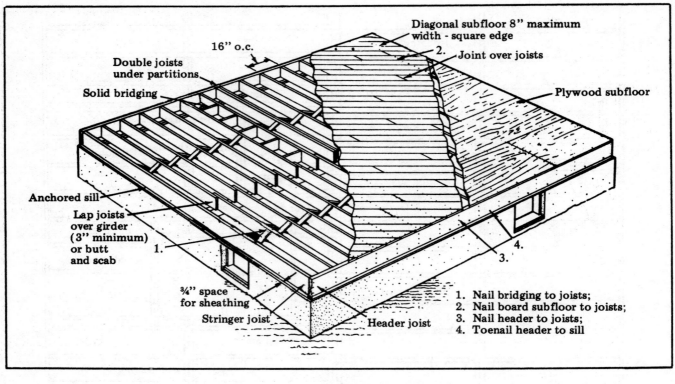

Floor framing
Figure 6-45

1. Nail bridging to joists;
2. Nail board subfloor to joists;
3. Nail header to joists;
4. Toenail header to sill

Figure 6-45 shows subflooring details and outlines the steps of floor framing. It shows two different kinds of subflooring. Many builders put down two layers of floor sheathing, besides the finished flooring. If you use 1/2''-thick plywood on joists spaced 16'' o.c., you can make a firm floor using underlayment of 5/8''-thick particleboard. You can put resilient floor covering or carpet right over the particleboard underlayment. Let's take a look at three ways to cover the floor framing.

Sturd-I-Floor panels— This method of flooring is more cost effective than double-layer floors. The APA-rated ''Sturd-I-Floor'' has all the features of panels that used to be called 2-4-1. This flooring method is especially useful with the plank-and-beam frame system. It's strong enough so you can lay it directly on top of the girders without using joists. You can also use it over joists to save on flooring costs.

Look at Figure 6-46 for details of installing Sturd-I-Floor over joists 32'' o.c. You can increase the joist spacing by using this kind of flooring. Allow enough ventilation, especially if you're flooring over a crawl-space. Also, make sure the panels are dry before you lay the finish floor. Unless a panel maker says otherwise, leave spacing of about 1/8'' at all end and edge joints.

Putting Sturd-I-Floor in a plank-and-beam frame system isn't much different. Figure 6-47 shows how to install the panels over 4-by girders spaced 48'' o.c. The supports can be either 2-by joists spiked together, 4-by lumber, lightweight steel beams, or wood-steel floor trusses. If the girders are doubled 2-by members, make the top edges flush, so that panel end joints will be smooth.

For a low profile with supports 48'' o.c., you can set beams in foundation pockets, or on posts supported by footings. This way the panels will bear directly on the sill. If you use 4-by lumber girders, air-dry them or set them higher than the sill to allow for shrinkage.

Now let's talk about two conventional types of subflooring.

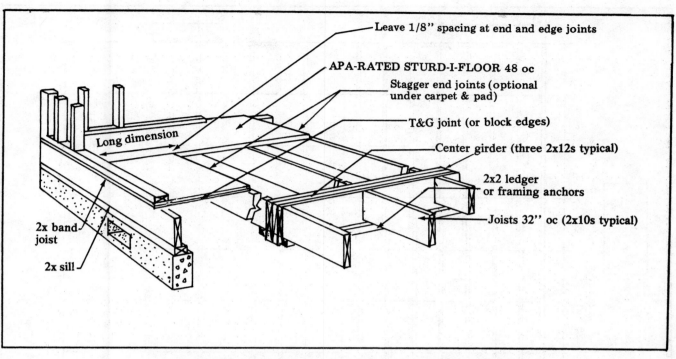

Leave 1/8" spacing at end and edge joints

APA-RATED STURD-I-FLOOR 48 oc

Stagger end joints (optional under carpet & pad)

T&G joint (or block edges)

Center girder (three 2x12s typical)

2x2 ledger or framing anchors

Joists 32" oc (2x10s typical)

Long dimension

2x band joist

2x sill

APA-rated Sturd-I-Floor 48 oc (2-4-1) (over supports 32" oc)
Figure 6-46

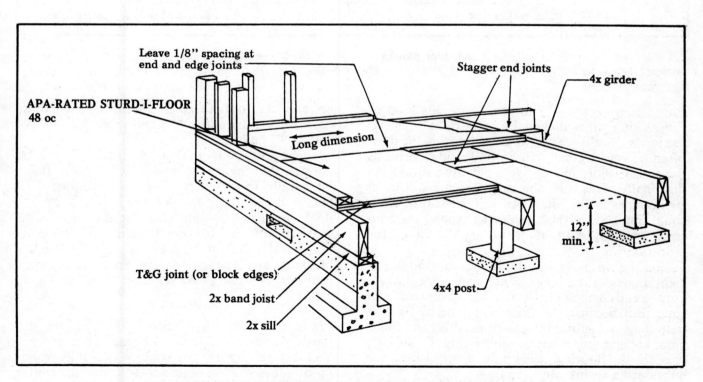

Leave 1/8" spacing at end and edge joints

Stagger end joints

4x girder

APA-RATED STURD-I-FLOOR 48 oc

Long dimension

12" min.

T&G joint (or block edges)

2x band joist

2x sill

4x4 post

APA-rated Sturd-I-Floor 48 oc (2-4-1) (over supports 48" oc)
Figure 6-47

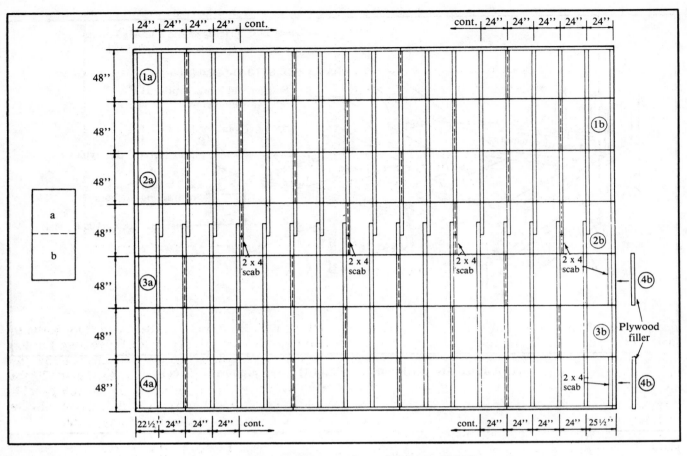

Subfloor panel layout for a 28' house depth
Figure 6-48

Board subflooring— You can apply this kind of subflooring either diagonally or at right angles to the joists. You'll see diagonal subflooring the most often. If you're going to use wood strip flooring as the finish floor, be sure you put the subflooring diagonally. This way, the strip flooring will be at right angles to the joists. Also, it gets rid of floor squeaking and lets the subflooring expand and contract normally without disturbing the fit of the strip flooring.

You'll save by laying the board subflooring at right angles to the joists. Since you don't have to cut the ends diagonally, it costs less in material and time. But because the floor won't be as rigid as with diagonal subflooring, you must lay the finish floor at right angles to the subflooring. Otherwise, the finish flooring will pull apart when the subflooring shrinks and expands.

End joints of the boards should rest directly over the joists, unless the boards are tongued and groov-ed. Nail the subflooring to each joist; use two 8d nails for widths under 8", and three 8d nails for widths 8" and up.

Space joists no more than 16" o.c. when you use parquet flooring, or when you lay the finish flooring parallel to the joists. You can make joist spacing up to 24" o.c. if the flooring is at least 25/32" thick *and* is laid at right angles to the joists.

When you use balloon framing, put blocking between the ends of the joists at the wall. Use this blocking as a nailer for the ends of diagonal subfloor boards. (Take another look at Figure 6-19.)

Plywood subflooring— Plywood comes in different grades, suitable for a broad range of uses. You can get any interior grade with a fully waterproof adhesive—the same adhesive used in exterior plywood. Use waterproofed interior grades in areas of high moisture, like underlayments or subfloors

Spacing	Edges (inches)	Ends (inches)
Underlayment or interior wall lining	1/32	1/32
Panel sidings and combination subfloor underlayment	1/16	1/16
Roof sheathing, subflooring, and wall sheathing (under wet or humid conditions, spacing should be doubled)	1/8	1/16

Plywood location and spacing
Figure 6-49

next to plumbing fixtures. Otherwise, standard sheathing grades will do the job.

Install plywood with the grain direction (of the outer plies) at right angles to the joists. For plywood 1/2" to 3/4" thick, use 8d box or 7d threaded nails to fasten the panels to the joists at each bearing. Space nails 6" apart along all edges and 10" apart along other members. If the plywood is to be both subfloor and underlayment, space nails 6" to 7" apart at all joists and blocking. In this case, use 8d or 9d box nails or 8d threaded nails.

Another way to save money on subflooring is to use the 4' house depth module. You can use full 48"-wide panels for 24', 28', and 32' house depths when joists are either lapped or in-line. Figure 6-48 shows a way to lay out panel subflooring in the 28' house depth using lapped joists. The subflooring panels are all staggered for extra strength, and the half pieces are used efficiently.

Remember to leave a space between the ends and edges of all adjoining plywood panels, just like you do with Sturd-I-Floor panels. Both interior and exterior plywood need this space. Figure 6-49 shows you how much space to leave. This spacing is based on field experience studies done by the American Plywood Association. Figure 6-50 lists minimum requirements for plywood subflooring.

Grades of plywood you can use for subflooring include: Standard, Structural I and II, and C-C Exterior. Each plywood panel should be stamped with a mark like those shown in Figure 6-51. This index

tells you what spacing of rafters and floor joists to allow for different thicknesses of plywood. For example, an index marking of 32/16 tells you that the maximum spacing for this plywood panel is 32" for rafters and 16" for floor joists. Since each panel is stamped with an identification index, you always know the strength capability of any panel.

Framing the Walls

Once you've framed the floor and completed the subfloor, you have a handy working platform for doing the wall framing. Wall framing includes: the studs and the members (sole plates, top plates, door and window headers) of any walls that support the ceilings, the upper floors, and the roof. The wall framing also gives you a nailing base for wall covering materials.

The exterior walls of most homes in the United States are framed with wood studs. As a builder, there are many times you'll be able to use wood-frame walls to lower your material and labor costs. By using wood-frame construction, modular planning, and tilt-up construction, you can cut your costs and increase your profits.

Look at Figures 6-52, 6-53, and 6-54. Figure 6-52 shows how modular planning can reduce your window framing costs. Figures 6-53 and 6-54 show the difference between modular and nonmodular wall framing. As you can see, your cost savings will be greatest when you use standard modular stud spacing to determine the size of the house and the size and location of wall openings.

Panel identification index (1), (2), (3), and (4)	Plywood thickness (inches)	Maximum span (5) (inches)	Common nail size and type	Nail spacing (inches)	
				Panel edges	Intermediate
30/12	5/8	12 (6)	8d	6	10
32/16	1/2	16 (7)	8d (8)	6	10
36/16	3/4	16 (7)	8d	6	10
42/20	5/8	20 (7)	8d	6	10
48/24	3/4	24	8d	6	10
1-1/8" groups 1&2	1-1/8	48	10d (9)	6	6
1-1/4" groups 3&4	1-1/4	48	10d (9)	6	6

Notes:
(1) These values apply for Structural I and II, Standard sheathing and C-C exterior grades only.
(2) Identification index appears on all panels except 1-1/8" and 1¼" panels.
(3) In some non-residential buildings, special conditions may impose heavy concentrated loads and heavy traffic requiring subfloor constructions in excess of these minimums.
(4) Edges shall be tongue and grooved or supported with blocking for square edge wood flooring, unless separate underlayment layer (¼" minimum thickness) is installed.
(5) Spans limited to values shown because of possible effect of concentrated loads. At indicated maximum spans, floor panels carrying identification index numbers will support uniform loads of more than 100 psf.
(6) May be 16" if 25/32" wood strip flooring is installed at right angles to joists.
(7) May be 24" if 25/32" wood strip flooring is installed at right angles to joists.
(8) 6d common nail permitted if plywood is ½".
(9) 8d deformed shank nails may be used.

Minimum thickness of plywood adequate for subflooring
Figure 6-50

Tilt-up wall framing cuts down on your labor costs and overall construction time. Also, it can mean better-quality construction. Consider preassembling all of your walls, including the framing, sheathing, siding, windows, and exterior trim. You can buy prefabricated wall panels, or you can assemble them flat on your subflooring deck.

Next, let's take a closer look at wall framing. We'll learn about ordering lumber, and about ceiling heights, platform and balloon wall framing, wall openings, corners, and intersections. We'll look at interior walls, non-bearing walls, nailers, fire cuts, blocking, and bracing.

Ordering Lumber
Wall framing lumber should be reasonably dry. A moisture content (MC) of about 15% is best. The maximum allowable moisture content is 19%. If you decide to use lumber with a 16% to 19% MC in your studs, plates, and headers, be sure to let the lumber reach in-service MC before you add any interior trim.

Before you order your wall framing lumber, look again at how much lumber you need. You can nearly always use 24" spacing for 2 x 4 studs in the exterior walls of one-story houses. There's no reason to space the studs any closer, as long as exterior and interior finish materials can span 24". Many times, you can also use 24" stud spacing for walls that support the upper floor and roof in a two-story unit.

Space studs at regular intervals, whether you use 16" or 24" spacing. This will mean less cutting and fitting of sheathing, insulation, vapor barriers, and finish materials.

Remember that you can leave out mid-height blocking between studs when the top and bottom plates serve as fire stops. There's also a way to leave out the blocking behind taped joints in gypsum wallboard (sheetrock). Space the studs 24"

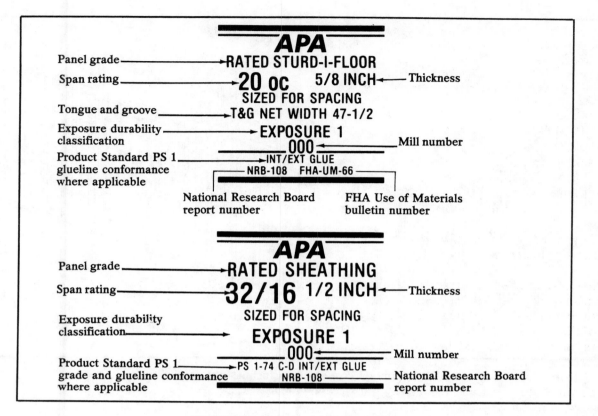

Typical APA registered trademarks
Figure 6-51

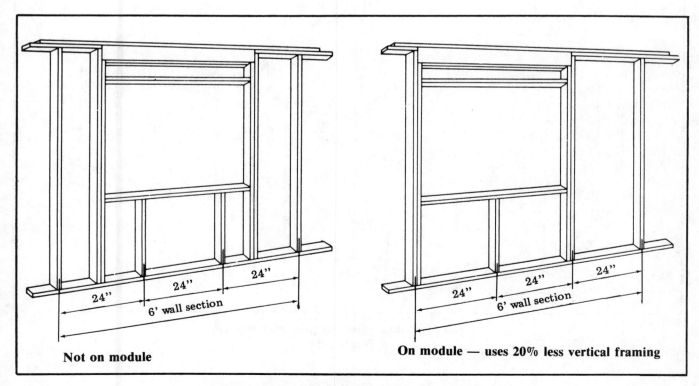

Windows located on module can save framing
Figure 6-52

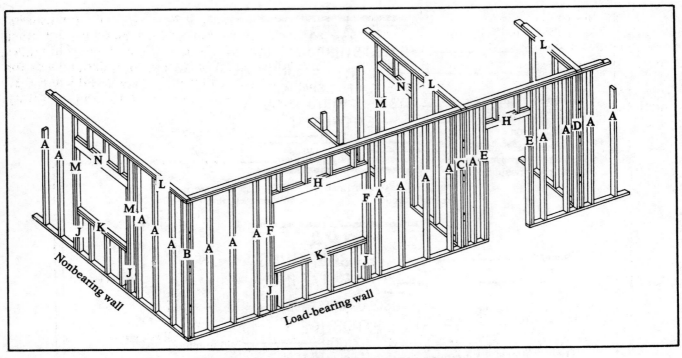

Wall framing: cost-saving principles not applied
Figure 6-53

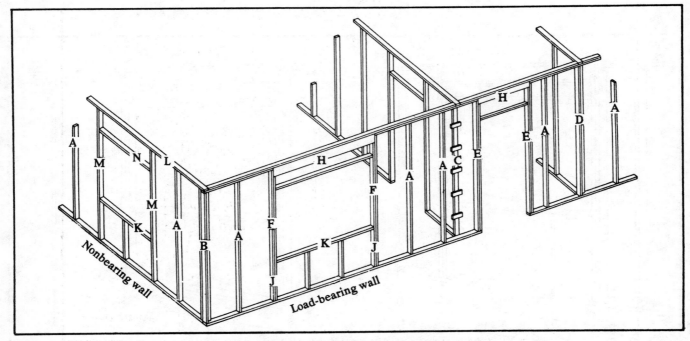

Wall framing: cost-saving principles applied
Figure 6-54

o.c. or less, and use 1/2'' thick sheetrock horizontally. The simple taped joint is strong enough to resist the normal loads placed on the board.

You can save more money simply by choosing your lumber carefully. Use the *lowest grade of lumber that will do the job*. Utility grade 2 x 4's are fine for supporting most wall loads. HUD accepts utility grade studs, spaced 24'' o.c., for exterior walls supporting roof and ceiling loads. For other residential bearing walls, standard and stud grades will do the job. They will support design loads for almost all residential uses. Figures 6-55 and 6-56 show lumber grade use in one- and two-story residential construction.

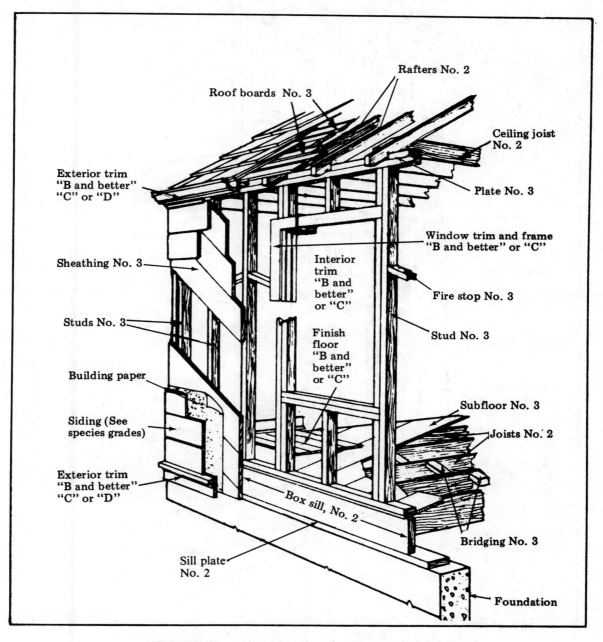

Grade use key for low-cost one-story construction
Figure 6-55

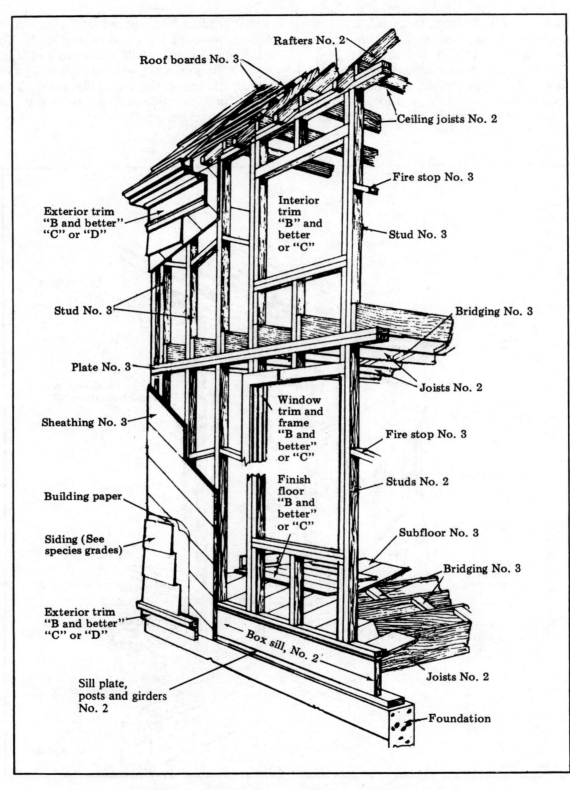

Rafters No. 2

Roof boards No. 3

Ceiling joists No. 2

Fire stop No. 3

Exterior trim "B and better" "C" or "D"

Interior trim "B" and better or "C"

Stud No. 3

Stud No. 3

Bridging No. 3

Plate No. 3

Joists No. 2

Sheathing No. 3

Window trim and frame "B and better" or "C"

Fire stop No. 3

Studs No. 2

Building paper

Finish floor "B and better" or "C"

Subfloor No. 3

Siding (See species grades)

Bridging No. 3

Exterior trim "B and better" "C" or "D"

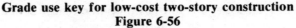

Box sill, No. 2

Joists No. 2

Sill plate, posts and girders No. 2

Foundation

Grade use key for low-cost two-story construction
Figure 6-56

Ceiling Heights

If the ceiling is unusually high or low, you'll probably spend extra money on labor and materials. Here's a look at how you can keep your costs in line by using standard ceiling heights.

First-floor ceiling— Normally the first floor ceiling is 8' high. Common practice is to rough-frame the wall (from the subfloor to the top of the upper plate) to a height of 8'1½''.

In platform construction, studs are often precut to a length of 7'8⅝''. Use these studs with a 1⅝''-thick plate. If dimension material is 1½'' thick, make the precut studs 7'9'' long. This way you can use 8' high wallboard panels or six courses of rock lath. And you still have enough clearance for floor and ceiling finish, or for plaster grounds, at the floor line.

In many cases, the precut studs are 7'8¾₆'', and the dimension material is 1½''. The sole plate and top plates are 1½'' thick. When a double top plate is used, this results in a ceiling height of only 8'1¹⁄₁₆'' from subfloor to ceiling joists. This takes into account the sole plate resting on top of the subfloor.

Second-floor ceiling— The second-floor ceiling height shouldn't be less than 7'6'' in the clear, except for the area under a sloping roof. A sloping ceiling should have a minimum height of 5', with one-half the floor area having at least a 7'6'' clearance.

Decide whether an 8' or 7'6'' second-floor ceiling is best for your construction methods. You may prefer an 8' ceiling for the second floor. This means that you can use the same size materials throughout the house. It saves you the time of cutting materials to fit nonstandard ceilings.

Wall Framing

The two most common methods of wall framing are platform construction and balloon frame construction. The platform method is well known for its simplicity. Balloon framing is often used where stucco or masonry is the exterior finish material in two-story houses.

Platform Construction— Look back again to Figure 6-2. This figure shows platform framing for a two-story house. Now look at Figure 6-57. Notice that the wall framing in platform construction sits

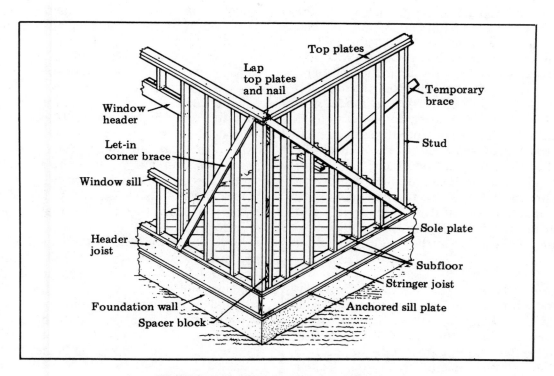

Wall framing with platform construction
Figure 6-57

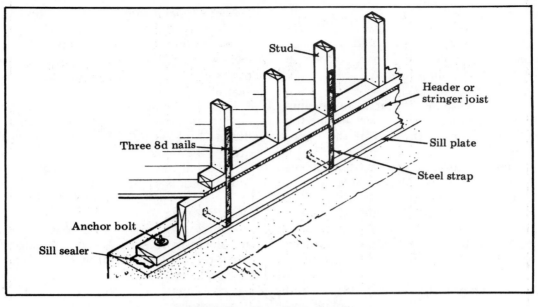

Anchoring wall to floor framing
Figure 6-58

on top of the subfloor. And the subfloor extends to all edges of the building.

You can use a combination of methods in a single-story house: platform construction for the first-floor side walls, and full-length studs for end walls extending to end rafters of the gable ends.

Here's how to preassemble a wall section. Lay out the precut studs, window and door headers, cripple studs, and window sills. Nail the first top plate and sole plates to all vertical members with 16d nails. Use 16d nails to fasten adjoining studs to headers and sills. Cut notches into the studs for let-in corner bracing, if you need it. Then tilt-up the entire wall section, plumb, and brace it.

In hurricane areas or areas with high winds, you should fasten wall and floor framing to the anchored foundation sill if the framing isn't tied together with sheathing. Figure 6-58 shows how you can use steel straps to anchor studs to the floor framing.

Balloon Construction— The main difference between balloon framing and platform framing is at the floor line. In balloon framing, the wall studs extend from the sill of the first floor to the top plate or end rafter of the second floor. In platform framing, the wall is complete at each floor. Take a look at Figure 6-59 for ways to arrange the studs at outside corners of the house frame.

In balloon construction, you'll frame the joists, bearing partitions, and outside wall sections before you apply subflooring or sheathing. Both the wall studs and the floor joists rest on the anchored sill, as shown in Figure 6-59. The studs and joists are toenailed to the sill with 8d nails and nailed to each other with at least three 10d nails.

The ends of the second-floor joists bear on a 1 x 4 or a 1 x 6 ribbon let into the studs. Use four 10d nails to fasten the joists to the studs at these connections. Nail the end joists of both the first and second floors to each stud.

In most areas, building codes require fire stops in balloon framing. Without fire stops, a fire can spread through the stud cavity. Use 2 x 4 blocking between the studs, as shown in Figure 6-60. But check your local code first. Some areas have special requirements.

Next look at the bracing shown in Figure 6-60. If you use diagonal braces instead of plywood or structural fiberboard, put them in the corners of the side wall of each story. Do this after the side walls have been erected and plumbed. You'll usually use 1'' x 4'' members for these braces. You have to cut a notch in each stud where the brace crosses it. This way, you can set the brace flush with the outside of the studs.

If you can't use diagonal bracing because of an opening near the corner of a building, use knee

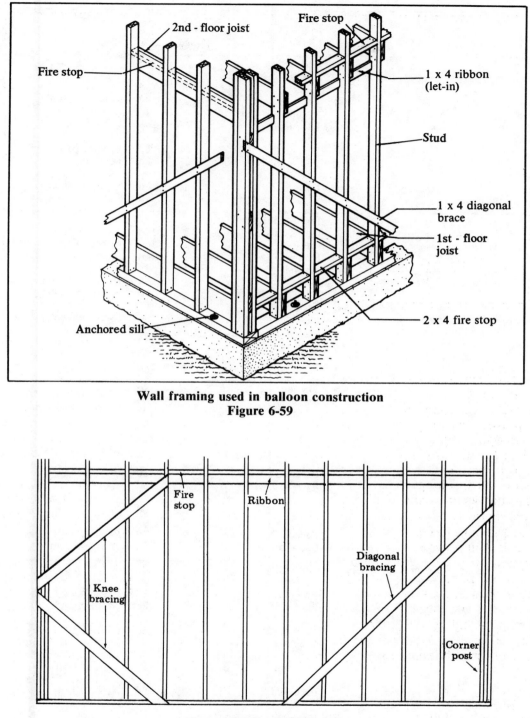

Wall framing used in balloon construction
Figure 6-59

Knee and diagonal bracing
Figure 6-60

braces, as shown in Figure 6-60. Install knee braces the same as you would diagonal bracing.

Take a look at Figure 6-61 to see the best methods for nailing the framing and sheathing. You'll see all the nailing details that you would need for framing a house.

Joining	Nailing Method	Number	Nails	
			Size	Placement
Header to joist	End-nail	3	16d	
Joist to sill or girder	Toenail	2	10d or	
		3	8d	
Header and stringer joist to sill	Toenail		10d	16-in. on center
Bridging to joist	Toenail ea. end	2	6d	
Ledger strip to beam, 2-in. thick	Face-nail	3	16d	At each joist
Subfloor, boards:				
1 x 6 and smaller	Face-nail	2	8d	To each joist
1 x 8	Face-nail	3	8d	To each joist
Subfloor, plywood:				
At edges	Face-nail		8d	6-in. on center
At intermediate joists	Face-nail		8d	8-in. on center
Subfloor (2 x 6-in., T&G) to joist or girder	Blind-nail (casing) & face-nail	2	16d	
Sole plate to stud, horizontal assembly	End-nail	2	16d	At each stud
Top plate to stud	End-nail	2	16d	
Stud to sole plate, upright assembly	Toenail	4	8d	
Sole plate to joist or blocking	Face-nail		16d	16-in. on center
Double studs	Face-nail, stagger		10d	16 in. on center
End stud of intersecting wall to ext. wall stud	Face-nail		16d	16-in. on center
Upper top plate to lower top plate	Face-nail		16d	16-in. on center
Upper top plate, laps and intersections	Face-nail	2	16d	
Continuous header, two pieces	Each edge		12d	12-in. on center
Ceiling joist to top wall plates	Toenail	3	8d	
Ceiling joist laps at partition	Face-nail	4	16d	
Rafter to top plate	Toenail	2	8d	
Rafter to ceiling joist	Face-nail	5	10d	
Rafter to valley or hip rafter	Toenail	3	10d	
Ridge board to rafter	End-nail	3	10d	
Rafter to ridge board	Toenail	3	10d	
Collar beam to rafter:				
2-in. member	Face-nail	2	12d	
1-in. member	Face-nail	3	8d	
1-in. diagonal let-in brace to each stud and plate (4 nails at top)		2	8d	
Built-up corner studs:				
Studs to blocking	Face-nail	2	10d	Each side
Intersecting stud to corner studs	Face-nail		16d	12-in. on center
Built-up girders and beams, 3 or more members	Face-nail		20d	32-in. on center, each side
Wall sheathing:				
1 x 8 or less, horizontal	Face-nail	2	8d	At each stud
1 x 6 or greater, diagonal	Face-nail	3	8d	At each stud
Wall sheathing, vertically applied plywood:				
⅜-in. and less thick	Face-nail		6d	6-in. edge
½-in. and over thick	Face-nail		8d	12-in. intermediate
Wall sheathing, vertically applied fiberboard:				
½-in. thick	Face-nail		1½-in. roofing nail	3-in. edge &
²⁵⁄₃₂-in. thick	Face-nail		1¾-in. roofing nail	6-in. intermediate
Roof sheathing, boards, 4-, 6-, 8-in. width	Face-nail	2	8d	At each rafter
Roof sheathing, plywood:				
⅜-in. and less thick	Face-nail		6d	6-in. edge and 12-in.
½-in. and over thick	Face-nail		8d	intermediate

Recommended procedures for nailing framing and sheathing of wood-frame house
Figure 6-61

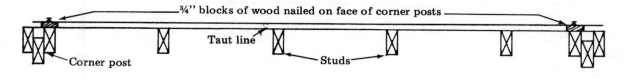

Taut gauge line stretched between posts
Figure 6-62

In some high-wind areas, you can use steel bands for more bracing. These bands are like the bands that hold bundles of framing lumber when they're delivered to the job site. In fact, sometimes you can even take the straps off the lumber bundles and use them for bracing. Here's how.

Apply the strap as you would diagonal bracing, but don't notch the studs. Predrill a hole in the steel at each stud location and at the sill and top plate. Use 16d nails for all fastening. Nail at the top plate first. Then, drive the nail at the sill at an angle so the strap will pull tight as the nail is driven. Last, nail the steel strap to the studs. Use this extra bracing system at corners, or anywhere the bracing won't interfere with wall openings.

Making Sure the Walls are Straight
When you're putting up the exterior walls, *make sure they're straight.* Here's how: For platform frame construction, snap a chalkline on the deck for placement of the sole plate. For balloon frame construction use a taut line to align the walls at the second-floor ribbon, and for top-plate alignment in all framing methods.

Tack a 3/4" block on the outside of each corner post near the ribbon. Figure 6-62 shows just where to mount the blocks. Stretch a line between the blocks. Put up temporary braces to hold the wall while you align it. Adjust each stud so that it's 3/4" from the line. An easy way to measure this distance is to slip a 3/4" block in between the line and the stud. When aligning the top plate, tack the 3/4" blocks to the top plate and follow the same procedure. Use temporary braces to help you adjust the top plate so that it's 3/4" from the taut line.

Corners and Intersections
There are four common framing methods for intersections of partitions and walls. Each one must give support and a nail base for interior or exterior wall materials. Figure 6-64 shows three methods used for exterior corners. Now let's take a look at the four methods shown in Figure 6-63.

Figure 6-63 A shows the traditional method. Three studs provide the support and nail base for interior and exterior wall materials. This arrangement means that you must insulate the space between the studs before putting on the sheathing. Look at Figure 6-64 A to see how to use this method at exterior wall corners.

Figure 6-63 B shows an alternate method. An extra stud provides the necessary insulation. It's effective insulation, but it's also more expensive. Take a look at B in Figure 6-64 to see the method on exterior wall corners.

Figure 6-63 C shows the best three-stud arrangement. You don't need any spacer blocks (often cut from full-length studs or 2 x 4's). There's enough backup support for wall materials, and you can insulate after the sheathing is on. This method is shown on an exterior corner in Figure 6-64 C.

An alternative is shown in Figure 6-63 D. You attach backup cleats to the partition stud. The cleats can be 3/8" plywood or 1" lumber. This way, you don't have studs that are only used as backup for interior wall material. The problem with cleats is that they are poor back-up nailers. They are often "springy," and it's easy to knock them loose. You're better off with a solid stud as a back-up nailer.

Wall Openings
Where you place the window and door openings will affect your wall framing costs. Always try to locate one side of each window or door rough opening at a regular stud position. You can use fewer studs this way. Look back to Figure 6-54, at details E, F, and M.

Choose window and door sizes so that the rough opening width is a multiple of the stud spacing. Both sides of the opening are then on the stud module. This reduces the number of studs you

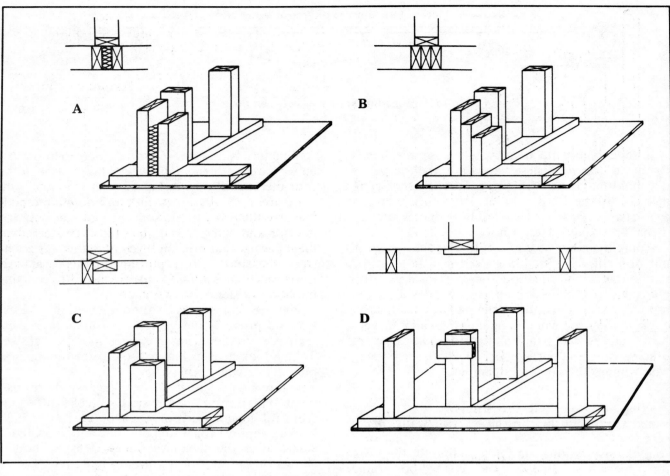

Wall and partition intersection framing
Figure 6-63

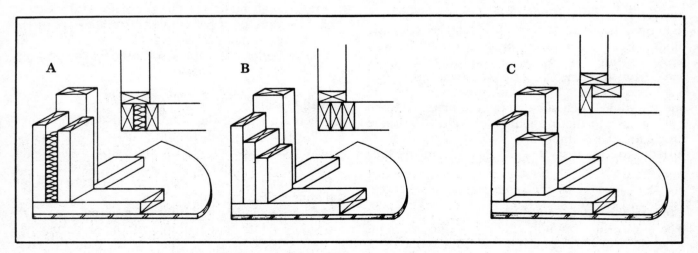

Stud arrangements at exterior corners
Figure 6-64

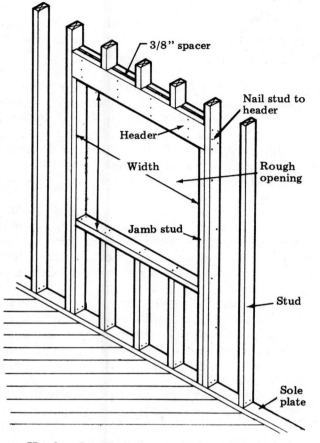

Headers for windows and door openings
Figure 6-65

same species of lumber for these headers as you use for floor joists. Here's a guide that should help you select the right size headers:

Maximum span (feet)	Header size (inches)
3½	2 x 6
5	2 x 8
6½	2 x 10
8	2 x 12

For other than light-frame construction, you may need specially-designed headers. Wider openings often need trussed headers. You may have to custom-design trussed headers for each individual opening. Figures 6-66 and 6-67 show two types of trussed openings. You can also use these methods for interior load-bearing partitions and over spans in projecting bays and porches.

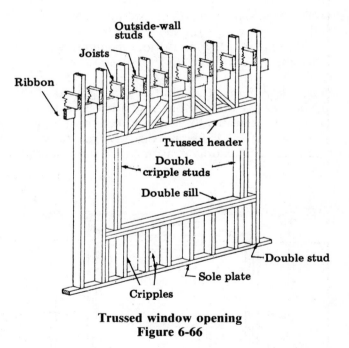

Trussed window opening
Figure 6-66

need. Also, you can use standard sheathing and siding without having to make special cut-outs for openings.

Headers, or lintels, are the members used over window and door openings. As the size of the opening increases, you have to increase the depth of the headers so they can support the ceiling and roof loads.

Figure 6-65 shows that a header is made up of two 2-by members. You can space the two members with a 3/8" lath or wood strip spacer. Nail the members and the spacer together. Some builders use 3/8" plywood as the spacer. They just cut the plywood to the same size as the header. Others use no spacer, leaving the 3/8" space open.

The ends of the headers are supported by the inner studs, called jamb studs or jack studs, as you can see in Figure 6-65. You must do this at exterior walls and at interior bearing walls. For normal light-frame construction, you can usually use the

Window openings are often the most over-built sections of wall framing. Builders often double the sills under windows and put short support studs at the ends of the sill.

Figure 6-68 shows how vertical loads at windows can be transferred downward by the studs that support the header. The sill and wall beneath the window carry only the load of the window; so you don't need a second sill and sill support studs. The load over the window is carried entirely by the

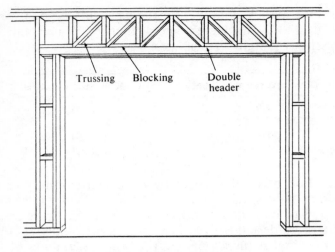

Trussed door opening
Figure 6-67

jamb studs. The ends of the sill are supported by end-nailing through the jamb stud.

For some window and door openings, you can move the header beam up to replace the lower top plate. Take another look at Figure 6-68. You'll need longer jamb studs, but you can get rid of the *cripple studs* (short, fill-in studs) between the top plate and the header. You don't need them because now the vertical loads are directly transferred to the header. The only reason you'll need cripple

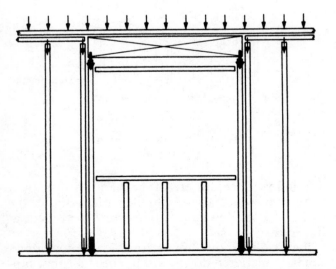

Load distribution through header and support studs at opening in load-bearing wall
Figure 6-68

studs below the header is that wall materials can't span the distance between the header beams and the rough opening top plate. The same is true of headers across doorway openings.

Openings in non-bearing walls— Not all exterior walls are load-bearing. You can eliminate lintels and headers over doors and windows *providing that there are no floor or roof loads acting downward over the opening.* In houses with gable-end roofs, the end walls are often framed so roof trusses and floor joists bear on front and rear walls rather than on end walls.

You can often frame these non-bearing exterior walls the same way as interior partitions. Since there are no loads to transfer, you can frame doors and windows with single members. A single top plate is enough if you use metal straps to tie the top plates together at the joints.

But remember: When you use single top plates in non-bearing walls, make the studs 1½" longer than the studs in bearing walls. Many builders use double top plates in all walls so they get enough tie-in of walls at the top plates without having to use metal straps.

Figure 6-69 shows how you can save on materials when you frame window and door openings in non-bearing exterior walls. On one or both sides of window or door openings, locate jamb studs on the module of the wall stud framing. If you don't locate the jamb studs on module, you'll need extra studs. You'll also have less waste in sheet facing materials when you locate the jamb studs on the stud spacing module.

Nailed and nail-glued plywood headers— You can substitute plywood headers for conventional headers in exterior wall openings in single-story houses, or in the top story of multi-story houses up to 36' wide. You can use plywood headers when you sheath the walls with 1/2"-thick material. Any roof and ceiling loads must be transferred to the exterior walls through trusses or rafters.

Figure 6-70 shows construction details and material requirements for a nailed 1/2"-thick plywood header. You can use this design for rough opening widths up to nominal 4' openings. Although the figure shows stud spacing 24" o.c. (for modular planning), you can use this method for smaller spacing, including 16" o.c. Take a look at the section X-X in the figure to get an idea of

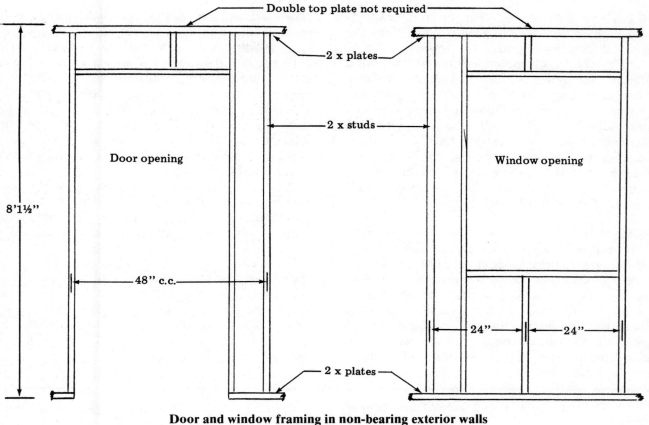

Door and window framing in non-bearing exterior walls
Figure 6-69

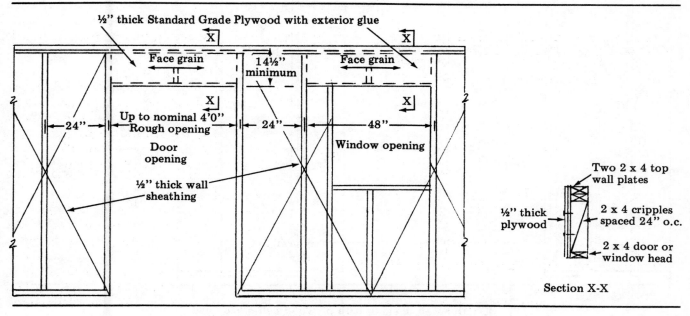

Exterior wall opening with plywood nailed header
Figure 6-70

how the plywood is put on. The plywood header is nailed with 8d wire nails spaced 4" o.c. along the edges and the cripple studs.

Figure 6-71 shows the construction details and material requirements for nail-glued plywood headers spanning more than 4', and up to nominal 6', openings. Use any elastomeric adhesive, or any other adhesive that meets the standards of American Plywood Association (APA) Specification AFG-01. Then nail the header 6" o.c. along the edges. Nail it to any studs in between with 8d common wire nails. Notice that there's a 14½" minimum for the header depth.

Interior Walls

Figures 6-1, 6-2, and 6-3 showed you how to frame load-bearing partitions. Make sure the intersection between interior and exterior walls is secure. And that it provides a nailing surface for the plaster base or drywall finish. Figure 6-72 shows another view of framing intersections. This figure shows a very important part of framing interior walls: connecting the interior to the exterior walls. Notice

how much less lumber (and labor) is needed for the method shown in Figure 6-72 A.

In a house with conventional joist and rafter construction, the interior walls normally are room dividers *and* bearing walls for the ceiling joists. Walls that are parallel to the direction of the joists are usually non-bearing. Nominal 2 x 4 studs are normally used throughout the structure, for both interior and exterior walls. But check your local code. In some areas you can use 2 x 3 studs for interior non-bearing walls.

Assemble and erect the interior walls the same way you do exterior walls. Use a single sole plate and double or single top plates. The upper top plate (or single top plate) ties intersecting walls to each other.

The location of walls, and the stud size and spacing, depend on the room size and type of interior finish material. The bottom chords of the trusses fasten and anchor crossing partitions. When partition walls are parallel to, and located between, trusses, fasten the walls to 2 x 4 blocks that are nailed between the lower chords.

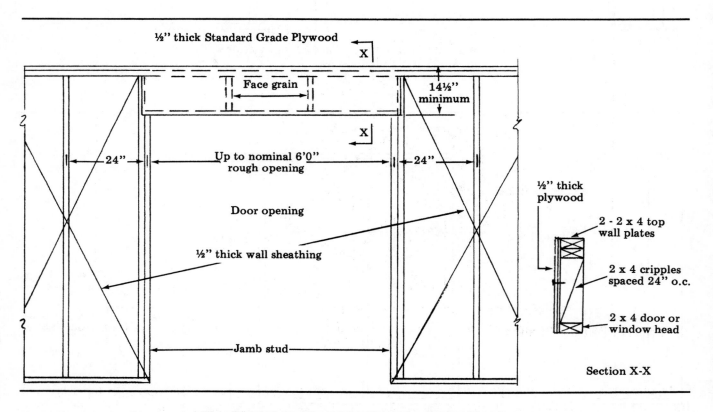

Exterior wall opening with plywood nail-glued header
Figure 6-71

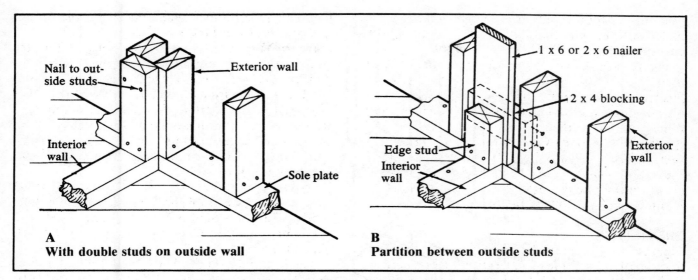

Intersection of interior wall with exterior wall
Figure 6-72

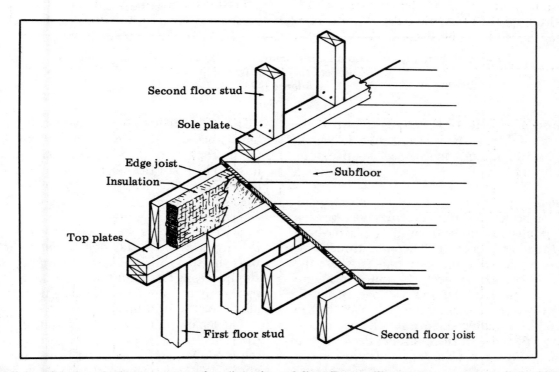

End-wall framing for platform construction (junction of first-floor ceiling and upper-story floor framing)
Figure 6-73

The framing for the end walls is different in platform and balloon construction. Figure 6-73 shows platform construction for the wall and ceiling for a 1½ or 2 story house with finished rooms above the first floor. Use 8d nails to toenail the edge floor joist to the top wall plate. Space the nails 16" o.c. Install the subfloor, sole plates, and wall framing in the same manner as for the first floor.

In balloon framing, the studs continue through the first and second floors. Take a look at Figure

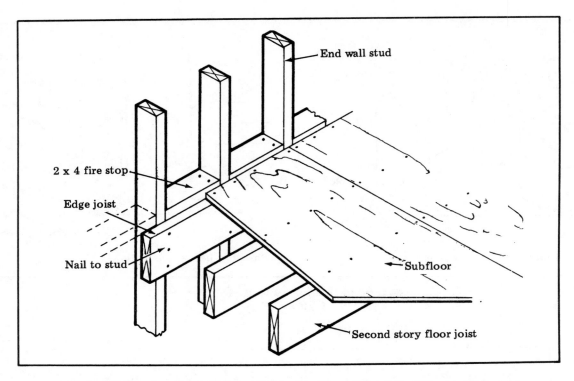

End-wall framing for balloon construction (junction of first-floor ceiling and upper-story floor framing)
Figure 6-74

6-74. The edge joist can be nailed to each stud with two or three 10d nails. However, many carpenters prefer to use 12d or larger nails. Use 2 x 4 fire stops between the studs.

Lath Nailers
Framing for both the wall and ceiling has to have fastening points for either drywall or plaster-base lath at all the inside corners. There are three common ways to put horizontal lath nailers where the wall and ceiling framing meet. Section A in Figure 6-75 shows double ceiling joists above the wall. Each of the two joists sticks out a little over the top plate, to make a nailing surface for the ceiling.

In section B, the wall runs parallel between two ceiling joists. Fasten a 1 x 6 lath nailer to the top plates. You could use a 2 x 6 or wider member instead of the 1 x 6. Here's where those pieces of scrap and short wood come in handy — they make ideal nailers. Put on the nailers before installing the roof sheathing. Insert backing blocks spaced 3' to 4' o.c. so the bottom of the joists are held level with the bottom of the lath nailer.

Now, look at C in Figure 6-75. When the partition wall is at right angles to the ceiling joists (in-

stead of parallel to them), you can let in 2 x 6 blocks between the joists. Nail the blocks directly to the top plate, then toenail them to the ceiling joists. The lip of the nailer hanging over either side of the top plate gives you a place to connect the edge of the ceiling.

Stairwells
When you frame a stairwell, support the trimmers with the studs for the stairwell walls. These studs should project beyond the inside face of the trimmer. When you place the trimmers, be sure to allow for the stud projection.

The length of a stairwell depends on the pitch of the stairs, the style of the stairs, and the headroom required. Measure headroom from the ceiling line, above the stairs where the headers are located, down to the top of the stairs. Headroom of 6'8'' to 7' is required when the risers are from 7½'' to 8'' each. Here's a formula for finding the location of the headers.

Divide the number of risers by two. Then add the width of the joists in inches. This gives you the length of the stairwell.

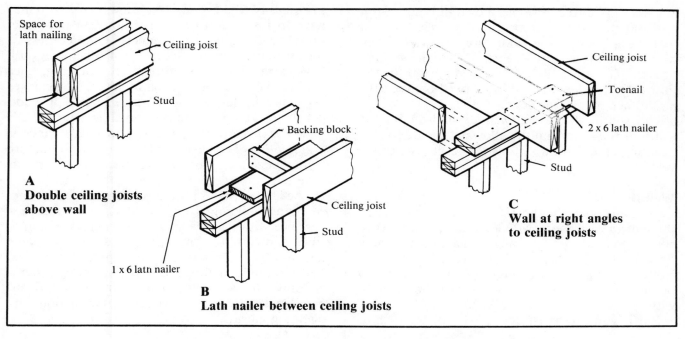

Horizontal lath nailers at junction of wall and ceiling framing
Figure 6-75

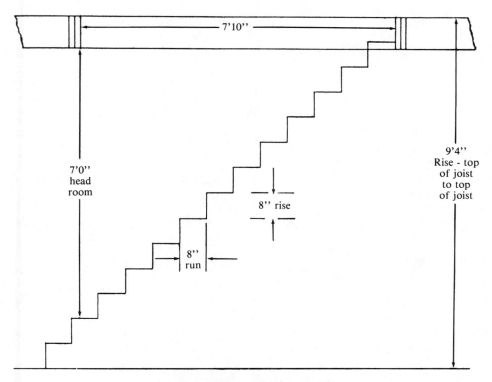

Finding headroom for a stairwell
Figure 6-76

Let's look at the example shown in Figure 6-76. Let's say the rise between the two floors is 9'4'' and that the joists between floors are 10'' wide. Then we'll figure 14 risers at 8'' each. Divide the number of risers (14) by 2 to get 7. Add the width of the joist (10). The length of this stairwell between the inside faces of the headers is 7'10''.

The width of the stairwell depends upon the width of the stairs and the allowance made for the partition studs and wall finish on each side of the well.

Wall Bracing

Last of all comes the sheathing, or wall bracing, on the outside of most wood frame homes. Most codes require exterior walls to have a minimum "racking" strength for stability under wind loads. You can usually add enough racking strength by bracing the walls with sheathing or siding panels, without any extra let-in corner bracing.

But if you use nonstructural sheathing, such as low-density fiberboard or gypsum board, you may have to install extra bracing. You can put structural sheathing or siding panels at the corners of the structure. Or, use let-in 1 x 4 bracing. You'll have extra labor costs with the 1 x 4 bracing, since you'll have to hire a skilled craftsman. You can see it's generally more cost-effective to use plywood panels for bracing whenever possible.

In rough carpentry, the key to cost savings is *preplanning*. The wise builder plans his job before layout begins and before the framing material is ordered.

The United States probably has the most advanced home building techniques of any nation. Experience with various sizes and spacings of joists, studs and rafters, along with standard 4 x 8 panel dimensions, has resulted in rapid, economical house framing methods. But your framing systems are only as good as the planning you put into them. You'll get more out of every construction dollar by planning ahead.

Chapter 7

Framing the Roof

In this chapter we'll discuss both conventional roof framing and the use of prefabricated and job-built trusses. Let's begin with a review of roof trusses.

Roof Trusses

The truss is a rigid framework of triangular shapes. Trusses can carry loads over long spans without any intermediate support. They can span from one exterior wall to the opposite exterior wall without support. Since you don't need bearing walls for support, you can use the entire house as one large workroom. You also have more flexibility when designing the interior: partitions can go anywhere.

Prefabricated roof trusses are used by many home builders. They're widely available in different styles and sizes. Or you can build your own trusses on the job site. Trusses save you money. They use less framing lumber, and need less labor to install, than conventional roof framing.

Trusses are usually designed with 24" spacing, rather than the 16" spacing used for conventional joists and rafters. Therefore you'll need thicker interior and exterior sheathing if you use roof trusses.

You can often buy prefabricated gable ends as part of the truss package. These look like a truss but have vertical members spaced at 24" o.c. instead of truss web members. You can nail sheathing right onto the vertical members.

Kinds of Trusses

The wood trusses common in houses include *W-type, king-post,* and *scissors.* These trusses are best for rectangular houses, where the constant width needs only one kind of truss. But you can also use trusses for L-shaped plans and for hip roofs. You can use special hip trusses for the ends and for valley areas. Let's take a closer look at the three kinds of trusses.

W-type truss— The W-type truss is the most popular. Note Figure 7-1 A. This truss uses three more members than the king-post truss, but the distance between connections is less. This means you can use a lower grade of lumber across greater spans.

King-post truss— The king-post truss is the simplest truss. It has upper and lower chords and a vertical center post, as shown in Figure 7-1 B. Allowable spans are less than for the W-type truss even when the same size lumber is used. This is because the upper chord of the king-post truss is only supported at the center. For short and medium spans, the king-post is probably more economical than other trusses because it has fewer pieces and goes up faster.

The span, however, isn't the only thing to consider when you decide what kind of truss to use.

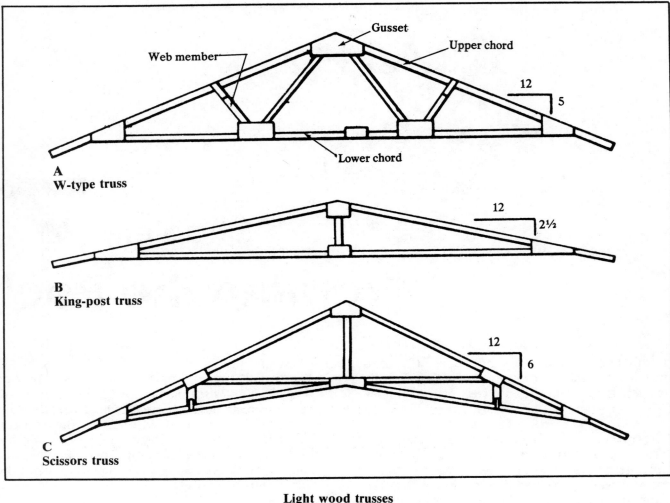

Light wood trusses
Figure 7-1

Consider the design load requirements for snow, ice, and wind. And be sure to check local availability and prices for each kind of truss.

Scissors truss— The scissors truss is shown in Figure 7-1 C. This truss is used for houses with a sloping living room ceiling. It's more complicated than the W-type truss, but it's a good choice for "cathedral" ceilings. Even scissors trusses use less material than conventional roof framing methods.

Truss Design and Fabrication

The design of a truss must allow for snow, ice, and wind loads. Figure 7-2 shows snow loads by geographical area. For example, builders in Maine must reinforce roofs to handle snow loads of up to 80 pounds per square foot. Notice that for the area stretching from the Northwest down to Texas the snow loads depend on the location. This is because of the great variety of weather from area to area.

Truss design must also allow for the weight of the roof itself. And it must take into account the slope of the roof. The flatter the slope, the greater the stress. A flat roof requires larger members and stronger connections. Keep all of these factors in mind when you're deciding what kind of truss to use.

Figure 7-3 shows maximum allowable spans for W-type trusses. The chart assumes 24" o.c. truss spacing with a 40 pounds per square foot total load. It gives spans for three types of lumber, depending on grade and roof pitch.

Most trusses are made with plywood gussets

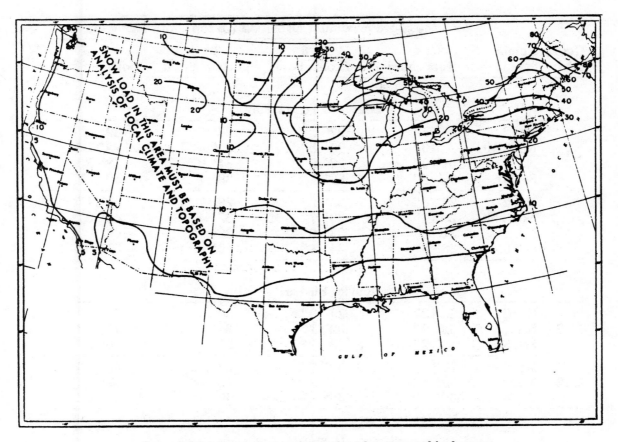

Snow load in pounds per square foot by geographical area
Source: Environmental Science Services Administration, Environmental Data Service
Figure 7-2

(nailed, glued, or bolted in place) or with metal gusset plates. Some trusses are made with split-ring connectors. Your local lumber dealer can usually provide the designs for standard W-type and king-post trusses with plywood gussets, and information on metal gussets. Many lumber dealers have complete trusses ready for installation.

Building a typical truss— Figure 7-4 shows how to design and build a typical wood W-type truss. This nail-glued gusset truss spans 26'; the slope is 4 in 12; and the spacing is 24'' o.c. The allowable roof load is 40 pounds per square foot, enough for moderate-to-heavy snow belt areas. You can generally use 2 x 4's for the upper and lower chords, though the upper chord should be a higher grade of material. Use wood with a moisture content of about 15% to 19%.

Use gussets on both sides of the truss. You can make plywood gussets from 3/8'' or 1/2'' standard plywood with exterior glue, or from exterior sheathing grade plywood. The cut-out size of the gussets and the general nail pattern for nail-gluing are shown in Figure 7-4.

In areas of normal-to-high relative humidity, like the Southern and Southeastern states, use a resorcinol glue for the gussets. But in dry areas you can use a casein or similar glue.

Gluing should be done under closely controlled temperature conditions. This is especially important if you're using resorcinol adhesives. Always follow the assembly temperatures recommended by the manufacturer. Spread glue on clean surfaces of both the gusset and truss members. Then use nails to apply pressure until the glue has set. Since both nails and glue are holding the gusset, the truss is much sturdier than if either was used alone.

Roof pitch	Lumber grade	2 x 4 top chord maximum spans		2 x 6 top chord maximum spans	
		2 x 4 bottom chord	2 x 6 bottom chord	2 x 4 bottom chord	2 x 6 bottom chord
		Southern Yellow Pine			
4/12	No. 2	26' 5"	27' 3"	27' 7"	39' 7"
	No. 2 MG	28'10"	29' 8"	30' 4"	43'11"
	No. 1	31'10"	32' 9"	34' 0"	48' 5"
	Sel Str.	33' 2"	33'11"	38' 7"	50' 3"
5/12	No. 2	27' 6"	28' 2"	29' 7"	40'11"
	No. 2 MG	29'11"	30' 7"	32' 4"	45' 3"
	No. 1	32' 3"	33' 0"	36' 1"	49' 0"
	Sel. Str.	33' 8"	34' 4"	40' 8"	51' 0"
		Douglas Fir-Larch			
4/12	No. 2	28'10"	29' 8"	30' 4"	43'11"
	No. 1	32' 0"	32'11"	34'10"	48' 5"
	Sel. Str.	33' 2"	33'11"	38' 2"	50' 3"
5/12	No. 2	29'11"	30' 7"	32' 4"	45' 3"
	No. 1	32' 6"	33' 2"	36'10"	49' 0"
	Sel. Str.	33' 8"	34' 4"	40' 4"	51' 0"
		Hem-Fir			
4/12	No. 2	25' 0"	25' 9"	26' 0"	38' 5"
	No. 1	28' 7"	29' 5"	29' 7"	42'10"
	Sel. Str.	30' 6"	31' 5"	33' 0"	46' 6"
5/12	No. 2	26' 0"	26' 8"	27'11"	39' 9"
	No. 1	29' 7"	30' 3"	31' 7"	44' 3"
	Sel. Str.	30'11"	31' 7"	35' 1"	47' 0"

Source: Span Tables for Light Metal Plate Connected Wooden Trusses, Truss Plate Institute, 1972.

Allowable maximum spans for W-type trusses
Figure 7-3

Use 4d nails, spaced 3" o.c., for plywood gussets up to 3/8" thick. Use 6d nails and 4" spacing for plywood 1/2" to 7/8" thick. For plywood less than 1/2" you can use staples instead of nails. When members are nominal 4" wide, use two rows of nails with a 3/4" edge distance. If the members are 6" wide, use three rows of nails. You can put the nails closer together, or in between the other nails, to make sure that the glue is squeezed evenly to all edges of the gusset.

Attaching trusses to the wall plate— Figure 7-5 shows how to fasten a truss to the wall plate. The method of anchoring is an important part of erecting trusses. Resistance to upward, as well as to downward, stresses must be considered. You usually need some sort of metal connector in addition to the toenailings. Plate anchors are available commercially or can be formed from sheet metal.

We've seen how quick, easy, and cost-efficient the use of prefabricated roof trusses can be. Keep

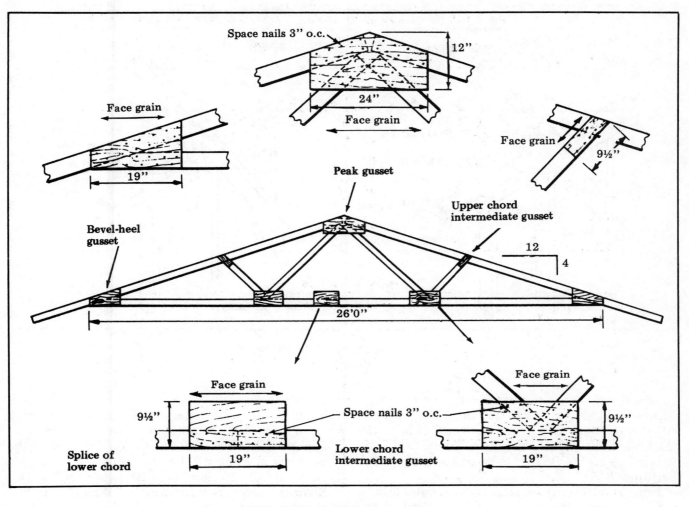

Construction of 26' W-type truss
Figure 7-4

in mind what you now know about trusses as we take an in-depth look at conventional roof framing. Conventional roofs are made with ceiling joists and rafters instead of trusses. We'll talk about roof slopes and about flat, gable, and hip roofs. Then we'll look at overhangs, and rafter run and rise.

Conventional Roof Framing

After all the walls are plumbed, braced, and have top plates, you can position and nail the ceiling joists. Usually they run across the width of the house, like the rafters. Try to place the partitions of the house so that you can use ceiling joists of even lengths (10', 12', 14' and 16' or longer) to

span from exterior walls to load-bearing interior walls without waste. The sizes of the joists depend on the span, wood species, spacing between joists, and the load on the second floor or attic. The correct size of joist is usually specified on the house plan.

Ceiling joists serve a number of purposes. They support ceiling finishes. They can be floor joists for second stories or attics. They tie exterior walls and interior partitions together. And they are tension members that resist the thrust of the rafters on pitched roofs.

Assemble ceiling joists as you would floor joists. Then, nail them securely to the top plate at outer and inner walls. Nail ceiling joists together where they cross or join at the load-bearing partition.

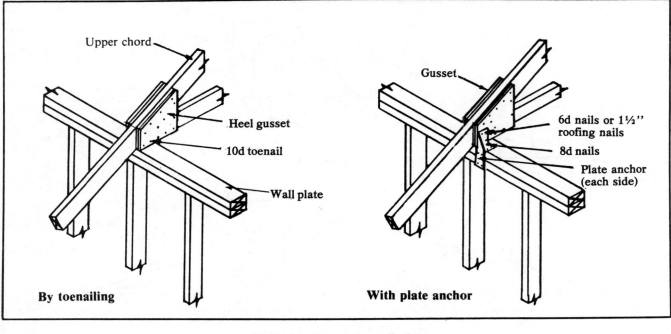

Fastening trusses to wall plate
Figure 7-5

Figure 7-6 A shows how to nail the joists together, either directly or with wood or metal cleats. Nail the joists directly to the rafters at exterior walls, as in B in Figure 7-6. Be sure to toenail at each wall or partition as well.

In areas of severe windstorms, it's a good idea to use metal strapping or some other method of anchoring ceiling and roof framing to the wall. Look back at Figure 7-6 B.

Framing a Chimney Hole
Figure 7-7 shows how a chimney should pass through a roof frame opening. The headers, trimmers, and studs should be kept 2'' from each surface of the chimney. This will keep the chimney's heat from passing to the wood frame. Fill the space between the woodwork and the chimney with some noncombustible material.

You usually figure the size of the chimney hole from the size of the flue lining. Follow along on Figure 7-8 while I explain how to do it. Figure the opening for a chimney with one 8'' x 8'' flue like this: the outside dimensions of an 8'' x 8'' flue are about 9¼'' x 9¼''. The brickwork surrounding the flue, in most cases, would be of bricks 8'' (nominal) long. Therefore, it takes two bricks to cover two opposite sides of the flue, and one brick to cover each of the other two sides, allowing 1/2'' for the joint between them. So, the size of the chimney opening will be 16'', plus a 2'' space on each side of the chimney, making a total of 20''. Since the chimney is square, the hole in the ceiling should be about 20'' x 20''. Follow this same procedure for larger flues and more bricks, always keeping the 2'' clearance between chimney and woodwork regardless of the chimney size.

Roof Slopes
The architectural style of a house often determines the kind of roof, and the amount of slope, you'll use. A contemporary design may have a flat or slightly pitched roof; a rambler or ranch-type, an intermediate slope; and a Cape Cod cottage a steep slope. A further consideration in choosing a roof slope is the kind of roofing you're going to be using. Shingles obviously won't work on a flat or very low-pitched roof. Instead, you'd use a built-up roof with a slope of up to 2/12. On sloped roofs with wood or asphalt shingles, you can get the slope as low as 3/12.

As you may remember from an earlier chapter, when we talk about roof slope (or pitch) we mean

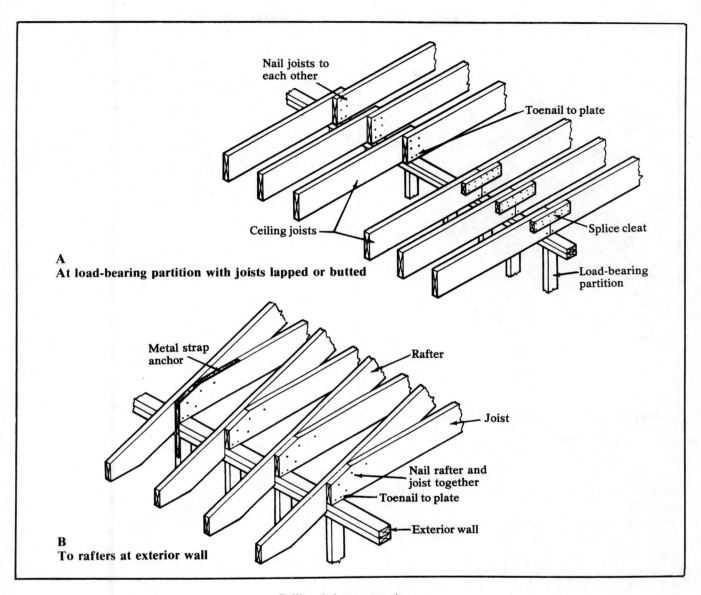

Ceiling joist connections
Figure 7-6

the number of inches of vertical rise in 12'' of horizontal run. The rise is given first: for example, 4 in 12 or 4/12.

We'll talk about the two basic types of roof: flat and pitched. Flat roofs (or slightly pitched roofs) have only one kind of member for roof and ceiling supports. Pitched roofs, however, need both ceiling joists and rafters. First we'll cover the flat roofs.

Flat Roofs

Flat or low-pitched roofs, sometimes known as shed roofs, can take a number of forms. Two kinds are shown in Figure 7-9. Roof joists for flat roofs are also the ceiling joists. They're laid level or with a slight pitch, with roof sheathing and roofing on top, and with the ceiling hung on the underside. You need larger-sized lumber for flat or low-pitched roofs than for steeper-pitched roofs

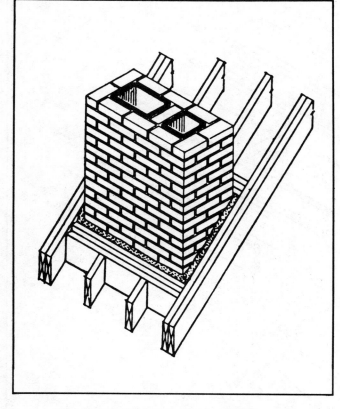

Headers and trimmers around chimneys
Figure 7-7

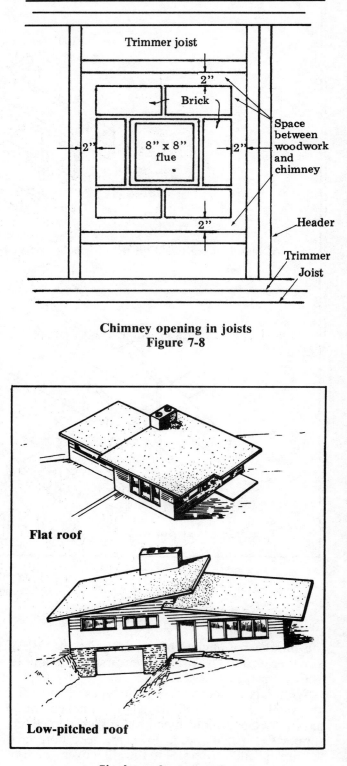

Chimney opening in joists
Figure 7-8

Flat roof

Low-pitched roof

Single-roof construction
Figure 7-9

because the lumber carries both roof and ceiling loads.

Sagging should never be a problem if you always use framing lumber of sufficient size and strength. Also, it's a good idea to have some roof slope to keep water from puddling on your flat roof. Provide a slight drainage slope by tapering the joists. To avoid the problem of condensation under the roof in winter, leave airways just under the sheathing when you install the insulation.

To provide overhang on all sides of the flat roof, use lookout rafters. Take a look at Figure 7-10 to see two common methods of overhang framing. Which method you choose depends on how much overhang you want. For an overhang of less than 3', nail lookout rafters to a double header and toenail them to the wall plate with 8d nails. The distance from the double header to the wall line is usually twice the overhang distance. For an overhang of more than 3' use the king rafter, as shown in Figure 7-10 B.

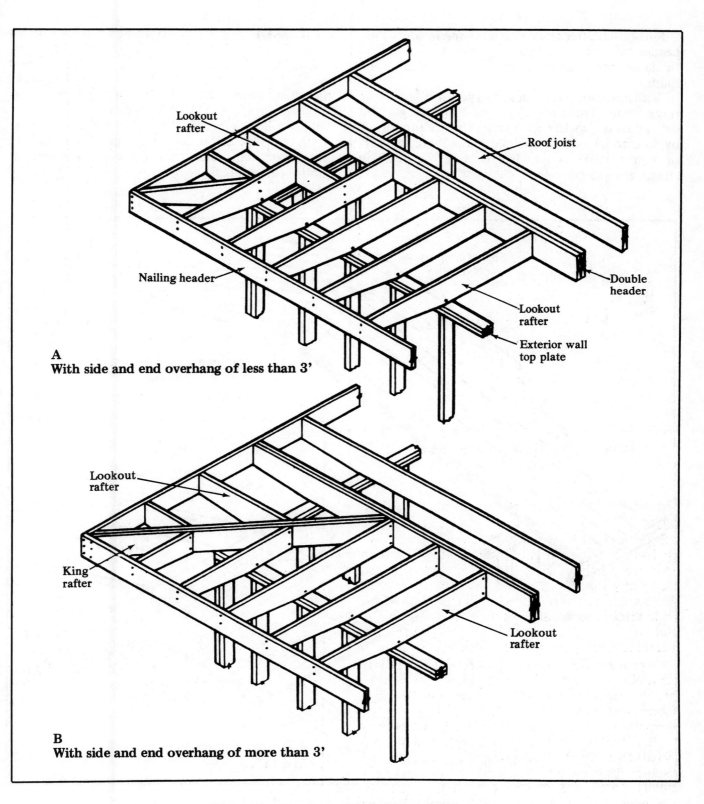

A
With side and end overhang of less than 3'

B
With side and end overhang of more than 3'

Typical construction of flat or low-pitched roof
Figure 7-10

Finish the rafter ends with a nailing header. This header can later be a fastening point for soffit and fascia boards. And be sure to build in enough ventilation.

With low-slope roof designs, you'll often use a ridge beam. These solid or built-up (with glue or nails) beams span the open area. They're supported by an exterior wall at one end and an interior wall or a post at the other. Just be sure the beam is strong enough to support the roof load.

There are several ways to frame a flat roof with a ridge beam. You can lay wood decking directly between the ridge beam and the top plate, just as you would for the plank-and-beam method. The decking then becomes both the ceiling and the roofing. Look at Figure 7-11 A. When you use a ridge beam and wood decking, they must be well anchored at both the ridge and outer wall. You should use long ring-shank nails and metal straps or angle irons at both bearing areas.

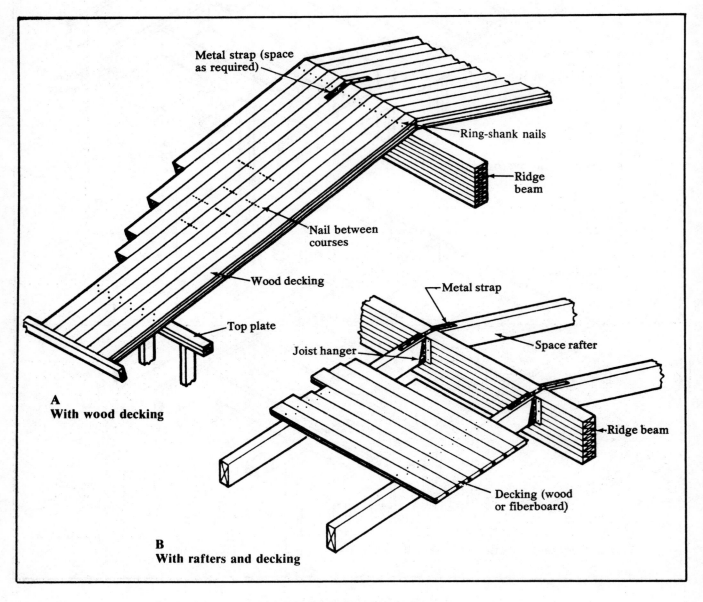

Ridge beam construction
Figure 7-11

Another method of flat roof framing uses large, spaced rafters called purlin rafters. These rafters are supported by metal hangers at the ridge beam, shown in Figure 7-11 B, and can extend beyond the outer walls to form an overhang. Wood or structural fiberboard decking is laid from rafter to rafter. Make sure you reinforce the roof with strapping or metal angles.

Gable Roofs

Perhaps the simplest form of pitched roof is the gable roof, shown in Figure 7-12 A. All rafters are cut to the same length and pattern, and erection is relatively simple.

Figure 7-13 shows gable framing and some details. In normal pitch-roof construction, you'll nail the ceiling joists in place after the interior and

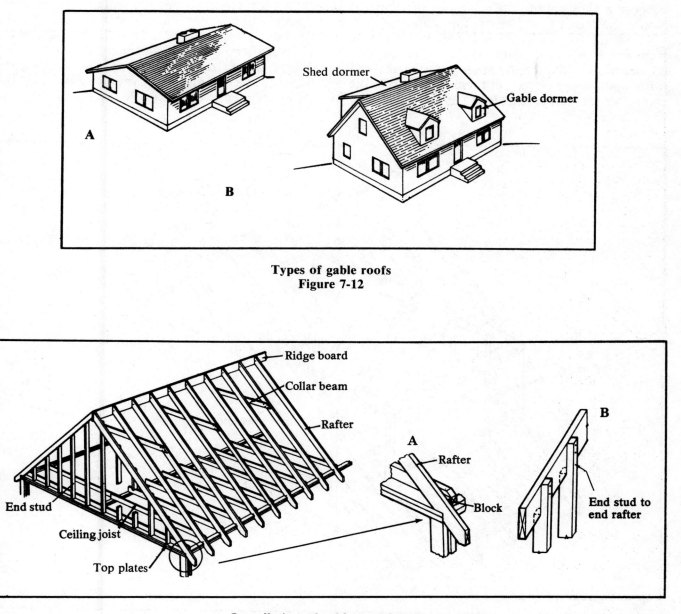

Types of gable roofs
Figure 7-12

Overall view of gable roof framing
Figure 7-13

exterior wall framing is complete. If you erect rafters before the ceiling joists are fastened in place, the thrust of the rafters will push out the exterior walls.

Rafters are usually cut to length before they're put up. Cut angles at the ridge and eave, and notches for the top plates, as shown in section A in Figure 7-13. Erect the rafters in pairs, one on each side of the ridge board. The ridge board may be a 1 x 8 or 2 x 8 member (for 2 x 6 rafters). It provides support and a nailing area for the rafter ends. Notice the use of collar beams for support.

The studs for the gable end walls are separate from the wall studs, unless you're using balloon construction. Connect the studs between the end rafter and the top plate. Cut them at the top so the end rafter can be inset, as B in Figure 7-13 shows.

Shed and gable dormers— These are variations of the basic gable roof, often used on Cape Cod or similar styles. Look at B in Figure 7-12. This is essentially a one-story house because most of the rafters rest on the first-floor plate. Space and light are provided on the second floor by the shed and gable dormers for bedrooms and bath. Roof slopes for this style can vary from 9 in 12 to 12 in 12 to provide the needed headroom.

Figure 7-14 shows how to frame a dormer in pitched roof construction. To build a small gable dormer, double the rafters at each side of the dormer. Then rest the side studs and the short valley rafters on these members. Tie the valley rafters to the roof framing at the roof with a header. You can also carry the side studs past the rafter so they bear on a sole plate nailed to the

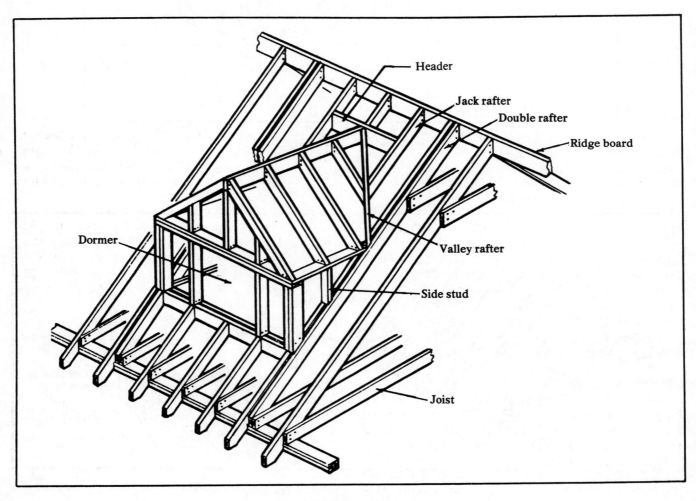

Typical dormer framing
Figure 7-14

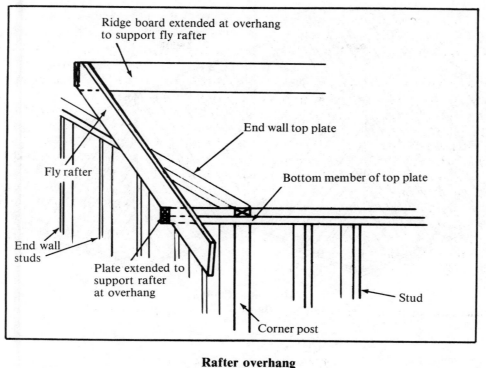

Ridge board extended at overhang
to support fly rafter

End wall top plate

Bottom member of top plate

Fly rafter

End wall
studs

Plate extended to
support rafter
at overhang

Stud

Corner post

Rafter overhang
Figure 7-15

floor framing and subfloor. Frame the side walls of shed dormers the same way.

With a gable (or rake) overhang, use a fly rafter beyond the end rafter. Fasten it with blocking and sheathing, or by extending ridge boards and top plates. Figure 7-15 shows an economical way to handle rafter overhangs. Extend the ridge and the bottom member of the side-wall top plate out to fit flush with the outside edge of the fly rafter. Then add a short top member at the extension.

Adding dormers to a home is usually money well spent. If there are dormers on the house, adding an attic room is relatively easy and inexpensive. This is a bonus point when selling the house, and should help it sell faster or at a higher price.

Hip Roofs

A third style in roof design is the hip roof. See Figure 7-16. Hip roofs are framed the same as gable roofs at a house's center section. Look at Figure 7-17 to understand hip roof framing.

Now look at Figure 7-18 for a top view of that same hip roof, which is actually two roofs intersecting. Notice the different names of rafters. Hip rafters extend from each outside corner of the wall to the ridge board at a 45-degree angle. Figure 7-19

shows a detail of a hip roof corner. See how the end of the hip rafter is cut so that a cornice can be nailed on. Also notice that the end joist isn't framed at the edge, but is set farther in so it won't block the hip jacks.

A *valley* is formed wherever two intersecting roofs meet. Valleys can be found in more than just hip roof construction, but let's talk about them here. The key member of valley construction is the valley rafter, as shown in Figure 7-20. When two equal-sized roof sections intersect, always double

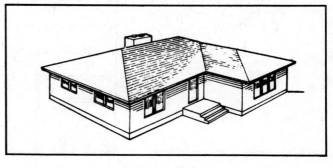

Hip roof
Figure 7-16

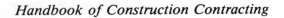

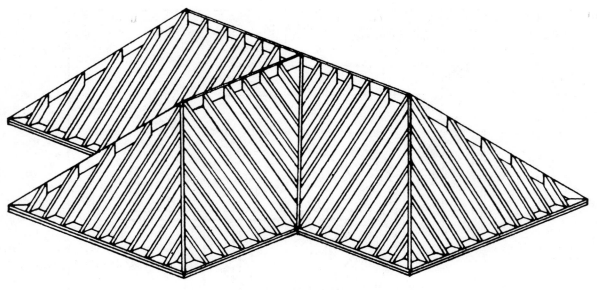

Hip roof framing
Figure 7-17

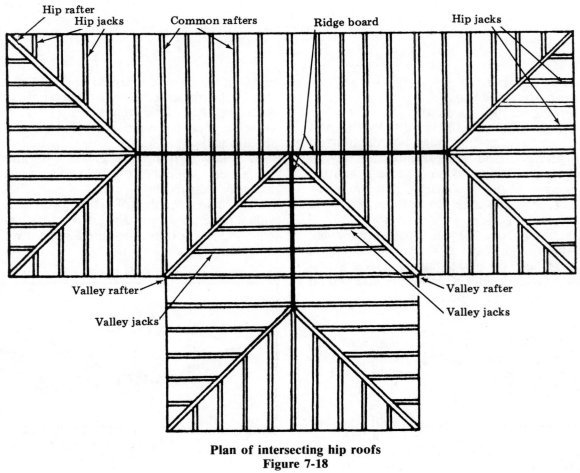

Plan of intersecting hip roofs
Figure 7-18

the valley rafter to carry the roof load. You should also make the valley rafter 2" deeper than the common rafter to provide full contact with the jack rafters. Nail the jack rafters to the ridge and toenail them to the valley rafter with three 10d nails.

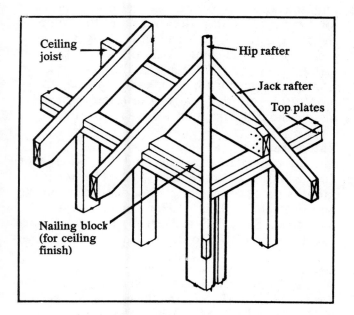

Framing at a hip roof corner
Figure 7-19

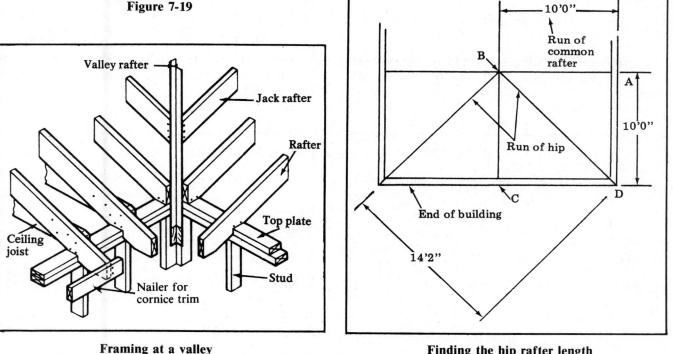

Framing at a valley
Figure 7-20

Collar beams are usually put between opposing rafters when roofs have long spans and flat slopes. Steeper slopes and shorter spans may also require collar beams, but only on every third rafter. In this case, use 1 x 6's for collar beams, nailed with 8d nails on every third pair of rafters. For 1½ story houses, 2 x 4 (or larger) members at each pair of rafters also act as ceiling joists for the finished rooms.

Figure 7-21 shows how to find the intersection of the hip rafters with the ridge board at point B. Notice that the distance AB is equal to BC. The two squares are equal in size. The distance DB is the diagonal of the square ABCD and shows the run of the hip rafter.

In regular hip roof construction, the length of the ridge is the building length minus the building width.

Example: A 40'-long building has a regular hip roof. The building is 20' wide. What's the length of the ridge?

Answer: Ridge = length minus width
= 40' − 20'
= 20'

Finding the hip rafter length
Figure 7-21

Now you can determine the length of the hip rafter. The hip rafter (Line BD, Figure 7-21) runs at a 45-degree angle to the sides of the building. It's longer than the common rafters, because of the extra distance from the corner of the building to the center of the span. There are several ways to find the length of the hip rafter. One of the simplest is to use the rafter table on the framing square.

Look ahead to the rafter table in Figure 7-41. If you know the rise, you can find the length of the hip rafter. Assume the pitch is 8/12, which is 8" rise for each 12" run. (We'll get into the details of the pitch a little later.)

Below the 8" mark on the body of the square is the "Length Common Rafters Per Foot Run," in this case 14.42". The next line, "Length Hip or Valley Rafter Per Foot Run," reads 18.76".

The run for the common rafter in Line AB, Figure 7-21, is 10'. To obtain the common rafter length, multiply the per-foot run (14.42") by 10' (the run).

$$14.42" \times 10' = 144.2" = 12.01'$$

We find the length of the hip rafter in the same way. From the framing square, we take the second figure under the 8" mark, 18.76, and multiply by 10' (the run of the *common* rafter).

$$18.76" \times 10' = 187.6" = 15.633' = 15'7\text{-}5/8"$$

These lengths for the common rafter and the hip rafter are the mathematical lengths: the length from the outside edge of the wall plate to the *center*

of the ridge board. The actual length is determined by subtracting *half* the thickness of the ridge board. This gives the true length of the rafter along the measuring line.

Rafter Run and Rise

Rafters make up the main body of the frame in all pitched roofs. As we have seen, rafters do the same job for a roof as joists do for a floor: they support sheathing and covering material. As with joists, there are several kinds of rafters — common rafters, headers, trimmers, etc. Each is similar in principle to the corresponding type of joist. Common rafters are the straight-run rafters which extend without interruption from eave to ridge.

Rafter Run

For an explanation of this and the other roof dimensions, look at Figure 7-22. The *run* of a rafter is the level, or horizontal, distance from the outside of the wall plate to the point straight under the ridge of the building. The run of a rafter is one-half the span of the roof.

Rafter Rise

The *rise* of a rafter is the vertical distance from the top of the wall plate to the measuring line on the ridge. Not all blueprints show the total rise of a rafter. However, you can find it by scaling or computation. To figure the rise, calculate the distance from the top of the plate line to the point where the measuring line (shown by the dotted line in Figure 7-22) meets the ridge board. The overall rafter length usually includes the overhang, as shown in Figure 7-23.

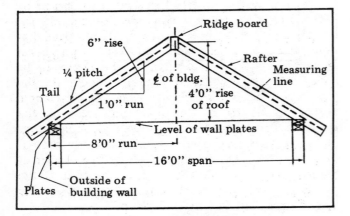

**Roof definitions
Figure 7-22**

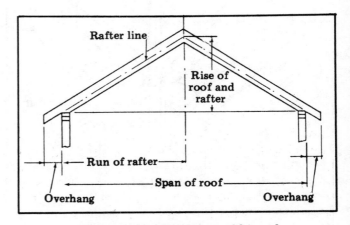

**Rise and run of rafter—gable roof
Figure 7-23**

Span

The roof *span* is the distance between the outer faces of the wall plates that support the rafters. It's always marked on plans. In a gable roof with a centered ridge, the span is twice the run of a rafter.

Figures 7-23, 7-24, and 7-25 show run, rise, and span for different types of roofs.

Here's a way to figure out the pitch: In Figure 7-22 the rise is 4' and the span 16'. The ratio of these measurements is equal to 4/16. Reduce this to a smaller fraction by dividing both numbers by a common denominator (the largest number that will go evenly into each of them). In this case, that number is 4, so we have 1/4 pitch.

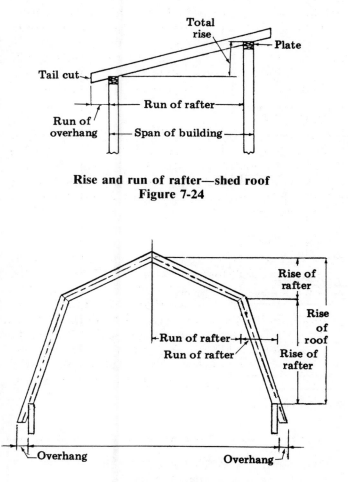

Rise and run of rafter—shed roof
Figure 7-24

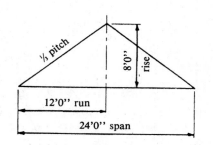

One-third pitch
Figure 7-26

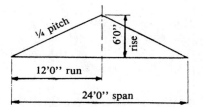

One-quarter pitch
Figure 7-27

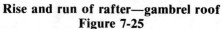

Rise and run of rafter—gambrel roof
Figure 7-25

Roof Pitch

The *pitch* of a roof is the slant, or slope, from ridge to plate. Because this slope changes if the rise (height of ridge) or span changes, we express it as a *ratio of the rise to the span.* For example, a "1/3 pitch" roof means the rise is 1/3 of the span, and "1/4 pitch" means that the rise is 1/4 of the span. See Figures 7-26 and 7-27.

Another way to describe the pitch of a roof is: *the increase in the rise for each foot of run.* For example, a slope may be 6" per foot. This means that the rafter rises 6" for every foot of *run* (horizontal distance covered). The pitch of the roof can be given by a third method: a small triangle like the one near the top of Figure 7-28. (The triangle doesn't have to be to the same scale as the drawing.) For example, a 6" rise roof means that the rafters are pitched to rise 6" for every foot of run, as illustrated in this figure. Note the small right triangle: the base represents one foot of the rafter run and the short vertical leg shows rafter rise. Simply put, it is a 6/12 pitch.

Here's some information relating to roof pitches that you might find useful:

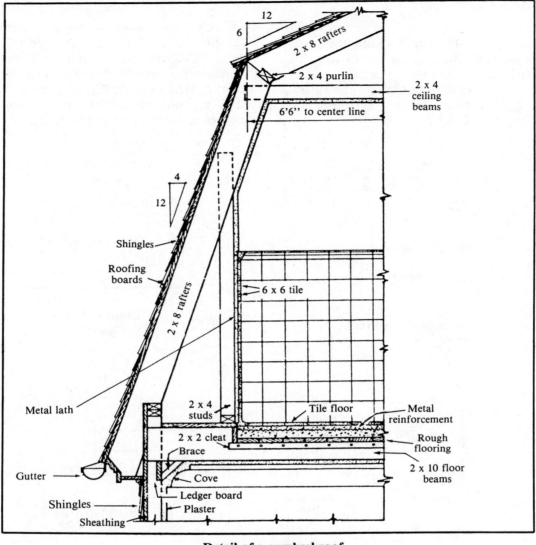

Detail of a gambrel roof
Figure 7-28

1) How many different ways can you express the rise of a roof rafter having a 1/3 pitch?

Answer: You can express the rise of the rafter as:

   (a) 1 foot for every 3 feet of *span.*
   (b) 1 foot for every 1½ feet of *run.*
   (c) 8 inches for 1 foot of *run.*

2) A simple way to find the rise of a roof when you know the pitch is: Multiply the pitch by the span.

Example: A building is 30' wide. The roof pitch is 1/3. What is the rise?

Answer: Rise equals 1/3 times 30', or 10'.

3) A simple way to find the pitch of a roof is: Divide the rise by the span.

Example: The total span of a building is 10'; the total rise of the roof is 3'4'' (3⅓'). What is the pitch?

Answer: Pitch equals 10' divided by 3⅓' which equals 1/3, or 1/3 pitch.

4) A simple way to find the number of inches of rise per foot of run is: Multiply the rise in feet by 12 to find the total number of inches in the rise; then divide the total number of inches in the rise by the length of the run in feet.

Example: Find the rise per foot of run of a roof when the span is 36' and the rise is 12'.

Answer: Since span is 36', the run is 1/2 of 36', or 18'. 12 times 12 equals 144, divided by 18 equals 8. Therefore the rise is 8" per foot.

The roof pitches in common use are shown in Figure 7-29. This figure shows a steel square (also called *carpenter's square* or a *framing square*). The square's body (blade) is up and the tongue is resting on a level surface. Figures on the body represent the inches of rise, and the 12 on the tongue represents 1 foot of run. A roof rises so many inches for each foot of run, and this determines the pitch. For example, 24" on the body and 12" on the tongue represent 1 pitch; 6" on the body would be 6/24 of 1 pitch, or 1/4 pitch; 8" on the body would be 8/24 of 1 pitch, or 1/3 pitch; 12" on the body would be 12/24 of 1 pitch, or 1/2 pitch; 18" on the body would be 18/24 of 1 pitch, or 3/4 pitch.

Rafter Overhang

The overhang (also called the eave or rafter tail) is the lower end of a rafter that extends beyond the building line. You must add its length to the calculated length of the rafter. If the finish for the overhang of a roof rafter isn't shown on the plans or mentioned in the specs, then you must decide what kind of design to use. If rafter ends will be enclosed by a cornice, you'll have to cut the ends accordingly.

Kinds of Rafters

Figure 7-30 is an isometric drawing, and Figure 7-31 is a framing plan of the same roof. The figures show several different kinds of rafters used in roof work.

In both figures, the letter "A" indicates the plate on which the rafters rest. The plate is important because its upper and outer edge is the line from which measurements are taken. "B" shows the ridge board; "C", a common rafter; "D", a hip rafter; "E", a valley rafter; "F", a hip jack rafter; "G", a valley jack rafter; and "H", a cripple jack rafter.

Since the dimensions are included, the framing plan in Figure 7-31 shows the location of the main framing members. The plan also shows the way to

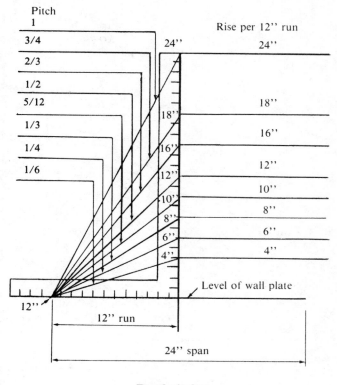

Roof pitches
Figure 7-29

build the intersections of hips and ridge, and of the valley and ridge.

A good set of plans should show the framing plan for the roof. If there isn't any roof framing plan, you have to improvise. First, draw the main members (the ridge, hips, valleys, and decks) on the second floor plan of the house. Then fill in the other small details. A simple roof plan like that will give even an inexperienced carpenter a general idea of how to lay out the roof.

Common rafters run all the way from the plate to the ridge, at right angles to both. They're called common rafters because they're common to all types of roofs. Hip rafters run diagonally from the plate to the ridge and form a hip in the roof. A valley rafter runs diagonally from the plate to the ridge and forms a valley at an inside corner in a roof. Jack rafters are short rafters that don't run full length from plate to ridge. They may run from ridge to hip, plate to hip, ridge to valley, or valley to plate. Jack rafters are, therefore, found on any kind of roof which is broken up in some way. Cripple jack rafters run between hip and valley rafters.

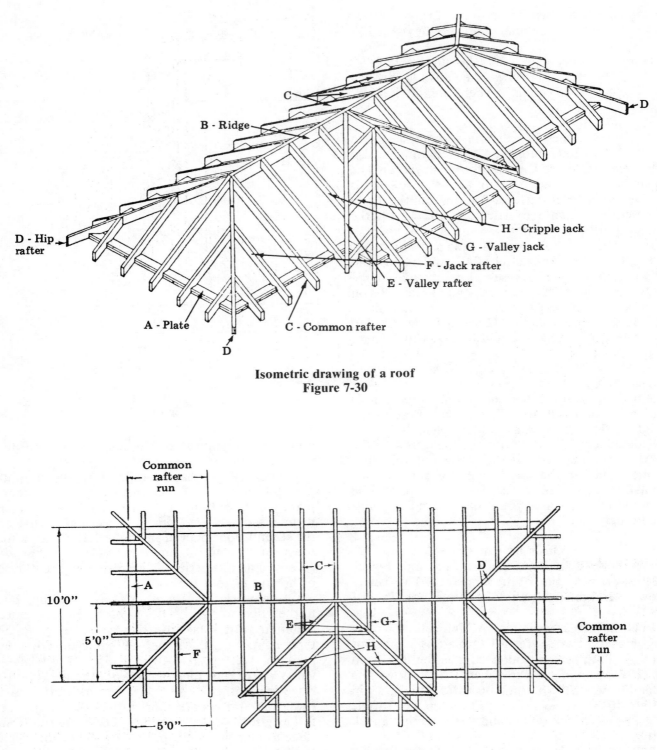

Isometric drawing of a roof
Figure 7-30

Framing plan of roof in Figure 7-30
Figure 7-31

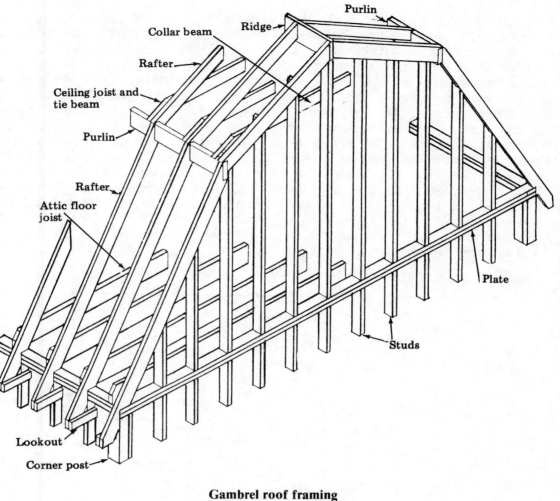

Gambrel roof framing
Figure 7-32

Gambrel Roof Rafters
The gable roof has two sloped surfaces running from ridge to plate; the gambrel roof has four such surfaces, two on each side of the ridge. The upper surface on each side is fairly flat, and the lower surface is fairly steep. The advantage of the gambrel roof is the additional space the gambrel gives, without any increase in the height of the ridge. Headroom, especially, is increased when you use a gambrel roof.

Figure 7-32 is a detail of the framing of such a roof. The only difference between gambrel roof framing and gable or hip roof framing is at the point where the rafters of the two slopes meet. You can frame this point two ways: with or without a purlin plate. You should use a purlin, as shown in

Figures 7-28 and 7-32, if you can support it with partition walls.

Cutting Rafters
There are several ways to find the length and cuts of roof rafters. However, the basic principle of all methods is geometric construction (trigonometry). Each of the three common methods — the graphic, the rafter table, and the step-off — works well. Use the method you find most convenient.

The architectural draftsman finds the graphic method easiest because the work can be done using drafting instruments. The carpenter finds the step-off method or the rafter table method convenient because the work is laid out with a framing square.

The step-off method of laying out rafters is the

most logical one for the carpenter. We'll also describe rafter tables, so you can check your work regardless of which method you use. However, to avoid confusion, learn only one rafter layout system at a time. When you've learned that system, then go on to learn another one.

Rafter layout is complicated. So always do the steps in the same order. Lay out all rafters in the same relative position. Crown each board by sighting along it to find the slight bow, or crown, near the middle of the board. The rafter will be stronger if the crowned edge is up. So hold the board with the crowned edge (top edge) toward you.

Hold the tongue (short leg) of the steel square in your left hand and the body (long leg) in your right hand. This way, the tongue will form the vertical, or top, cut of the rafter and the body will form the level, or seat cut, as in Figure 7-33.

Finding The Rough Length of a Common Rafter

You can use your square as a calculator to find the approximate length of a common rafter. The rise of the rafter (in feet) is represented on the tongue of the square. The run (in feet) is represented on the body. Measure the length of the diagonal between these two points. This measurement, expressed in feet, is the rough length of the rafter. For example, assume that the total rise of a rafter is 9' and the run is 12'. Find 9 and 12 on the square in Figure 7-34 and measure the diagonal. This will be 15'.

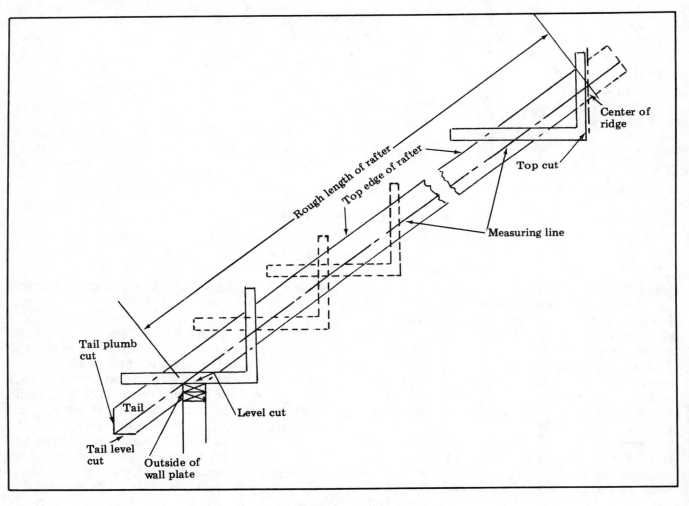

Common rafter
Figure 7-33

If you need cornice overhang, remember to add the overhang length. In the above example the rafter length is 15' without overhang. If the overhang will be 12'', you'll need 16' stock. Figure 7-35 is a conversion diagram you can use to find rough rafter lengths.

Locating the Measuring Line on a Rafter
Make the first pattern rafter out of straight stock of the right rough length. Lay this piece flat across two sawhorses and place the square near the right-hand end. See Figure 7-36.

The 12'' mark on the outside edge of the body (representing the run of the roof), and the inch mark on the outside edge of the tongue that represents the rise of the roof, should both come at the edge of the rafter. This is the top edge. See points A and F of Figure 7-36. Draw the line AB on

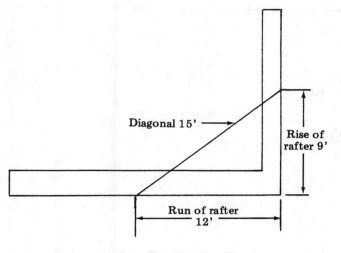

Rough length of rafter
Figure 7-34

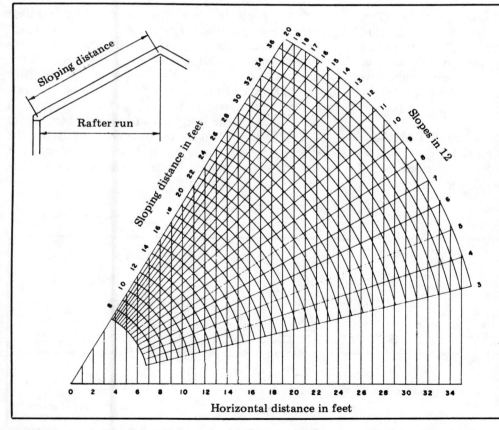

To use the diagram select the known horizontal distance and follow the vertical line to its intersection with the radial line of the specified slope, then proceed along the arc to read the sloping distance. In some cases it may be desirable to interpolate between the one foot separations. The diagram also may be used to find the horizontal distance corresponding to a given sloping distance or to find the slope when the horizontal and sloping distances are known.

Example: With a roof slope of 8 in 12 and a horizontal distance of 20 feet the sloping distance may be read as 24 feet.

Conversion diagram for rafters
Figure 7-35

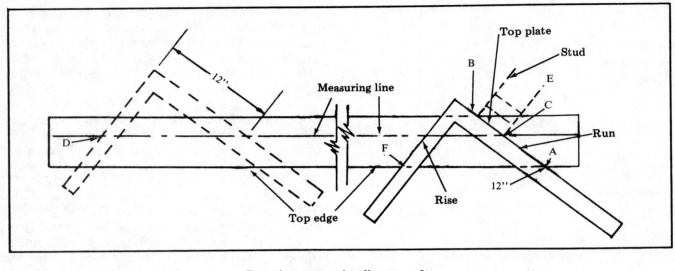

Locating measuring line on rafter
Figure 7-36

the plank to mark the top of the wall plate. Then measure 3⅝" along this line, from B to C, to find the outside top corner of the plate. This should be far enough from the right-hand end of the rafter to allow for the tail (overhang). Then gauge the measuring line (CD) parallel to the edge of the rafter.

Laying Out a Common Rafter (Step-Off Method)
When using the step-off method to lay out a rafter, it helps to fasten a wooden fence or small metal clamps to the tongue and body of the square. We'll use a fence in our examples. Many skilled carpenters don't use a fence or clamps. Instead, they carefully place the square at each step.

Our sample building is 24' wide. The pitch of the rafter will be 3/8, or 9" of rise on 12" of run. Here's how to find the exact length of the rafter:

Lay the square on the stock so the 12" mark on the outside of the body is at point C, as in Figure 7-37. Put the 9" mark on the outside edge of the tongue on the measuring line at E. It's important that these two marks are *exactly* on the measuring line for the following step.

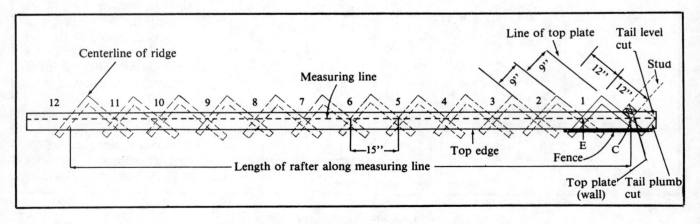

Laying out a common rafter (step-off method)
Figure 7-37

Put the fence against the rafter stock, as shown at the far right of Figure 7-37. Keep the square from moving and adjust the fence so its edge lies against the top edge of the rafter. Tighten the fence on the square. Mark along the outside edges of the square's tongue and body. Now you have point E on the measuring line. Slide the square to the left until the 12" mark is over E (position 2 on Figure 7-37) and again mark along the tongue and body. Do this as many times as there are feet run of the rafter (12 in this case). The successive positions of the square are shown by the numbers 1 through 12 on Figure 7-37.

When you reach the last position, draw a line along the tongue across the rafter to mark the centerline of the ridge board. Do each step in this whole process carefully, and make all marks clear. Even a small mistake will greatly affect the fit and length of the rafter.

Laying Out a Rafter for an Odd Span

Assume that a building span is an odd number, 25' wide. The pitch of the rafter is 3/8, or 9" rise to 12" run (9/12 pitch). The run of the rafter will be one half the width of the building, or 12'6".

You'll lay out this rafter the same as the one just described, *except* that at the end you'll take an extra half step for the added 6" of the rafter. After the twelfth step is marked, put the square on the top edge of the rafter as shown in Figure 7-38. Then move the square until the 6 on the outside of the body is right over line A from the 12th step. Draw a line along the outside edge of the tongue. This line shows the centerline of the ridge board.

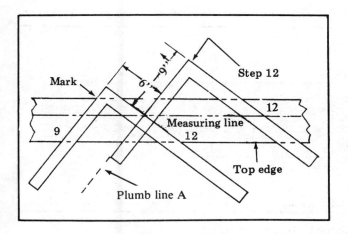

Additional half-step
Figure 7-38

Allowance for Ridge Board

The last line you marked on the rafter shows where you'd cut the rafter, *if* all rafters were to be butted against each other without a ridge board. Since most roofs have a ridge board, the rafter you just laid out will be a little too long. You have to cut a piece off the end of the rafter. This piece will be equal to half the thickness of the ridge board, as shown in Figure 7-39.

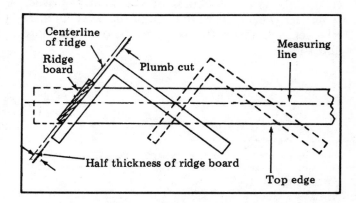

Allowance for ridge board
Figure 7-39

To lay out this cutting line, measure half the thickness of the ridge board at a right angle to the last line (position 12 in Figure 7-37). Then slide the square back for this distance. Keep the fence tight against the rafter. When you've moved the square back half the thickness of the ridge board, mark the plumb cut along the edge of the tongue. This line should be inside of, and parallel to, the original line marking the end of the rafter.

Bottom or Seat Cut

The bottom, or seat, cut of the rafter (called "birdsmouth" by some carpenters) is a combination of level and plumb cuts. The level cut (BC, Figure 7-36) rests on the top face of the side wall plate. The plumb cut (CE) fits against the outside edge of the wall plate. Lay out the plumb cut by squaring a line from line AB, Figure 7-36, through point C. This line CE then represents the plumb cut.

Cutting the Tail

If the rafter tail is the kind shown in Figure 7-37, make the plumb cut along the measuring line as shown. Find the level cut at the end of the tail (see Figure 7-33) by sliding the square toward the tail. Mark the rafter where the body crosses both the plumb and measuring lines.

For a box cornice as shown in Figure 7-40, you have to cut the rafter tail differently. Lay out the level and plumb cuts as before. Then make allowance for the thickness of the sheathing, since it extends into the rafter notch up to the top of the plate. To do this, lay out line AB parallel to line CD. Make it the thickness of the sheathing away from CD. Then extend the line of the level cut at D to meet AB. The rafter will now fit over the plate and sheathing.

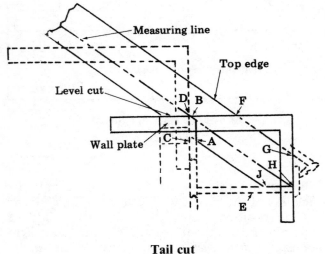

Tail cut
Figure 7-40

Now, assuming that a 12″ piece will be used at E, continue the line of the level cut through B to F on the top edge of the rafter. Lay the body of the square along this line. Put the 12″ mark of the outside edge directly over point B. Mark a line GH along the outer edge of the tongue across the rafter. The tip of the rafter is at the spot where this line meets the measuring line at H. Square the line HJ across the rafter from line GH to locate J.

Rafter Tables

Many steel squares have rafter tables stamped on them. The manufacturer of the square you use probably offers a booklet of instructions on how to use these tables. However, you can figure correct rafter lengths even without an instruction book. Here's how: Once you know the rise and run of the rafter, you can find the length of the rafter, along the measuring line, from the rafter table. See Figure 7-41. Use the inch marks on the outside edge of the square to show rise per foot of run.

For example, the figure 8 (the 8″ mark) means 8 inches of rise per foot of run. Right below each inch mark is the length of main rafters per foot of run. In this case, 14.42″. If the rafter run is 10′, multiply the 14.42 by 10. This gives 144.20″, or 12.01′, the length of the rafter from the center of the ridge board to the outside surface of the top plate. Now you can lay out the rafter cuts as discussed in the step-off method.

For a complete manual on framing roofs, get a copy of *Roof Framing,* published by Craftsman Book Company. There's an order form at the back of this book. This manual shows how to use a

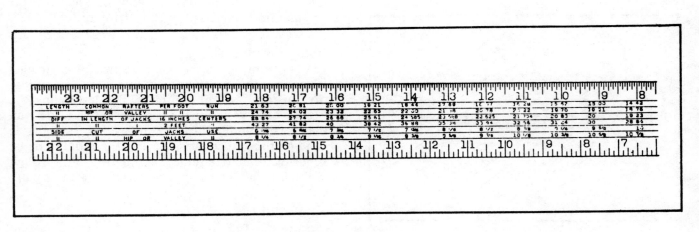

Rafter table
Figure 7-41

hand-held calculator to figure lengths and cuts for all types of roofs.

Roof Sheathing

Roof sheathing is the covering over the rafters or trusses. It should be thick enough to carry the weight of finish roofing between supports. Plywood is usually used, although 2''-thick wood roof planking or fiberboard roof decking might be used on some flat or low-pitched roofs with post and beam construction.

Plywood Roof Sheathing

Lay plywood roof sheathing so the long dimension is perpendicular to the rafters. Figure 7-42 shows how. Make end joints over the center of the rafter and stagger them by at least one rafter. Standard sheathing-grade plywood is usually specified. However, in humid areas you should use a standard sheathing grade with exterior glue line.

Nail plywood at each bearing, 6'' o.c. along all edges and 12'' o.c. along intermediate members. Unless plywood has an exterior glue line, the raw

edges shouldn't be exposed to the weather at the gable end (rake) or at the cornice. Protect the edges with the trim or metal drip cap. Leave 1/8'' edge spacing and 1/16'' end spacing.

Use the minimum allowable thickness of plywood that will effectively span roof framing members. Figure 7-43 shows the minimum plywood thickness for roofs covered with asphalt shingles, wood shingles, shakes, and built-up roofing if you use the metal clips discussed below. For slate and similar heavy roofing materials, 1/2'' plywood is the minimum for 16'' rafter spacing.

Plywood thickness (inches)[2]	Plywood Panel Identification Index[1]	Maximum span (inches)[3,4]
5/16	12/0	12
5/16	16/0	16
5/16	20/0	20
3/8	24/0	24
5/8	30/12	30
1/2	32/16	32
3/4	36/16	36
5/8	42/20	42
3/4	48/24	48

Notes:

(1) Applies to Standard, Structural I and II and C-C grades only conforming to U.S. Commerce Dept. PS 1-66.

(2) Use 6d common smooth, ring-shank or spiral-thread nails for ½-inch thick or less, and 8d common or 8d ring-shank or spiral-thread for plywood 1-inch thick or less.

(3) These spans shall not be exceeded for any load conditions.

(4) Provide adequate blocking, tongue and groove edges, or other suitable edge support such as metal fasteners. Use two metal fasteners for 48 inches or greater spans.

Plywood sheathing spans
Figure 7-43

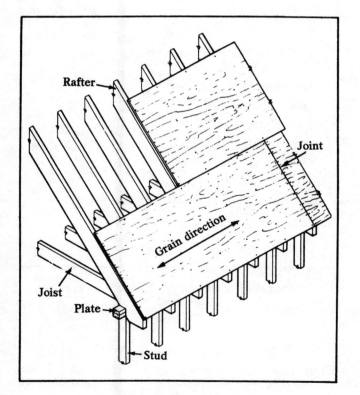

Plywood roof sheathing
Figure 7-42

Unsupported plywood edges that are perpendicular to the roof framing may be connected with special metal clips. Figure 7-44 shows how to use these clips. If the plywood is 1/8'' thicker than the minimum thickness for the span and grade, no clips or blocking are needed.

Eliminate unnecessary waste and cutting by planning your plywood sheathing layout in advance.

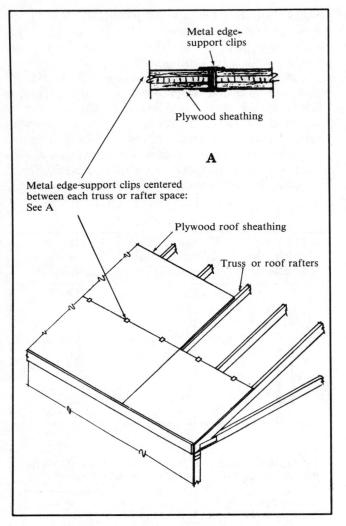

Eliminating plywood edge blocking
Figure 7-44

Fiberboard Roof Decking

Fiberboard roof decking is used the same way as wood decking. Planking is usually supplied in 2 x 8 sheets and with T&G edges. Thickness of the plank and spacing of supports ordinarily comply with the following schedule:

Minumum thickness (inches)	Maximum joist spacing (inches)
1½	24
2	32
3	48

Manufacturers of some types of roof decking recommend that you use 1⅞" thickness for 48" support spacing.

Use corrosion-resistant nails to fasten the fiberboard to the wood members. Don't space them more than 5" o.c. They should be long enough to penetrate the joist or beam at least 1½". Flat and low-pitched roofs with wood or fiberboard decking usually have built-up roofing.

Spaced Sheathing

When wood shingles or shakes are used in damp climates, the roof boards are usually spaced. Figure 7-45 illustrates both spaced and closed board sheathing. Spacing lets air circulate under the roof to dry the shingles or shakes.

Space 1 x 3 or 1 x 4 wood nailing strips the same distance o.c. as the shingles are to be laid to the weather. For example, if shingles are laid 5" to the weather over 1 x 4 strips, there would be spaces of 1⅜" to 1½" between each board for ventilation.

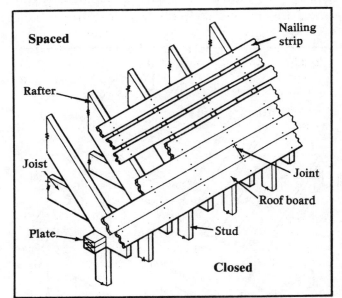

Board roof sheathing: both spaced and closed
Figure 7-45

Roof Coverings

Roof coverings should give a durable, waterproof finish to protect the building and its contents from rain, snow, and wind. Many time- and weather-tested materials are available.

Wood shingles, asphalt shingles, and built-up roofs are the most commonly used roof coverings. Less common materials (including asbestos, slate, tile, metal, and others) may need specialized applicators. Several new materials such as plastic films and coatings are showing promise for future moderate-cost roof coverings. These materials are more expensive than currently popular materials. However, they are likely to be used more in the near future.

Wood, asphalt, asbestos shingles, tile, and slate are all used for pitched roofs. Sheet materials like roll roofing, galvanized iron, aluminum, copper, and tin are also used. The most common covering for flat or low-pitched roofs is the built-up roof with a gravel topping or a cap sheet. Your choice of roofing materials will be influenced by the cost, local code requirements, house design, or your own past experiences.

In shingle application, the exposure distance is important. The amount of exposure depends on the roof slope and the material used. This may vary from a 5'' exposure for standard-size asphalt and wood shingles on a moderately steep slope to about 3½'' for flatter slopes. However, asphalt shingles can be used on low slopes with double underlay and triple shingle coverage. Built-up construction is used mainly for flat or low-pitched roofs but can be adapted to steeper slopes by the use of special materials and methods.

Roof underlay material is usually 15- or 30-pound asphalt-saturated felt. Use it on moderate and low-slope roofs covered with asphalt, asbestos or slate shingles, or tile roofing. It isn't common with wood shingles or shakes.

To install underlayment in a valley, first center a 36''-wide strip of 15-pound asphalt-saturated felt in the valley. Use only enough nails to hold it in place. Then, trim the horizontal courses of felt underlayment applied on the roof so they overlap the valley strip by at least 6''.

Ice dams can occur in areas of moderate to heavy snowfalls if the cornices aren't properly protected. Ice dams form when snow melts, runs down the roof, and freezes at the colder cornice area. Gradually the ice forms a dam that backs up water under the shingles, as shown in section A in Figure 7-46. You can reduce the chance of water backing up and entering the wall: Use an undercourse (26'' width) of 50-pound or heavier smooth-surface roll roofing along the eave line as flashing, as in Figure 7-46 B. However, the best way to eliminate ice

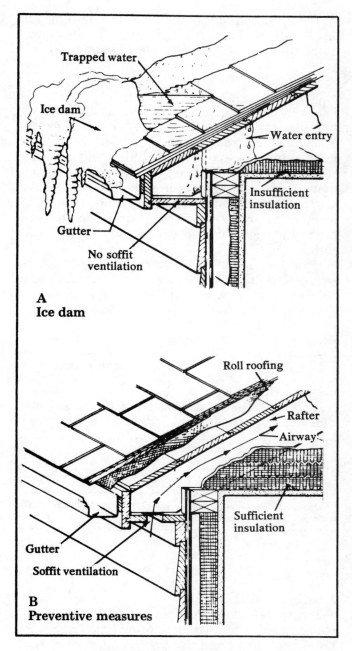

Winterizing the cornice
Figure 7-46

dams is with proper attic ventilation and good ceiling insulation.

Metal roofs (tin, copper, galvanized iron, or aluminum) are sometimes used on flat decks of dormers, porches, or entryways. Make joints watertight. And flash the deck properly where it meets the house. Nails should be of the same metal as the roof, with one exception. Steel nails may be used with tin roofs. All exposed nailheads in low-slope tin roofs should be soldered with rosin-core solder.

Wood Shingles

No. 1 grade wood shingles are most common on house roofs. They are all-heartwood, all-edge-grain, and tapered. Second grade shingles make good roofs for secondary buildings and excellent sidewalls for main buildings. Shingles come in random widths, the narrower shingles being in the lower grades. Western red cedar, cypress, and redwood are the main commercial shingle woods: their heartwood has high decay resistance and low shrinkage.

Four bundles of 16" shingles laid 5" "to the weather" will cover 100 square feet. Recommended exposure for the standard shingle sizes are shown in Figure 7-47.

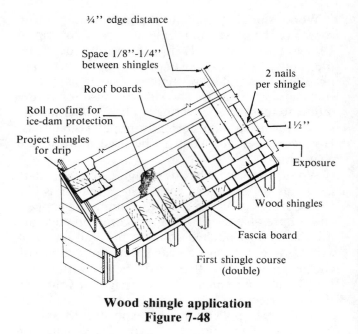

Wood shingle application
Figure 7-48

		Maximum exposure	
Shingle Length (inches)	Shingle Thickness (Green)	Slope Less[1] than 4 in 12 (inches)	Slope 4 in 12 and over (inches)
16	5 butts in 2"	3¾	5
18	5 butts in 2¼"	4¼	5½
24	4 butts in 2"	5¾	7½

[1] Minimum slope for main roofs — 4 in 12. Minimum slope for porch roofs — 3 in 12.

Recommended exposures for wood shingles
Figure 7-47

Figure 7-48 shows the right way to apply a wood-shingle roof. Underlay or roofing felt is necessary only in ice-dam areas, for protection. Spaced roof boards under wood shingles are most common, particularly in damp areas. In other areas, spaced or solid sheathing is optional.

In some areas, wood shingles increase fire insurance rates, and local codes may not even allow wood shingles in areas with high fire risk.

Rules for applying wood shingles:

1) Shingles should extend about 1½" beyond the eave line and about 3/4" beyond the gable (rake) edge.

2) Use two rust-resistant nails in each shingle; space them about 3/4" from the edge and 1½" above the butt line of the next course. Use 3d nails for 16" and 18" shingles and 4d nails for 24" shingles. I recommend a ring-shank nail (threaded) if plywood roof sheathing is less than 1/2" thick.

3) Double the first course of shingles. In all courses, allow 1/8" to 1/4" space between each shingle for expansion when wet. The joints between shingles should be offset at least 1½" from the joints between shingles in the course below. Further, space the joints in succeeding courses so they don't directly line up with joints in the second course below.

4) On a hip roof, shingle away from the valleys, selecting and precutting wide valley shingles.

5) A metal edging along the gable end will help guide water away from the sidewalls.

6) When laying No. 1 all-heart, edge-grained shingles, you don't need to split wide shingles. Top-quality wide shingles are less likely than lower-quality shingles to split due to normal expansion and shrinkage.

Wood Shakes

Wood shakes are applied much the same as wood shingles. To create a rustic appearance, lay the butts unevenly.

Because shakes are longer than shingles, they have greater exposure. Exposure distance is usually 7½" for 18" shakes, 10" for 24" shakes, and 13" for 32" shakes. Shakes are thick, and the longer shakes have thicker butts. So you'll need to use long galvanized nails.

Shakes aren't smooth on both sides (faces). In areas with wind-driven snow, you'll need to build in some extra protection. Use solid sheathing under the shakes. Also, it's essential to use underlay between each course. Put a 36" wide starting strip of 30-pound asphalt felt at the eave line. Use an 18" wide layer of the felt between each course. Put the bottom edge of the felt above the butt edge of the shake at a distance equal to double the weather exposure.

Asphalt Shingles

Asphalt shingles are the most popular roof covering for homes. They're available with an organic or glass fiber base. The trend is toward shingles with the glass fiber base.

The asphalt strip shingle is available in many styles:

• Self-sealing random-tab multi-thickness shingle weighing 285 to 390 pounds per square

• Self-sealing random-tab single-thickness shingle weighing 250 to 300 pounds per square

• Self-sealing square-tab three-tab shingle weighing 215 to 300 pounds per square

• And many other configurations, including individual interlocking shingles. Asphalt strip shingles are also available without self-sealing, factory-applied adhesives.

There are 66 to 90 shingles to the square, depending on weight. Shingles are packed three to five bundles per square, depending on the type of shingle. A square covers 100 square feet. Follow the manufacturer's instructions (usually printed on the wrapper) for storage, handling, and application. Strip shingles are usually laid with 4" to 6" exposure.

First, cover the roof sheathing with 15-pound asphalt-saturated felt underlayment. (Some local codes may require 30-pound felt underlayment.) Apply the felt when the deck is dry. Do not use coated felts, tar-saturated materials, or laminated waterproof papers. Any of these could act as a vapor barrier to trap moisture or frost between the covering and the roof deck. In areas where ice-dams are likely, flash the eave area as previously discussed.

Applying asphalt shingles— Put a starter strip under the first course. The starter strip protects the roof by filling in the spaces under the cutouts and joints of the first course of shingles. You can use a row of shingles trimmed to the shingle manufacturer's recommendations, or a strip of mineral-surfaced roll roofing at least 7" wide. It should overhang the eaves and rake edges by 1/4" to 3/8". See Figure 7-49.

The first course is the most critical. Be sure it is laid perfectly straight. Check the alignment often against a horizontal chalk line. You can mark a few vertical chalk lines, aligned with the ends of the first-course shingles, to insure proper alignment of cutouts.

Start the first course with a full shingle. But start succeeding courses with portions removed according to the style of shingle and the pattern you want. Save the pieces cut from the first shingle in each course. If they're full tabs you can use them for finishing the opposite end of the course, and for hip and ridge shingles.

Align each shingle carefully. Drive nails straight so that the edge of the nail head doesn't cut into the shingle. Also, drive the nail heads flush with the shingle surface, not sunk into it. Don't nail into or above factory-applied adhesives.

Start nailing from the end nearest the shingle last laid, then work across. This prevents buckling. Try to ensure that no cutout or end joint is less than 2" from a nail in an underlying course. *Never* try to re-align a shingle by shifting the free end after two nails are in place.

To get the correct exposure for square-tab strip shingles, align the butts with the top of the cutouts in the course below. For correct exposure for no-cutout shingles and those with variable butt lines, follow the manufacturer's directions.

Three-tab strip shingles— The most popular strip shingle is the three-tab square butt. The shingles measure 12" to 13¼" wide and 36" to 40" long, depending on the weight.

There are three methods of applying these shingles: the 6'' method, 5'' method, and the 4'' method. The names refer to the amount removed from the first shingle in each successive course to get a desired pattern. By removing different amounts from the first shingle, cutouts in one course of shingles don't line up directly with those in the course below.

Figure 7-49 shows the 6'' method, Figure 7-50 the 5'' method, and Figure 7-51 the 4'' method.

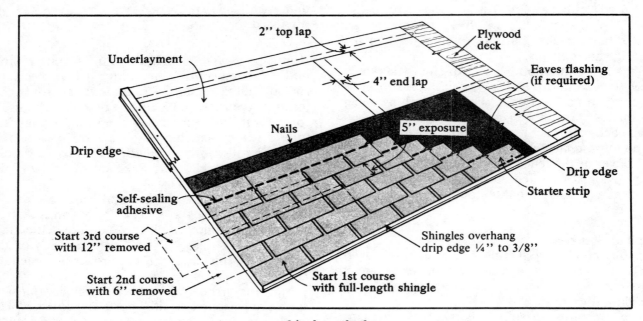

6-inch method
Figure 7-49

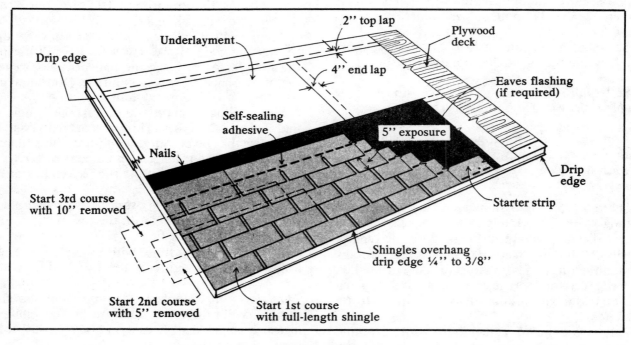

5-inch method
Figure 7-50

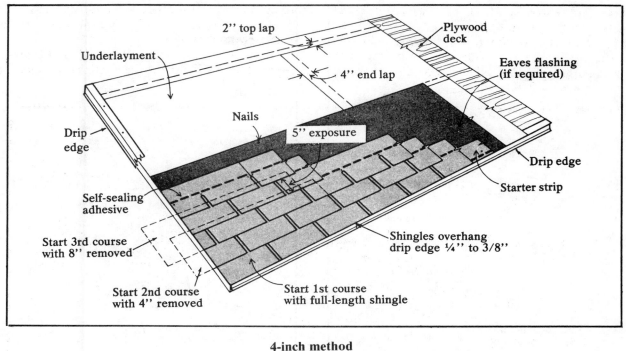

4-inch method
Figure 7-51

Valley Flashing

Valleys are formed where the sloping roof planes join at an angle. The sloping planes direct water toward the valley. Thus, drainage is concentrated along the joint, which makes it especially likely to leak. Therefore, one of the most important parts of roof installation is proper valley flashing.

There are several different ways of shingling valleys, including the open, woven, and closed-cut methods. Woven or closed-cut valleys are the best treatment for strip shingles. For all methods, valley flashing should be finished before the shingles are put on, except for open valleys around dormers. There, the valley flashing must overlap the top courses of shingles along the dormer sidewalls. Figure 7-52 shows how to install flashing around a dormer.

Open Valleys

The flashing material I recommend is 90-pound mineral surface roll roofing. Choose a color that either matches the shingles or is neutral, because the flashing will show. Apply the valley flashing in two layers. Refer to Figure 7-53 for details.

Center the first layer, 18" wide, in the valley, with the mineral surfacing *down*. (See Figure 7-53

again.) Trim the lower edge flush with the eaves' drip edge. Install this first layer up the entire length of the valley. If you need two or more strips of roll roofing, lap the upper piece over the lower so that drainage will be carried over the joint, not into it. The overlap should be 12" and fully bonded with asphalt plastic cement. Nail along a line 1" from each edge. Use only enough nails to hold the strip in place. Start at one edge and work all the way up. Then return to nail the other side, pressing the flashing strip firmly into the valley at the same time.

After you've nailed the 18" strip, center a second strip 36" wide in the valley over the first strip. This time have the mineral surface facing *up*. Nail the strip in place the same way you did the one under it. Overlaps should be 12" and cemented. The valley will be completed as the shingles are installed.

Woven and Closed-Cut Valleys

Cover both types of valleys with 36" wide mineral or smooth-surfaced roll roofing, 50 pounds or heavier. Center the strip in the valley, over the underlayment. Use only enough nails to hold it in place. Nail the strip along a line 1" from the edges,

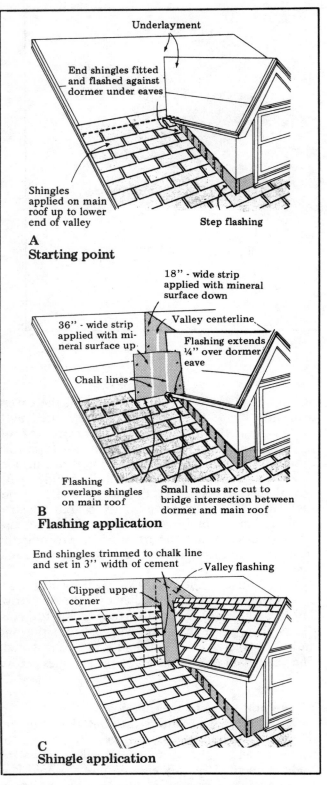

A
Starting point

Underlayment

End shingles fitted and flashed against dormer under eaves

Shingles applied on main roof up to lower end of valley

Step flashing

B
Flashing application

18" - wide strip applied with mineral surface down

Valley centerline

36" - wide strip applied with mineral surface up

Flashing extends ¼" over dormer eave

Chalk lines

Flashing overlaps shingles on main roof

Small radius arc cut to bridge intersection between dormer and main roof

C
Shingle application

End shingles trimmed to chalk line and set in 3" width of cement

Valley flashing

Clipped upper corner

Open valley flashing at dormer roof
Figure 7-52

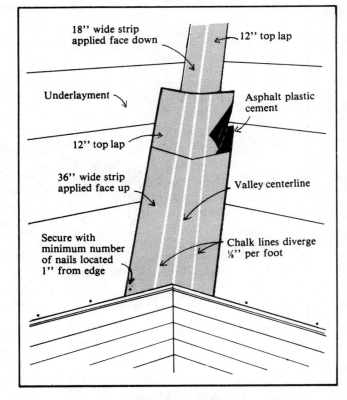

18" wide strip applied face down

12" top lap

Underlayment

Asphalt plastic cement

12" top lap

36" wide strip applied face up

Valley centerline

Secure with minimum number of nails located 1" from edge

Chalk lines diverge ⅛" per foot

90-pound roll roofing as open valley flashing
Figure 7-53

first on one edge all the way up, then on the other, while pressing the flashing strip firmly and smoothly into the valley. Laps should be 12" and cemented. The valley will be completed as the shingles are installed.

Figure 7-54 shows how to apply shingles in the woven-valley method, and Figure 7-55 shows the closed-cut method. Figure 7-56 shows three ways to finish a roof at the ridge. Or you can use a ridge vent instead of shingles or a metal ridge.

Built-Up Roofs

Built-up roofing is installed by roofing contractors who specialize in this work. Roofs may have three, four, or five layers of roofer's felt, each hot-mopped down with tar or asphalt. The final surface is coated with asphalt. Then it's either covered with gravel embedded in the asphalt or in tar, or covered with a cap sheet. Built-up roofs are called 10, 15, or 20-year roofs, depending upon the method of application.

For example, a 15-year roof over a wood deck (see Figure 7-57 A) may have a base layer of

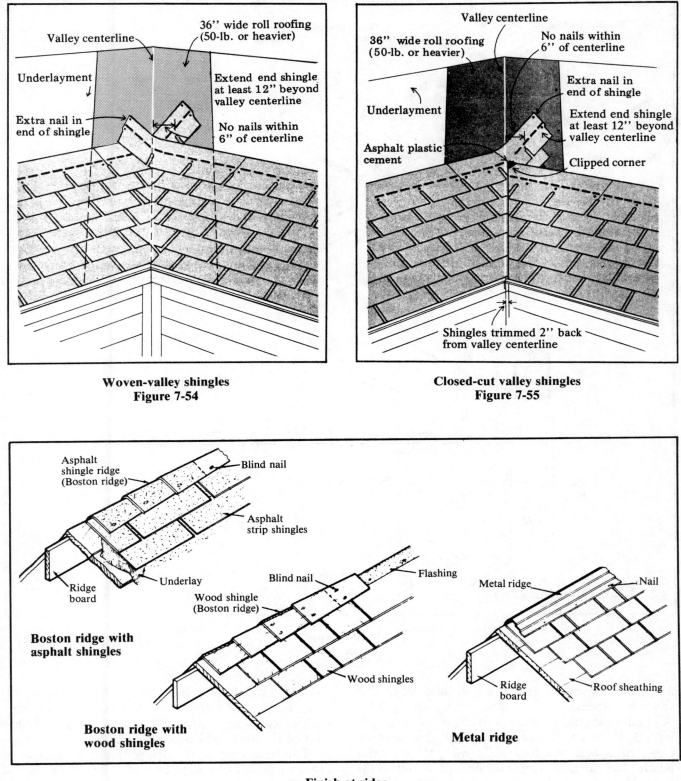

Woven-valley shingles
Figure 7-54

Closed-cut valley shingles
Figure 7-55

Finish at ridge
Figure 7-56

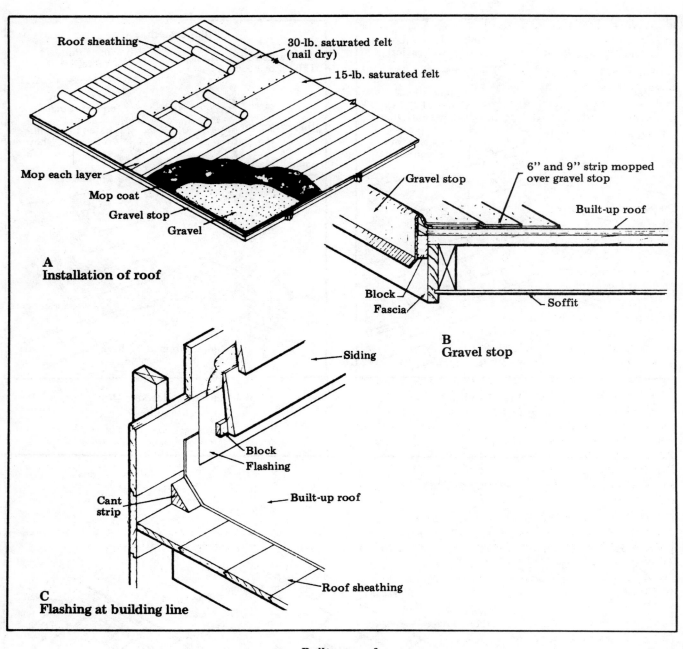

Built-up roof
Figure 7-57

30-pound saturated roofer's felt laid dry. The edges are lapped and held down with roofing nails. All nailing should be done with either roofing nails with 3/8" heads driven through 1" diameter tin caps, or special roofing nails with 1" diameter heads. The dry sheet keeps tar or asphalt out of the rafter spaces. This sheet is covered with three layers of 15-pound saturated felt. Each layer is mopped on with hot tar rather than nailed. Finish the cornice or eave line of projecting roofs with metal edging or flashing, which acts as a drip.

The final coat of tar or asphalt is covered with roofing gravel or a cap sheet of roll roofing. If the roof is covered with gravel, use a metal gravel strip along with the flashing at the eaves, as shown in B in Figure 7-57.

When built-up roofing is finished against another wall, the roofing is turned up on the wall sheathing over a cant strip. Often it's also flashed with metal as in Figure 7-57 C. This flashing usually extends up about 4'' above the bottom of the siding.

Flashing Around Chimneys

To prevent problems that uneven settling can cause, chimneys are usually built on a separate foundation from that of the main structure. This doesn't eliminate possible differences in settling between the chimney and the main structure. But it does free the chimney from the stresses and distortions that would occur if both were on the same foundation.

Because of the difference in settling, the flashing where the chimney projects through the roof must be built to allow movement without damage to the water seal. To do this you must fasten the base flashings to the masonry of the chimney. If movement occurs, the cap flashing slides over the base flashing without affecting water runoff.

Chimneys that go through the roof surface should have a cricket (or wood saddle) installed where the back face of the chimney and the roof deck meet. The cricket is important for the life of the flashing. It prevents the build-up of ice and snow at the rear of the chimney and diverts water runoff around the chimney.

Apply shingles up to the front edge of the chimney before installing any flashings. Also, put a coat of asphalt primer on the chimney's brickwork to seal the surface. The primer will also improve the adhesion of the asphalt plastic cement you'll apply later.

Base flashing— Start with 26-gauge corrosion-resistant metal base flashing between the chimney and the roof deck on all sides. Apply the base flashing to the front first, as shown in Figure 7-58. Bend the base flashing so that the lower section extends at least 4'' over the shingles and the upper section extends at least 12'' up the face of the chimney. Work the flashing firmly and smoothly into the joint between the shingles and chimney. Set both the roof and chimney overlaps in asphalt plastic cement placed over the shingles and on the chimney face. You can use one or two nails driven into the mortar joints to hold the flashing against the chimney until the cement sets.

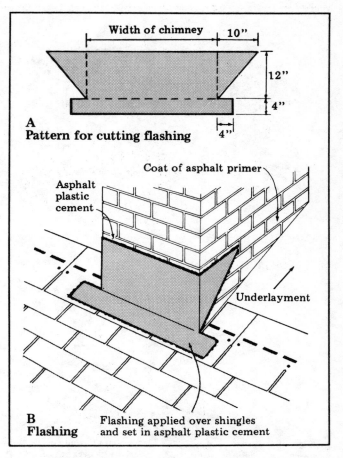

Base flashing at front of chimney
Figure 7-58

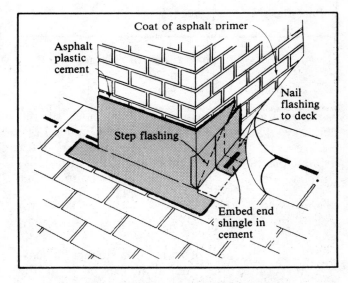

Base flashing at side of chimney
Figure 7-59

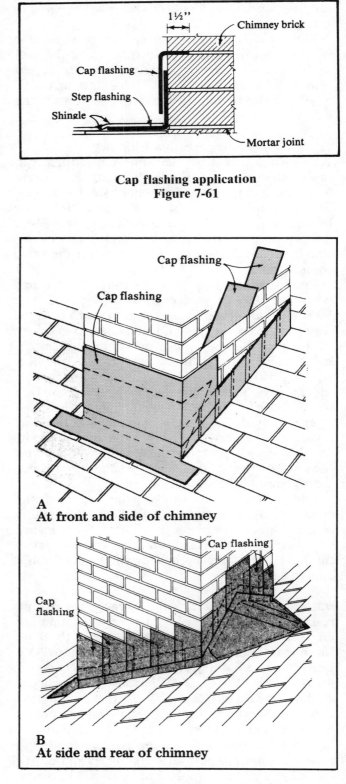

Coat of asphalt primer

Cricket

Asphalt plastic cement

Nail flashing to deck

Corner flashing laps step flashing

A
Base flashing at corner

Base flashing extends up chimney at least 6''

Nail flashing to deck

Asphalt plastic cement

Base flashing cut to fit over cricket and extend onto roof at least 6''

B
Base flashing over cricket

Base flashing at rear of chimney
Figure 7-60

1½''

Chimney brick

Cap flashing

Step flashing

Shingle

Mortar joint

Cap flashing application
Figure 7-61

Cap flashing

Cap flashing

A
At front and side of chimney

Cap flashing

Cap flashing

B
At side and rear of chimney

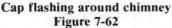

Cap flashing around chimney
Figure 7-62

Use metal step flashing for the sides of the chimney. Place the pieces as shown in Figure 7-59. Secure each flashing piece to the masonry with asphalt plastic cement and to the deck with nails. Embed the end shingles in each course that overlaps the flashing in asphalt plastic cement.

Place the rear base flashing over the cricket and the back of the chimney as shown in Figure 7-60. Cut and bend the metal base flashing to cover the cricket and extend onto the roof surface at least 6''. It should also extend at least 6'' up the brickwork and far enough to the side to lap the step flashing on the sides.

If it's big enough, you can cover the cricket with shingles. Otherwise, apply the rear base flashing, then bring the end shingles in each course up to the cricket and cement them in place.

Cap flashing— Now you have to put cap flashing over all the base flashing to keep water out of the joints. Begin by setting the metal cap flashing into the brickwork, as shown in Figure 7-61. Rake out a mortar joint to a depth of 1½'' and insert the bent edge of the flashing into the cleared joint. Since it has a small amount of spring tension, the flashing won't come out easily. Refill the joint with mortar. Finally, bend the flashing down so it covers the base flashing and lies snug against the masonry.

Use one continuous piece of cap flashing on the front of the chimney as shown in Figure 7-62 A. On the sides and back of the chimney use several pieces of similar-sized flashing, trimming each to fit the particular location of brick joint and roof pitch. See Figure 7-62 B. Start the side pieces at the lowest point and overlap each at least 3''.

Roof framing is probably the most complex element of house construction. A solid roof is the key to quality building: it locks the entire structure together.

So, study this chapter again. Because roof framing is so complex, you'll need more than one reading to study all the details.

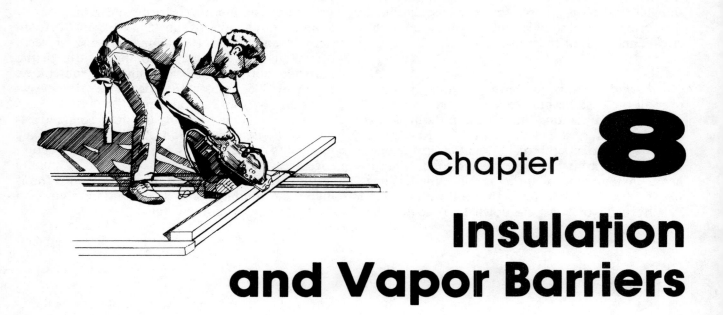

Chapter 8

Insulation and Vapor Barriers

All professional builders have special obligations in the area of insulation. It's in the country's interest and your interest to provide quality insulation in every home you build. Let's look first at what the law requires. Then we'll go into selecting and using insulation in your projects.

The Federal Trade Commission Rule titled "Labeling and Advertising of Home Insulation" requires you to give a potential buyer certain information about a home's insulation. This information includes the R-value and thickness of the various kinds of insulation used in the home. The information must be provided before the sale, and must be included in the sale contract for the house.

You must list the R-value of each material separately (air spaces or reflectivity not included) at a mean temperature of 75 degrees F. There are only three acceptable test methods for finding this R-value. If you include reflective R-values, you must list them separately. You must also explain exactly how the values were found. Fortunately for builders, the insulation manufacturers have done nearly all the work for you. All you have to do is pass along the information that the insulation manufacturer is more than happy to supply.

Written ads mentioning R-values must include this statement: *The higher the R-value, the greater the insulating power. Ask your seller for the fact*

sheet on R-values.

Written ads or promotional materials that state or imply fuel or dollar savings must include the statement: *Savings vary. Find out why in the seller's fact sheet on R-values. Higher R-values mean greater insulating power.*

Here's a direct quotation from the trade rule:

If you say or imply that a combination of products can cut fuel bills or use, you must have a reasonable basis for the claim. You must make the statement about savings in sub-section (b). Also, you must list the combination of products used. They may be two or more types of insulation; one or more types of insulation and one or more other insulating products, like storm windows or siding; or insulation for two or more parts of the house; like the attic and walls. You must say how much of the savings come from each product or location. If you cannot give exact or appropriate figures, you must give a ranking. For instance, if your ad says that insulation and storm doors combined cut fuel use by 50%, you must say which one saved more.

The FTC Rule has the effect of federal law. Compliance is mandatory. Passing on information from a supplier which doesn't comply with the rule can result in a $10,000 per day fine per occurrence. Your best bet is to deal only with reputable companies who supply the correct information you need to comply with the regulation.

Write the Federal Trade Commission for a copy of the Rule at:

Public Reference Branch
Federal Trade Commission Headquarters
6th Street and Pennsylvania Ave. N.W.
Washington, D.C. 20580

Insulation

Building insulation materials are those developed specifically to reduce heat transfer. The materials are generally very lightweight. They're available in five forms: batts or blankets, loose fill or granulated, rigid boards, reflective, and foamed-in-place.

Insulation Materials

Common thermal insulating materials include:

- Mineral fibers, made from glass, rock, slag, and asbestos

- Wool

- Vegetable fibers, such as wood, cane, cotton, and redwood bark

- Expanded mineral granules, such as perlite and vermiculite

- Vegetable granules, such as ground cork

- Foamed materials, both glass and synthetic resins like styrene and urethane

- Aluminum foil

Insulating batts, blankets, and boards are made from fibers or granules. These fibers or granules are mixed with binders, then formed into useful widths, lengths, and thicknesses. The binders make the insulation more resistant to water and mildew. They also make it stronger. Vegetable fibers usually need chemical treatment to make them fire-resistant. Foamed glass and synthetic resins are formed into blocks, sheets, or boards of various sizes. The foaming process is carefully controlled to get the needed cell-size and density.

Forms of Insulation

As you saw in the section above, insulation comes in a great variety of forms. These forms include: batts, blankets, reflective insulators, and loose fill.

Batts or blankets are put between the joists or studs. You can get them in widths to fit standard joist and stud spacing. You can get them unfaced, or faced on one or both sides with paper, aluminum foil, or plastic. If batts or blankets are faced on one side, that side is less permeable to moisture, and so acts as a vapor barrier.

Paper-faced batts or blankets have continuous paper flanges along the long edges. You can easily nail or staple these flanges to studs and joists.

Reflective insulators are made with aluminum foil. This foil is usually reinforced with paper backing. Insulators generally have two or more reflective surfaces, with air space between the layers. Like batts or blankets, these insulations have flanges (for nailing or stapling), and are put between studs or joists.

Loose fill insulation is made from mineral or vegetable fibers, or from mineral or vegetable granules. It's applied either by blowing or by hand-packing. You'll use it mainly in masonry cavity walls and over ceilings in attic spaces. You can, though, blow it into the spaces between wall studs.

Insulation-blowing is best done by a specialist who has the necessary equipment. It's important that you check on the finished blown insulation work. Make sure that all areas have the right amount of insulation. Also check that the insulation is evenly distributed over the entire area, and that it doesn't block any vents.

Ratings

An insulation's effectiveness can be specified in two ways. One is the resistance offered by a material or materials to the flow of heat under known conditions. This "resistance" is stated as "R-value." The higher the R-value, the better the insulating ability.

The second specification is called the U-value. It's the amount of heat, measured in Btu's, that moves through one square foot of construction in one hour from air on one side to the air on the other side at a temperature differential of 1 degree F. The lower the U-value, the better.

Most insulation manufacturers use the R rating and stamp it on their product for your convenience. For example, a 3-inch batt of one insulation might have an R-10 rating, whereas a 3-inch batt of another type might have an R-12 rating. On the other hand, a 2-inch thick rigid foam board like Thermax can have an R-14.4 rating.

Paper-faced batts or blankets have continuous paper flanges along the long edges. You can easily nail or staple these flanges to studs and joists.

Reflective insulators are made with aluminum foil. This foil is usually reinforced with paper backing. Insulators generally have two or more reflective surfaces, with air space between the layers. Like batts or blankets, these insulations have flanges (for nailing or stapling), and are put between studs or joists.

These simple ratings make it easy to compare costs of insulation. If one insulation rated R-11 meets your needs and costs 25 cents a square foot, and another rates the same but costs 30 cents a square foot, your choice is clear.

Also, the ratings help you determine the thickness and type of insulation you need for a particular R-value for any kind of construction. You can find the total R-value for a wall built of several materials if you know the R-value of each material. Simply add the values together. For example, R-4 and R-5 total R-9. Another way you can use the rating is to find the values of different thicknesses of materials. If two inches of a certain insulation rates at R-8, then four inches of it would rate R-16.

U-values, however, can't be added directly. As an example, suppose you have U-values of the materials in a wall and want to find the total insulating value. You must convert to R-values, then add. To convert, divide 1 by the U-value to get the R-value. Like this: if the U-value is 0.5, the R-value is 1 divided by 0.5, which equals 2.

Other Factors to Consider

Some features of insulation besides R-values and cost could influence your choice of materials for particular conditions. Here are a few to consider:

• Structural requirements must be considered when you want to use rigid insulation for such things as sheathing, plaster base, interior ceiling or wall finish, and roof decks.

• Fire resistance of any insulating material should be considered regardless of where it is used in the house.

• Effects of moisture on an insulation material sometimes determine whether it's used for jobs like perimeter insulation under concrete slabs. The insulation value of any material decreases to almost nothing when saturated with water.

• Vermin resistance is important, but it's relative. No material is absolutely vermin proof.

Reflective Insulation

Most materials reflect some radiant heat, and some reflect a great deal of heat. Materials with great reflective ability include aluminum foil, sheet metal with tin coating, and paper products coated with a reflective oxide compound. You can use these materials in enclosed stud spaces, attics, and similar areas to slow down heat transfer by radiation. Remember that these materials have little or no insulating value: they work only when the reflective surface faces an air space of at least 3/4 inch. The insulating properties are lost when the reflective surface contacts another surface.

How Much Insulation

Figure 8-1 shows suggested insulation values for different areas of the U.S. as recommended by Owens-Corning. The figures take into account national weather data, energy costs and projected increases, and insulation costs. As you can see, most areas require R-19 wall insulation.

A double layer of 6-inch mineral fiber (made from glass, rock, slag, or asbestos) would give a ceiling an R-38 rating. The second layer of insulation shouldn't have a vapor barrier. If it does, either remove the barrier or tear holes in it at 16 to 24-inch intervals. This will keep moisture from being trapped between the two layers.

When you estimate batt or blanket insulation needs, first calculate the total area of the ceiling, floor, or wall to be covered. Then, find 95% of this total area. (About 5% of the total is taken up by the thickness of the framing members.) Thus, 1000 SF of floor area needs 950 SF of insulation. Allow 160 staples per 100 SF of insulation.

Estimate insulating panels or sheathing needs the same way you do regular fiberboard or plywood sheathing panels. One 4' x 9' insulating panel covers 36 SF. Apply it with the length parallel to the studs.

Here's how you can work with various insulation materials and thicknesses to get the R-value you want.

Ceilings, double layers of batts:

R-38: Two layers of R-19 (6'') mineral fiber
R-33: One layer of R-22 (6½'') and one layer of R-11 (3½'') mineral fiber

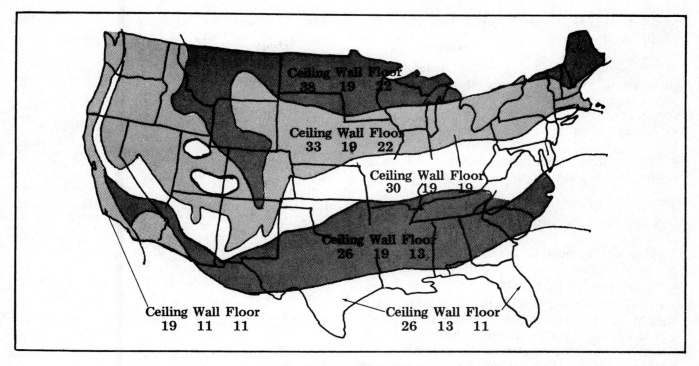

Recommended R-values
Figure 8-1

R-30: One layer of R-19 (6'') and one layer of R-11 (3½'') mineral fiber
R-26: Two layers of R-13 (3⅝'') mineral fiber

Ceilings, loose fill mineral wool and batts:

R-38: R-19 (6'') mineral fiber and 20 bags of wool per 1,000 SF (8¾'')
R-33: R-22 (6'') mineral fiber and 11 bags of wool per 1,000 SF (5'')
R-30: R-19 (6'') mineral fiber and 11 bags of wool per 1,000 SF (5'')
R-26: R-19 (6'') mineral fiber and 8 bags of wool per 1,000 SF (3¼'')

Walls, using 2 x 6 framing:

R-19: R-19 (6'') mineral fiber batts

Walls, using 2 x 4 framing:

R-20: R-11 (3½'') mineral fiber batts and 3/4'' Thermax sheathing
R-19: R-13 (3⅝'') mineral fiber batts and 1'' plastic foam sheathing
R-11: R-11 (3½'') mineral fiber batts

Foam Sheathing
Plastic foam sheathing panels, like Thermax and Styrofoam, are available in 4' x 8' or 4' x 9' sheets, and are easy to install.

Thermax consists of glass-fiber-reinforced polyiscolyanurate foam plastic core with aluminum foil faces. You can use it for conventional 2 x 4 framing. Styrofoam is expanded polystyrene. Neither Thermax or Styrofoam is a structural material. You must provide diagonal bracing with 1 x 4 let-ins, or with metal straps (whichever your building code allows).

Recommended uses— Foam plastic sheathing is generally concealed in the walls in residential construction. You then provide an interior finish acceptable to your building code. Examples of concealed uses are:

• High-performance insulation sheathing in new frame wall construction.

• Thin-profile cavity wall insulation in new masonry construction.

• High-performance, insulating, vapor-barrier

undercourse behind new interior wall or ceiling finish material.

• Thin-profile, insulating underlayment under roof shingles in vaulted, cathedral-type ceilings and in "A-frame" construction.

• Under slab or perimeter thermal insulation.

• Thin-profile, insulating underlayment installed behind new exterior siding.

• Thin-profile, insulating undercourse behind new interior wallboard.

Foam plastic sheathing is not intended for exposed use in living quarters. Follow the manufacturer's instructions concerning this.

When you handle and install foam plastic sheathing, be careful not to puncture the foil face of the board. After installation, this foil protects the insulation from ultraviolet light which would harm the foam. If the sheathing is accidentally damaged, repair it with aluminized tape.

Installing foam sheathing— If you need to trim insulation boards to fit, use a utility knife and a straightedge.

Install foam panels vertically. Have the long edges in moderate contact where they meet, and have them bearing directly on the framing members. Avoid horizontal joints, unless they bear on a framing member or occur over vent strips.

Use galvanized roofing nails with 3/8" heads, and long enough to penetrate at least 3/4" into the framing. Or use 16-gauge wire staples, with a minimum crown of 3/4", and legs long enough to penetrate at least 1/2" into the framing.

Apply staples with the crowns running parallel to the direction of the framing. Don't over-drive the heads, or you'll tear the facer sheets.

Vapor Barriers

Vapor barriers always go on the warm side (toward the home's interior) of the insulation, regardless of the type or location of the insulation. Vapor barriers prevent the moisture in warm air from entering the insulation in walls, ceilings, and floors. Insulation loses its effectiveness if it gets wet. Enough moisture will also cause rot in the structure, and will cause paint problems on exterior wood siding.

Vapor barriers are required in areas with an average winter temperature of 35 degrees F or lower.

If you rip or tear the vapor barrier, repair it by stapling vapor barrier material over the tear or by taping the torn barrier back into place.

Installing Insulation and Vapor Barriers

For slabs on grade, you can use perimeter insulation of 1-inch styrofoam or urethane (R-5 to R-6), as shown in Figure 8-2. Install it at the edge of the slab, and 18" to 24" down the foundation or under the slab. This insulation will *not* support the slab's weight on the lip. The slab must rest directly on the lip, at sufficient intervals, on all sides. After applying the 18" to 24" under-slab insulation, cut out a 6-inch wide section over the lip, at 3-foot intervals. When the slab is poured the mix will fill the cut-outs onto the lip. This provides solid bearing for the slab.

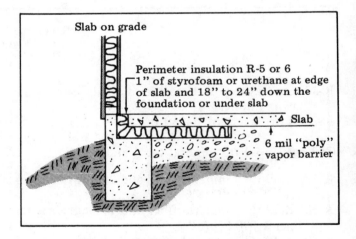

Perimeter insulation under slabs
Figure 8-2

Floors

Water pipes installed in the floor joists in unheated crawl spaces can freeze in the winter in cold climates. You can wrap the pipe with insulation to prevent freezing. But if the pipes are installed close to the subfloor, it's easier to put insulation batts with attached vapor barrier between the floor joists. You can fasten them by placing the tabs over the edge of the joists before installing the subfloor, but only if the vapor barrier is strong enough to support the insulation batt. But there's a good chance that the insulation will get wet before the subfloor is installed and the house enclosed or "dried in." I recommend the following alternative installation method:

Install friction-type batt insulation from the crawl space after the house is enclosed and dried in. Batts are made to fit tightly between the joists. Use small dabs of mastic adhesive to insure that the insulation stays in place against the subfloor. If a vapor barrier isn't part of the insulation, use a separate film between the subfloor and the underlayment.

Walls

Flexible insulation in blanket or batt form is usually made with a vapor barrier. These vapor barriers have tabs at each side, designed to be stapled to the frame. The best way to do this is to staple the tabs over the edge of the studs, as shown in Figure 8-3. This minimizes vapor loss and possible condensation problems. However, many builders don't follow this procedure because it's harder to make a smooth wall with paper stapled to the face of the

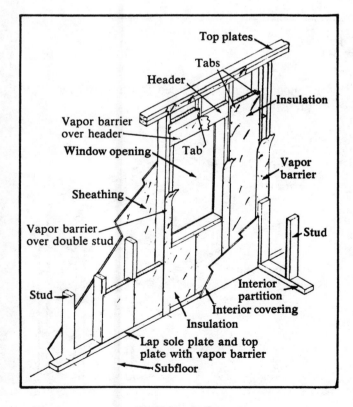

Wall insulation
Figure 8-3

studs. Instead, they fasten the tabs to the inner sides of the studs. This usually leaves some openings along the edge of the vapor barrier, which allows vapor to escape through the wall. If you use this method, it's good practice to install a plastic (polyethylene) vapor barrier over the entire wall.

For spaces of non-standard width, cut the insulation and the vapor barrier an inch or so wider than the space to be filled. Staple the uncut tab as usual. Pull the cut side of the vapor barrier to the other stud, compressing the insulation behind it, and staple through the vapor barrier to the stud. For unfaced blankets, cut them slightly oversize and wedge into place.

Extra protection is needed around window and door areas when you use flexible insulation with an integral vapor barrier. If the vapor barrier on the insulation doesn't cover doubled studs and header areas, add additional vapor barriers for protection. See Figure 8-3. Most conscientious builders include such details when they apply insulation.

Where interior partitions and exterior walls meet, be sure to cover the intersection with some kind of vapor barrier. For the best protection, insulate the space between the double exterior wall studs and apply the vapor barrier before assembling the corner post, as in Figure 8-3. At the very least, the vapor barrier should cover the stud intersections at each side of the partition wall.

Enveloping— You can practically eliminate condensation problems in walls by *enveloping*. This means wrapping a vapor barrier around the whole wall. Here's how:

Use unfaced friction-type insulation without vapor barrier. It's semi-rigid and made to fit tightly between frame members spaced 16 or 24" o.c. After insulation, rough wiring, ductwork, and window frames are installed, staple vapor barrier over the entire wall. As you can see in Figure 8-4, the window and door headers, top and bottom plates, and other interior wall framing are completely covered. This kind of vapor barrier should be 4-mil or thicker polyethylene and in 8-foot wide rolls. After you install the rock lath plaster base or drywall finish, trim the vapor barrier around window and door openings.

Put insulation behind pipes, ducts, and electrical boxes. You can pack the space with loose insulation. Or you can cut a piece to size and fit it into place. Pack small spaces between rough framing and door and window headers, jambs, and sills

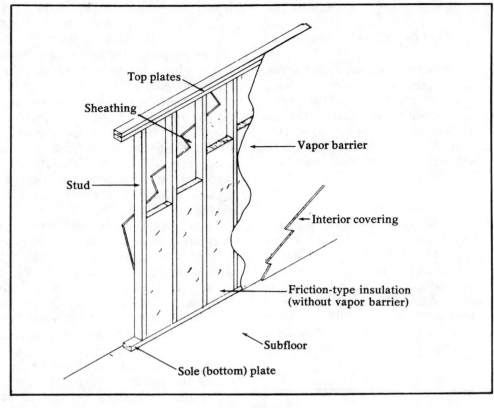

"Enveloping"
Figure 8-4

with pieces of insulation. Staple insulation vapor barrier paper or polyethylene over these small spaces.

Masonry walls— The only way to insulate solid masonry walls such as brick, stone and concrete is to apply insulation to the interior, even though some space is lost. One way to do this is to glue insulating board directly to the interior surface. Then you can plaster it, leave it exposed, or cover it with any finish material. Another way to insulate the inside of masonry walls is to attach 2 x 2 or 2 x 4 furring strips 16 or 24" o.c., as in Figure 8-5. Then staple the insulation's vapor barrier tabs to the faces of the furring strips. A third way is to use 1 x 2 furring with unfaced masonry wall blankets (normally 1" thick) or 1" thick rigid-foam board insulation between the furring strips. In cold climates insulation blankets are often used with 2 x 4 furring strips, as in conventional framing. The vapor barrier may be a separate polyethylene barrier over the insulation, or foil-backed gypsum board used as the interior finish wall. A separate vapor barrier

isn't needed with foil-faced foam-type rigid board insulation.

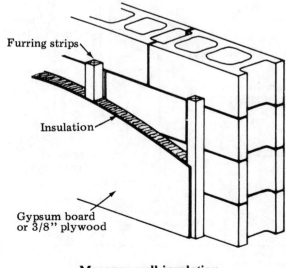

Masonry wall insulation
Figure 8-5

SAMPLE CALCULATION FOR COMPARISON WALL R VALUES USING ¾" THERMAX SHEATHING
R-5.4 AT 75°F MEAN TEMPERATURE

Illustrated below are the calculations for one design based on a 19% framing factor.

	Winter Conditions	
	"R" Values Thru Frame	"R" Values Thru Cavity
Inside surface film	0.68	0.68
½" (12.7 mm) Gypsum wallboard	0.45	0.45
6 mil (0.15 mm) Poly film	negl.	negl.
Wood framing 2 x 4's 16" (406.4 mm) o.c.	4.35	—
3⅝" (92.1 mm) Friction fit glass batt compressed	—	12.68
¾" (19.1 mm) Thermax Sheathing (75°F mean)	5.4	5.4
¾"x10" (19.1 mm x 254 mm) Lapped wood siding	1.05	1.05
Outside surface film 15 mph wind	0.17	0.17
'R''s at Section	12.10	20.43
"U''s at Section	.0826	.0489

$$\text{Total Design "U"} = \frac{.19}{12.10} + \frac{.81}{20.43} = .055$$

$$\text{Total Design "R" Value} = \frac{1}{.055} = 18$$

Per ASHRAE 1977 "Handbook of Fundamentals" Methods. Note: Standard rules for rounding numbers applied.

Sample R-value calculation
Figure 8-6

In cavity wall construction, you can put rigid board insulation between the concrete block wall and the brick or stone veneer exterior wall. Some builders prefer to use blown-in foam or loose-fill insulation.

The R-20 wall— Many builders in colder parts of the U.S. insulate walls to a minimum value of R-20. You can do this simply and economically with conventional 2 x 4 framing. Just substitute Thermax (or similar) sheathing for the regular exterior sheathing material. No major changes in house design, framing lumber, or trim are needed.

Thermax sheathing gives you design freedom with a wide range of materials. For example, the basic components of an R-20 wall system, starting from the interior of the wall, are:

1/2 inch gypsum wallboard (sheetrock)
6-mil polyethylene vapor barrier
Nominal 2 x 4 inch wood framing
Stud cavity insulation
Thermax sheathing thickness
Exterior siding material

Figure 8-6 shows a sample R-value calculation for an R-20 wall system. (The R-20 wall is calculated at the cavity between studs, and excludes windows and doors. The total-design R-value is therefore slightly lower.)

In all frame construction, no matter what amount or kind of insulation is put between the studs, there is still the underinsulated area of these studs to consider. This framing area, known as the framing factor, varies between 18% and 27% of the total opaque exterior wall area, depending on construction.

In Figure 8-6, substituting 7/8" thick Thermax would give a value of R-21.33 at section; 1" Thermax would result in R-22.23 at section.

Finished Basement Rooms
Treat walls of finished rooms in basements the same as framed walls with respect to vapor barriers and insulation. See Figure 8-7.

However, if you have a full masonry wall you need to consider several factors: (a) if drainage in the area is poor and soil is wet, or if the basement has a history of dampness, install drain tile on the outside of the footing to remove excess water; (b) in addition to an exterior wall coating, apply a waterproof coating to the interior surface to insure a dry wall, as in Figure 8-7.

Furring strips (2 x 2 or 2 x 3's) on the wall provide space for blanket insulation with attached vapor barrier and nailing surfaces for interior finish. Look at Figure 8-7 for details. You could also use 1'' or 1½'' thick friction-type insulation

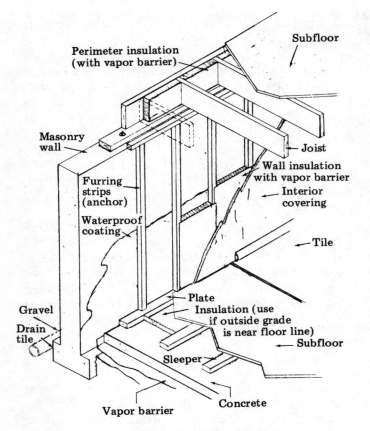

**Vapor barrier in finished basement
Figure 8-7**

Second Stories

Builders often overlook the need to insulate and put vapor barriers around the edges of the second-floor joists in two-story houses. Protect the space between the joists at the header and along the stringer joists with sections of batt insulation with vapor barrier, as shown in Figure 8-8. Be sure that the sections fit tightly so both the vapor barriers and the insulation fill the joist spaces. Install insulation and vapor barriers in exposed second-floor walls the same way as in single-story walls.

Sometimes, part of the second story of a house projects beyond the first. Be sure that every exterior surface in your homes is enclosed with vapor barriers and insulation.

with a plastic-film vapor barrier of 4-mil polyethylene.

Rigid insulation such as expanded polystyrene or Thermax is also used over masonry walls.

If you didn't use a vapor barrier under the concrete slab, you should put a vapor barrier of some kind over the slab itself before applying the sleepers. One such system for unprotected in-place slabs uses treated 1 x 4 sleepers fastened to the slab with mastic. Next comes vapor barrier, followed by a second set of 1 x 4 sleepers placed over and nailed to the first set. Finally, apply the subfloor and finish floor over the sleepers.

To prevent heat loss and minimize escape of water vapor, use blanket or batt insulation with attached vapor barriers around the edge of the floor framing above the foundation walls. Put the insulation between joists, or along stringer joists, with the vapor barrier facing the basement side. Make sure the vapor barrier fits tightly against the joists and subfloor.

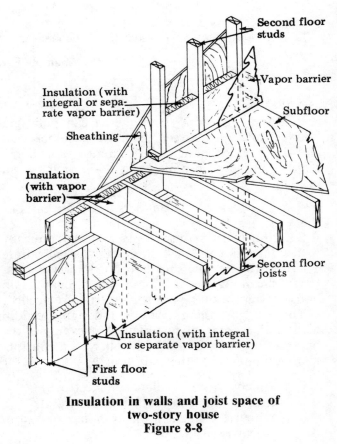

**Insulation in walls and joist space of
two-story house
Figure 8-8**

In 1½-story houses with bedrooms or other occupied rooms on the second floor, insulate: (a) in the first-floor ceiling area, (b) at the knee walls (partial walls extending from the floor to the rafters), and (c) between the rafters. Figure 8-9 shows how.

Put insulation batts, with the vapor barrier facing down, between joists, from the outside wall plate to the knee wall. Fill up the entire joist space directly under the knee wall with insulation, as shown in Figure 8-9. When putting the insulating batt at the junction of rafter and exterior wall, be sure to allow an airway for attic ventilation.

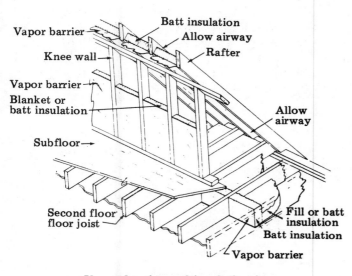

Vapor barrier and insulation in knee-wall areas of 1½ story house
Figure 8-9

Insulate the knee wall with blanket or batt insulation, either with integral vapor barrier or with a separate vapor barrier as described for first and second floor walls.

Use batt or blanket insulation between the rafters at the sloping portion of the room. As with all insulation, the vapor barrier should face the inner, or warm, side of the room. Always leave an airway between the top of the insulation and the roof sheathing at each rafter space. Make the airway at least a 1-inch clear space, without obstructions (such as might occur with solid blocking). This way air can move behind the knee wall and into the attic area above the second-floor rooms.

Ceilings

Most houses have an accessible attic with exposed ceiling framing, so you can apply any kind of insulation easily. If you use batt or blanket type, get the width that matches joist spacing — 16 or 24 inches. You could also use loose-fill insulation: simp-

ly dump or blow it between joists and screed it off to the thickness you want.

Pay attention to the places in the attic where air can leak in and add to the heating/cooling load of a home. Weatherstrip attic access doors. In addition, cut a piece of insulation board to the size of the attic door and tack it to the attic side of the door. This improves the insulating ability of a panel or hollow core door. If there is an attic scuttle hole, weatherstrip it and insulate the back of the scuttle closure panel. Insulate around pipes, flues, or chimneys in the attic space, especially in cold climates.

When insulating a ceiling, extend the insulation over the top plate, as in Figure 8-10. There are three ways to install mineral-fiber blankets in ceilings: You can staple the vapor barrier tabs from below; place unfaced pressure-fit batts between joists; or lay blankets from above after the ceiling is in place. Insulate the side of pipes and ducts that will be cold in winter.

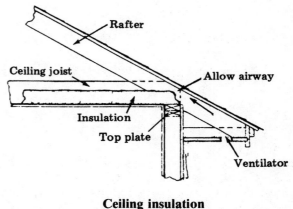

Ceiling insulation
Figure 8-10

In unfinished attics, you can simply empty bags of pouring wool evenly between ceiling joists. Follow the manufacturer's recommendations for proper thickness and coverage per bag. Then level the wool with a wood slat or garden rake. Be sure that you don't block the eave ventilation openings.

Hand-pack small openings, such as those around a chimney, with noncombustible insulation. Cover all cavities, drops and scuttles with insulation. But be sure that recessed lighting fixtures or exhaust fan motors protruding through the ceiling are *not* covered with insulation.

Many builders have insulation blown into attic areas by insulation contractors. If you use a subcontractor, always inspect the work to be sure that the job is properly done.

Summary

Today's homebuyers are knowledgeable about insulation, and they rightly insist on well-insulated homes with proper vapor barriers. The successful builder never skimps in these areas. Codes in many parts of the country specify insulation types and thicknesses to be used. We all benefit from properly constructed and insulated buildings.

Inspection is the key word. *Always* check your insulators' work. Make sure they do the job right.

Chapter 9

Designing and Building Stairs

A well-designed stairway has graceful, well-proportioned, perfectly-matched parts made of quality materials. It reflects quality construction and fine craftsmanship.

You can buy most standard stairways precut and designed to fit your job. All you do is assemble the pieces on site. That makes quality stair building easy because complete directions and all the hardware are included. Most stair fabricators have catalogs that show the many different designs of balusters, rails and newels they offer. In many communities there are stair specialty shops that act as subcontractors for the more difficult stair jobs.

But most residential contractors still build main stairways, porch, basement and exterior stairs on the job from lumber and plywood. That usually saves money — if you have a qualified stair builder on your payroll.

This chapter is intended to sharpen your stair building skills and serve as a reference when you or your stair builder has a question on a particular job. Most of the details in this chapter apply to carpenters, but general contractors, masons, and structural iron workers who build stairs follow the same design and layout principles.

Stairway Terms

Everyone who builds stairs should understand the following terms. Figure 9-1 illustrates some of the terms.

A *staircase* is the whole set of stairs. It includes landings, winders, and stairs leading from one story to another; also several stories connected by flights of stairs one above the other. "Staircase" is often used with the same meaning as the word "stairs."

A *flight of stairs* is a series of unbroken steps leading from one landing to another.

The *wellhole* is the framed opening in the floor of a building through which a stairway passes. The wellhole is guarded with balusters and a handrail if the stairway is open.

The *riser* is the upright member between two treads.

The *tread* is the horizontal top surface where people step when going up or down the stairway. Usually the tread projects a short distance in front of the riser to form the *nosing,* as in Figure 9-2. A tread of a finished set of stairs has two parts: the run and the nosing. The nosing, however, is not considered when making stair calculations.

The *rise* of a step is the height from the top of one tread to the top of the next tread.

The *run* of a step is the horizontal distance from the face of one riser to the face of the next riser. The run equals the width of a tread, not including the nosing.

The *total rise* of a flight of stairs is the height

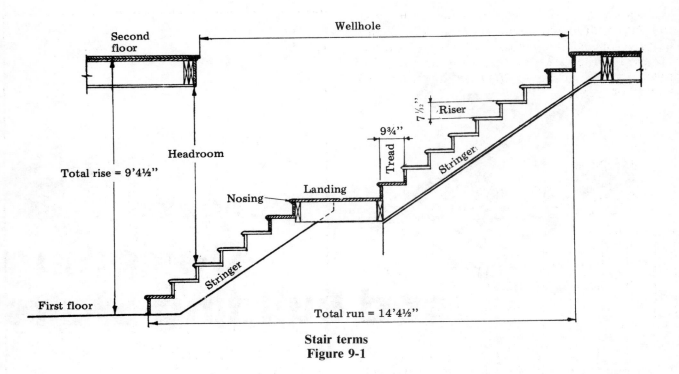

Stair terms
Figure 9-1

from finish floor to finish floor. It's the basic measurement for all stair layouts. It must be accurate, so you have to establish the measurement from the top of the finish flooring of both floors. Since only the rough flooring will be laid out when you begin stair construction, consult the specifications for floor thicknesses. They may not be the same.

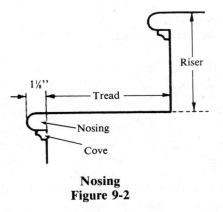

Nosing
Figure 9-2

The *total run* of a flight of stairs is the horizontal distance from the face of the first riser to the face of the last riser in the same flight. It equals the sum of the treads.

Stringers support the treads just as joists support the floor. The average small stairway has two stringers, one on each side of the treads. Three stringers, however, give better support.

Headroom is the clearance above a stair. It's measured from the lowest point of the open end of the wellhole down to the outside corner of the tread and riser directly below.

The *handrail* is a plain or decorative piece that acts as handrest and guide when climbing or descending. It's parallel to the stringer. Handrail is also used as a guard around the wellhole.

Balusters are the uprights put between the treads and the rail. They support the handrail and provide a side enclosure for the stairway. They also act as a guard at the outside of an exposed stairway and around the wellhole. Balusters can be plain or decorated, square or round. A tread with 10" to 12" of run should have two evenly-spaced balusters, as in Figure 9-3.

A *newel post* is the finished post located at the bottom tread, at the top tread, and at the corners of the wellhole. Newels may be either solid or built-up, square or round, plain or fancy. Platform stairs have at least one newel at the landing. The handrail is always fitted to the newel posts.

Occasionally a stairway is designed with a continuous handrail and no newel posts. Special curved pieces called *easements, goosenecks,* or *ramps* are used at the various turns of the stairway. The lower end of the railing may be finished with a wreath or spiral.

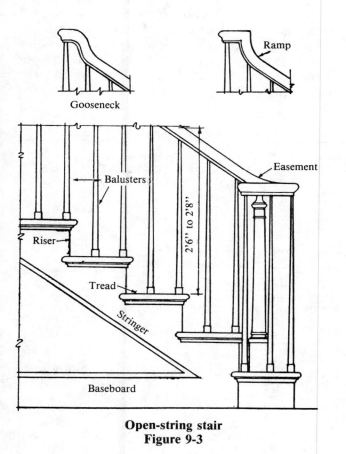

Open-string stair
Figure 9-3

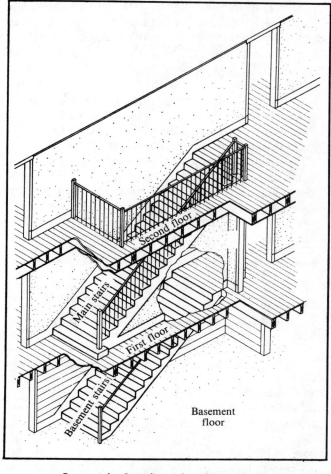

Isometric drawing of stairway from
basement to second floor
Figure 9-4

How Stairs Are Shown on Plans

You can get the information you need about the stairs in a building from the floor plans, though sometimes it's hard to interpret. This section shows you how. Figure 9-4 is an isometric view of a stairway leading from the basement to the second floor. Figure 9-5 shows the same stairs on several floor plans.

The *basement plan* shows everything below a horizontal section taken through the basement, one or two feet below the first-floor joists and just above the grade line. The basement stairs, except for the last few steps at the top, are shown on the basement plan. The top steps are shown broken off by a shaded line called a *break line*. These top steps

are then shown on the first-floor plan.

As you can see in Figure 9-5, stairways are shown as a series of parallel lines representing the risers. Two lines drawn close together indicate the handrails, with the small squares at the ends marking the newel posts. You'll usually see an arrow on the stairs with the word *up* or *down* to show whether the stairs ascend or descend.

The *first-floor plan* shows everything below a horizontal section cut through the windows and doors about three or four feet above the floor. The first-floor plan shows six or eight steps of the main stairway. As in the basement plan, a break line is shown in the stairway. The rest of the stairway is shown on the second-floor plan.

These flights of stairs are directly below one

another (Figure 9-4). The first-floor plan shows the top part of the basement stairs at the point where the main stairway is broken off. To see this better, lay a pencil or draw a line at the point about 3 or 4 feet from the first floor of Figure 9-4. Assume that you're looking down from this line. Everything you see would be shown on the first-floor plan. Remember, though, that since you can see the lower steps of the first-floor stairs, you can only see the upper steps of the basement stairs.

The *second-floor plan* shows everything below a horizontal section cut through the windows and doors 3 or 4 feet above the second-floor line. It

shows the wellhole for the stairs, and as many of the steps as are directly under the wellhole. Any part of the main stairs not directly under the wellhole isn't shown on the second-floor plan.

Figure 9-6 is a sectional drawing of the first- and second-floor plans of the stairs shown in Figures 9-4 and 9-5. Line B-B of Figure 9-6 is the position of the horizontal section for the first-floor plan shown at the bottom of the figure. You can see seven steps of the first-floor stairway below this line. If you project the lines to the first-floor plan at the bottom of Figure 9-6, you'll see seven steps on the plan. The break line in the seventh step

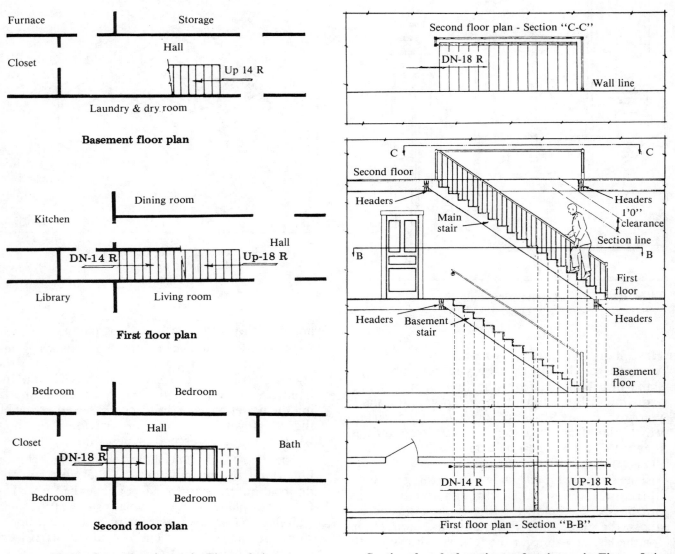

Floor plans of stairway in Figure 9-4
Figure 9-5

Sectional and plan views of stairway in Figure 9-4
Figure 9-6

shows that the stair plan stops here.

Below section B-B you'll also see the eight top steps of the basement stairs. These steps must also be shown on the first-floor plan. Therefore, the first-floor plan shows the bottom seven steps of the main stairs and the top eight steps of the basement stairs. Between these two is a break line to show that the steps don't all belong to the same flight of stairs.

Line C-C is the section line from which the second-floor plan is taken. If you could look down from this line, you'd see the outline of the wellhole, the railing around it, and the top 14 steps of the main stairs.

Symbols on Stair Plans

An arrow on the plan shows the direction the stairway leads. Usually a note says how many risers there are in the flight. Treads are always numbered from the bottom to the top. (The bottom riser is number 1, the next number 2, and so on.) This numbering system is important for locating the stairway platform.

The plan of the basement stairs in Figure 9-5 indicates: "Up 14 risers". However, on the first-floor plan the basement stairs are shown with an arrow and a note "Down 14 risers". The main stairs on the first-floor plan have an arrow and a note "Up 18 risers". The second-floor plan shows the main stairs with an arrow and a note "Down 18 risers". The abbreviations used are: "Dn" for *down* and "R" for *risers*.

Some stairways are harder to show on plans than the stairs in these figures. But if you can imagine yourself looking at the building from the section line drawn through the plan, you'll understand the stairs on any set of plans.

Railing, baluster and newel post details aren't shown on floor plans. These are given on detailed drawings in a larger scale. The floor plan only shows a little square or circle to mark the newel post and two lines to indicate the railing.

Location of a Stairway

The architect decides where to put a stairway, and decides such details as the shape, number and thickness of treads, stringer construction, and thickness of the finished floors. This last dimension isn't strictly a part of the stairway, but it's important: it affects stairway height from floor to floor.

Many stairways begin at a front hall in the center

of the house so people can use the stairs without going through any room. This arrangement, though, calls for a large entrance hall. The average home doesn't have space for that much hallway. Therefore, many stairs are put in living rooms. These stairs end in the hall of a second floor, never in a room.

It saves space if stairs to the basement are directly below the main stairs. But make sure people can reach the top of the basement stairs from the kitchen without passing through any other room.

Sometimes attic stairs are directly over the main stairs. This setup means the stairs and part of the attic space are visible from the hall below. Because they're visible, these attic stairs need to be made of more expensive materials. It's better to have a door at the foot of attic stairs. Use a little space on the second floor for a separate, narrow, attic stair. The second-floor space above the main stairs is then left open, with a balustrade around the landing. In a small house, however, you may not be able to waste even this much space.

Kinds of Stairways

The two most common kinds of stairways are *straight* and *platform*. See Figures 9-7 and 9-8. The straight-run stair is the simplest and least expensive to build because it leads from one floor to the next without turns or landing. Straight stairs can be tiring and more dangerous to use: with no landing, people must climb 13 or 14 steps without a rest. A fall on straight stairs is a long fall.

Closed-string stairs are straight stairs with a wall on each side. Straight stairs with a wall on one side

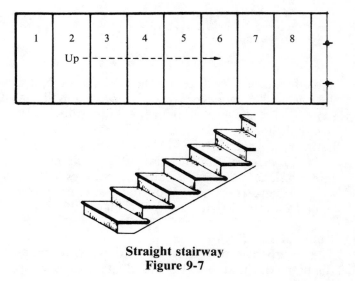

Straight stairway
Figure 9-7

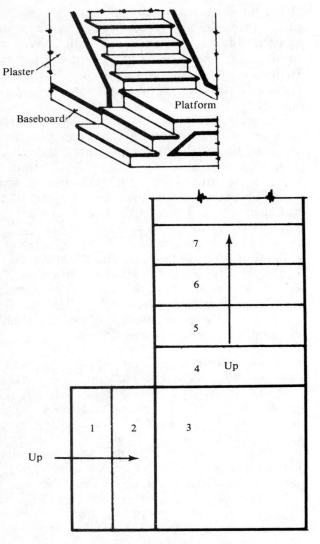

Platform stairway
Figure 9-8

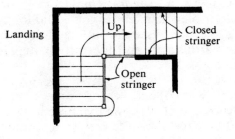

Open- and closed-string stairway
Figure 9-9

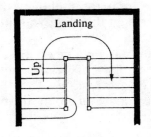

U-shaped stairway
Figure 9-10

or the bottom, the stair is called a *long L stair* or a *quarter-space stair*. The plan of the stairs is L-shaped. In going up, a person makes a 90-degree turn (one-quarter of a circle), and so is facing a different way at the top than at the bottom. Figure 9-8 shows a long L stair with the landing near the bottom.

If the space for the staircase is limited, the stairs often go to a landing then turn 180 degrees, so the second flight continues in the opposite direction from the first flight, as in Figure 9-10. The landing must be at least twice the width of the stairs. It's usually at the middle of the staircase, but can be closer to the top or to the bottom. This kind of stair, where the two flights have little space between them, is called a *narrow U stair,* a *platform stair returning on itself,* or a *half-space stair.*

and a handrail or baluster on the other, as in Figure 9-3, are called *open-string stairs*. Occasionally, a straight-run stairway is open on both sides, with two balusters. Straight stairs often need a long hallway, a disadvantage in a small house.

The *platform stairway* includes a change of direction part way up the stairs. Sometimes it starts as an open-string stair and changes after a few steps to a closed-string stair. See Figure 9-9.

Kinds of Landings
The number and location of landings determines the name of the stairs. If the landing is near the top

Winders and Their Arrangement
Some homes are too small for a stair with landings. In this case, you need to use steps called *winders* for changing direction. Figure 9-11 shows a common three-winder stairway.

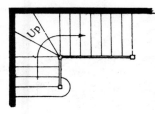

Winders
Figure 9-11

Most building codes prohibit this kind of stairway: the wide outside and narrow inside makes them dangerous. The main objection to winders is that there's little or no tread for foot support at the inner corner, where all the winders meet. Also, the risers all come together at the inner corner, making a very steep descent. It's easy to slip and fall.

Overcome the problem by using a stair with more winders, as shown in Figure 9-12, and laying them out so they don't come to a point. This arrangement has a tread width at the line of travel that's closer to the tread width of a normal step. People walk about 14" or 16" from the inside when making the turn on a stairway with winders. The tread width at this point is 9" or 10".

Figure 9-13 shows geometrical stairs — all steps

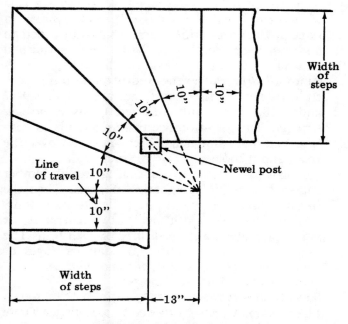

Preferred winder stairway
Figure 9-12

have the same tread width at the line of travel. Many ornamental stairs are shaped like part of a circle or an ellipse, as shown in the drawings, and are designed with landings to make them easy to ascend. This kind of staircase has a baluster or handrail following the curve, but no newel post at the turn of the stairs.

Geometrical stairs
Figure 9-13

The Width of a Flight of Stairs

The width of a flight of stairs is measured between handrails, or between the wall and a handrail. Width should be proportional to the greatest number of people that may use the stairs at one time. In general, a flight of stairs should be wide enough so two people can pass comfortably. Stairways between floors should be 32" wide for comfortable passage of one person, or 42" for two people side by side.

Most cities have building codes regulating the width, rise, and treads of stairs for various kinds of buildings. For example, the minimum stair width allowed by the National Board of Fire Underwriters is 44", except in buildings occupied by less than 40 people. There the width may be 36".

Stairways are also used for moving furniture between floors. The width needed for furniture moving depends on the kind of furniture you'd reasonably expect will need to be moved. Stairs that are open on one side, including open-well stairs, make moving large furniture easy because the pieces can be raised up over the handrails and newel posts.

Headroom

If there's more than one flight of stairs in the same stairwell (attic stairs over the main stairs, for example), the design must allow enough headroom

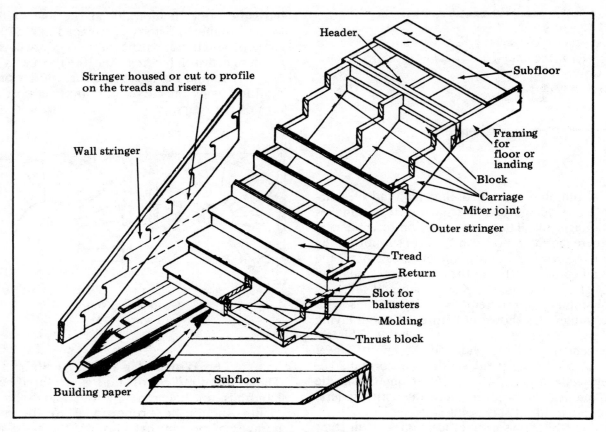

Stringer housed or cut to profile on the treads and risers

Wall stringer

Header

Subfloor

Framing for floor or landing

Block

Carriage

Miter joint

Outer stringer

Tread

Return

Slot for balusters

Molding

Thrust block

Subfloor

Building paper

Supports for stairs
Figure 9-14

underneath the upper stair. Look at Figure 9-1 again to see what we mean by "headroom." It's measured vertically from the top of a tread. Headroom will vary with the steepness of the stair, but is generally from 7'4" to 7'7". This means a person can swing an arm overhead without hitting anything. The minimum headroom needed is 6'8".

The main staircase, when carefully designed and built, is an attractive feature of a home. But it also takes considerable skill to erect the many different parts.

All stairways, no matter how simple or complicated, have two parts in common: the treads and the stringers.

Kinds of Stringers

Rough structural members support the finished stairs, as shown in Figure 9-14. They're called *stringers, strings, horses,* or *carriages.* These terms all mean the same thing.

Often, finished boards are used to cover the car-

riage on the open-string stairs. It's usually called an *outer stringer.* See Figure 9-14. When a finished board is used on the wall side of the stairs, it's called a *wall stringer.* When a finished board is grooved to receive the riser and tread, as in Figures 9-17 and 9-18, it's usually referred to as a *housed stringer.* This flip-flop use of terms tends to confuse everyone, but it has crept into use over the years and it looks as if we're stuck with it.

A carriage is usually made of 2" rough lumber cut so that it makes a series of steps. Then the finished risers and treads are nailed to it. A stairway of average width needs two carriages or stringers; wider stairs need three or even four. The two common constructions of stair carriages or rough stringers are: *sawed-out* and *built-up.*

Sawed-out Carriages

Figure 9-15 A shows a sawed-out carriage. These are used for porch, basement, and attic stairs. The different steps are sawn out of a piece of framing lumber that's wide enough to give support after it's cut. The two standard widths are 2 x 10 and 2 x 12. If you use this kind of carriage against a wall on attic or basement stairs, first nail wallboard against the wall, then nail the carriage against the wallboard. See Figure 9-15 B.

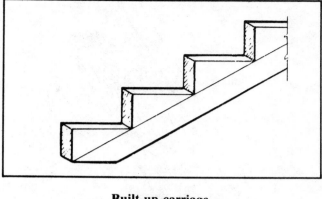

Built-up carriage
Figure 9-16

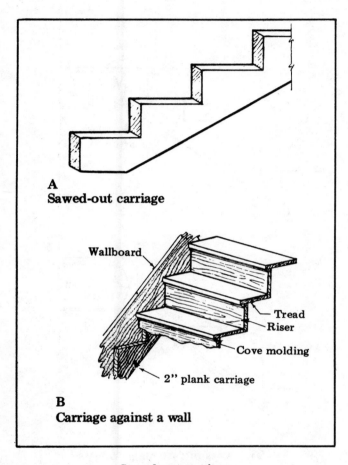

A
Sawed-out carriage

B
Carriage against a wall

Sawed-out carriage
Figure 9-15

shape as a carriage, or cut on a miter for the riser cuts. Sometimes they're grooved, or *housed.*

Housed stringers— This is the best way to build stairs. In high-quality housed stairways, the risers are rabbeted into the treads, then glued and nailed, as shown in Figures 9-17 and 9-18. The rabbeting and wedges make a stairway that won't squeak or allow dust and dirt to sift through the joints.

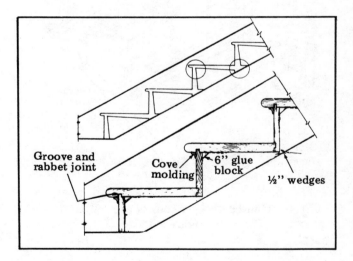

Housed stringer
Figure 9-17

Built-up Carriages
These are made by sawing triangular blocks from a narrower piece of framing lumber and nailing them on 2 x 4 or 2 x 6 stock. The two triangle sides that are at right angles to each other are equal to the riser and tread of the stairway. See Figure 9-16.

Finished Stringers
These are similar to a stair carriage but are made from finished stock. They're cut exactly the same

As you can see in these figures, the stringers are rabbeted with tapered grooves into which the treads and risers fit. When you assemble the housed stairway, you'll cut the risers and treads to the right length, then put them into the grooves. Fasten

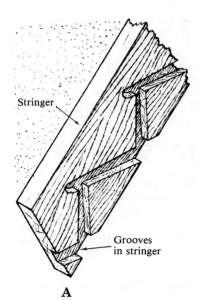

A

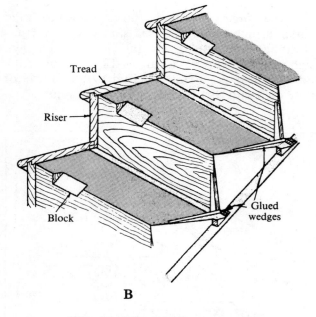

B

Housed stringer stair construction
Figure 9-18

One way to do it is to nail a piece of 2 x 4 scantling to the side of the trimmer joists. Then, cut the stringer so the top of the highest step of the stringer is flush with the top of the joist. See Figures 9-14 and 9-19 A. Figure 9-19 B shows another way to fasten the stringer at the top. In this case, the trimmer joist is placed so it supports the back of the top riser.

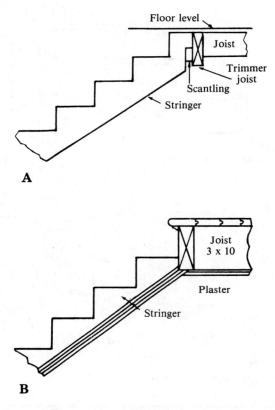

A

B

Two ways of fastening stringers
Figure 9-19

them in place with glued wedges driven into the grooves under the treads and against the back side of the risers. Figure 9-18 B shows how. Hold the tread and riser together with several triangular blocks, about 4" to 6" long, glued along the middle of the steps on the underside of the stairway.

Fitting the top of the stringer or carriage against the floor joists of the landing can be a problem.

Framing Around Stairs

Figure 9-20 shows how to frame around a platform stairway. Reinforce the openings by doubling the headers to receive the tail joists. Figure 9-21 A is a section through the stringers of the first-floor open-stringer stair. It shows the balusters and rough carriage. It also shows a center carriage, which you should use for rigidity. Space carriages of wider-than-normal (2'10" to 3'0") stairs not more than 24" o.c. Use a 3"-thick carriage for open-stringer stairs. The second-floor stair is a typical closed-stringer stair.

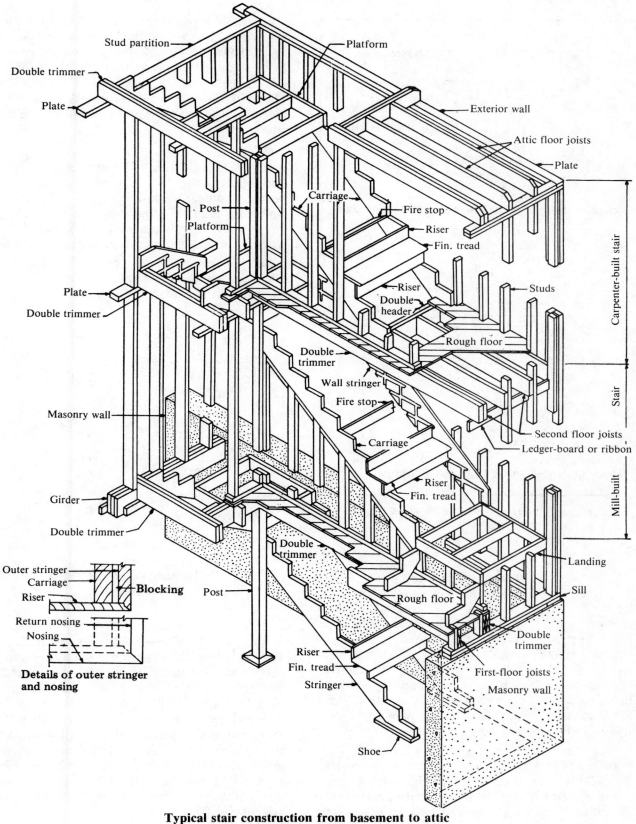

Typical stair construction from basement to attic
Figure 9-20

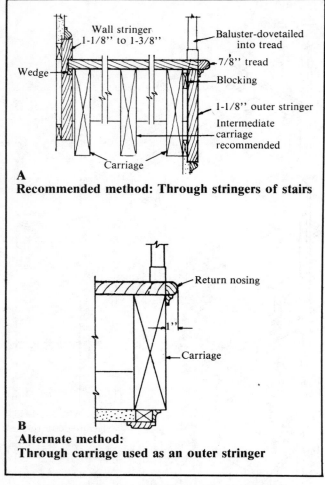

A
Recommended method: Through stringers of stairs

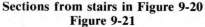

B
Alternate method:
Through carriage used as an outer stringer

Sections from stairs in Figure 9-20
Figure 9-21

Treads and Risers

The treads of a residential stairway should be at least 1⅛" thick. Usually 7/8" risers are thick enough. In a medium-priced house, make the treads and the risers from the same material. If you want to save money, however, you can make the risers from a less expensive material. If the stairway isn't carpeted, finish the treads and risers with a different stain so they're easy to tell apart in a dim light. If the stairway is to be carpeted, plywood treads and risers will be acceptable.

Stairs that are either too steep or too shallow are both tiring to walk on and dangerous, especially for older people. Risers over 8" high are too steep; risers less than 6" high result in stairways that take up too much room. It's best to have risers just over

7" high, with the exact height depending upon the total rise. The risers may not vary in height by more than a 1/4". Therefore, divide the total rise by the number of risers that will give about a 7" rise for each stair. For example, if the total rise is 114", 17 risers give a rise of 6.71" and 16 risers give a rise of 7.13". The best choice would be 16 risers.

Stair Angles

As a general rule, the angle between the floor and a line touching the riser edge of all treads should be between 20 degrees and 50 degrees. But you won't find many stairways that steep or shallow. The best angle is between 30 and 35 degrees. The ideal stairway has an angle close to 32 degrees 30 minutes.

See Figure 9-22 for preferred angles. This diagram was prepared by the National Worker's Compensation Service to help prevent accidents.

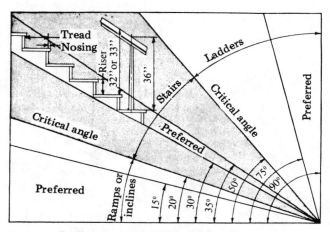

Preferred and critical angles for ramps,
inclines, stairs and ladders
Figure 9-22

Ramps are angled surfaces without steps. They're better than stairs when a large number of people will use them, as in airports and stadiums. They can replace elevators in multi-story parking garages and are common where access to the handicapped is essential. Ramps take up much more space than stairs because they can't be as steep. In fact, foot ramps should have a slope of 1 in 8 or less. A slope of 1 in 10 or less is even better. Figure 9-22 shows that the angles for ramps can be up to 20 degrees, but that angles up to 15 degrees are best.

Rules of Proportion

There's a range of proportions between risers and runs that's best for comfortable walking. For this reason, grand staircases often have shallow risers and wide runs. Basement stairs, with a minimum run available, have high risers and narrow runs.

The relationship between the risers and treads is especially important for safety and ease of travel. The accepted rules are:

Rule 1: The sum of two risers and one tread (in inches) should be between 24" and 25". Thus, a 7" to 7½" riser with a 10" to 11" tread is acceptable.

Rule 2: The sum of the width of one run and the height of one riser (in inches) should equal 17" to 18". Thus, a 10" tread and a 7" to 8" riser, or an 11" tread and a 6" to 7" riser are acceptable.

Rule 3: The product of multiplying the height of the riser by the width of the tread (in inches) should not exceed 75. Thus, either a 10" tread and 7" riser or a 9" tread and an 8" riser is acceptable. Another rule is that the product should equal 66. This would eliminate a 9" tread with an 8" riser, since these dimensions don't produce a good stair. They are, though, acceptable for attic or basement stairs.

Double-checking— Use Rule 3 as a check on Rule 2. If you do this, you'll eliminate the 10" tread with 8" riser. This is because the product of multiplying the two numbers, 80, is too much. You wouldn't eliminate the 11" tread and 6" riser because the product, 66, is acceptable, though small. This last proportion would be appropriate for outside stairs to a public building.

Figure 9-23 shows the best dimensions and angles for stairs. Though it's not always possible to use just these dimensions, stay as close as possible. That guarantees a safe stairway.

Planning Stairs

Calculating Run

Let's use Figure 9-1 as a practical example for illustrating the rules above. The total rise is 9'4½", or 112.5". The total run is 14'4½", or 172.5". This is the total space that you can allow for the stairs. It must include the landing, of course, which will usually be about as long as it is wide. Assuming a 3'0" stair width, you also have a 3'0" landing. Subtract the landing width from the total stair run. Now you have 172.5" less 36", giving 136.5", or

Angle with horizontal		Riser (inches)	Tread (inches)	
Degrees	Minutes			
22	00	5	12½	
23	14	5¼	12¼	
24	38	5½	12	
26	00	5¾	11¾	
27	33	6	11½	
29	03	6¼	11¼	
30	35	6½	11	
32	08	6¾	10¾	These are preferred
33	41	7	10½	
35	16	7¼	10¼	
36	52	7½	10	
38	29	7¾	9¾	
40	08	8	9½	
41	44	8¼	9¼	
43	22	8½	9	
45	00	8¾	8¾	
46	38	9	8½	
48	16	9¼	8¼	
49	54	9½	8	

Angles, risers, and treads for stairs
Tread + Riser = 17½"
Figure 9-23

11'4½" for the run of the stairs, not including the landing.

Calculating Risers

Now your problem is to design these stairs so the rise and the run will make a convenient step. Since 7" is the ideal riser height, start out by using seven as a trial divisor. Divide the total rise, 112.5", by 7": 112.5" divided by 7" equals 16.07. Assume, therefore, that 16 risers is the right number for this stair.

Now, since the 7" you used above was only an approximate number, you need to find the exact height of each riser. So, divide the total rise by the number of risers. The total rise, 112.5", divided by 16 equals 7.031", or 7¹⁄₃₂" for the exact height of one riser.

Steps for Finding Riser Height:

1) Look on the plan to find the distance from one floor to the top of the next. Convert that dimension to inches.

2) Look on the plan for the number of risers. It will usually be in a note.

3) Divide the distance from one floor to the top of the next by the number of risers. Now you have the height of one riser.

Calculating Tread Width

Your next step is to find the width of each tread. Remember that there is always one more riser than there are treads. This is because the upper floor is considered as the last tread. But it isn't counted as a tread when you lay out the stairway. So subtract one tread. Now you have 15 treads. The landing doesn't count in this calculation, so you have 14 treads. You already subtracted the width of the landing (when you figured run) and got a run of 11'4½'', or 136.5''. Divide this by 14 treads: 136.5 divided by 14 equals 9.75'', or 9¾''. Therefore, each tread is 9¾''.

Steps for Finding Tread Width:

1) Find the length of the total run, less any landing, and change feet to inches.

2) Find the number of treads by subtracting one from the number of risers shown. For example, if five risers are shown, there would be four treads; if 15 risers, there would be 14 treads. Then subtract one for any landing.

3) Divide the answer you found in step 1 by the number of treads. Now you have the width of one tread.

Now, double-check your answers by testing them against the three rules. The height of two risers plus the width of one tread is: 7.031 plus 7.031 plus 9.75, equaling 23.812, or about 24. This checks with Rule 1. The rise plus the run is: 7.031 plus 9.75 equals 16.781, or 16²⁵⁄₃₂'', about 17''. This checks with Rule 2. The height of one riser times the width of one tread is: 7.031 times 9.75 equals 68.55. This checks with Rule 3.

Laying Out Stairs

So far, we've only considered planning stairs. Now I'll describe how to lay out the framing for a straight flight of stairs. As an example, we'll use a flight of exterior stairs. Figure 9-24 shows a plan of this stair.

The distance between the floors will be 10'4'', as taken from the architect's plan. The plan calls for 16 risers. The width and length aren't listed, but by scaling the plan we discover that the total run of the stairs is 12'0''. Some plans show the tread height. This one doesn't, so we'll have to figure it ourselves.

First figure the height of risers and the width of

treads. As the total height is 10'4'', or 124'', the height of one riser is: 124 divided by 16, which equals 7.75, or 7¾''. Next, divide the total run by 15: 12'0'', or 144'', divided by 15 equals 9.6'', or 9¹⁹⁄₃₂''.

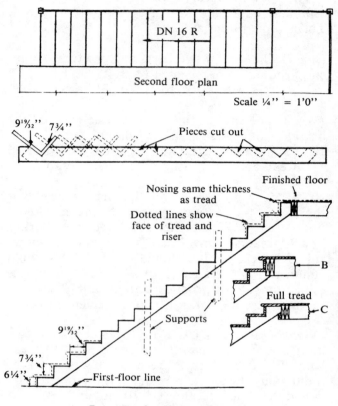

Layout of a flight of stairs
Figure 9-24

Rough Length of Stringer

Now that you know the height of each riser and the width of each tread, you can calculate the stringer length. Be sure to select stock that's long enough. Otherwise all the time you spend doing the layout will be wasted. You won't know the *actual* length of the stringer until you lay it out. But you can find the rough length quite easily with a steel square.

To find the rough length, read the total run, given in feet, as inches on the blade of the steel square. Then read the total rise, given in feet, as inches on the tongue of the square. The distance in inches between these two points gives you the approximate length, in feet, of the stringer. (Remember? You did the same thing to find the rough length of rafters.)

In this case, you'll measure the distance between 12 (the run) on the blade and 10⅓ (the rise) on the tongue. It's about 16''. You can, therefore, use a 16-foot length of either 2 x 10 or 2 x 12 for the stringer.

Laying Out the Stringer

Now lay out the stringer. Rest the stringer on a pair of sawhorses. Start at the left end and work from the top edge. Make sure the outside point of the corner of the square is toward you, as shown in Figure 9-25. Let the number on the square's blade stand for the width of one tread. In this case, 9¹⁹⁄₃₂. Let the number 7¾ on the square's tongue stand for the height of one riser.

Lay out the steps as in Figure 9-25. Each time you move it, the square shows the rise and tread of one step. Mark these and move the square to the next step, as shown by dotted lines. Number each riser as soon as you've marked it so you'll know when you've laid out enough risers.

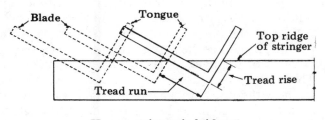

How a stringer is laid out
Figure 9-25

Use a *pitchboard* to make your stringer layout easier, faster, and more accurate. A pitchboard is a triangular pattern laid out to the exact rise and run of the stairway. See Figure 9-26. If it's carefully made, a pitchboard reduces the chance of an error. Make a pitchboard from 1'' thick material. Plane the edges smooth and square with the surface. Mark all cutting lines with a sharp pencil or a knife.

Now cut the stringer along the lines you marked. When you cut out the bottom step, allow for the thickness of one tread, as shown in the lower part of Figure 9-24. If the treads will be 1½'' thick, make the first riser of the stringer 1½'' less than the other risers. By cutting off this amount from the bottom of the stringer, you'll bring the stairs 1½'' lower and to the exact location for the stringer.

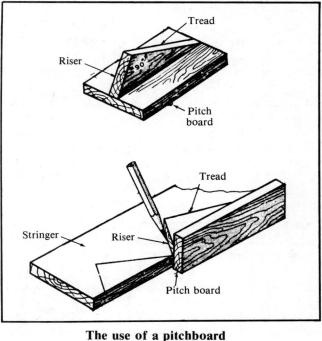

The use of a pitchboard
Figure 9-26

Figure 9-24 shows you all the dimensions. At "B" and "C" you can see different methods of fastening the stringer board to the floor framing above.

Winding Stairs

There's an easy way to build any winder stair: first draw a full-sized layout on the floor. Show the size and shape of treads, the lengths of risers, and all angle cuts of both treads and risers. (The risers should be nailed against stringers in a butt joint.) If you do the drawing on paper, use a scale of 3'' equals 1'0''.

Figure 9-27 shows how to get the different dimensions. Find on the plan the distance between floors and the number of risers. Then find the height of one riser and the width of one tread from the plan. In this figure, the height of a riser is 7½'' and the width of a tread is 10''.

After you draw the plan of the stairway, you can project points (as shown in Figure 9-27) to draw the elevation of the stairs. From the elevation you can project further to get the cuts for the stringer.

Use your steel square to calculate the straight run of stairs. The numbers on the steel square again represent the run and rise of a tread. (In this case 7½'' and 10'')

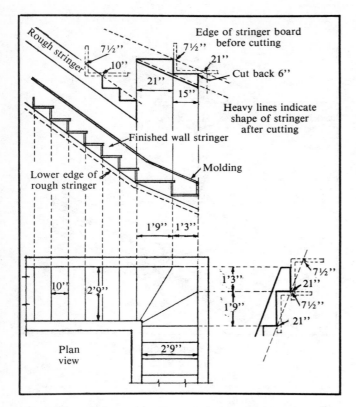

Laying out the stringer for a winding stair
Figure 9-27

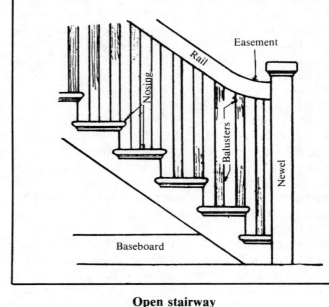

Open stairway
Figure 9-28

You also use the steel square to find the length of the stringer for a winding stairway. Take the height of the riser on the tongue and the width of the tread at the wall line on the blade of the square. The heavy lines in Figure 9-27 show the shape of the rough stringers after they are cut. If you're making a housed stringer, the square shows the edge of the line to be mortised out to receive the treads and risers.

Newel Posts

Here are some hints on installing newels and balusters, whether made by you or bought from a mill shop. Figure 9-28 shows part of an open stairway with the rail, balusters, nosings, and the newel. The curved part of the rail where it joins the newel is called an *easement*. Often, however, the rail joins the newel without an easement.

Diagram A in Figure 9-29 shows an enlarged detail of the first two steps in Figure 9-28. You can see part of the newel, just above the first step. It's ready to be put in place. The dotted lines in B show how to cut the inside corner of the bottom of the

newel so it will slip over the corner of the first step. The downward arrows show how to place the newel.

One of the balusters is also shown ready to be put in place. The heavily shaded housing for the dovetail of the baluster is shown directly below it. There, the return nosing has been cut out enough to show the housing. The arrows indicate how the baluster is fitted into place. Also, notice the return nosing and the two balusters already installed on the second tread.

Figure 9-29 A shows a partial plan view of the two steps in 9-29 B. The part marked "X" is the corner you'll set the newel over. The nosing and the return nosing of this tread have to be cut so they'll tie into the newel.

Figure 9-29 C shows a platform newel in place but not fastened. Set the newel before laying the finish flooring of the platform. Mark any necessary cutting lines carefully with a steel square. Cut them just as carefully. You'll also need to cut the nosings to let the newel in. You can usually leave the return nosings off until the newel is permanently fastened. In this drawing, the return nosing is shown in place. Notice that the pendant (the ornamental detail at the bottom of the newel) fits against the ceiling of the platform.

Above the 2 x 12 framing you can see the wedge holding the newel. When the newel is perfectly aligned and plumb, fasten it permanently. Fill any

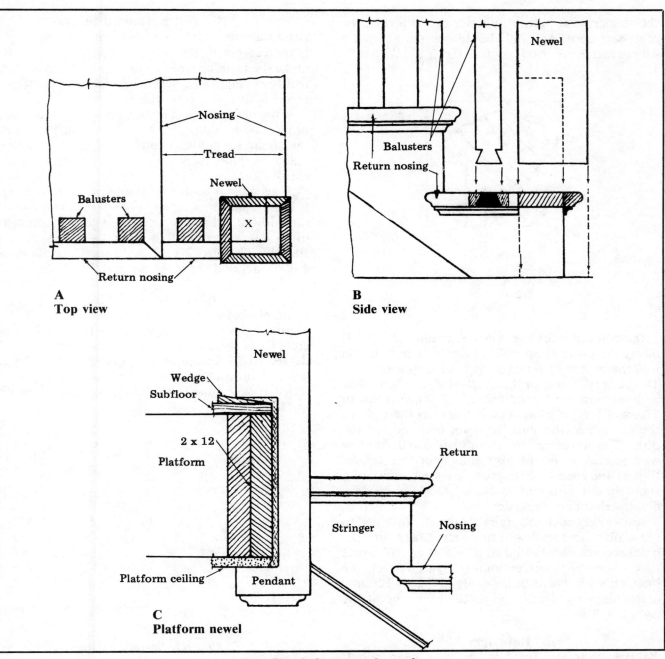

A
Top view

B
Side view

C
Platform newel

Installing balusters and newel
Figure 9-29

open space around the wedge with blocking that's fastened securely so it can't work out of place. Sometimes you can use 2 x 10 or even 2 x 8 framing pieces instead of 2 x 12, depending on the job.

Handrail Height

All open stairs need rails, and rails must go around the edges of open stairwells. Make rails 32" or 33" above the treads along the slope of the stair, depending on the pitch. Look again at Figure 9-22. Rail height on landings and around open wells should be 34".

Handrail design is a matter of taste. It should always complement the interior finish. Figure 9-30 shows a cross section of a handrail. As a rule, the cross-section of any rail should contain at least 9 square inches. A good handrail size is 3½″ x 3½″.

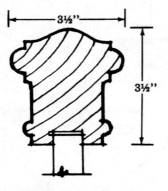

Cross section of a handrail
Figure 9-30

In ordinary stairs the rail is generally straight. It meets the posts at an oblique angle. The rails will join the post at different heights on opposite sides. To get around this problem, the rail is often made with a ramp, or "gooseneck," as shown back in Figure 9-3. The gooseneck is made so the ends of the rail are at the same level on each side of the post. The lower end of the rail is often finished with an easement, as also shown in this figure. Ramps and easements improve the appearance of a stairway but also add to the cost. Many modern designs eliminate these details.

Rails along enclosed stairs are usually fastened to the walls with metal wall brackets. Make sure the brackets are substantial and less than 10′ apart unless the rail is exceptionally rigid. Attach the brackets with screws or lag-bolts to the wall frame, or to blocking firmly spiked to the frame at the right height.

Balusters

Balusters support the handrail and keep people from falling over the ends of the steps. They're also ornamental. They can be any convenient size, but should be made of hardwood. A common style is square at the ends and turned at the middle. Others are square or round and tapered from 1¾″ to 1⅛″. For basement stairs the balusters can be cut from 7/8 x 3 material.

You'll usually put two balusters on each step: one flush with the face of the riser and the other halfway between the risers. See Figure 9-28.

Attic and Basement Stairs

Attic stairs are similar to other stairs, but are usually made from lighter and less costly material. But make the framing strong enough so the stairs won't spring or sag when several people are carrying a heavy load down from the attic.

If the basement will be used as a recreation room or playroom, it deserves a stairway nearly as nice as the main stairway. You'll build most basement stairs of pine with simple cut 2″ stringers and plank treads.

A very plain basement stair doesn't have risers. It has a small, square, solid newel post and a handrail without balusters. Since basement headroom is often low, the run of the stairs will be short and straight. Use two 2 x 12 stringers if the stairs are 3′0″ wide, and three if they are 4′0″ wide.

Other Kinds of Stairs
Concrete Basement Stairs

Figure 9-31 shows forming details for exterior concrete basement steps. Cut or fill the ground to a 30 degree to 35 degree angle. (Look back to Figure 9-22.) Undisturbed soil is best as a base for the steps. If you must use fill dirt, compact it well by tamping.

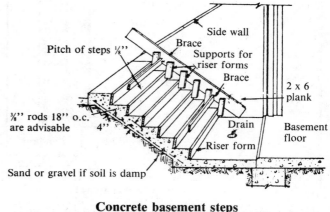

Concrete basement steps
Figure 9-31

Determine the rise and tread of the steps the same way we've done it for stairs. Use a 2 x 6 along each side of the steps and brace it to a side wall or with stakes driven into the ground.

When the steps will butt against a side wall, mark the location of the riser and tread on the wall. Nail supports for the riser forms to the 2 x 6 at the marked locations. The riser form (board) should be

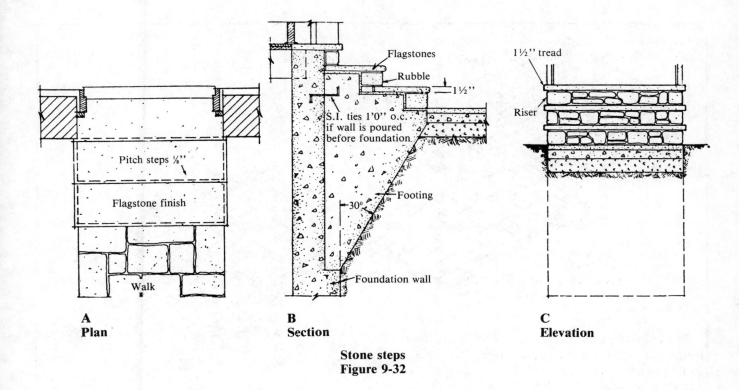

Stone steps
Figure 9-32

the same width as the height of the riser. Install all riser forms and brace them so that they'll hold when you pour the concrete.

Pour the steps directly on the ground unless there's a dampness problem. If the soil is damp, pour a 1" layer of dry sand, gravel or crushed stone to help drain excess water. Place 3/8" reinforcement rods parallel to treads 18" on center.

Avoid using a mix that's too wet. You want the concrete stiff enough to prevent excess slumping. Be sure to work the concrete against the face of the riser forms for a good finish. Give the treads a broom finish to prevent a slippery surface.

Provide a drain at the bottom of the steps. The floor at the base of the steps must be lower than the basement floor to prevent water from flowing into the basement.

Install an isolation joint where the steps join the building. Use 1/2"-thick asphaltum-impregnated felt.

Pitch treads 1/8" downward toward the front to insure proper drainage. When concrete has set, remove the forms.

Stone and Brick Steps
Figure 9-32 shows the plan, vertical section, and elevation of exterior stone steps. The flagstone treads have a rise of about 6". You can change the footing material according to conditions. As with any exterior steps, pitch the treads slightly forward about 1/4" per foot. This provides good drainage. Figure 9-33 illustrates three styles of exterior brick steps. Here are the basic rules for laying brick steps:

- Always lay the brick on a firm base.
- Make the treads at least 12" wide, or they'll be dangerous when covered with ice or snow.
- Pitch steps forward with a slope of 1/4" per foot.
- Never slope the concrete base: step it off horizontally.

Pour the concrete thick enough to prevent cracking. A 4"-thick slab is usually sufficient. Where the sub-base isn't firm, reinforced brick construction is an economical solution. For best results in exterior step construction, use a vapor barrier between the ground and concrete base.

Always fill joints in steps with cement mortar and point it with a "thumb" joint, which is a broad, slightly concave joint thoroughly rubbed with a steel jointing tool. Lay the front of the treads with full length headers. Never use half brick in this position.

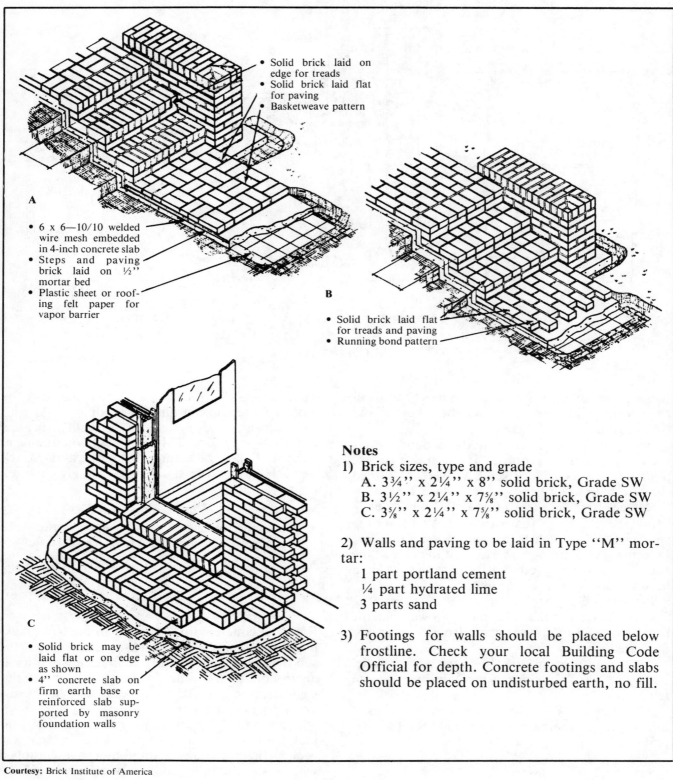

Notes on the figure:

• Solid brick laid on edge for treads
• Solid brick laid flat for paving
• Basketweave pattern

A
• 6 x 6—10/10 welded wire mesh embedded in 4-inch concrete slab
• Steps and paving brick laid on ½" mortar bed
• Plastic sheet or roofing felt paper for vapor barrier

B
• Solid brick laid flat for treads and paving
• Running bond pattern

C
• Solid brick may be laid flat or on edge as shown
• 4" concrete slab on firm earth base or reinforced slab supported by masonry foundation walls

Notes

1) Brick sizes, type and grade
   A. 3¾" x 2¼" x 8" solid brick, Grade SW
   B. 3½" x 2¼" x 7⅞" solid brick, Grade SW
   C. 3⅝" x 2¼" x 7⅞" solid brick, Grade SW

2) Walls and paving to be laid in Type "M" mortar:
   1 part portland cement
   ¼ part hydrated lime
   3 parts sand

3) Footings for walls should be placed below frostline. Check your local Building Code Official for depth. Concrete footings and slabs should be placed on undisturbed earth, no fill.

Courtesy: Brick Institute of America

**Brick steps
Figure 9-33**

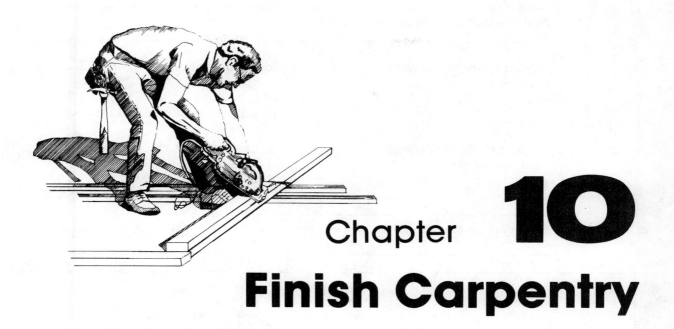

Chapter 10

Finish Carpentry

Here's a good rule to remember: It takes a magician to sell a house that shows the handiwork of a poor finish carpenter. Use the wrong materials, have a slapdash finish carpenter hurry through the job, leave key items "almost done," don't take the trouble to correct obvious defects, and you've got an albatross on your hands.

Nothing shows like finish work. And finish work *must look good.* There's no substitute for selecting the right materials and having them installed by skilled craftsmen.

A house with a poor finish will sell eventually, of course. But it will probably sell at a loss, and losses won't keep you in the construction business.

I'll give you a second rule: the less trim, the better. If you keep exterior and interior trim to a minimum, you reduce the construction costs. You'll also eliminate a lot of maintenance costs for the owner.

Sheathing

The house is framed and roofed. Your next step is to apply the rough outside covering called *sheathing.* Sheathing is normally considered "rough carpentry," but we'll discuss it here briefly since it serves as the base for exterior finish materials such as siding, stone or brick veneer. And sheathing has several other functions:

• It makes the whole building stiffer.

• It makes the walls more resistant to winds.

• It helps insulate the building.

You can install sheathing horizontally before the walls are in place, or after the framing is completed.

In some parts of the country, you can eliminate the sheathing and apply the siding directly to the studs. Look at Figure 10-1 to see how to apply panel siding. Normally, you'll install the panels vertically, butting the edges over a stud. Begin at the left corner of the wall and plumb the first panel. Nail as recommended by the manufacturer or as shown in Figure 10-1. Avoid nailing in grooves. Leave a gap at butt joints. Use an inside corner board 1⅛" x 1⅛" and outside corner boards 1⅛" x 3" and 1⅛" x 4". Treat horizontal joints as illustrated in Figure 10-1 to provide proper drainage. Even when it isn't required, most builders use sheathing to improve the wall's insulating value.

Figure 10-2 shows how to install sheathing. The sheathing in this figure is rigid fiberboard. The nailing specs shown are for a house without diagonal corner bracing. You could increase the nail spacing to 8" at intermediate framing members, and 4" at the edges of boards over studs. If you're going to use masonry veneer, use insulating sheathing directly on the studs.

Nail size and spacing requirements

When racking requirements must be met	6d or 8d galvanized box nail* 3/8'' in from edges 4'' o.c. along all edges 8'' o.c. along intermediate supports
When racking requirements need not be met	6d or 8d galvanized box nail* 3/8'' in from edges 6'' o.c. along all edges 12'' o.c. along intermediate supports

*Use 6d box nails only for direct panel-to-stud applications or for panel over wood or plywood sheathing applications.

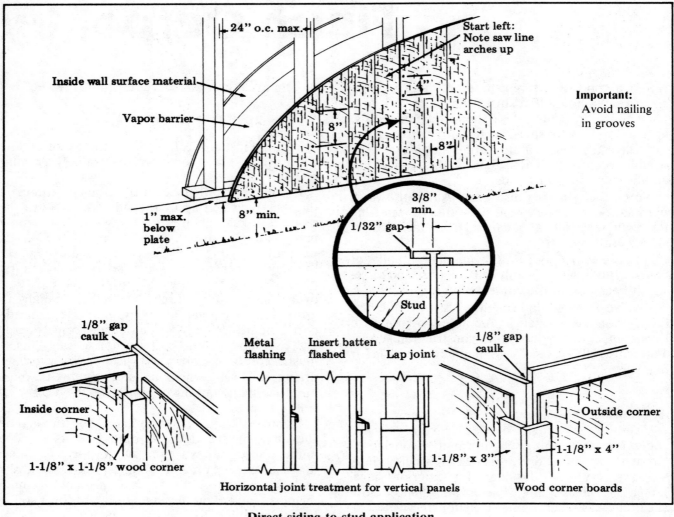

Direct siding-to-stud application
Figure 10-1

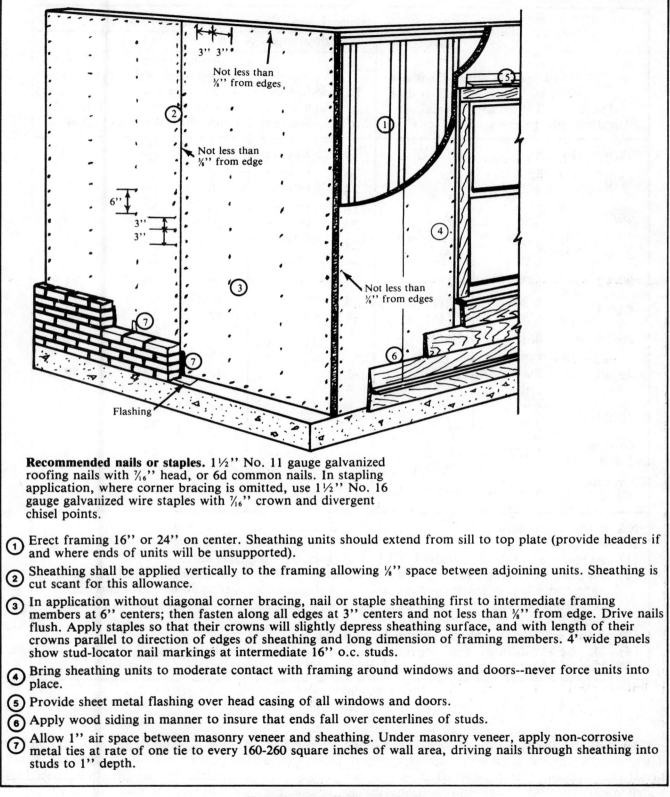

Recommended nails or staples. 1½'' No. 11 gauge galvanized roofing nails with 7⁄16'' head, or 6d common nails. In stapling application, where corner bracing is omitted, use 1½'' No. 16 gauge galvanized wire staples with 7⁄16'' crown and divergent chisel points.

① Erect framing 16'' or 24'' on center. Sheathing units should extend from sill to top plate (provide headers if and where ends of units will be unsupported).

② Sheathing shall be applied vertically to the framing allowing ⅛'' space between adjoining units. Sheathing is cut scant for this allowance.

③ In application without diagonal corner bracing, nail or staple sheathing first to intermediate framing members at 6'' centers; then fasten along all edges at 3'' centers and not less than ⅛'' from edge. Drive nails flush. Apply staples so that their crowns will slightly depress sheathing surface, and with length of their crowns parallel to direction of edges of sheathing and long dimension of framing members. 4' wide panels show stud-locator nail markings at intermediate 16'' o.c. studs.

④ Bring sheathing units to moderate contact with framing around windows and doors--never force units into place.

⑤ Provide sheet metal flashing over head casing of all windows and doors.

⑥ Apply wood siding in manner to insure that ends fall over centerlines of studs.

⑦ Allow 1'' air space between masonry veneer and sheathing. Under masonry veneer, apply non-corrosive metal ties at rate of one tie to every 160-260 square inches of wall area, driving nails through sheathing into studs to 1'' depth.

Sheathing installation methods
Figure 10-2

Panel identification	Thickness (inches)	Subfloor and roof used as walking surface (1)	Maximum spacing of supports (inches)		Roof sheathing (1)		
			Wall sheathing		Asphalt, wood shingles and shakes, built-up roofs Edges (2)		20 year bonded roofs Edges (2) blocked
			Exterior finish nailed to				
			Studs	Sheathing	Blocked	Unblocked	
12/0	5/16	NA	16	16 (3)	12	12	NA
16/0	5/16 3/8	NA	16 24	16 (3) 16 (5)	16 16	16 16	NA
20/0	5/16 3/8	NA	16 24	16 (3) 16 (5)	20 20	20 20	NA
24/0	3/8 1/2	NA	24 24	16 (5) 24	24 24	20 24	16 16
30/12	5/8	12	24	24	30	26	24
32/16	1/2 5/8	16 (4) 16 (4)	24 24	24 24	32 32	28 28	24 24
36/16	3/4	16 (4)	--	--	30	30	24
42/40	5/8 3/4 7/8	20 (4)	--	--	42	32	32
48/24	3/4 7/8	24 24	-- --	--	48 48	36 36	48 48
Groups 1 & 2	1 1/8	48	--	--	72	48	72
Groups 3 & 4	1 1/4	48	--	--	72	48	72

NA = Not acceptable
Notes:
(1) Applicable to C-D, Structural I and II, and C-C Exterior grades only.
(2) Blocking of unsupported edges of roof sheathing shall be by means of wood blocking, approved T & G edges or special corrosion resistant metal H clips designed for this purpose. Use one clip for each space less than 48" wide, and two clips for spaces 48" and wider.
(3) Apply grain of face ply perpendicular to studs.
(4) May be 24" if 25/32 wood strip flooring is applied perpendicular to joists.
(5) May be 24" if face grain is perpendicular to stud.

Plywood subfloor and sheathing
Figure 10-3

Figure 10-3 shows specs for plywood sheathing. These specs cover a wide range of uses. In conventional construction, 32/16 panels 1/2" thick are widely used for floor, roof and wall sheathing. (The panel identification *3/16* means it can be used on rafters spaced up to 32" o.c. and on floor joists spaced up to 16" o.c.) Note that 1/2" plywood is suitable for use on studs spaced 24" o.c. Figure 10-4 gives the specs for using gypsum sheathing, fiberboard sheathing, and fiberboard

Material Size Corner bracing (3)	Gypsum sheathing (2)			Fiberboard roof deck (1)	
	½" x 2' x 8' Installed	½" x 4' x 8' (or longer) Installed	Omitted	2" NA	3" NA
Max. framing spacing	24"	24"	16"	32"	48"
Nail size	1½"x 11 ga.	1½"x 11 ga.	1½"x 11 ga.	Adequate to penetrate 1½" into framing	
Nail pattern:					
Edge	8"	8"	4"	4½" at main supports 8" ridge and openings	
Intermediate	8"	8"	8"		
Installation	Horizontal	Horizontal or vertical	Vertical	Long dimension perpendicular to framing. Stagger end joints	

Material Size Corner bracing (3)	Fiberboard sheathing (2)							
	½" x 2' x 8' Installed	Omitted	½" x 4' x 8' Installed	Omitted	25/32" x 2' x 8' Installed	Omitted	25/32" x 4' x 8' Installed	Omitted
Max. framing spacing	24"	NA	24"	16"	24"	NA	24"	16"
Nail size	1½" roofing nail or 6d common nail				1¾" roofing nail or 8d common nail			
Nail pattern:								
Edge	4"	NA	4"	3"	4"	NA	4"	3"
Intermediate	8"	NA	8"	4"	8"	NA	8"	6"
Installation	Horizontal	NA	Vertical	Vertical	Horizontal	NA	Vertical	Vertical

Notes:
NA = Not applicable.
(1) Structural fiberboard insulating roof deck, without vapor barrier, used as combined roof deck and ceiling finish shall be limited to areas where the outside design temperature is more than 10 F.
(2) Sheathing, unless identified as Nail Base, shall not be used as a nailing base. Shingle type exterior finish materials may be supported by sheathing when special methods of attachment designed for this use are used.
(3) Omission of corner bracing is dependent upon compliance with criteria for resistance to racking.

Gypsum sheathing, fiberboard sheathing and
fiberboard roof deck
Figure 10-4

roof deck. Follow the manufacturer's directions and the specs given in Figure 10-4 when you install these sheathings.

Insulation Sheathing
If you want to build truly energy-efficient homes, use insulating sheathing. Figure 10-5 shows why. Compare the R-values of the different sheathing materials.

The best insulator shown in Figure 10-5 is Ther-

max. (Chapter 8 described Thermax in more detail.) Its uniform closed-cell structure is exceptionally efficient. You can buy Thermax in standard 4 x 8 or 4 x 9 sheets, and from 1/2" to 3" thick.

Wood Boards and Planks
Most builders use plywood or insulation board sheathing, but wood board sheathing is sometimes useful. The stud spacing specs for wood board sheathing are:

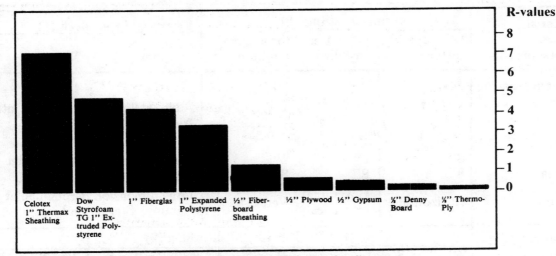

Source ASHRAE 1977 Fundamental Handbook and published product literature

Thermo-Ply—Reg. trademark of Simplex Industries, Adrian, MI
Denny Board—Reg. trademark of Denny-Corp., Calwell, OH
Expanded Polystyrene is sold by many different manufacturers
Styrofoam—Reg. trademark of Dow Chemical manufacturers
Fiberglas—Reg. trademark of Owens-Corning Fiberglas Corporation, Toledo. OH

**Comparative R-value (at 70⁰ mean temperature)
of sheathing
Figure 10-5**

• Studs 24" o.c. for 1" solid board.

• Studs 16" o.c. for 3/4" net spaced boards (If boards are 4" or more wide, studs can be 24" o.c.).

• Studs 24" o.c. for 3/4" solid board.

• Studs 16" o.c. for 5/8" net spaced boards. (Again, studs can be 24" o.c. if boards are 4" or wider.)

Wood boards can be square-edged, shiplap, or T&G. Unless you use end-matched boards, put the joints over studs. Make sure that no two successive joints are on the same stud. Use at least two 8d nails for 6" wide boards. And use three 8d nails for 8" widths.

You can nail the boards horizontally, as shown in section A in Figure 10-6, or diagonally, as shown in Figure 10-6 B. Horizontal sheathing uses less material and fewer manhours than diagonal sheathing, but it isn't as strong and rigid.

Siding

Popular exterior finishes include wood siding, plywood panels, hardboard panels, and vinyl or aluminum siding. See Figure 10-7 for a short course on estimating how much siding to buy.

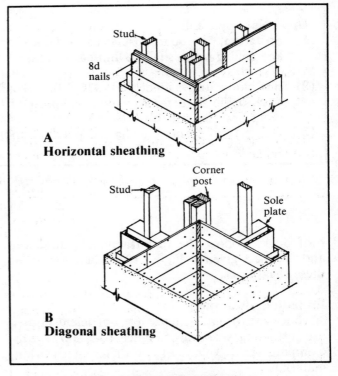

**Board sheathing application
Figure 10-6**

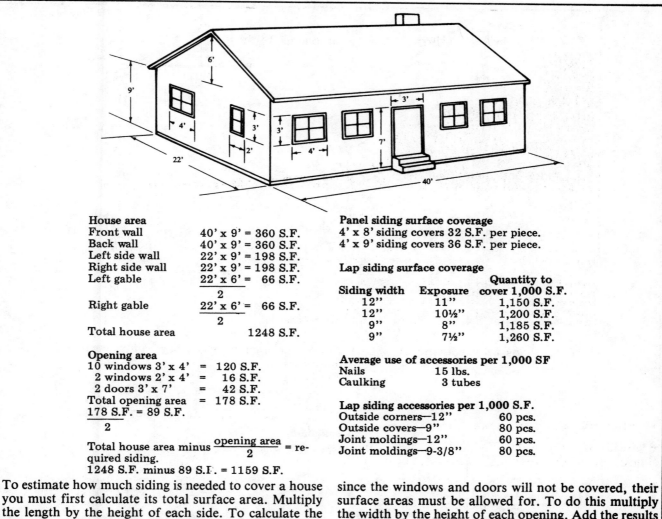

House area

Front wall	40' x 9'	= 360 S.F.
Back wall	40' x 9'	= 360 S.F.
Left side wall	22' x 9'	= 198 S.F.
Right side wall	22' x 9'	= 198 S.F.
Left gable	$\frac{22' \times 6'}{2}$ =	66 S.F.
Right gable	$\frac{22' \times 6'}{2}$ =	66 S.F.
Total house area		1248 S.F.

Opening area

10 windows 3' x 4'	=	120 S.F.
2 windows 2' x 4'	=	16 S.F.
2 doors 3' x 7'	=	42 S.F.
Total opening area	=	178 S.F.

$\frac{178 \text{ S.F.}}{2}$ = 89 S.F.

Total house area minus $\frac{\text{opening area}}{2}$ = required siding.

1248 S.F. minus 89 S.F. = 1159 S.F.

Panel siding surface coverage
4' x 8' siding covers 32 S.F. per piece.
4' x 9' siding covers 36 S.F. per piece.

Lap siding surface coverage

Siding width	Exposure	Quantity to cover 1,000 S.F.
12"	11"	1,150 S.F.
12"	10½"	1,200 S.F.
9"	8"	1,185 S.F.
9"	7½"	1,260 S.F.

Average use of accessories per 1,000 SF

Nails	15 lbs.
Caulking	3 tubes

Lap siding accessories per 1,000 S.F.

Outside corners—12"	60 pcs.
Outside covers—9"	80 pcs.
Joint moldings—12"	60 pcs.
Joint moldings—9-3/8"	80 pcs.

To estimate how much siding is needed to cover a house you must first calculate its total surface area. Multiply the length by the height of each side. To calculate the area of a gable, multiply the length by the height and then divide the result in half. Add these square foot figures together plus the areas of any other elements that will be sided such as dormers, bays and porches.

You now have the total surface area of the house. But since the windows and doors will not be covered, their surface areas must be allowed for. To do this multiply the width by the height of each opening. Add the results together. Since some waste is involved in cutting the siding to fit around these openings, the total area of the openings is divided in half. This figure is then subtracted from the total area of the house to arrive at the total square footage of siding required to cover the house.

Estimating siding requirements
Figure 10-7

Wood Siding

Wood siding comes in various patterns: plain, beveled, or matched. It's made in sizes ranging from 1/2" x 4" to 3/4" x 12". The woods commonly used for siding include western red cedar, Idaho white pine, cypress, redwood, and Douglas fir. Since siding is exposed to the weather, durability and paint-holding ability are important con-

siderations in choosing the material. Make sure the wood siding you use is well-seasoned.

Plain siding boards are dressed on four sides. Nail them over the sheathing horizontally or vertically. Figure 10-8 A shows how. You can tongue and groove the joints, or cover them with battens about 1/2" x 1¾". Or lap the boards over one another in horizontal courses, as shown in the

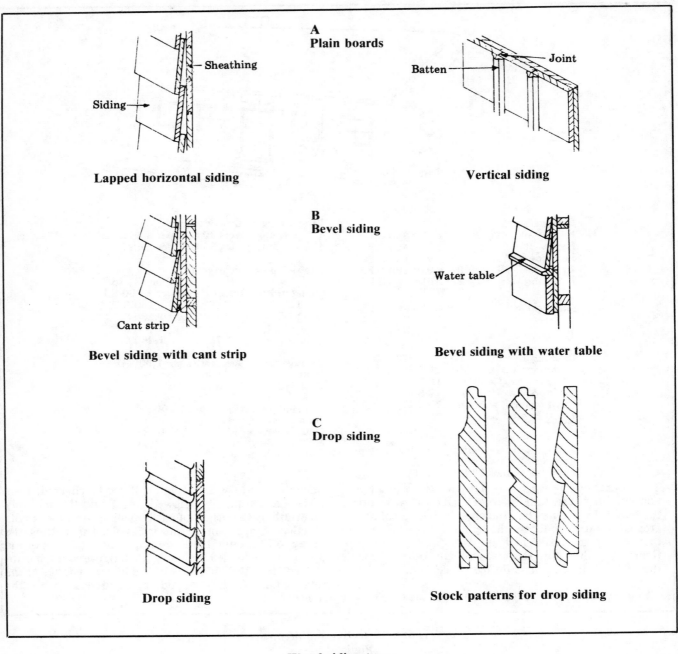

A
Plain boards

Sheathing

Siding

Batten — Joint

Lapped horizontal siding

Vertical siding

B
Bevel siding

Cant strip

Water table

Bevel siding with cant strip

Bevel siding with water table

C
Drop siding

Drop siding

Stock patterns for drop siding

Wood siding types
Figure 10-8

illustration.

Bevel siding boards are tapered, or beveled, so they're thinner on the upper edge than on the lower edge. Always use sheathing under bevel siding. Nail the siding over the sheathing in lapping courses as shown in section B of Figure 10-8. Make the lap at least 1''. Bevel siding 8'' or wider is often called *bungalow* or *colonial siding*.

Drop siding, as shown in Figure 10-8 C, has a tongue-and-groove or other matched joint. It's made in many different designs. Figure 10-8 C shows some stock patterns. You can use drop

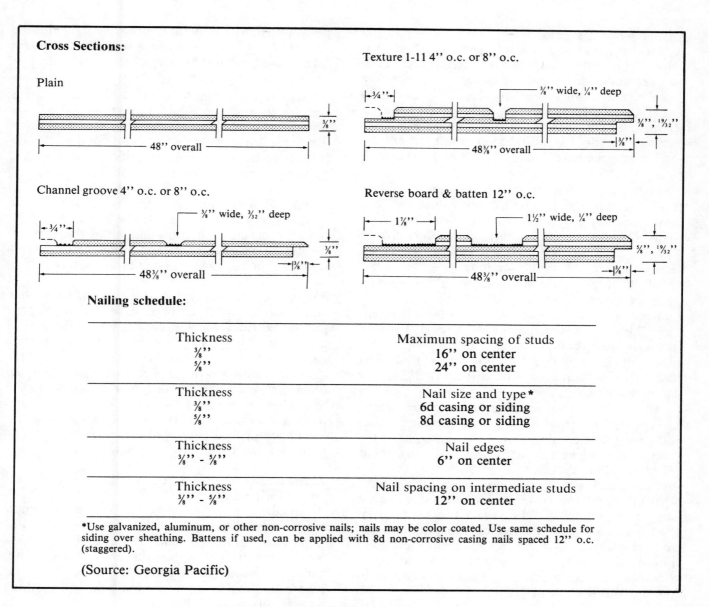

Plywood siding types
Figure 10-9

siding with sheathing, or without it, to save money. Drop siding over sheathing makes a very strong, well-insulated outside wall. But on farm out-buildings, garages, summer homes, and similar buildings, just nail the siding right to the studs.

Plywood Siding

Plywood siding is a popular exterior covering. It's weather resistant, easily installed, and cuts labor costs. In many cases, it can be applied directly to studs, as we saw in Figure 10-1. Figure 10-9 gives specifications for stud spacing and nailing.

Plywood siding includes APA Sturd-I-Wall panels, panel siding and lap siding. Many styles are available: plain, textured, grooved, board and batten, rough-sawed and more.

Stack plywood siding flat in a dry enclosure until you're ready to install it. If an enclosure isn't available, stack the siding on boards and cover them with polyethylene to keep them dry. Plywood treated with a water repellent or wood preservative may react chemically with foam sheathing if it's installed when it's wet.

Most plywood siding is installed over sheathing. Hot-dipped or hot-tumbled galvanized steel nails are recommended for most applications. For best results, use stainless steel or aluminum nails. On panels, space 6d nails 7'' at edges and 12'' at intermediate points. On lap siding, use 4d nails at vertical butt joints and 6d along the bottom edge and at intermediate points. Use 8d nails if siding is wider than 12''.

Plywood panels are normally installed vertically, but they can also be applied horizontally. When applying siding over rigid foam sheathing, drive nails flush with the siding surface. Don't overdrive the nails. Overdriven nails will dimple the siding and compress the foam sheathing.

When installing plywood siding, it's sometimes hard to maintain the uniformly flat appearance required for a finished wall. Straight studs are essential. Leave the recommended spacing between panel edges. This leaves room for small adjustments if the panel gets out of plumb. It also lets the panels expand slightly without causing the siding to buckle.

The nailing sequence also affects the appearance of the finished wall. To install plywood panel siding without compressing the panels, install the first panel flush at a corner. Use a level to plumb it. Then position the second panel, leaving the required gap. Lightly tack the panel at each corner. Drive the first row of nails along the edge nearest the preceding panel. Nail from top to bottom. Remove tacking nails. Then nail along the first intermediate stud. Continue by nailing at the second intermediate stud, and then along the outside edge. Complete the installation by nailing the siding to the top and bottom plates.

Hardboard Siding

Hardboard siding is popular because it costs less than most other siding, and is easy to install without special tools. It has several advantages over wood: nails don't split it; there are no knots; and the thickness, density, and appearance are uniform. Hardboard siding is made from logs, chips, and sawmill byproducts. Since it's wood, it must be maintained like all wood siding. Termites and weather affect it. It doesn't, however, have any grain to raise or check.

Protect hardboard from the weather until you install and prime it. Stack it evenly. Protect it from grease and dirt. And handle it carefully. Use a pro-

per vapor barrier (1 perm or less) like polyethylene film or foil-backed gypsum board, on the warm side of the building. This keeps damaging condensation from forming within the walls.

You can use siding over sheathed or unsheathed walls, with studs spaced not more than 16'' o.c. Be sure the wall is adequately braced with corner bracing or sheathing. Whether you apply siding directly to the studs or put it over wood sheathing, use building paper or felt directly under the siding.

Cutting— Whenever possible, do all cutting and marking on the back side. If the finished surface must be cut, use clean tools and put heavy paper or cardboard under the saw. Use a fine-toothed hand saw, or a power saw with a combination blade.

Nailing— Use 8d galvanized box nails for siding and outside metal corners. Use 6d or longer, galvanized siding or nonstaining box nails for the heavy-gauge vinyl strip and the inside metal corners.

Finish nail heads of exposed nails with matching touch-up paint. Or use color-matched nails. Special color-matched nails, caulk, and touch-up paint are all available. If you use colored nails to install prefinished panels, use the plastic hammerhead cap that comes with the nails.

Joints and corners— Butt joints should always be on a stud, whether the wall is sheathed or unsheathed. Use metal joint molding at all butt joints: insert it from the top into the gap at the butt joint. Use prefinished metal outside and inside corners and trim. The wood trim around doors and windows should be at least 1⅛'' thick. Look at Figure 10-10 to see how to handle siding at joints, corners, doors and windows.

Finishing— All exposed areas must be painted. Panels must be primed before you can paint them. First, use a good quality, exterior-grade, oil-based primer. Then use two coats of paint over the primed surface. For finish paint, you can use any good-quality exterior alkyd, oil, or latex paint. Follow the manufacturer's recommendations about using special primers or undercoats, the rate of spread, and application procedures. Paint primer-coated panels within 120 days of installation. If they've been exposed to the weather any longer than that, reprime before painting.

Use stain only on siding with a wood-like tex-

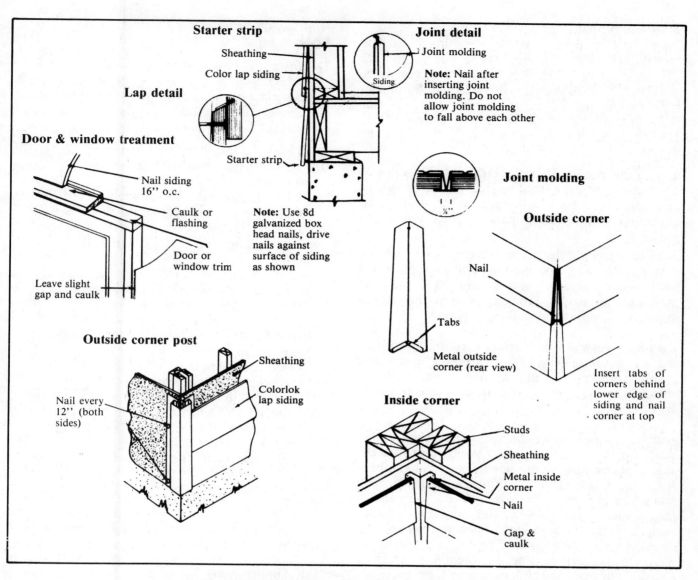

Prefinished metal inside and outside corners and trim
Figure 10-10

ture. Use good-quality exterior, opaque, acrylic latex stains. Don't use oil-base stains. Several coats of stain will make the siding last longer. If you prefer, buy prestained panel siding that doesn't need further staining.

Installing Hardboard over Foam Sheathing
You can apply hardboard over foam plastic sheathing like Thermax. If you do, follow these special application and construction techniques:

1) Make sure the wall is braced well.

2) Use longer nails to account for the greater thickness of Thermax. Nail carefully to keep from crushing the sheathing. For 3/4'' foam-lap siding, use 10d nails. For 3/4'' foam and panel siding, use 8d nails. Use 12d nails for 1'' foam-lap siding and use 10d nails for 1'' foam and panel siding.

3) *Be sure* to use a continuous, unbroken vapor barrier (such as 6-mil polyethylene film) on the interior face of the studs. This lessens the possibility of condensation inside the wall. In some cases, you may even have to vent these cavities to the outside.

Hardboard Shakes

Apply shakes over sheathing or directly to the studs. Use corner bracing if you put the shakes directly on the studs. Lower edges should always be at least 8" above a grade with adequate drainage. Apply vapor barriers on the room side of exterior walls, regardless of the kind of sheathing. Use them even if there is no sheathing. Omitting the barrier may cause buckling.

You can also use shakes on framing that's sloped less than 15 degrees from vertical (45/12 pitch). In this case, first put sheathing over the studs, then cover it with building paper. Then apply the shakes.

Unless your building code requires it, building paper isn't necessary over sheathing in a vertical wall. When building paper is required, put it over the studs if you're applying shakes directly to studs. Then apply the shakes.

Cutting— As when cutting any hardboard product, use a fine-tooth hand saw or a power saw with a combination blade. Cut into, or toward, the finish side. And always prime the cut edge before painting.

Nailing— Apply siding horizontally, over framing spaced up to 16" o.c. Shim to avoid deforming the siding above and below windows or other openings, and to avoid deformation caused by uneven walls. Do not over-drive nails. When you nail the shakes into place, use 8d or 10d corrosion-resistant nails. Use nails long enough to penetrate at least 1" into the studs. Nails should be 16" o.c., on a line 1" from the bottom edge. Set them 3/8" in from the side edge of the shiplap. Use at least two nails on any small trimmed pieces.

Joints and corners— Always leave a slight gap at joints, junctions, corners, and openings. Never spring pieces into place. Leave a 1/32" gap between shiplap edges, and a 1/16" gap at joints. On horizontal runs of 50' or more, use expansion joints or other means of breaking up the continuity of the wall.

Finishing— Use a good quality, non-hardening sealant wherever the siding meets windows, doors, or vertical and horizontal trim. Don't seal shiplap edges. Where siding rests on a wood or concrete sill, make sure there's a good slope for drainage. And seal the joint with a permanent sealant.

Application of hardboard shakes:
Look at Figure 10-11 as you read the steps below.

1) Install a starter strip of wood lath, 3/8" x 1½". Use a chalk line to position the strip at the bottom plate, so it's parallel to the top plate or soffit. Use 6d galvanized siding nails to nail the starter strip to the bottom plate at studs.

2) Level and install the first course of siding (with the bottom edge at least 1/2" below the starter strip) as follows:

   a) Install inside corner members of 1⅛" x 1⅛" wood. And install outside corners of 1⅛" x 3", and 1⅛" x 4" wood members. Textured metal outside corners may also be used.

   b) Leave a 1/16" gap between the siding and the corner boards. Begin at the lower left corner. If necessary, trim the first board on its left-hand edge, so that its right edge falls over a stud. Then, use full pieces for the rest of the first course until you reach the next corner. At the corner, trim the siding to again leave a 1/16" gap.

   c) Begin the second course with a piece trimmed 16" shorter than the first course. Trim it from the left edge. Overlap the first course a minimum of 1½".

   d) Begin the third course with a piece 32" shorter (from the left edge) than the first course. Overlap the second course a minimum of 1½".

3) Repeat steps *b, c,* and *d* for the fourth and all the rest of the courses.

Half-Timber Work

Half-timber work is a decorative treatment that's popular with some architectural styles. Originally it wasn't decorative at all. Many years ago in Europe the supporting timbers in a wall were left exposed as part of the finished wall. Homes were built by first erecting heavy horizontal, diagonal, and vertical timbers. This frame formed the sills, girts, plates, and corner posts. Homes often had secondary vertical and horizontal timbers that framed every door and window opening. Figure 10-12 A shows an example. Many diagonals stiffened the frame. Timbers were all the same thickness through the wall, and were mortised and tenoned together. They formed a strong skeleton to carry the entire load of the building. Spaces between timbers were filled with brickwork or stucco.

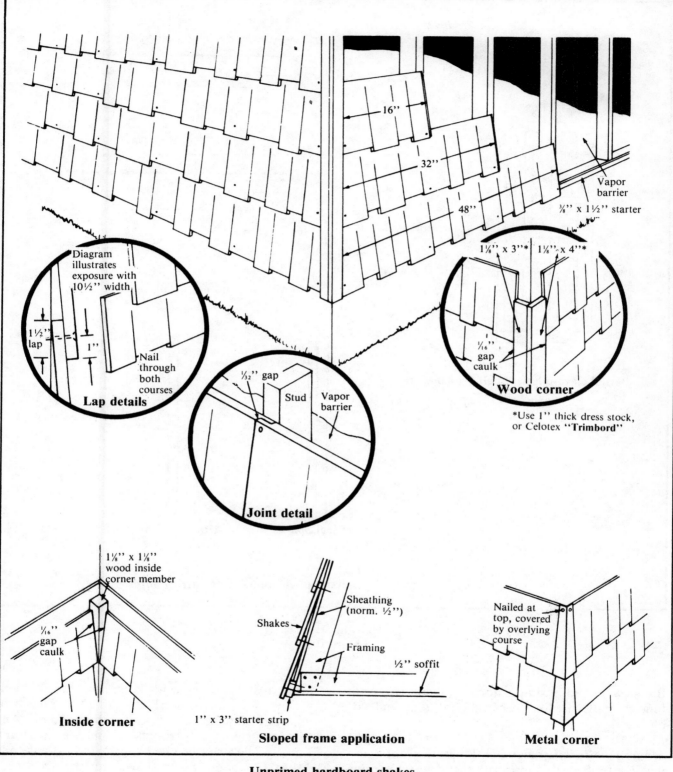

16"

32"

48"

Vapor barrier

⅜" x 1½" starter

Diagram illustrates exposure with 10½" width

1½" lap

1"

Nail through both courses

Lap details

1⅛" x 3"*   1⅛" x 4"*

1/16" gap caulk

Wood corner

*Use 1" thick dress stock, or Celotex "**Trimbord**"

1/32" gap

Stud

Vapor barrier

Joint detail

1⅛" x 1⅛" wood inside corner member

1/16" gap caulk

Inside corner

Sheathing (norm. ½")

Shakes

Framing

½" soffit

1" x 3" starter strip

Sloped frame application

Nailed at top, covered by overlying course

Metal corner

Unprimed hardboard shakes
Figure 10-11

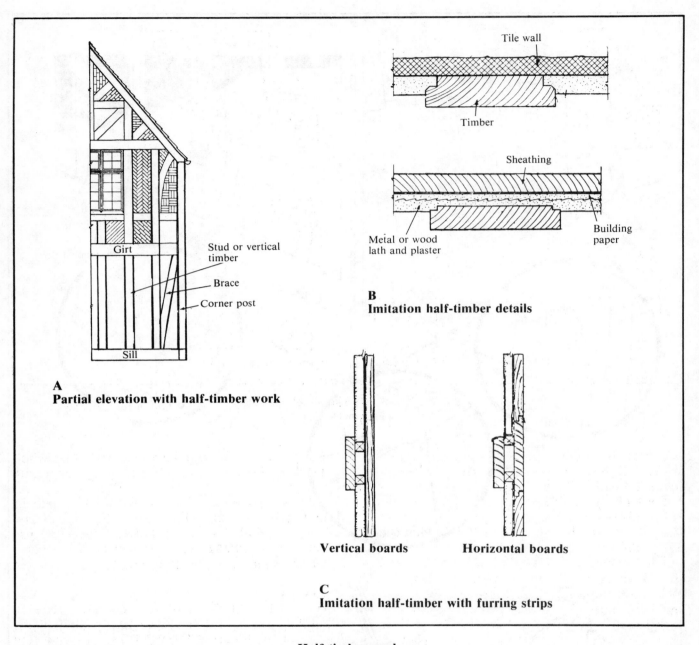

Tile wall

Timber

Sheathing

Metal or wood
lath and plaster

Building
paper

**B
Imitation half-timber details**

Stud or vertical
timber

Girt

Brace

Corner post

Sill

**A
Partial elevation with half-timber work**

Vertical boards   **Horizontal boards**

**C
Imitation half-timber with furring strips**

**Half-timber work
Figure 10-12**

Bricks were laid in simple horizontal courses, or in fancier patterns like herringbone, diagonal, or block.

Modern half-timber work— Today we don't use massive timbers to support the roof and to frame windows and doors. But we imitate the older style by applying rough boards to the wall exterior.

Modern half-timber work is mainly stucco panels with wooden strips. See Figure 10-12 B. The strips are used only on the outside of the building. And they're usually left rough to imitate structural timber. The strips can be from 1'' to 2'' thick, and 4'' to 12'' wide. Sizes depend on the architect's design.

There are several ways to fasten the half-timber

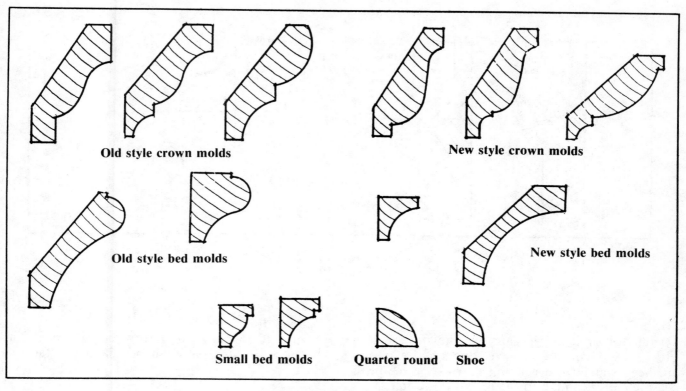

Old style crown molds

New style crown molds

Old style bed molds

New style bed molds

Small bed molds   **Quarter round**   **Shoe**

Common exterior moldings
Figure 10-13

boards to the building wall: On a tile wall, place the strips right next to the tile. On other walls, apply the metal lath and one coat of stucco first.

Look at Figure 10-12 C to see an easy way to do half-timber work. Nail a pair of furring strips to the wall, wherever the half-timber members will be. Tack flashing to the sheathing, wherever the horizontal boards will be. (Later, bend the flashing over the horizontal boards.) Next, stucco the wall. Wait until the plaster has hardened. Finally, nail the half-timber boards over the furring strips. The boards must be at least 3/4'' thick. They also should project out 3/4'' from the face of the wall. There are several ways to do this: You can use ordinary, square-edged boards, and you apply the boards after applying the stucco.

Outside Trim

Outside trim, moldings, cornices, window and door casings, and porch material are more than merely decorative. They cover joints and make the building more weather resistant. They also make the building's exterior look better. Eliminate exterior trim when possible. But when necessary, take

it seriously. Select attractive materials that won't shrink or discolor with age and apply them with care.

All these items of trim or millwork should be made from a wood with a soft, uniform texture. Idaho white pine is a good choice. If the wood also has a low shrinkage value, it will keep its shape, and the joints will remain tight.

Moldings

Keep trim simple. Use it sparingly. Exterior building trim consists mostly of molding. Older molding often have indentations and projections of various shapes (often parts of circles and ellipses) worked on the edges of wood. They produce a variation of light and shade, and give an ornamental and artistic appearance. Modern molding is usually much more plain.

Figure 10-13 shows common moldings used on cornices. The moldings in the top row are crown molds — the old standard forms, and then the later forms of the same moldings. The shapes in the second row are bed molds. The bottom row shows small bed molds, quarter round and shoe.

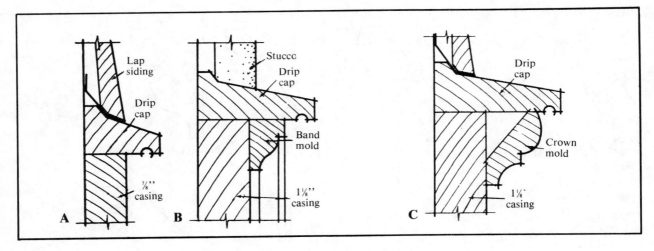

Sections of drip caps
Figure 10-14

Drip Caps

Details of several drip caps are shown in Figure 10-14. Drip caps are put over doors and windows to keep water from running down over the head casing and seeping into the top of the frame.

The top piece projects 2″ or 2½″ from the face of the casing. Use a band or crown mold under the cap as shown in Figure 10-14. If a house has plain trim, use the drip cap in section A. The drip caps in B and C are often used in more expensive construction. In the best quality work, flash the tops of all openings where they're exposed to the weather: See the heavy black lines in sections A and C.

Cornices

The cornice is formed where the eaves of the roof meet the side walls. In gable roofs it's on each side of the house. And in hip roofs it's continuous around the house. In flat or low-pitched roofs, it's usually formed by the extension of the ceiling joists (which also act as rafters).

There are three basic kinds of cornices: the *box cornice,* the *close cornice* (it has no projection), and the *open cornice.*

The box cornice is the most common. It adds a finished look. It also helps protect the side walls from rain. Use an open cornice to protect side walls of low-priced homes at a reasonable cost. Also, use open cornices with exposed beams in contemporary or rustic designs. The close cornice, with its small overhang, doesn't provide very much protection.

Narrow Box Cornice

Figure 10-15 shows a narrow box cornice. The rafter projection is also a handy nailing surface for

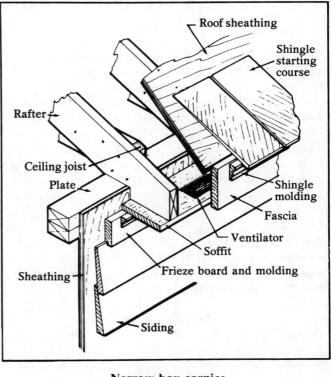

Narrow box cornice
Figure 10-15

the soffit board and fascia trim. Depending on roof slope and rafter size, the extension will vary between 6" and 12" (or more). The soffit is a good place to put inlet ventilators.

Use a *frieze board* or a simple molding to finish off the siding at the top of the wall.

Box Cornice with Lookouts

This cornice style is common on hip-roofed houses. Extra boards must often be added to a wide box cornice so the soffit can be attached. So, toenail lookout boards to the wall and face-nail them to the ends of the rafter extensions. Figure 10-16 shows how. Add a nailing header to the ends of the rafters for the soffit and fascia trim. If the cornice extension is moderate, you can rabbet the fascia instead of using a nailing header. Put ventilators (often continuous narrow slots) in the soffit area.

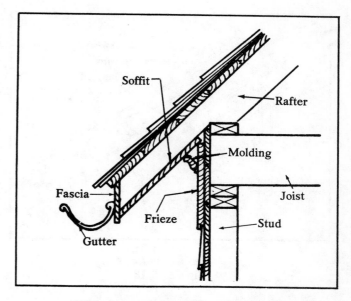

Wide box cornice without lookouts
Figure 10-17

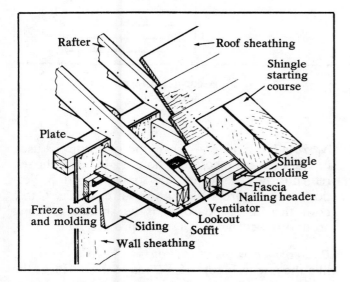

Wide box cornice with horizontal lookouts
Figure 10-16

Soffits can be of lumber, plywood, paper-overlaid plywood, hardboard, medium-density fiberboard, or other sheet material. Soffit thickness depends on the distance between the supports. For 16" rafter spacing, use 3/8" plywood or 1/2" fiberboard.

The cornice with lookouts shouldn't stick out too far from the wall. If it does, you won't be able to put a frieze or frieze molding above the top casing of the windows. If there's a steep slope and a wide projection, the soffit will be too low. So in

this case, use a box cornice without lookouts, as in Figure 10-17.

Box Cornice without Lookouts

A wide box cornice without lookouts has a sloped soffit. It's often used on houses with wide overhangs, as in Figure 10-17 and in the second diagram in Figure 10-21, and is sometimes called a closed cornice. Nail the soffit material directly to the underside of the rafter extensions. In hip-roofed houses, extend this sloping soffit around the roof extension at each end. Install ventilators, either individual ones or a continuous screened slot, in the soffit area.

Open Cornice

The open cornice in Figure 10-18, made with or without a fascia board, saves you money in low-cost construction.

Close Cornice

A close cornice has no rafter projection beyond the wall. See Figure 10-19. Carry the sheathing to the ends of the rafters and ceiling joists. Use just a frieze board and shingle molding to edge the roof. This cornice is easy to build, but it's not very attractive. And it doesn't give the side walls much weather protection, or provide space to put ventilators.

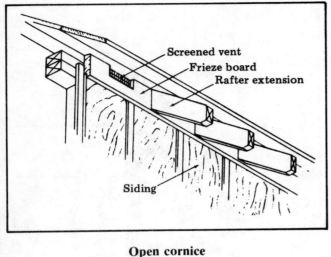

Open cornice
Figure 10-18

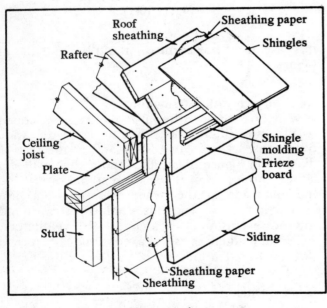

Close cornice
Figure 10-19

Sidewalls are less expensive to maintain where there is ample roof overhang — less painting, replacement, and window deterioration. Even brick lasts longer. And eave ventilators are the best thing you can do for an attic. They provide a natural air current, which is good in any season.

Cornice Return

In hip and flat roofs, the cornice usually continues around the entire house. But in a gabled house it must be joined with the gable ends. So, the cornice return is used to finish the end of the cornice. Figure 10-20 shows a box cornice return. The detail you use will depend on the style of cornice, and how far the gable roof projects past the end wall.

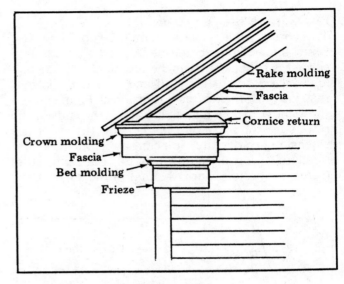

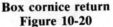

Box cornice return
Figure 10-20

Narrow box cornice— This kind of cornice is used in Cape Cod or colonial houses. It has a boxed return if the rake section has some projection. Section A in Figure 10-21 shows an example. Carry both the fascia board and the shingle molding around the corner of the rake projection.

Box cornice without lookouts— For this cornice, like the one in Figure 10-17, make the soffit of the gable-end overhang at the same slope as the cornice soffit. Join them as shown in Figure 10-21 B. Use this simple system when there are wide overhangs at both sides and ends of the house.

Close rake— This is a gable end with little projection. It's used with a narrow box cornice or a close cornice. In this style, the siding butts into the gable-end frieze board. This frieze board joins the cornice frieze board or fascia, as Figure 10-21 C shows.

Rake or Gable-End Finish

The *rake section* is the extension of a gable roof beyond the end wall of the house. The two kinds of

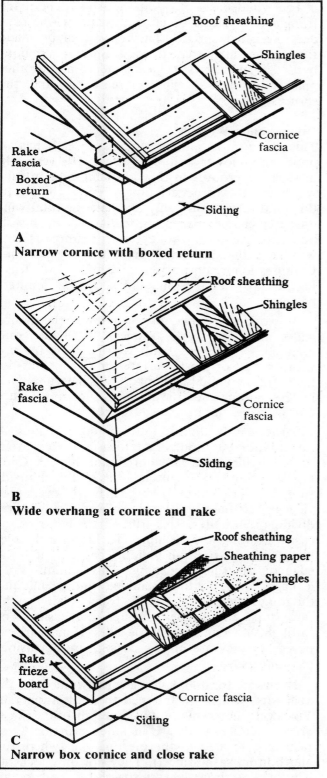

A
Narrow cornice with boxed return

Labels on figure A: Roof sheathing, Shingles, Cornice fascia, Rake fascia, Boxed return, Siding

B
Wide overhang at cornice and rake

Labels on figure B: Roof sheathing, Shingles, Rake fascia, Cornice fascia, Siding

C
Narrow box cornice and close rake

Labels on figure C: Roof sheathing, Sheathing paper, Shingles, Rake frieze board, Cornice fascia, Siding

Cornice returns
Figure 10-21

rake are the close rake with little projection, and the boxed or open rake 6" to 12" or wider. The rake must project enough to protect the side walls from the weather. This projection also protects the paint.

Rake of 6" to 8"— With this style of rake, you can nail the fascia and soffit to a series of short lookout blocks. Look at Figure 10-22 A to see how. Fasten the fascia further by nailing through the projecting roof sheathing. Finish with a frieze board and moldings.

Rake of 8" to 20"— With this amount of rake, use both extended sheathing and a *fly rafter* to support the rake section. See Figure 10-22 B. The fly rafter (sometimes called a floating rafter) extends from the ridge board to the nailing header across the rafter ends.

Use roof sheathing all the way from the inner rafters to the end of the gable projection. This makes the gable-end strong and rigid. Nail the sheathing to the fly rafter and the lookout blocks. This sheathing will help support the rake section. Nail the soffit to the fly rafter and lookout blocks.

Windows

The basic parts of any window are the *frame* and the *sash.* The frame encloses and supports the sash. It includes the *side jambs* (one on each side), the *head jamb* at the top, and the *sill* at the bottom of the frame.

The sash holds the glass in the frame opening. It includes the vertical *stiles* on each side of the glass and the *rails* at top and bottom. The bottom rail is usually wider than the other sash parts. Sashes can be *single-lite* or *multi-lite.* In multi-lite sashes, *bars* (vertical or horizontal pieces running the full length of the sash) or *muntins* (short bars) divide the pieces of glass. A snap-out grille can be used to give a single-lite sash the look of a multi-lite sash. It's easy to remove the grille for cleaning and painting.

The common kinds of windows are *double-hung, casement, stationary, awning,* and *horizontal sliding.* Frames and sashes can be made of wood or metal. Remember, though, that heat loss is greater through metal frames.

These windows can be combined in many different ways. A *picture window* is a large, stationary unit. It's often surrounded by other kinds of windows. A *bow-bay window* is a series of windows in

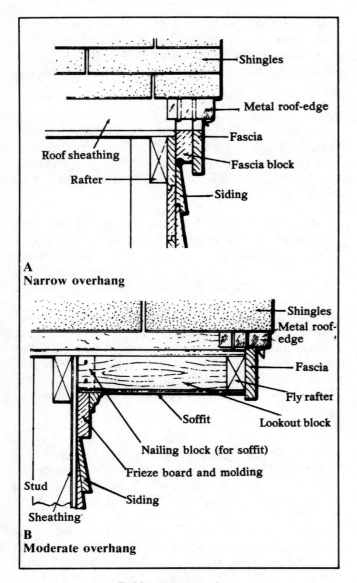

**A
Narrow overhang**

**B
Moderate overhang**

**Gable-end extensions
Figure 10-22**

the shape of an arc. An *angle bay window* has its windows arranged in the shape of a polygon. A *stacked window* consists of several awning, hopper, or casement windows combined to form a large glass area. Opening and stationary units can be included.

Insulated glass can be used in both stationary and movable sashes. It's made of two or more sheets of glass, spaced apart, with hermetically-sealed edges. It's very resistant to heat loss, and so can be used without a storm sash.

The wood parts of a window should be made from a clear grade of all-heartwood stock, of a decay-resistant species. They may also be made from preservative-treated wood. Species commonly used for windows include ponderosa and other pines, the cedars, cypress, redwood, and the spruces.

Double-hung Windows
This is the most familiar window style. It has an upper and a lower sash. The sashes slide vertically in separate grooves in the side jambs, or in full-width metal weatherstripping (Figure 10-23). In this kind of window, only one-half the total window area can be opened for ventilation. Each sash has springs, balances, or compression weatherstripping to hold it in place at any position. Compression weatherstripping keeps air from leaking in, provides tension, and acts as a counter-balance.

In wooden double-hung windows, the jambs (sides and top of the frames) are made from nominal 1" lumber. Jamb width allows for use with drywall or plastered interior finish. Sills are made from nominal 2" lumber. Sills have about a 3 in 12 slope for good drainage, as shown in Figure 10-23. Sashes are normally 1⅜" thick.

Sashes can be divided into a number of lites. A ranch-type house looks best with top and bottom sashes divided into two horizontal lites. A Cape Cod or colonial house usually has each sash divided into six or eight lites. Some manufacturers provide grilles that snap in place over a single lite, dividing it into lites. This grille can be removed for easier painting and cleaning. To install a double-hung window, first put building paper around the edge of the rough opening. This minimizes air leakage. Then, put the assembled window unit into the rough opening. Plumb the frame. Nail it to the side studs and header through the casings, or through the blind stops at the sides. If nails are exposed, as they are on the casing, use corrosion-resistant nails.

Hardware for double-hung windows includes the sash lifts, which are fastened to the bottom rail. You don't need these lifts if the rail has a finger groove. Other hardware includes sash locks, or fasteners, located at the meeting rail (where top and bottom sashes meet). They not only lock the window; but draw the sash together for a good air seal.

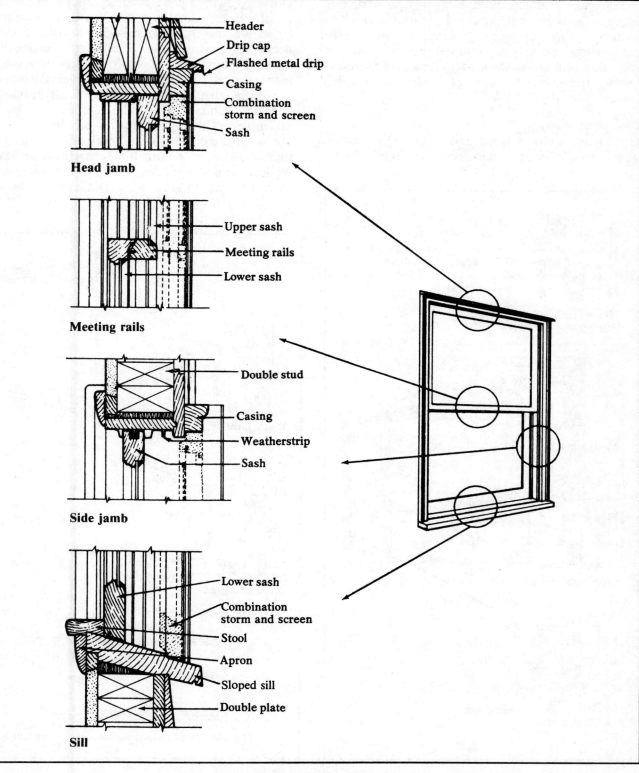

Head jamb

- Header
- Drip cap
- Flashed metal drip
- Casing
- Combination storm and screen
- Sash

Meeting rails

- Upper sash
- Meeting rails
- Lower sash

Side jamb

- Double stud
- Casing
- Weatherstrip
- Sash

Sill

- Lower sash
- Combination storm and screen
- Stool
- Apron
- Sloped sill
- Double plate

Double-hung window
Figure 10-23

You can arrange double-hung windows several ways: as a single unit, doubled (or mullion), or in groups of three or more. A window wall is often made with one or two double-hung windows on each side of a large, stationary, insulated window. Frame these large openings with headers big enough to carry the load.

Casement Windows

Casement windows consist of side-hinged sashes, which usually swing outward. See Figure 10-24.

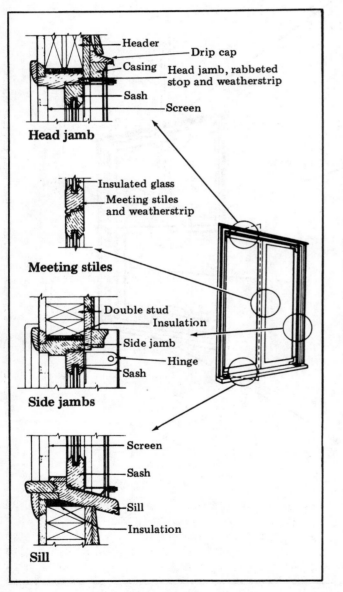

Head jamb
Header
Drip cap
Casing
Head jamb, rabbeted stop and weatherstrip
Sash
Screen

Meeting stiles
Insulated glass
Meeting stiles and weatherstrip

Side jambs
Double stud
Insulation
Side jamb
Hinge
Sash

Sill
Screen
Sash
Sill
Insulation

Outswinging casement window
Figure 10-24

Outward-swinging casements can be made more weathertight than inward-swinging ones. Screens go inside the outward-swinging windows. Storm windows, or insulated glass in the sashes, give winter protection. The whole window area of casement windows can be opened for ventilation.

Weatherstripping is available for this kind of window. Units come from the factory entirely assembled. The hardware is already in place. Closing hardware consists of a rotary operator and sash lock. As in double-hung units, casement sashes can be used several ways: as a pair, or in combinations of two or more pairs. Use divided lites to vary the

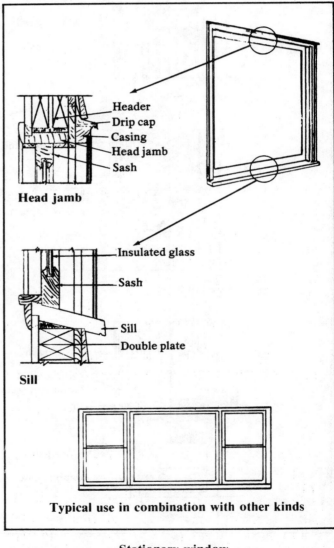

Head jamb
Header
Drip cap
Casing
Head jamb
Sash

Sill
Insulated glass
Sash
Sill
Double plate

Typical use in combination with other kinds

Stationary window
Figure 10-25

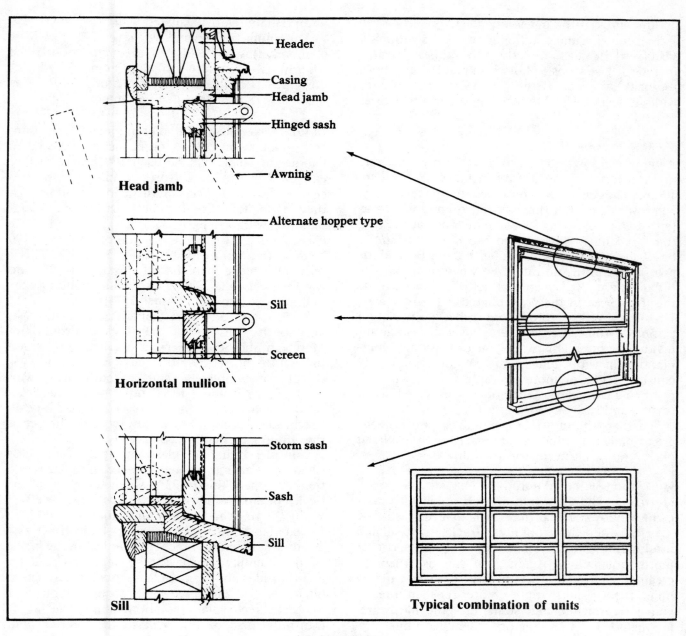

Header
Casing
Head jamb
Hinged sash
Awning

Head jamb

Alternate hopper type

Sill

Screen

Horizontal mullion

Storm sash

Sash

Sill

Sill

Typical combination of units

Awning window
Figure 10-26

style. Or use snap-in grilles for a multi-pane look for traditional homes.

Metal sash windows are sometimes used, but they have low insulating value. They must be installed carefully to prevent condensation and frosting on the insides during cold weather. In cold climates, add a full storm window to eliminate this problem.

Stationary Windows

Stationary windows consist of a wood sash with a large single lite of (usually) insulated glass. They're permanently fastened to the frame. Look at Figure 10-25. They're used alone, or in combination with double-hung or casement windows. These windows can be as much as 6' to 8' wide, so use 1¾'' thick sashes for strength. This size is needed because of the thickness of the insulating glass.

You can also make stationary windows without a sash: set the glass directly into rabbeted frame members. Use stops to hold it in place. Putty both the back and face of the glass (as you should do with any window sash unit). This assures moisture-resistance.

Awning Windows

Awning windows consist of a frame in which one or more opening sashes are installed. See Figure 10-26. They're often combined to make a large window wall that's three or more units wide and tall. Awning-type sashes swing outward at the bottom. A similar kind, called the *hopper window*, has sashes that swing inward at the top. Both kinds give protection from rain even when open.

Jambs are 1¹⁄₁₆'' or more thick. They need to be this thick because they're rabbeted. Sills are at least 1⁵⁄₁₆'' thick when two or more sashes are used in a complete frame. Or, you can give each sash an individual frame. This way, you can use any combination in width and height. Window units may consist of one or more operable sashes, with the rest being stationary. Operable sashes have hinges, pivots, and sash-supporting arms.

Weatherstripping, storm windows, and screens are available. If windows are glazed with insulated glass, you don't need storm windows.

Horizontal-sliding Windows

Horizontal-sliding windows look much like casement windows. The difference is that the sashes — in pairs — slide horizontally in tracks in the sill and head jamb. Multiple window openings have two or more single units. Again, you can use them to create a window wall. As in most modern window units, the weatherstripping, water-repellent preservative treatments, and sometimes the hardware, are included in these factory-assembled units.

Energy-efficient Windows

In winter— Heat loss per unit area through windows (and exterior doors) is much greater than through most wall materials. If you assume that a single pane of glass resists heat transfer on the order of one unit, an insulated wall offers 10 units of resistance. Double-glazed windows have over three units of resistance. If you take solar heat gain into account, a double-glazed window on the south wall of a building (located as far north as Canada) can have about the same net heat loss over a winter

as an equal area of insulated wall.

Aluminum-frame windows have 25% heat loss. Compare this to 13% loss through wood-frame windows. A 10-square-foot aluminum-framed window facing south (the best direction) can lose as much as 425 Btu's per day in winter.

In summer— During summer in much of the country, about a ton of cooling capacity is needed to handle the solar load on each 100 square feet of east or west windows. North-facing windows (except in winter) and south-facing windows (if they're shaded) are the best. If a rectangular building with a length 2½ times its width, 50% glass, and an east-west orientation is turned 90 degrees, the cooling load is reduced by 30%. This is only 50% more load than if the building had no windows at all. Shade trees surrounding the building can provide another 25% savings.

Windows of any appreciable size in east, west, and south walls should be shaded. From mid-March to mid-September, an overhead projection just a little shorter than the height of the window is enough to completely shade a south-facing window. If this window faced only 30 degrees from south (toward either east or west), it would take an overhead projection more than twice the height of the window to shade it completely.

Consider tilting the glass out at the top. Glass tilted at 78 degrees reflects 45% of the solar radiation. Compare this to 23% when the glass is vertical. For an 8' window, this tilt has the same effect as does a 16'' shade above the window.

Venetian blinds have the same shade effect as a building projection. Since the blind is on the inside of the building, however, the heat absorbed remains inside. For example, a light-colored venetian blind, set at a 20-degree slant, absorbs about half of the direct solar radiation falling on it. It transmits most of this heat to the interior, and reflects only about 35%. The color of the blind determines how much heat is reflected. A two-color blind could be used, with the dark side out in winter to absorb heat, and the light side out in summer to reflect heat and light.

Rough Opening Sizes

You'll buy new windows as complete units which include sash, frame, and exterior trim. Screens are usually available at extra cost. As a general rule, find rough opening sizes for these windows as follows:

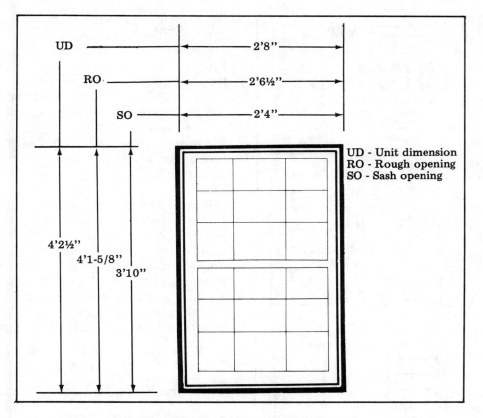

Rough opening determined by sash openings
Figure 10-27

• Double-hung windows (single unit): The width equals the glass width plus 6''. Height equals glass height plus 10''.

• Casement windows (two sashes): Width equals total glass width plus 11¼''. Height equals total glass height plus 6⅜''.

• Another way to figure the rough opening is based on the sash opening size. Figure 10-27 shows how.

Doors

Exterior Doors
Many different door and entry designs are used in contemporary houses. And both exterior doors and combination (storm) doors are available in various designs. Match the style to the style of the house. If a house has an entry hall, use a main door with some glass, unless there's another provision for natural light.

Exterior doors are 1¾'' thick, and not less than 6'8'' high. Main entrance doors are 3' wide, and side or rear service doors are 2'8'' wide.

Traditional doors have panels, as in the top row of Figure 10-28. The doors consist of *stiles* (solid vertical members), *rails* (solid cross members), and *filler panels* in a number of designs. Glazed upper panels are combined with raised wood or plywood lower panels.

Flush doors have thin plywood faces over a framework of wood, with a woodblock or particleboard core. You can get many designs, from plain to fancy with raised panels and glazed openings. See the middle row of Figure 10-28. But be sure to use solid-core, not hollow-core, exterior flush doors. This minimizes warping during winter. (Warping is caused by a difference in moisture content on the exposed and unexposed faces.)

Combination doors (storm and screen) also come in many styles, and are available in wood or metal. The screen and storm inserts are usually in the upper part of the door. You can buy doors with self-storing features, like window combination units. As with windows, heat loss through metal

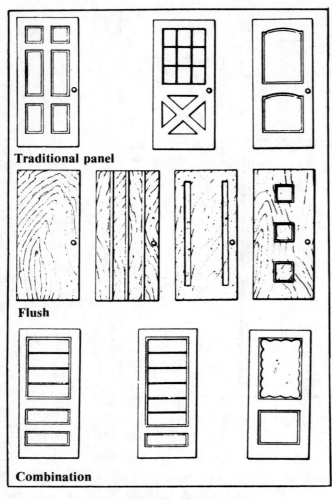

Traditional panel

Flush

Combination

Exterior door types
Figure 10-28

combination doors is greater than through wood doors.

Installing exterior doors— Prehung door units are already hinged to the frame and weatherstripped. The exterior casing is fastened to the frame.

Exterior door frames are made of 1⅛'' or thicker wood. Rabbeting on the side and head jambs provides stops for the door, as Figure 10-29 shows. The wood sill is often made of oak for wear resistance. If you use a softer wood for the sill, add a metal nosing and wear strips. As with many window units, the outside casing is wide enough for the 1⅛'' combination or screen door.

Nail the frame to the studs and headers of the rough opening through the outside casing. Normal-

ly, the sill can rest on the subfloor. Butt the 5/8''-thick underlayment, or wood flooring, to the sill. After finish flooring is in place, cover the joint between floor and sill with a hardwood or metal threshold.

The exterior trim around the main entrance door can vary, from a simple casing to a molded pilaster with a decorative head casing. Of course, any decorations should be in keeping with the architecture of the house. You can buy millwork adaptable to any style.

Interior Doors
Interior doors should open in the direction of natural entry. If you can, place them to open against a blank wall. And make sure other opening doors won't get in the way. The two kinds of interior doors are the flush door and the panel door. Folding or sliding doors can be flush or louvered.

Standard interior doors are 1⅜'' thick. And interior doors have standard minimum widths:

- Doors to bedrooms and other habitable rooms are at least 2'6'' wide.

- Doors to bathrooms are at least 2'4'' wide.

- Doors for small closets and linen closets are at least 2' wide.

The standard minimum height for interior doors is 6'8'', though doors above the first floor are sometimes 6'6'' tall. Door sizes, especially widths, vary a lot from these minimums, however.

Sliding doors, folding doors, and similar doors are often used for wardrobes. They might be 6' wide, or more. In most cases, though, the jamb, stop, and casing parts are used to frame and finish the opening.

A *flush interior door* has a hollow core with a light framework in it. Over this is a thin plywood, or hardboard, face. Look at Figure 10-30 to see how these doors are made. The plywood face can be of birch, gum, mahogany, oak, or other species. Most woods can take a natural finish, unless they're non-select grades. These are usually painted, as are hardboard-faced doors.

Interior *panel doors* have the same parts as exterior panel doors. Figure 10-31 shows the different styles. Five-cross and Colonial panel doors are the most common. The louvered door in the lower left of the figure is also popular. It's often used for closets, because it provides ventilation even when closed.

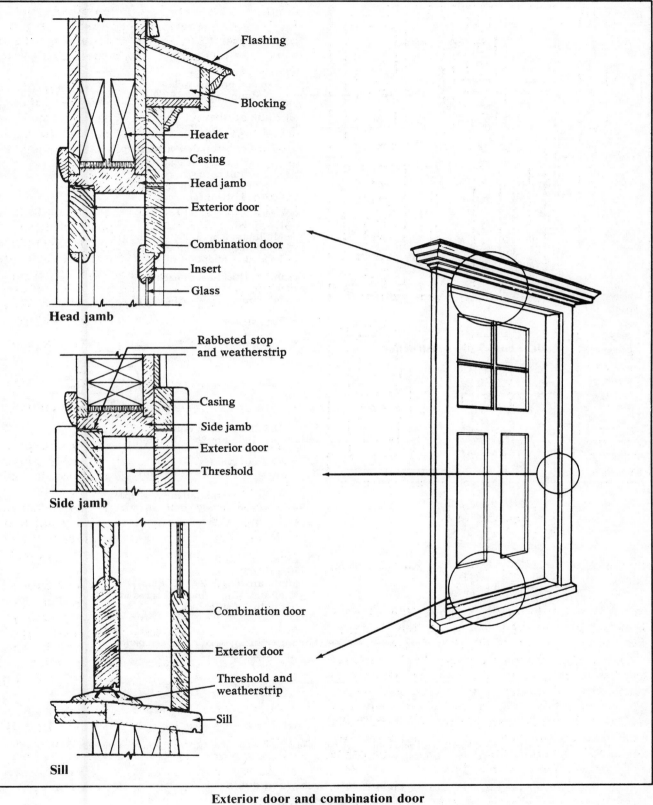

Head jamb

Flashing

Blocking

Header

Casing

Head jamb

Exterior door

Combination door

Insert

Glass

Side jamb

Rabbeted stop and weatherstrip

Casing

Side jamb

Exterior door

Threshold

Sill

Combination door

Exterior door

Threshold and weatherstrip

Sill

Exterior door and combination door
Figure 10-29

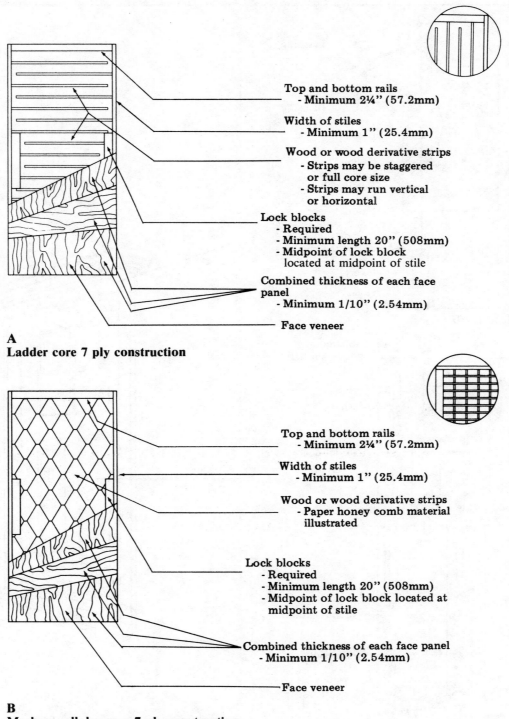

Top and bottom rails
- Minimum 2¼" (57.2mm)

Width of stiles
- Minimum 1" (25.4mm)

Wood or wood derivative strips
- Strips may be staggered or full core size
- Strips may run vertical or horizontal

Lock blocks
- Required
- Minimum length 20" (508mm)
- Midpoint of lock block located at midpoint of stile

Combined thickness of each face panel
- Minimum 1/10" (2.54mm)

Face veneer

A
Ladder core 7 ply construction

Top and bottom rails
- Minimum 2¼" (57.2mm)

Width of stiles
- Minimum 1" (25.4mm)

Wood or wood derivative strips
- Paper honey comb material illustrated

Lock blocks
- Required
- Minimum length 20" (508mm)
- Midpoint of lock block located at midpoint of stile

Combined thickness of each face panel
- Minimum 1/10" (2.54mm)

Face veneer

B
Mesh or cellular core 7 ply construction

High pressure laminate, hardboard, and composition face panels may be used as alternates to the face panels illustrated above.
Source: National Woodwork Manufacturers Association

Hollow-core flush wood doors
Figure 10-30

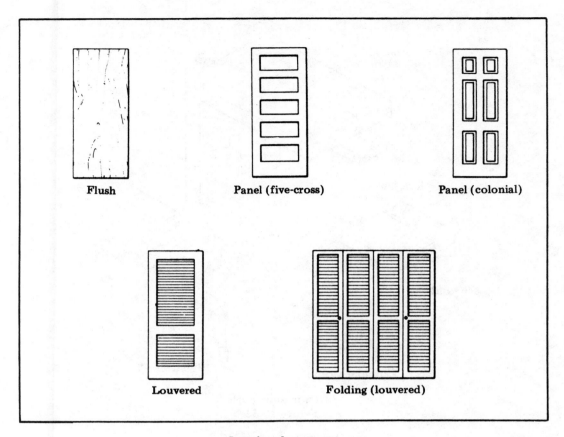

Interior door types
Figure 10-31

Finish large wardrobes with sliding or folding doors, or with flush or louvered doors. These doors are usually 1⅛'' thick.

Interior door frames— Interior door frames are made of two side jambs and a head jamb. They include the stop molding upon which the door closes. A common jamb is the one-piece style in Figure 10-32 A. Jambs come in standard 5¼'' widths for plaster walls, and 4⅝'' widths for walls with 1/2'' drywall.

The two- and three-piece adjustable jambs (Figures 10-32 B and C) are also common. And many builders prefer them. Their main advantage is that they're adaptable to a variety of wall thicknesses. Prehung units of these types usually come with the casings installed.

Frame the rough openings for interior doors to be 3'' more than the door height, and 2½'' more than the door width. This gives room for plumbing

and leveling the frame.

Installing interior doors— In general, install interior doors after the finish floor is in place. At this time also install cabinets, built-in book cases, fireplace mantels, and other millwork pieces.

Some contractors install the interior door frames before the finish floor is in place if the jambs will act as plaster grounds. They leave space for the flooring at the bottom of the jambs. This isn't good practice. Frames installed too soon absorb excessive moisture from the plaster. Also, the edges of the jambs are often marred by the tradesmen.

Installing Prehung Doors
Prehung doors go directly into the rough opening. Even prehung double-door units are installed this way. So it's important to know the exact rough opening dimension. Check the rough opening door size *before* framing the wall. Once you've got the

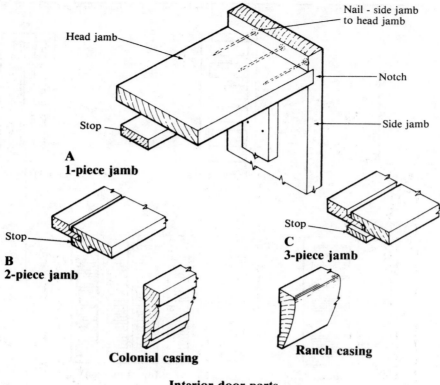

Interior door parts
Figure 10-32

right size door to fit the opening, half the job is done.

Figure 10-33 shows how a prehung exterior door unit fits into the rough opening. Remember that a prehung exterior or interior door is installed as a complete unit. *Never remove the door from the hinges during installation.*

Figure 10-32 shows the two-piece and three-piece adjustable split jambs commonly found in prehung units. Another type of interior split jamb uses machine-fitted steel dowels to connect the front and reverse jambs. The dowels fit into countersunk receiving holes. This makes it easy to separate and reassemble the front and reverse jamb parts.

Here's how to install the interior prehung door unit:

1)   Remove the packing nails connecting the keeper jamb to the door. See Figure 10-34. Separate the halves of the door frame.

2)   Place the *front* half of the frame into the opening, as shown in section A of Figure 10-35. (The

front half of the frame has the hinged door attached to it.) Plumb the jamb on the hinge side of the frame. Nail the hinge side trim to the wall. Use 8d casing nails spaced 24'' o.c. Don't drive the nails all the way into the wall.

3)   Press the keeper side of the jamb against the spacer tabs. Be sure the head jamb fits uniformly at the top of the opening. Nail the keeper side trim to the wall. Use 8d nails spaced 24'' o.c. Don't drive the nails all the way into the wall.

4)   With the door closed and spacers still attached, shim behind the jambs.

5)   Open the door and nail through the jambs and shims. Use 8d casing nails. Remove all spacers.

6)   Take the *reverse* half of the frame and slide it into position, as shown in Figure 10-35 B. Insert the top first. Then press the sides firmly into position against the wall. Nail through the trim into the wall. Use 8d casing nails spaced 24'' o.c. Don't drive the nails all the way into the wall.

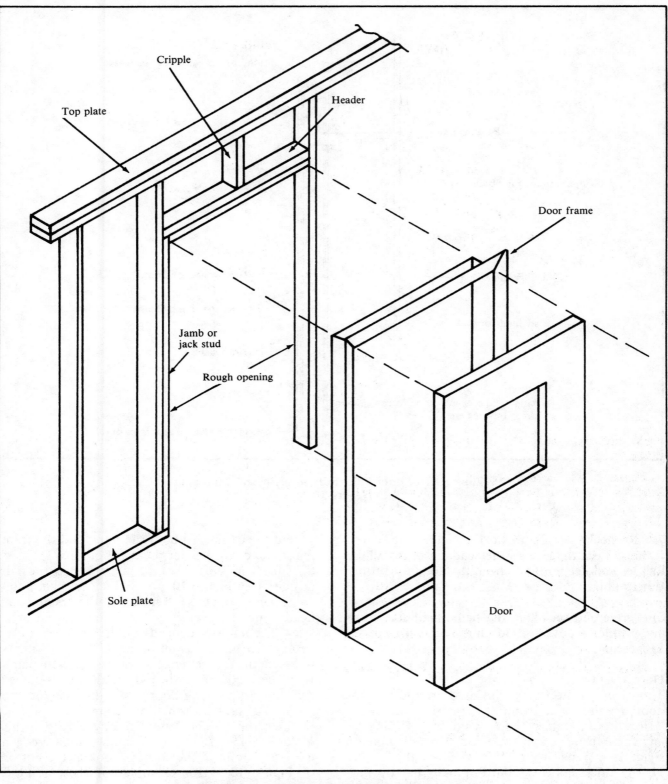

Fitting arrangement—exterior prehung door unit
Figure 10-33

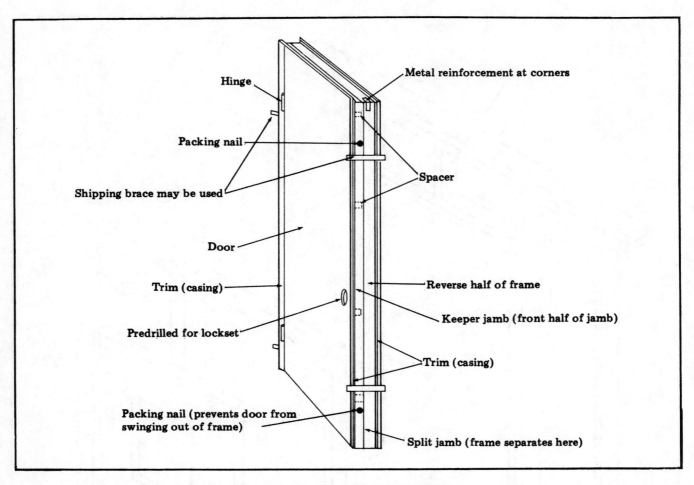

Prehung interior door unit ready to be installed in opening
Figure 10-34

7) Recheck the plumb on the hinge side of the frame. Check the level of the head jamb. Install the lockset and keeper, and check the door for proper functioning.

8) Drive and set all casing nails. Nail side stops and header stops. Use 3d finishing nails, spaced 12'' o.c.

Door Casings

The *casing* is the finish trim used around interior door openings and around the interior side of exterior door frames and windows. Casing is anywhere from 2¼'' to 3½'' wide, depending on the style. It can be from 1/2'' to 3/4'' thick. The 11/16'' thickness is standard in many narrow patterns. Two common casing patterns are shown in Figure 10-36.

Nail the casings to both the jamb and the framing studs, or the header. Leave about a 3/16'' edge distance from the face of the jamb. Locate the nails in pairs (see Figure 10-36), and space them about 16'' apart along the full height of the opening and head jamb. Use finish or casing nails in 6d or 7d sizes. The nail size depends on the thickness of the casing. Use 4d or 5d finishing nails (or 1½'' brads) to fasten the thinner edge of the casing to the jamb. If the casing is of hardwood, predrill it to prevent splitting. In prehung door units, the casing comes glued and stapled to the door frame.

Miter the corner joints of any casing with a molded shape. But when casing is square-edged, use a butt joint at the corner. Look at the bottom of Figure 10-36 for examples of these joint fit styles. In quality homes, of course, butt joints are seldom used.

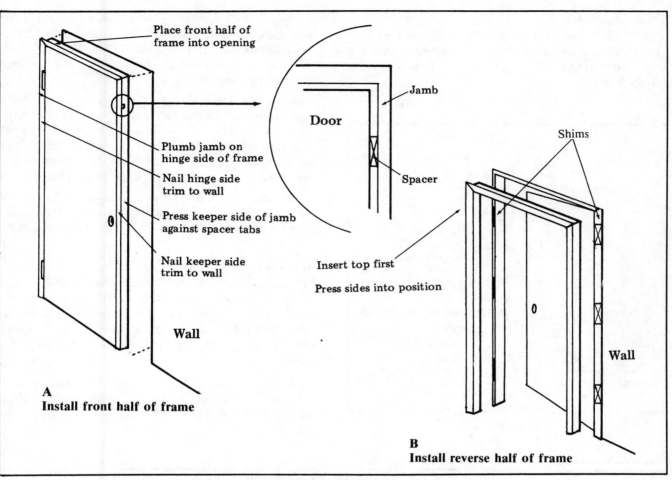

Place front half of
frame into opening

Jamb

Door

Spacer

Plumb jamb on
hinge side of frame

Nail hinge side
trim to wall

Press keeper side of jamb
against spacer tabs

Nail keeper side
trim to wall

Shims

Insert top first

Press sides into position

Wall

A
Install front half of frame

B
Install reverse half of frame

Wall

Installing the prehung door unit
Figure 10-35

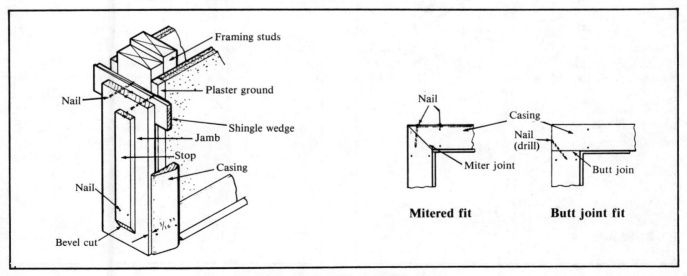

Framing studs

Plaster ground

Nail

Shingle wedge

Jamb

Stop

Casing

Nail

Bevel cut

3/16"

Nail

Casing

Nail
(drill)

Miter joint

Butt join

Mitered fit

Butt joint fit

Door frame and casing detail
Figure 10-36

Door Hardware

Figure 10-37 shows door details. It also shows how to install the strike plate and where to locate the door stops. Figure 10-38 shows how to install door hardware.

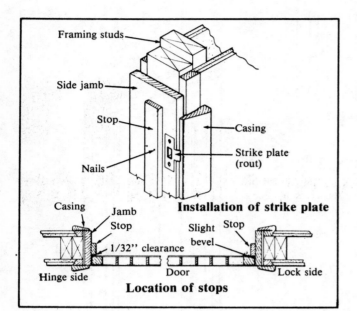

Door details
Figure 10-37

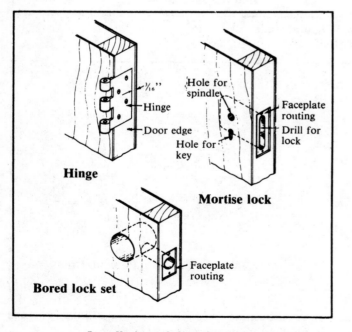

Installation of door hardware
Figure 10-38

Figure 10-39 illustrates the normal clearances for both exterior and interior doors. Leave a 1/2'' or more gap under interior doors. This allows for better air circulation in homes having central heating/cooling plants.

Don't stockpile doors on the job. Don't have them delivered to the job until you're ready to install them. They'll be better protected against moisture, damage and theft if they're left at the lumber yard until it's time for installation. When they're delivered, set them upright on end.

Prehung doors are hinged to their frames at the mill. The holes for the lockset are pre-drilled. Be sure to use a quality lockset on all exterior doors.

Summary

Finish carpentry is, of course, the "finish." If a house is poorly finished, it will show — and whether you're building a spec or custom house, your reputation as a quality builder will suffer, even if the rest of the house is the best-built in the world.

Quality builders make sure that qualified craftsmen do the finish work on their houses. Poor workmanship here will inevitably result in a slow sale or a sale at a reduced price.

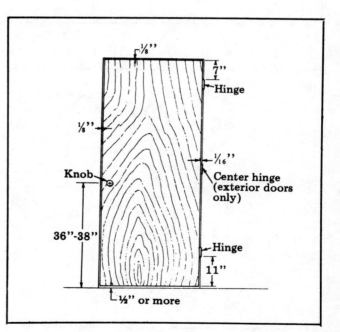

Door clearances
Figure 10-39

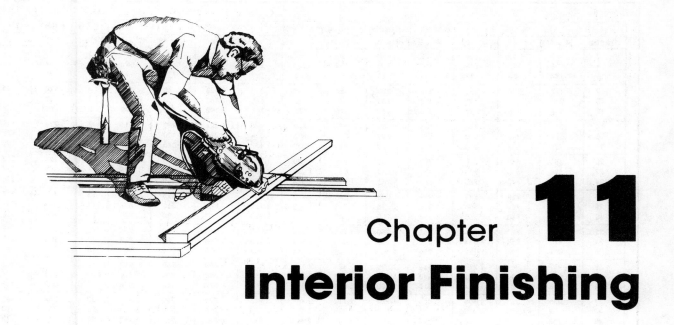

Chapter **11**
Interior Finishing

Interior Walls and Ceilings

Every professional builder should take pride in the quality of the interior finish in the homes he builds and remodels. Nothing makes a job look more amateurish than poor interior finish. And, unfortunately, many good tradesmen just don't have the ability or interest to do good quality finish work. It's your job to distinguish between professional interior finish and slapdash work that's suitable only for temporary construction. Know the difference and tolerate only what you and your tradesmen can be proud of.

Interior walls should have a smooth, even surface. Plaster walls should be free of flaws like waves, dips, or humps. Drywall, brick, stone, and other masonry-type walls should also be smooth.

The term *drywall* includes gypsum board (often called sheetrock), hardboard, wood paneling, plywood, and fiberboard. Apply drywall to framing, or to furring strips over framing or masonry walls. Figure 11-1 is a footage table to help you figure how much drywall to buy. Figure 11-2 is a guide to thickness for drywall applied to framing or furring strips on 16" and 24" centers.

Gypsum Board

Gypsum board is probably the best thing that ever happened to builders. Here's why:

- It's an attractive wall or ceiling finish.

- It's easy to maintain.

- It has a long life.

- It goes up ten times faster than lath and plaster.

- Most builders can hang drywall. Plastering takes skilled craftsmen.

- It's classified as a noncombustible material by building and fire insurance companies.

Gypsum board is made of gypsum filler faced with paper. Side edges are recessed or tapered to permit a smooth "even-wall" application of tape and joint compound. The sheets are 4' wide and available in 8', 10', or 12' lengths. Various thicknesses are available. Use the 3/8" thickness over old walls and ceilings. Be sure to protect gypsum wallboard from high humidity or moisture until it's hung.

Vertical application— When you use this method of application, you'll put the length of the gypsum board parallel to the framing. When you use the 1/2" or 5/8" thickness in a vertical application, the maximum frame spacing is 16" o.c.

Pieces	Board Products — Square Feet											16" x 48" Lath		
	4' x 6'	4' x 7'	4' x 8'	4' x 9'	4' x 10'	4' x 12'	4' x 14'	4' x 16'	2' x 8'	2' x 10'	2' x 12'	Bdls.	3/8"	1/2"
2	48	56	64	72	80	96	112	128	32	40	48	1	32	21.33
4	96	112	128	144	160	192	224	256	64	80	96	2	64	42.67
6	144	168	192	216	240	288	336	384	96	120	144	3	96	64.00
8	192	224	256	288	320	384	448	512	128	160	192	4	128	85.33
10	240	280	320	360	400	480	560	640	160	200	240	5	160	106.67
12	288	336	384	432	480	576	672	768	192	240	288	6	192	128.00
14	336	392	448	504	560	672	784	896	224	280	336	7	224	149.33
16	384	448	512	576	640	768	896	1,024	256	320	384	8	256	170.67
18	432	504	576	648	720	864	1,008	1,152	288	360	432	9	288	192.00
20	480	560	640	720	800	960	1,120	1,280	320	400	480	10	320	213.33
22	528	616	704	792	880	1,056	1,232	1,408	352	440	528	11	352	234.67
24	576	672	768	864	960	1,152	1,344	1,536	384	480	576	12	384	256.00
26	624	728	832	936	1,040	1,248	1,456	1,664	416	520	624	13	416	277.33
28	672	784	896	1,008	1,120	1,344	1,568	1,792	448	560	672	14	448	298.67
30	720	840	960	1,080	1,200	1,440	1,680	1,920	480	600	720	15	480	320.00
32	768	896	1,024	1,152	1,280	1,536	1,792	2,048	512	640	768	16	512	341.33
34	816	952	1,088	1,224	1,360	1,632	1,904	2,176	544	680	816	17	544	362.67
36	864	1,008	1,152	1,296	1,440	1,728	2,016	2,304	576	720	864	18	576	384.00
38	912	1,064	1,216	1,368	1,520	1,824	2,128	2,432	608	760	912	19	608	405.33
40	960	1,120	1,280	1,440	1,600	1,920	2,240	2,560	640	800	960	20	.640	426.67
42	1,008	1,176	1,344	1,512	1,680	2,016	2,352	2,688	672	840	1,008	21	672	448.00
44	1,056	1,232	1,408	1,584	1,760	2,112	2,464	2,816	704	880	1,056	22	704	469.33
46	1,104	1,288	1,472	1,656	1,840	2,208	2,576	2,944	736	920	1,104	23	736	490.67
48	1,152	1,344	1,536	1,728	1,920	2,304	2,688	3,072	768	960	1,152	24	768	512.00
50	1,200	1,400	1,600	1,800	2,000	2,400	2,800	3,200	800	1,000	1,200	25	800	533.33
52	1,248	1,456	1,664	1,872	2,080	2,496	2,912	3,328	832	1,040	1,248	26	832	554.67
54	1,296	1,512	1,728	1,944	2,160	2,592	3,024	3,456	864	1,080	1,296	27	864	576.00
56	1,344	1,568	1,792	2,016	2,240	2,688	3,136	3,584	896	1,120	1,344	28	896	597.33
58	1,392	1,624	1,856	2,088	2,320	2,784	3,248	3,712	928	1,160	1,392	29	928	618.67
60	1,440	1,680	1,920	2,160	2,400	2,880	3,360	3,840	960	1,200	1,440	30	960	640.00
62	1,488	1,736	1,984	2,232	2,480	2,976	3,472	3,968	992	1,240	1,488	31	992	661.33
64	1,536	1,792	2,048	2,304	2,560	3,072	3,584	4,096	1,024	1,280	1,536	32	1,024	682.67
66	1,584	1,848	2,112	2,376	2,640	3,168	3,696	4,224	1,056	1,320	1,584	33	1,056	704.00
68	1,632	1,904	2,176	2,448	2,720	3,264	3,808	4,352	1,088	1,360	1,632	34	1,088	725.33
70	1,680	1,960	2,240	2,520	2,800	3,360	3,920	4,480	1,120	1,400	1,680	35	1,120	746.67
72	1,728	2,016	2,304	2,592	2,880	3,456	4,032	4,608	1,152	1,440	1,728	36	1,152	767.00
74	1,776	2,072	2,368	2,664	2,960	3,552	4,144	4,736	1,184	1,480	1,776	37	1,184	789.33
76	1,824	2,128	2,432	2,736	3,040	3,648	4,256	4,864	1,216	1,520	1,824	38	1,216	810.67
78	1,872	2,184	2,496	2,808	3,120	3,744	4,368	4,992	1,248	1,560	1,872	39	1,248	832.00
80	1,920	2,240	2,560	2,880	3,200	3,840	4,480	5,120	1,280	1,600	1,920	40	1,280	853.33
82	1,968	2,296	2,624	2,952	3,280	3,936	4,592	5,248	1,312	1,640	1,968	41	1,312	874.67
84	2,016	2,352	2,688	3,024	3,360	4,032	4,704	5,376	1,344	1,680	2,016	42	1,344	896.00
86	2,064	2,408	2,752	3,096	3,440	4,128	4,816	5,504	1,376	1,720	2,064	43	1,376	917.33
88	2,112	2,464	2,816	3,168	3,520	4,224	4,928	5,632	1,408	1,760	2,112	44	1,408	938.67
90	2,160	2,520	2,880	3,240	3,600	4,320	5,040	5,760	1,440	1,800	2,160	45	1,440	960.00
92	2,208	2,576	2,944	3,312	3,680	4,416	5,152	5,888	1,472	1,840	2,208	46	1,472	981.33
94	2,256	2,632	3,008	3,384	3,760	4,512	5,264	6,016	1,504	1,880	2,256	47	1,504	1,002.67
96	2,304	2,688	3,072	3,456	3,840	4,608	5,376	6,144	1,536	1,920	2,304	48	1,536	1,024.00
98	2,352	2,744	3,136	3,528	3,920	4,704	5,488	6,272	1,568	1,960	2,352	49	1,568	1,045.33
100	2,400	2,800	3,200	3,600	4,000	4,800	5,600	6,400	1,600	2,000	2,400	50	1,600	1,066.67
200	4,800	5,600	6,400	7,200	8,000	9,600	11,200	12,800	3,200	4,000	4,800	60	1,920	1,280.00
300	7,200	8,400	9,600	10,800	12,000	14,400	16,800	19,200	4,800	6,000	7,200	70	2,240	1,493.33
400	9,600	11,200	12,800	14,400	16,000	19,200	22,400	25,600	6,400	8,000	9,600	80	2,560	1,706.67
500	12,000	14,000	16,000	18,000	20,000	24,000	28,000	32,000	8,000	10,000	12,000	90	2,880	1,920.00
600	14,400	16,800	19,200	21,600	24,000	28,800	33,600	38,400	9,600	12,000	14,400	100	3,200	2,133.33
700	16,800	19,600	22,400	25,200	28,000	33,600	39,200	44,800	11,200	14,000	16,800	200	6,400	4,266.67
800	19,200	22,400	25,600	28,800	32,000	38,400	44,800	51,200	12,800	16,000	19,200	300	9,600	6,400.00
900	21,600	25,200	28,800	32,400	36,000	43,200	50,400	57,600	14,400	18,000	21,600	400	12,800	8,533.33
1,000	24,000	28,000	32,000	36,000	40,000	48,000	56,000	64,000	16,000	20,000	24,000	500	16,000	10,666.67

Drywall footage table
Figure 11-1

Material	Minimum material thickness (inches) with framing spaced 16" o.c.	24" o.c.
Gypsum board	3/8	1/2
Hardboard	1/4	--
Wood paneling	1/4	1/2
Plywood	1/4	3/8
Fiberboard	1/2	3/4

Drywall thickness and spacing
Figure 11-2

Horizontal application— With this method of application, you'll put the length of the gypsum board at right angles to the framing. Then, the maximum frame spacing is 24" o.c.

Apply boards with square or beveled edges vertically. You can apply the tapered-edge board either vertically or horizontally. Bring panels to moderate contact, but don't force them together. Stagger the joints so that the corners of four boards don't meet at one point. When covering the opposite sides of partitions, make sure perpendicular joints don't fall on the same studs.

Floating angle application— This method reduces stress and strain on the gypsum board panels when the framing settles. It eliminates some of the nails or screws at interior angles — where ceiling and sidewalls meet, and where sidewalls intersect. We'll talk more about this a little later. Use conventional fastening in the remaining ceiling or wall areas. Fit the wallboard snugly into all corners.

Nailing— To make gypsum wallboard fasteners work right, you must have framing grade 2 x 4's that are straight, dry, and uniform in dimension. Figure 11-3 shows the recommended fasteners for single and double layer gypsum board application to wood framing. Figure 11-4 has more instructions for applying gypsum wallboard on wood framing.

Fasten gypsum panels by nailing from the center toward the ends and edges. On intermediate and end supports, space nails 6" to 8" apart on walls; 5" to 7" apart on ceilings. Keep nails 3/8" in from the ends and edges of panels. Dimple-set the nails. Be careful not to tear or puncture the paper.

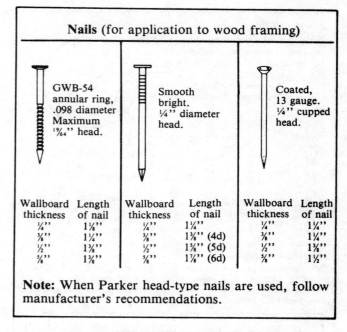

Nails (for application to wood framing)

 GWB-54 annular ring, .098 diameter Maximum 19/64" head.		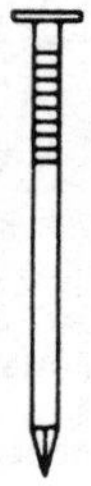 Smooth bright. 1/4" diameter head.		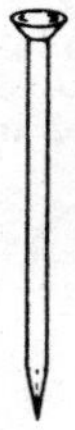 Coated, 13 gauge. 1/4" cupped head.	
Wallboard thickness	Length of nail	Wallboard thickness	Length of nail	Wallboard thickness	Length of nail
1/4"	1 1/8"	1/4"	1 1/4"	1/4"	1 1/4"
3/8"	1 1/4"	3/8"	1 3/8" (4d)	3/8"	1 1/4"
1/2"	1 3/8"	1/2"	1 5/8" (5d)	1/2"	1 3/8"
5/8"	1 3/8"	5/8"	1 7/8" (6d)	5/8"	1 1/2"

Note: When Parker head-type nails are used, follow manufacturer's recommendations.

Recommended fasteners for single and double layer wallboard
Figure 11-3

Finishing— Apply finishes to gypsum wallboard only when the temperature is above 55 degrees. Provide continuous, controlled heating for at least 24 hours before you apply the finish. If the humidity is high, be sure to provide adequate ventilation.

Ceilings— Always install ceilings first. Wood furring on ceilings should be no smaller than (nominal) 2 x 2. Space the furring no more than 24" o.c. (But if ceiling insulation will rest on 1/2" thick wallboard, space the framing not more than 16" o.c. and apply horizontally). You may want to put a vapor barrier between the gypsum board and the insulation. And make sure that unheated spaces above wallboard ceilings are well ventilated.

If a ceiling will support insulation, or if it will have a sprayed-on textured coating, you can use 5/8"-thick gypsum board in a horizontal application, over framing spaced either 16" or 24" o.c. For vertical application of 5/8"-thick gypsum board, the framing must be 16" o.c. Use 1/2"-thick gypsum board horizontally over 16" o.c. framing. Apply a pigmented primer-sealer to the wallboard before applying the texture. When you apply gypsum board horizontally (across ceiling joists), use conventional nailing where the ends

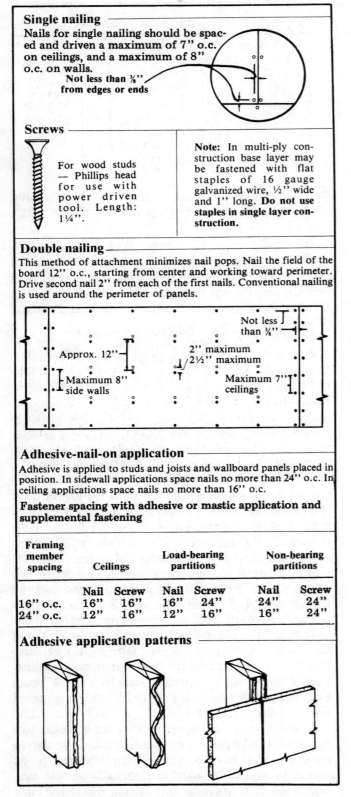

Single nailing

Nails for single nailing should be spaced and driven a maximum of 7" o.c. on ceilings, and a maximum of 8" o.c. on walls.

Not less than ⅜" from edges or ends

Screws

For wood studs — Phillips head for use with power driven tool. Length: 1¼".

Note: In multi-ply construction base layer may be fastened with flat staples of 16 gauge galvanized wire, ½" wide and 1" long. **Do not use staples in single layer construction.**

Double nailing

This method of attachment minimizes nail pops. Nail the field of the board 12" o.c., starting from center and working toward perimeter. Drive second nail 2" from each of the first nails. Conventional nailing is used around the perimeter of panels.

Adhesive-nail-on application

Adhesive is applied to studs and joists and wallboard panels placed in position. In sidewall applications space nails no more than 24" o.c. In ceiling applications space nails no more than 16" o.c.

Fastener spacing with adhesive or mastic application and supplemental fastening

Framing member spacing	Ceilings		Load-bearing partitions		Non-bearing partitions	
	Nail	Screw	Nail	Screw	Nail	Screw
16" o.c.	16"	16"	16"	24"	24"	24"
24" o.c.	12"	16"	12"	16"	16"	24"

Adhesive application patterns

Wood frame application methods
Figure 11-4

of the boards meet the wall. At the long edges of the board, which run parallel to the intersection, put the first nail about 7" from the wall joint. Figure 11-5 shows horizontal ceiling application with floating angles.

When you apply gypsum board vertically (parallel to ceiling joists), use conventional nailing where the long edges of the board meet the wall. At wall intersections, where the ends of the board meet the wall, put the first nail about 7" from the joint. Look at Figure 11-6. It shows vertical ceiling application with floating angles.

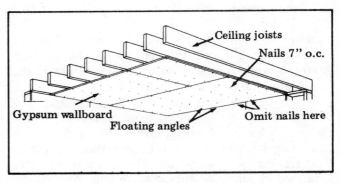
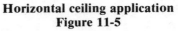

Horizontal ceiling application
Figure 11-5

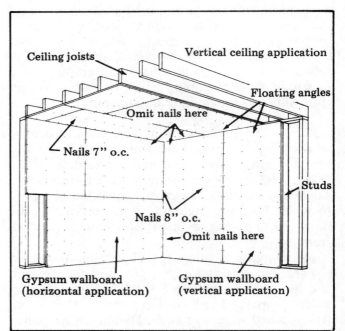

Wall and vertical ceiling application
Figure 11-6

Walls— On walls, make sure all the wallboard makes firm contact at the ceiling line. It has to support the ceiling boards that are already in place.

As always when framing walls, be sure the studs are accurately spaced 16'' or 24'' o.c. and that they are set straight and true. Provide nailing members in both planes of interior and exterior angles. Figure 11-6 shows both horizontal and vertical wall application.

Omit the nails directly below the intersection of ceiling and wall. Put the first nail about 8'' from this intersection. Also omit the corner nails on the first board you apply to an inside wall corner; the board on the other wall will overlap at the corner. Look at Figure 11-7. Then, nail the overlapping board conventionally, 8'' o.c.

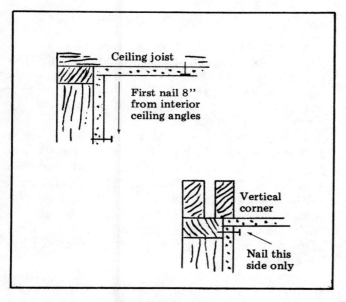

Nailing at corners
Figure 11-7

If you're finishing curved walls, you can bend dry gypsum wallboard to the radii shown in Figure 11-8. If you need curves with shorter radii, thoroughly dampen both sides of the gypsum wallboard. Wait for the moisture to dampen the core, then bend. The board will resume its original hardness when dry.

Water Resistant Gypsum Board
Use a special backer board (B/B) in shower stalls, tub enclosures, and other high-moisture areas. It's

Wallboard thickness	Bending radii (dry)	
	Width	Lengthwise
1/4''	15'	5'
3/8''	25'	7½'
1/2''	---	20'

Bending radii for dry wallboard
Figure 11-8

also used as a base for gluing on ceramic, metal, and plastic wall tile.

Backer board consists of an asphalt-treated core covered with a heavy water-repellent back paper and then an ivory face paper. You don't need to seal the surface before you apply tile and you don't need to tape joints. Don't, however, use backer board for ceilings or soffits.

Provide appropriate headers or supports for the tub, plumbing fixtures, soap dishes, grab bars, and towel racks. Install the tub, shower receptor or pan before you install the gypsum board. Reinforce interior angles to make rigid corners.

Look at Figure 11-9 to see how to install backer board. Tubs, shower receptors, or pans should have a lip or flange that's 1'' higher than the water drain (or the threshold of the shower). Use furring around the tub enclosure and shower stall to make the inside face of the fixture lip flush with the face of the backer board. Make the top of the furring even with the upper edge of the tub or shower pan.

If studs are spaced more than 16'' o.c. and walls have ceramic tile larger than 5/16'' thick, put one row of blocking about 1'' above the top of the tub or shower receptor. Then put another row midway between the fixture and the ceiling.

Apply the backer board horizontally to eliminate butt joints. Leave a 1/4'' space between the lip of the fixture and the edge of the board. Unless ceramic tile over 5/16'' thick will be used, space nails 8'' o.c., and space screws 12'' o.c. If you'll use ceramic tile over 5/16'' thick, space nails 4'' o.c. and screws 8'' o.c.

In areas that will be tiled, first treat joints and angles with a waterproof tile adhesive. Don't use regular joint compound and tape. Then caulk openings around pipes and fixtures with a waterproof, non-setting caulking compound.

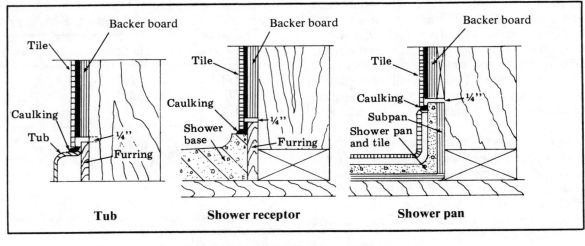

Backer board installation procedures
Figure 11-9

Next, apply tile down to the top edge of the shower floor surfacing material, to the return of the shower pan, or over the tub lip, as shown in Figure 11-9. Grout all joints completely and continuously.

Fill the space between fixture and tile with non-setting caulking, or with tile grout compound.

Plywood and Hardboard Panels
Plywood and hardboard panels are made in 4' by 8' sheets. They're finished to look like boards, or planks, of equal or random widths. Plywood and hardboard are also made with a durable plastic finish. And hardboard is made with vinyl coatings in different patterns and colors.

Plywood panels are made in a variety of wood species and finishes. Prices vary widely, depending on the species, the finish, and the quality. Even panels from the same lot may vary in color and texture. So always stand the panels up in the room, and arrange them in the best color or texture pattern for the room. Then mark (or restack) the panels according to the order in which they're to be installed.

Hardboard panels printed with a wood-grain pattern are the least expensive of the hardboard panels. High-quality hardboard paneling uses a photograph of wood to give the wood-grain effect. Therefore, all the panels have the same wood-grain pattern. (Real wood panels all have different grain patterns.)

Measuring and marking— To measure for window and door openings, often one person can hold the panel in place while someone else marks the opening on the back of the panel. Then return the panel to the workbench. Adjust the saw line, allowing for jamb thickness, and cut.

To mark wall outlets and switch boxes, draw along the edges of the box with a pencil or chalk. Put the panel in position, then tap it against the box with your hand. The box will make a mark on the back side of the panel, so you'll know the correct position. Return the panel to the workbench, with the back side up. Next, drill starter holes at each corner of the box mark. Then saw out the piece with a keyhole or saber saw.

Cutting— Use a hand or power saw to cut plywood and hardboard panels. Use a combination hollow-ground blade on power saws. Put the finish side of the panel up when cutting with a hand saw; put the finish side down when cutting with a power saw. Always cut *into* the face of the panel. If, for some reason, you must use a power saw to cut with the face up, first score the saw line with a knife to prevent splintering. Keep the blade on the waste side of the line.

Installing— You can either nail or glue plywood and hardboard panels to the framing. A vapor barrier, such as polyethylene, can be applied against the studs before the panels are installed.

The simplest installation procedure is to start at one corner and work around the room. After the first sheet is plumb and nailed or glued in place, butt the next sheet to it. Paneling edges should touch very lightly. Sometimes the edge of a panel doesn't exactly fit the edge of the next panel, so the stud shows. To fix it, use a black felt marker to mark the stud where the panels butt. Do this before you install the panel.

Space nails 8" to 10" apart on the panel edges and at intermediate supports. You can get color-matched nails for most paneling.

Finish Floors
Carpeting and sheet vinyl flooring are probably the most popular floor coverings today. Besides being wear-resistant, they're easy to maintain, and make any spec or custom-built house look better.

Sheet Vinyl Flooring
Polyvinyl chloride (PVC) is the main ingredient of vinyl flooring, which also contains resin binders with mineral fillers, plasticizers, stabilizers, and pigments.

Vinyl can be filled or clear. The clear style has a layer of opaque particles or pigments bonded to vinyl, or polymer-impregnated asbestos fiber, or resin-saturated felt. A wearing surface of clear vinyl tops it off. This style has good wear resistance.

Filled vinyl is made of chips of various shapes and colors in a clear vinyl base. Base and chips are bonded together with heat and pressure.

Most floor covering suppliers have installers that will lay sheet vinyl and carpet at a reasonable cost.

Installing Sheet Vinyl
If you decide to have your crew install the sheet vinyl, here's how. First, sweep the floor area and drive down all protruding nail heads. Unroll the floor covering in another room.

1) Select the wall against which you'll put one of the factory edges of the sheet vinyl. (Factory edges run the entire length of the flooring, not the width of the roll.) In general, the factory edge should run along the longest straight wall in the room.

2) Measure out from this wall a given distance — to about the center of the room — at each end of the room. (See *#2* in Figure 11-10.)

3) Snap a chalk line through these two points.

4) Now mark a second line at a 90-degree angle to the first line at a point where the second line can run the entire width of the room. It's very important that these two lines be exactly perpendicular. They now become your reference points by which you determine how to mark and cut the vinyl sheet. (See #4 in Figure 11-10.)

5) On a sheet of paper (preferably graph paper), draw the two reference lines from the floor. Sketch the room roughly around the intersection of the two lines. Indicate all cabinets and closets.

6) Measure from the reference lines on the floor out to the walls or cabinet every two feet around the room, including at least two measurements for every offset. Record these measurements on your paper layout. *Be careful!* Walls are seldom exactly straight. Measure every two feet. (See #6 in Figure 11-10.)

Now transfer the measurements from the paper to the flooring. Position your floor-plan sketch on the sheet vinyl so that your first reference line is parallel to a factory edge of the flooring. Using the measurement recorded on your floor plan, measure in from the factory edge on the flooring the distance from your first reference line to the wall. Mark this distance at both ends on the vinyl flooring and snap a chalk line through the points.

To establish a second reference line on the vinyl flooring, measure along the first reference line from the edges to where the floor plan indicates your reference lines intersect. Add 1½ inches and snap a chalk line through the points at an exact 90-degree angle to the first line. This second line should run the entire width of the floor.

You now have reference lines on the floor covering which correlate to those in the room. Measure out from these lines to all walls and cabinets at the same intervals as shown on the floor plan.

Connect these marks with a chalk line, being as accurate as possible. These outside lines become your cut lines. Look at Figure 11-10 again.

Cutting the flooring— Don't make any cuts until all room measurements have been transferred and rechecked.

Using a straight-blade utility knife and a metal straightedge as a guide, cut along all outside cut lines. (If you're cutting on a concrete floor, put a scrap of plywood or cardboard under the cut lines or you'll ruin your blade.)

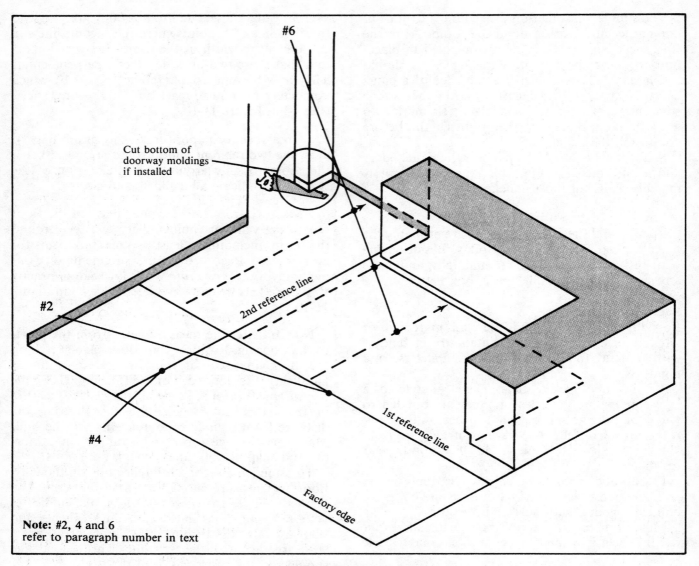

Cut bottom of
doorway moldings
if installed

#6

#2

#4

2nd reference line

1st reference line

Factory edge

Note: #2, 4 and 6
refer to paragraph number in text

Establishing reference points
Figure 11-10

One-piece installations— Before carrying the sheet vinyl into the room where it's to be installed, roll the material face in, so it rolls out into the room along the longest straight wall.

Don't force the covering under offsets or cabinets, but unroll enough to make sure that it's in the correct position.

Fold half of the covering back onto itself, being careful not to move the other half out of position.

Now you're ready to spread the adhesive. Be sure to follow instructions on the container. Spread the adhesive evenly, using a trowel with notches 1/16'' wide, 1/16'' deep and 3/32'' apart. Don't let the adhesive dry more than 15 minutes before placing the material onto it.

Fold back the other half of the covering and repeat the process. Use a heavy roller to ensure a good bond and remove any air pockets. Roll from the center out toward the edges. Finally, cap all doorways and openings with a metal threshold strip.

Two-piece installations— Where you'll need a seam in the flooring because the room is wider than

12 feet, follow procedures used for one-piece installations with the following exceptions:

1) Before transferring measurements on the floor covering, overlap the two pieces at the seam area, making certain that the pattern is matched.

2) Place strips of masking tape across the overlap to hold pieces together.

3) Proceed to transfer measurements and cut the covering to the size and shape of the room. (Don't cut the overlapped seams.)

4) Untape the seams before moving the floor covering to the room where it's to be installed.

5) Move and position one piece at a time.

6) Fold back the top piece half way and draw a pencil line on the subfloor along the edge of the second piece (at the seam area).

7) Fold back the second piece and spread adhesive to within 12 inches of either side of the line. (See Figure 11-11)

8) After placing the flooring in the adhesive, repeat the preceding two steps for the other half of the room.

9) Using a straight-blade utility knife and a metal straightedge as a guide, cut through both pieces where they overlap.

10) Fold back both seam edges, spread the adhesive, and place the covering into position.

Seam sealing is your last step when installing most vinyl sheet flooring. However, there are some coverings in which seam sealing is one of the first steps. Your supplier should be able to advise you here. Follow the seam sealing instructions on the container.

Carpeting

Wall-to-wall carpeting can do wonders for a room, or for a whole house. It helps insulate and sound-proof cold floors. It makes small rooms look larger. In general, rooms look more "finished." Carpeting offers comfort that's unavailable with the other types of flooring, and most carpeting is easy to maintain.

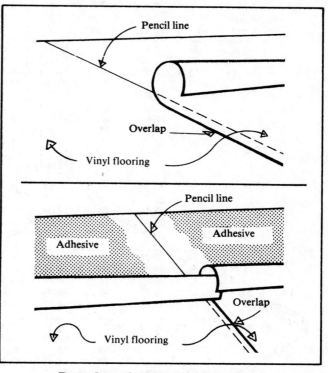

Procedures for two-piece installation
Figure 11-11

Carpet is available in outdoor, indoor, indoor/outdoor, and artificial grass. All carpets consist of a surface pile and a backing material. The surface pile may be wool, cotton, nylon, acrylic, polyester, or polypropylene. Each fiber has its advantages and disadvantages. Consult a reputable dealer for the type and grade of carpet best suited for different rooms or areas. Most carpet fibers offer good to excellent service when installed according to the manufacturer's recommendations.

Most carpets come in 12-foot widths. They're available in any length. Wall-to-wall carpeting is usually sold by the square yard. To find the square yards needed to cover a floor, multiply the length by the width (both in feet) and divide by 9, or use Figure 11-12, a carpet yardage calculator.

One advantage of carpeting is that it can be installed over almost any floor. The base floor only needs to be relatively smooth and free of serious imperfections. And carpeting is available for every room in the house, including the kitchen. Kitchen carpeting has a very close weave. This keeps spills from penetrating the surface too quickly.

Padding or an underlay extends carpet life, increases soundproofing, and adds underfoot com-

Length (feet)	12' SY	15' SY	Length (feet)	12' SY	15' SY	Length (feet)	12' SY	15' SY	Length (feet)	12' SY	15' SY
2	2.67	3.33	27	36.00	45.00	52	69.00	86.67	77	102.67	128.33
3	4.00	5.00	28	37.33	46.67	53	70.67	88.33	78	104.00	130.00
4	5.33	6.67	29	38.67	48.33	54	72.00	90.00	79	105.33	131.67
5	6.67	8.33	30	40.00	50.00	55	73.33	91.67	80	106.67	133.33
6	8.00	10.00	31	41.33	51.67	56	74.67	93.33	81	108.00	135.00
7	9.33	11.67	32	42.67	53.33	57	76.00	95.00	82	109.33	136.67
8	10.67	13.33	33	44.00	55.00	58	77.33	96.67	83	110.67	138.33
9	12.00	15.00	34	45.33	56.67	59	78.67	98.33	84	112.00	140.00
10	13.33	16.67	35	46.67	58.33	60	80.00	100.00	85	113.33	141.67
11	14.67	18.33	36	48.00	60.00	61	81.00	101.67	86	114.67	143.33
12	16.00	20.00	37	49.33	61.67	62	82.67	103.33	87	116.00	145.00
13	17.33	21.67	38	50.67	63.33	63	84.00	105.00	88	117.33	146.67
14	18.67	23.33	39	52.00	65.00	64	85.33	106.67	89	118.67	148.33
15	20.00	25.00	40	53.33	66.67	65	86.67	108.33	90	120.00	150.00
16	21.33	26.67	41	54.67	68.33	66	88.00	110.00	91	121.33	151.67
17	22.67	28.33	42	56.00	70.00	67	89.33	111.67	92	122.67	153.33
18	24.00	30.00	43	57.33	71.67	68	90.67	113.33	93	124.00	155.00
19	25.33	31.67	44	58.67	73.33	69	92.00	115.00	94	125.33	156.67
20	26.67	33.33	45	60.00	75.00	70	93.00	116.67	95	126.67	158.33
21	28.00	35.00	46	61.33	76.67	71	94.67	118.33	96	128.00	160.00
22	29.33	36.67	47	62.67	78.33	72	96.00	120.00	97	129.33	161.67
23	30.67	38.33	48	64.00	80.00	73	97.33	121.67	98	130.67	163.33
24	32.00	40.00	49	65.33	81.67	74	98.67	123.33	99	132.00	165.00
25	33.33	41.67	50	66.67	83.33	75	100.00	125.00	100	133.00	166.77
26	34.67	43.33	51	68.00	85.00	76	101.33	126.67			

Carpet yardage calculator
Figure 11-12

fort. Common kinds of padding are soft- and hard-backed vinyl foam, sponge-rubber foam, latex (rubber), and felted cushions made either of animal hair or a combination of hair and jute. Latex and vinyl foams are probably the most practical. Their waffled surface helps to hold the carpet in place. Standard padding is 4½ feet wide.

One-piece and cushion-backed carpeting need no extra padding or underlay. Foam rubber-backed carpeting is popular. It's mildew-proof and is unaffected by water. You can lay it in basements and below grade. You can lay it directly on unfinished concrete floors. The backing is non-skid. And, since it's a heavy material, it stays in place without being glued or tacked.

Indoor/outdoor carpeting is a good choice for installation over concrete and tile floors. This carpet's backing is made of a closed-pore vinyl or latex foam that keeps out moisture.

Don't carpet over vinyl or asbestos floors. These floors accumulate moisture when covered with carpet. The moisture soaks through into the carpet and eventually causes a musty odor and mildew stains.

Wood Floors

Both hardwoods and softwoods are used for flooring. The hardwoods most used for flooring are oak, hard maple, beech, birch, and pecan. Most residential wood floorings are made of oak. Softwoods (the harder ones) used for flooring are yellow pine, redwood, Douglas fir, spruce, and Western hemlock. Softwoods don't wear as well as hardwoods. But good-quality, edge-grained softwood flooring is excellent for porches and other exposed places.

To eliminate warping, wood flooring is kiln-dried. Keep it in dry storage and lay it only after all masonry and plaster are thoroughly dry. You can also buy prefinished floors, with the face brought to a hard gloss. Prefinished flooring saves a lot of time, but you must handle it with care to protect the finish.

Hardwood flooring is classified according to the

method used in sawing it from the log. When sawed perpendicular to the annual rings, it's called quarter-sawed, vertical- or edge-grained, as shown at the left in Figure 11-13. Flooring sawed parallel to the annual rings is called plain sawed, flat-grained, or flat-sawed. The right side of Figure 11-13 shows flat-grained flooring. Oak is the only flooring that's ordinarily quarter-sawed. This method of sawing makes it less likely to swell and buckle from moisture.

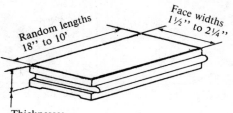

End-matched hardwood flooring

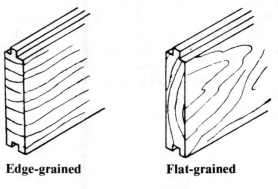

Edge-grained   Flat-grained

Hardwood flooring
Figure 11-13

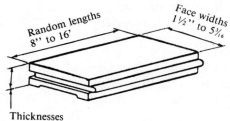

Softwood flooring

Strip flooring
Figure 11-15

Wood flooring is available in three basic styles, as shown in Figure 11-14. You may also hear of a style called "block flooring." This is simply flooring blocks, pre-assembled at the factory for parquet installation.

Strip flooring— This flooring consists of pieces cut in narrow strips, all the same width.

Hardwood strip flooring is usually tongued and

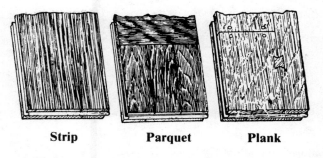

Strip   Parquet   Plank

Types of wood floors
Figure 11-14

grooved and end-matched for tight jointing. The backs of the strips are hollowed or grooved to prevent rocking and squeaking after installation. Common thicknesses range from 1/2" to 25/32". Common widths are 1½", 2", 2¼", and 3¼". Lengths are random. The most popular size is 25/32" thick and 2¼" wide. The top section of Figure 11-15 shows this kind of flooring.

Softwood strip flooring isn't always end-matched. If it's not, the floor-layer must square the ends. The trend recently is toward end-matched softwood flooring because it's faster to lay. Common thicknesses range from 25/32" to 1⁵⁄₁₆". Common widths are 1½", 2⅜", 3¼", 4¼", and 5³⁄₁₆". Lengths are random. The bottom section of Figure 11-15 shows softwood strip flooring.

Parquet flooring— Also called pattern flooring, it consists of equal short lengths installed in patterns like herringbone, squares, or rectangles. The length of the pieces is generally some multiple of the width. Parquetry is tongued and grooved and end-matched. It's usually 25/32" thick.

Over subfloor   In mastic over concrete

**Parquet flooring
Figure 11-16**

Lay each piece separately, either by nailing or by setting in a suitable mastic. Figure 11-16 shows how.

Plank flooring— It consists of random-width boards, so it looks like an early American heavy plank floor. Plank flooring is available with or without beveled edges. Use the beveled-edge style for a more authentic-looking hand-hewn effect.

Fasten the planks with countersunk screws. Then glue wood plugs in the holes on top of the screws for the look of colonial wood-peg fastening.

Installing Strip Flooring
Have flooring delivered only during dry weather. Store it in the warmest and driest area in the house so it won't absorb moisture. Moisture absorbed on-site is a common cause of the open joints that appear between flooring strips several months into the heating season. The recommended average moisture content for flooring (and other finish woodwork) at time of installation varies in different areas of the United States. The moisture content map of Figure 11-17 outlines these recommendations.

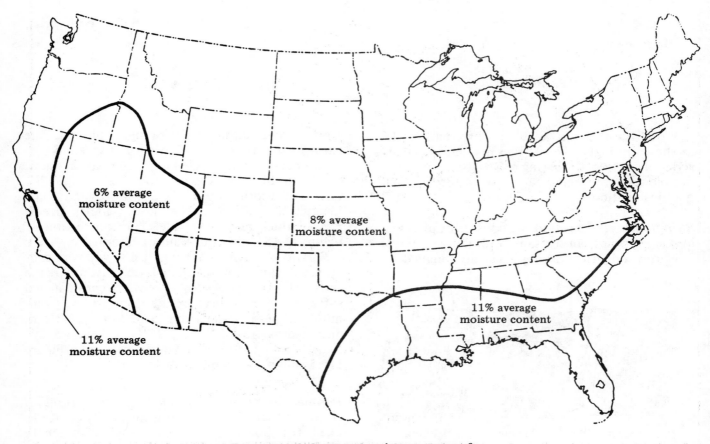

**Recommended average moisture content for
interior finish woodwork
Figure 11-17**

Lay the flooring after the plastering or other interior wall and ceiling finish is complete and has dried out, after the windows and exterior doors are in place, and after the interior trim (except base, door jambs, and casing) are installed. Thus, the flooring won't be damaged by getting wet or by construction activity.

Ensure that subfloors are clean, level, and covered with sound-deadening felt or heavy building paper. In addition to deadening sound, the felt or paper stops some dust. If the house has a crawl space, the felt or paper will also help insulate the floor.

Lay strip flooring crosswise to the floor joists, as in Figure 11-18 A. Do this whether the subfloor is of board or of plywood. In conventional houses,

the floor joists span the width of the building, over a center supporting beam or wall. So you'll lay the finish flooring of the entire floor area of a rectangular house in the same direction. Floors with "L" or "T" shaped plans will usually have the flooring change direction at the wing. This depends on the joist direction. Since joists usually span the short dimension of a wing, you'll lay the flooring lengthwise in the wing. This makes for a good-looking floor, and also reduces shrinkage and swelling in the floor during seasonal changes.

Floor squeaks are caused by boards moving against each other. Reasons for this movement are:

- Floor joists aren't heavy enough, so the floor bends too much.

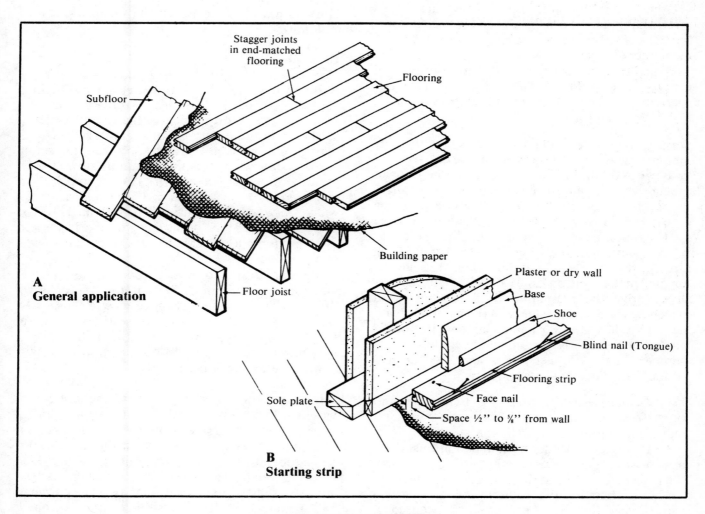

Application of strip flooring
Figure 11-18

- Sleepers over concrete slabs aren't held down tightly.

- Tongues of the boards are loose-fitting.

- The nailing is poor. Proper nailing minimizes squeaks.

Nailing— Different kinds of nails are used for different flooring thicknesses. For 25/32'' flooring, use 8d flooring nails. For 1/2'' flooring, use 6d flooring nails. And for 3/8'' flooring, use 4d casing nails. Blind nail in all these cases.

For thinner square-edge flooring, use 1½'' flooring brads. Face-nail every 7'' with two nails. Drive one brad into the subfloor near each edge of the strip.

Other kinds of nails, like the ring-shank and screw-shank, have been developed for use on flooring. Check the manufacturer's instructions for the recommended nails.

Probably the best way to nail strip flooring is to use a floor power nailer. The machine uses specially-designed nails. Flooring installation is much easier and faster.

Starting strip— Figure 11-18 B shows how to apply the starting strip. Nail this first strip 1/2'' to 5/8'' away from the wall. This space allows the floor to expand when its moisture content increases. Drive the nails straight down through the board, at the groove edge. Put the nails near enough to the edge so they'll be covered by the base or shoe molding.

You could also nail the first strip of flooring through the tongue. If you blind-nail through the tongue, make it at an angle of 45 to 50 degrees. Don't drive the nail quite flush so you avoid damaging the flooring edge with the hammer. Nailing devices like a power nailer eliminate this problem. When properly positioned, one blow of the hammer on the plunger drives and sets the nail without any damage to the flooring.

Second and succeeding courses— For the second course from the wall, choose pieces carefully. You want the butt joints to be well separated from those in the first course. Under normal conditions, drive each board tightly against the previous board. When you finish laying the floor, leave a 1/2'' to 5/8'' space between the wall and the last flooring strip. Because this strip is so close to the wall, face-nail it so the base or shoe molding covers the set nail heads.

Ceramic Tile

Ceramic tile and similar floor coverings come in many sizes and patterns for bath, lavatory, and entry areas. These floorings can be installed with the cement-plaster method or with adhesives. The cement-plaster method requires a concrete-cement setting bed. The cement base is reinforced with woven wire fabric or expanded metal lath.

Ceramic-tile installation requires skilled tradesmen. Most builders subcontract this work to professionals who have the equipment and know-how to do quality work.

Moldings

Use moldings at the junctions of walls and floor, and at the junction of walls and ceiling. These moldings make the house look "finished." Figure 11-19 shows where to install the various kinds of moldings.

Base Moldings

Use base molding to finish the joint between the wall and the floor. You can get base molding in a variety of shapes and widths. *Two-piece base* consists of a baseboard topped with a small base cap, as shown in the upper left of Figure 11-20. If the plaster on a wall isn't straight and true, the small base molding will match the variations more closely than the wider base alone will. A common size for this kind of baseboard is 5/8'' by at least 3¼'' wide. *One-piece base* varies in size from 7/16'' x 2¼'' to 1/2'' x 3¼'', or more, wide. The lower left section of Figure 11-20 shows narrow and wide one-piece base.

It's a good idea to have a wood member at the junction of the wall and floor (as a protective "bumper"). This piece is called a *base shoe.* It's 1/2'' x 3/4'', and is shown in all sections of Figure 11-20. A single-base molding without the shoe is sometimes used at the wall-floor junction, especially when carpeting will be used.

Installing base molding— Install square-edged baseboard with a butt joint at inside corners, and a mitered joint at outside corners. The right section of Figure 11-20 shows how. Nail the molding to each stud with two 8d finishing nails.

Install molded single-piece base, base moldings, and base shoe with a coped joint at inside corners, and a mitered joint at outside corners. A coped joint is one in which the first piece is square-cut against the wall, and the second piece is coped to fit

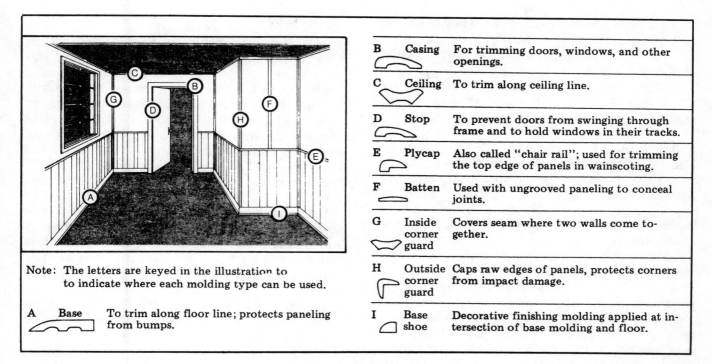

B	Casing	For trimming doors, windows, and other openings.	
C	Ceiling	To trim along ceiling line.	
D	Stop	To prevent doors from swinging through frame and to hold windows in their tracks.	
E	Plycap	Also called "chair rail"; used for trimming the top edge of panels in wainscoting.	
F	Batten	Used with ungrooved paneling to conceal joints.	
G	Inside corner guard	Covers seam where two walls come together.	
H	Outside corner guard	Caps raw edges of panels, protects corners from impact damage.	
I	Base shoe	Decorative finishing molding applied at intersection of base molding and floor.	

Note: The letters are keyed in the illustration to to indicate where each molding type can be used.

A Base To trim along floor line; protects paneling from bumps.

Where to install moldings
Figure 11-19

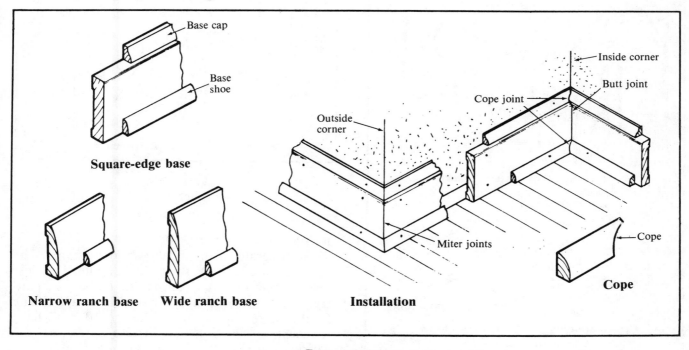

Base moldings
Figure 11-20

the first piece. Do this by sawing a 45-degree miter cut, then trim the molding with a coping saw along the inner line of the miter, as the lower right section of Figure 11-20 shows.

Nail the base shoe into the floor, not into the baseboard. Thus, if there's a small shrinkage of the joists, no opening will occur under the shoe.

Ceiling Moldings

These moldings are sometimes used at the junction of wall and ceiling for an architectural effect. They're also used to "finish" drywall paneling of gypsum board or wood. Figure 11-21 shows how to install ceiling moldings. As with base moldings, cope-joint the inside corners. This ensures a tight joint and retains a good fit even if there are minor moisture changes.

A cutback edge at the outside of the molding will partly conceal any unevenness of the plaster. It will also make painting easier, especially where there are color changes. For gypsum drywall construction, you might use a small, simple molding like the one in the lower right section of Figure 11-21. Drive finish nails into the upper wallplates.

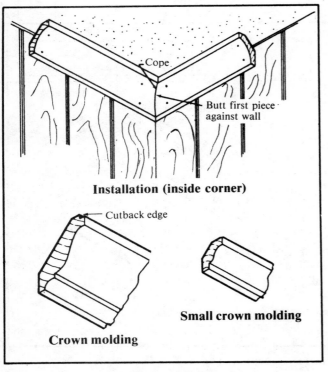

Ceiling moldings
Figure 11-21

Kitchen Layout

The kitchen is probably the most-studied part of the house. Equipment makers have hired architects, engineers, and home economic agencies to study homemakers working in the kitchen. They're looking for the most efficient arrangement of equipment. And they want to work out floor plans that get the most use from the least amount of space. As a successful construction contractor you know that well-planned, well-equipped kitchens in the homes you build are one of your best sales tools.

Complete kitchen-planning services are available to help you plan economical kitchens. Manufacturers of kitchen equipment, as well as associations interested in kitchens, provide these services. You can save yourself a lot of time and expense by using them.

Figure 11-22 shows a modern kitchen arrangement. This kitchen has, as all kitchens do, three main work areas:

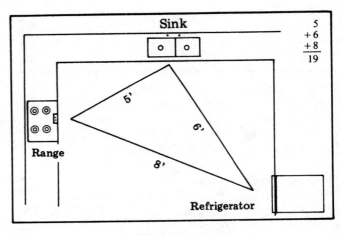

Work triangle
Figure 11-22

Refrigerator/Storage Center

The refrigerator is the focal point of this area. The cabinets next to the fridge store canned goods, cereals, and other nonperishables. If there's a pantry, put it near the fridge also. This way, all the food is stored in one area. The cook doesn't have to chase all over the kitchen to gather up the various ingredients when preparing a meal.

Locate the fridge near the kitchen entrance door, or back door. The idea is to have a short distance from car to storage area when bringing in the groceries. Provide counter space next to the fridge. And have the fridge's door handle next to the counter top, so there's a direct route from fridge to counter. Your best bet is to put the fridge at the end of the cabinets, since this large appliance gives visual balance to the kitchen.

Cleaning/Preparation Center
This includes the sink and the adjacent countertop area. There's nothing more awkward than working at a sink with too little counter space on each side. Provide at least 30 inches of counter space on each side of the sink.

You also need storage space in this area for items like cleansers, sponges, and scrub brushes. Also, put a storage area for a garbage pail near the sink. Finish the bottom of the storage area with the same material as the countertop to make cleanup of spills easier. And be sure to provide some ventilation for the compartment. A decorative screen or air holes will allow free air circulation and prevent accumulation of moisture and odors.

Put the dishwasher near the sink, not more than one step away from it. This way, dishes can easily be moved from sink to dishwasher. If you're not installing a dishwasher now, plan a compartment for later installation. Make the compartment at least 24'' wide, so the appliance will fit. For the moment, you can install temporary shelves in this space. Just remember that the dishwasher has to sit on the floor. Design the cabinet accordingly. You'll avoid a lot of trouble later.

Cooking/Service Center
This area holds the range, oven, microwave oven, and any other pots, pans, and plug-in gadgets designed for cooking. Provide lots of storage space. Include a separate storage area for the microwave, and the special dishes and items it requires.

The cooking/service center is also a serving center. That is, food is served from here. So provide storage for toasters, waffle irons, and similar items.

The Work Triangle
The path connecting the three kitchen work centers usually forms a triangle: the "work triangle." The length of the sides of the work triangle is a measure of the kitchen's efficiency. The three sides of the triangle should total no more than 22'. The

distances are measured from the fronts of the three major appliances: refrigerator to sink is 4' to 7'; sink to range is 4' to 6'; and range to refrigerator is 4' to 9'. The kitchen in Figure 11-22 has a good work-triangle arrangement: the sum of the sides is 19'.

Normal traffic lanes are the paths that people, other than the cook, use to walk through the kitchen or around in it. Traffic lanes shouldn't pass through the work triangle. If they do, the cook's efficiency is reduced.

Kitchen Shapes
There are four time-tested, basic kitchen floor plans. Of course, many variations are possible based on these four plans. You can have a combination of two types, or variations in each type.

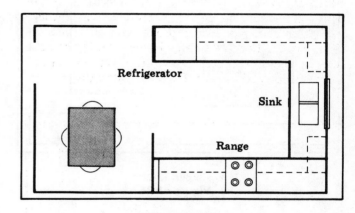

**U-shaped kitchen
Figure 11-23**

The U-shaped Kitchen— This is probably the best kitchen arrangement. Look at Figure 11-23 to see how it works. It's closed on three sides, so it eliminates through traffic. And it eliminates wasted steps because the work area is a true triangle. Three walls provide adequate counter space. One of the walls can have a pass-through, which can be a decorative focal point.

Make sure U-shaped kitchens are at least 10' wide at the base. If they're any less, the work space at the sink is cramped, and efficiency suffers.

The L-shaped kitchen— If the walls aren't too long, the L-shape is a good arrangement. Watch the location of the doors, though. If the wall space is broken by a door on each wall, you'll have traffic interfering with triangle efficiency. Locating continuous counters and appliances on two adjoining walls leaves comfortable space for eating in the kitchen. However, not as much cabinet space is available in the L as in the U. Figure 11-24 shows an L-shaped kitchen.

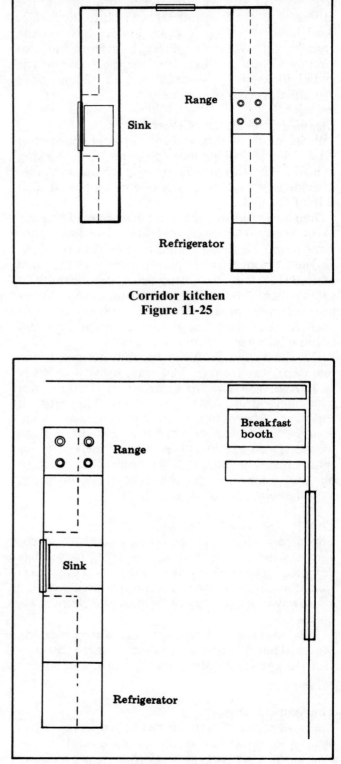

Corridor kitchen
Figure 11-25

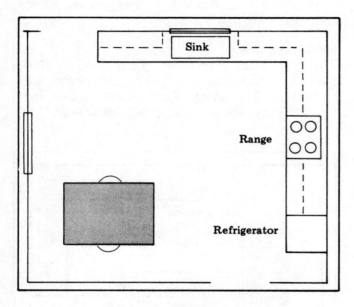

L-shaped kitchen
Figure 11-24

The corridor kitchen— The corridor kitchen is similar to the U-shape, but occupies only two walls. Figure 11-25 shows a corridor kitchen. It's a great step-saver, and is ideal for a long, narrow room because it takes up very little floor space. The many good points of this design are offset if traffic goes through the kitchen. Watch the door location. Through traffic is no problem if there's only one way in and out. Cabinet space is generous in this plan.

Be careful not to close up the room too tightly. Allow 5' of space (at floor level) between the two cabinet walls. A greater distance will reduce efficiency.

The sidewall kitchen— The sidewall, or single wall, kitchen is usually part of another room, as in Figure 11-26. While cabinet and storage space are

Sidewall kitchen
Figure 11-26

obviously sacrificed in this plan, you can combine the functions of the kitchen with the functions of the entire room. The work centers are in a line against one wall. Efficiency is good, but counter work space is limited. The sidewall is ideal for small houses and apartments.

The kitchen sells the house in many cases. So always be alert for ways to improve or beautify this area. If I had a couple of hundred dollars "extra" to spend on a house, I'd put it into the kitchen. Many builders spend a few dollars more than needed in the kitchen to get a quick sale.

Kitchen Storage

The amount of storage space needed in the kitchen is determined by the number of people expected to live in the house. The number of people is determined by the number of bedrooms in the house: one person for each bedroom, plus an additional person for the master bedroom.

Studies show that six square feet of wall-cabinet storage space is required for each permanent resident of a house. Add to this amount 12 square feet for entertaining. So, you need six square feet of wall cabinet space for each of the three people who will live in a two-bedroom house; or 18 square feet. This, plus the 12 square feet for entertaining, totals 30 square feet of wall cabinet storage space.

Wall and Base Cabinets

Measure base cabinet storage by the linear foot. As a general rule, fill all space under wall cabinets not already taken by major appliances with base cabinets. If you build enough wall storage, you'll usually have enough base storage.

Sometimes, windows limit the amount of wall space you can use, though you have plenty of base cabinet space. To solve this problem, install another wall cabinet, a tall shelf cabinet, or a storage closet elsewhere in the room, without a base cabinet below. It's important to have enough wall storage space above the counter. Lack of wall storage space reduces efficiency, makes for poor housekeeping, and creates awkward storage arrangements.

Cabinet proportions— Figure 11-27 illustrates the proper proportions for kitchen cabinets. Have about 18" clearance between a counter surface and the bottom of wall cabinets. Base cabinets are 36" high, to match the established heights of ranges

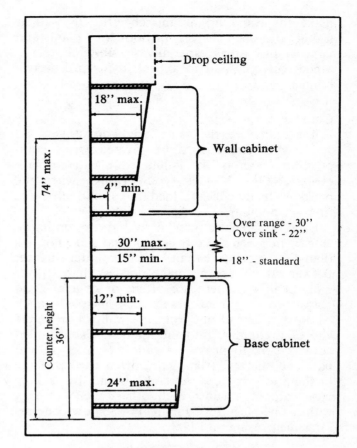

Kitchen cabinet proportions
Figure 11-27

and sinks. Wall cabinets over the range and sink need clearances of 22" to 30".

Locate doors and windows away from corners of the room, and so there's room for equipment to be installed. Appliances and cabinets are from 24" to 28" deep. To clear such equipment, put door openings no less than 30" from a corner. If a range is next to a door, use a clearance of 48". Don't put window openings less than 15" from a corner. This leaves you room to install the standard flat wall cabinets, which are about 13" deep, all the way to the corner.

Because cabinets are 36" high, with a backsplash at the rear of the countertop that varies from 3" to 8" high, make sure the undersides of window sills are no less than 44" from the floor.

Good kitchen planning requires more than the basic layout of work areas and equipment. Because people spend so much of their time in the kitchen, make the room bright and cheerful, as well as useful. Use open shelves for pottery or plants and large window areas, especially over the sink, to add a cheerful note and to appeal to the prospective homebuyer.

Counter Areas

Countertops are the same height as stoves and sinks. Most countertops are covered with a composition material like laminated plastic or with ceramic tile. Molded countertops, with the backsplash and edges formed into a solid one-piece top, are popular.

Each kitchen counter area has an optimum length. Between the refrigerator and sink: not less than 4'6'', nor more than 5'6'' of counter surface. Between the sink and range: not less than 3', nor more than 4'. There are often, of course, good reasons to vary from these basic principles. Snack bars often extend out from the wall, or from the end of a counter, into the kitchen. Use rounded corners here to provide a radius countertop and base-end shelves. Islands are often used in large kitchens to save steps. Assemble islands by backing base cabinets against each other, then covering with a one-piece top. Figure 11-28 shows one arrangement using an island.

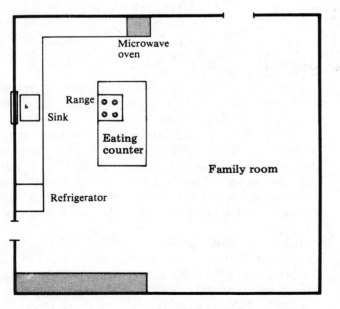

Work island/room divider
Figure 11-28

Cabinet Styles

Figure 11-29 shows sizes and styles of different wood kitchen cabinets. The cabinets have drawers and shelves, and often have special compartments. Often they have bread boards and chopping boards which slide into slots under the counter. Notice that the bottoms of the cabinets are recessed about 3'', and the bottom shelves are raised off the floor about 4''. This provides toe space and lets people working at the counter get closer to it, increasing their comfort.

Cabinet shelves— Shelves in the best-made cabinets are housed into the sides of the cabinet with dovetail joints (such as those shown in Figure 11-30) which are not nailed. In less expensive work, shelves are supported on cleats fastened to the sides and backs of cabinets. Shelves in kitchen cabinets are usually fixed in place. But you can make them adjustable with movable brackets that fit into slotted standards fastened to the back or sides of the unit. Ordinarily, 3/4'' stock is used for shelves.

Cabinet drawers— To prevent sagging, sticking, and loss of shape, build drawers carefully. Dovetail sides to the front piece, and house the back piece in slots cut into the side pieces. Make the bottoms of plywood or other rigid material. Insert bottoms into slots cut in the front and sides. Don't merely nail them on. The best drawers are both glued and screwed together. And make sure the drawers are straight and true. Drawers should run on rollers for ease of operation.

Cabinet doors— These may be paneled, have glass lites, or be flush. Flush doors with flush or lipped edges are popular because they're easy to keep clean. They have a smooth, easy-to-wipe surface. Flush doors have a plywood or lumber core, with veneer or some other hard facing, and solid edging strips. The plywood doors are easier to make, and can be cut to size from large sheets. Paneled doors have rails and stiles 3/4'' to 1'' thick, and have molding stuck directly to the rails and stiles. Panels may be plywood or glass. They're seldom made of solid lumber.

Unless you're a cabinetmaker, subcontract all your cabinet work. A cabinet shop can build and install professional-looking cabinets faster and less expensively than you or your crew can.

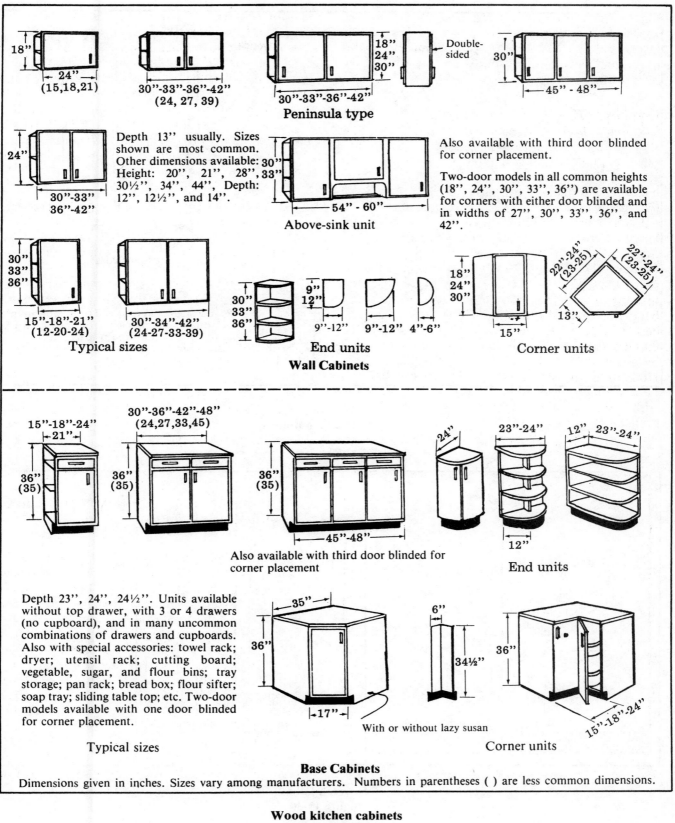

Peninsula type

Depth 13" usually. Sizes shown are most common. Other dimensions available: Height: 20", 21", 28", 30½", 34", 44", Depth: 12", 12½", and 14".

Above-sink unit

Also available with third door blinded for corner placement.

Two-door models in all common heights (18", 24", 30", 33", 36") are available for corners with either door blinded and in widths of 27", 30", 33", 36", and 42".

Typical sizes

End units

Corner units

Wall Cabinets

Also available with third door blinded for corner placement

End units

Depth 23", 24", 24½". Units available without top drawer, with 3 or 4 drawers (no cupboard), and in many uncommon combinations of drawers and cupboards. Also with special accessories: towel rack; dryer; utensil rack; cutting board; vegetable, sugar, and flour bins; tray storage; pan rack; bread box; flour sifter; soap tray; sliding table top; etc. Two-door models available with one door blinded for corner placement.

Typical sizes

With or without lazy susan

Corner units

Base Cabinets

Dimensions given in inches. Sizes vary among manufacturers. Numbers in parentheses ( ) are less common dimensions.

Wood kitchen cabinets
Figure 11-29

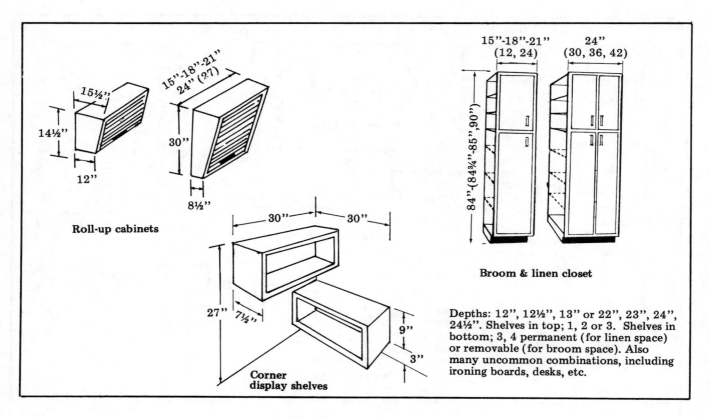

Roll-up cabinets

Corner display shelves

Broom & linen closet

Depths: 12", 12½", 13" or 22", 23", 24", 24½". Shelves in top; 1, 2 or 3. Shelves in bottom; 3, 4 permanent (for linen space) or removable (for broom space). Also many uncommon combinations, including ironing boards, desks, etc.

Wood kitchen cabinets
Figure 11-29 (continued)

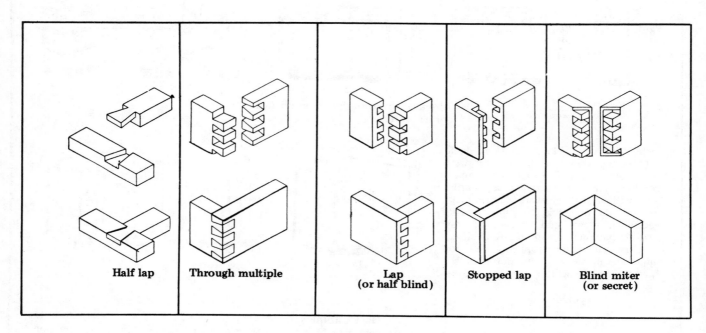

| Half lap | Through multiple | Lap (or half blind) | Stopped lap | Blind miter (or secret) |

Dovetail joints
Figure 11-30

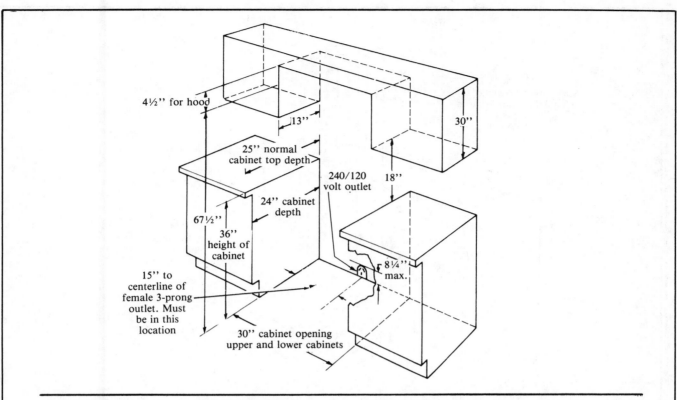

4½" for hood

13"

25" normal cabinet top depth

240/120 volt outlet

30"

18"

24" cabinet depth

67½"

36" height of cabinet

8¼" max.

15" to centerline of female 3-prong outlet. Must be in this location

30" cabinet opening upper and lower cabinets

Overall range			Microwave or electric upper oven interior			Lower oven interior			Approx. shipping wt. (lbs.)	Total connected load (KW)
Width	Height	Depth	Width	Height	Depth	Width	Height	Depth		
30"	67½"	26-5/8"	21"	12"	13"	22"	15"	16"	290	13.0KW at 120/240V 9.8KW at 120/240V

Overall range			Microwave or electric upper oven interior		
Width	Height	Depth	Width	Height	Depth
30"	67½"	26⅝"	21"	12"	13"

Lower oven interior			Approx. shipping wt. (lbs.)	Total connected load (KW)
Width	Height	Depth		
22"	15"	16"	290	13.0KW at 120/240V 9.8KW at 120/208V

**Cabinet specifications upper-lower cooking center
30-inch width
Figure 11-31**

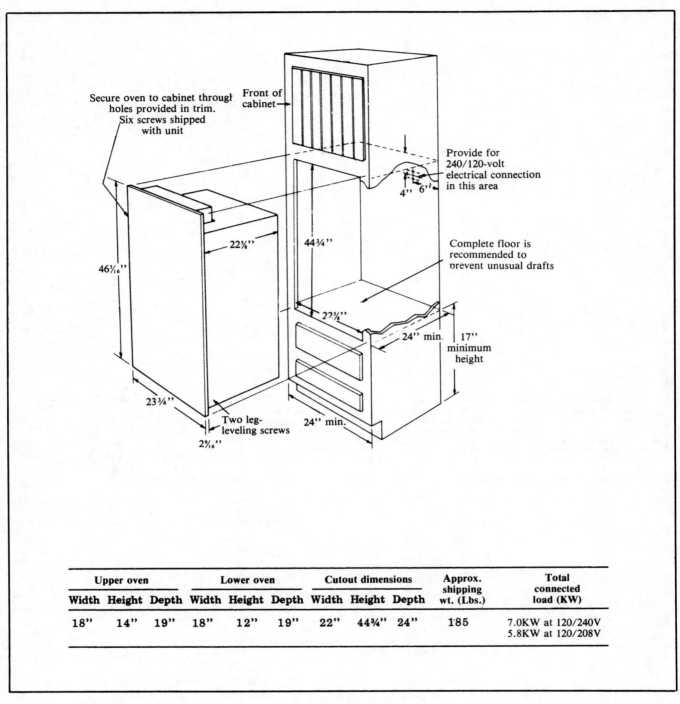

Electric wall ovens
Figure 11-32

Upper oven			Lower oven			Cutout dimensions			Approx. shipping wt. (Lbs.)	Total connected load (KW)
Width	Height	Depth	Width	Height	Depth	Width	Height	Depth		
18"	14"	19"	18"	12"	19"	22"	44¾"	24"	185	7.0KW at 120/240V 5.8KW at 120/208V

Built-in Appliances

Figure 11-31 gives cabinet specifications for installing an upper-lower cooking center, with either a microwave or an electric oven on the top.

Figure 11-32 shows specifications for installing an electric wall oven.

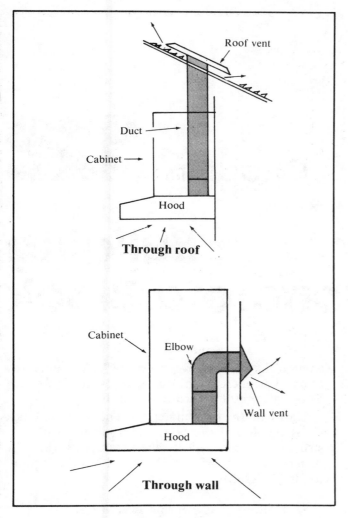

Range hood venting
Figure 11-33

Kitchen Ventilation

Kitchens are ventilated with a range hood, or with an exhaust fan in the wall over the range. But it's not the size of the stove that determines the amount of ventilation needed, it's the size of the kitchen. A 160-square-foot kitchen with an 8' ceiling requires a 320 cfm fan. (Fan capacity is measured in cubic feet per minute — cfm.)

Range hoods are either vented or ductless. The vented type moves the air to the outside either up through the ceiling and roof or out through the wall, as shown in Figure 11-33. When you install a vented hood, don't end the duct in the attic. You'll create a fire hazard. Figure 11-34 shows how to install a typical vented range hood. A ductless hood

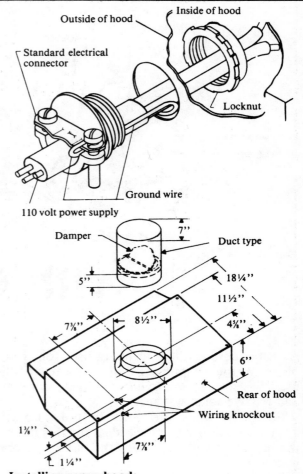

Installing range hood:

1. Pull electrical cable through cutout on wall or cabinet.

2. Attach a proper electrical connector to the cable just below the cabinet level.

3. Knock out appropriate knockout in junction box on hood. Remove junction box cover, and lift hood into position while feeding cable through the knockout.

4. Attach hood to cabinet.

5. Attach electrical connector securely to hood and make ground connection as shown.

6. Replace junction box cover.

7. Insert damper, for friction fit in duct pipe, 5 inches from bottom of pipe as shown.

Range hood installation
Figure 11-34

is not vented to the outside. The air circulates through a charcoal filter, then back into the room. Most ductless hoods also have an aluminum grease filter.

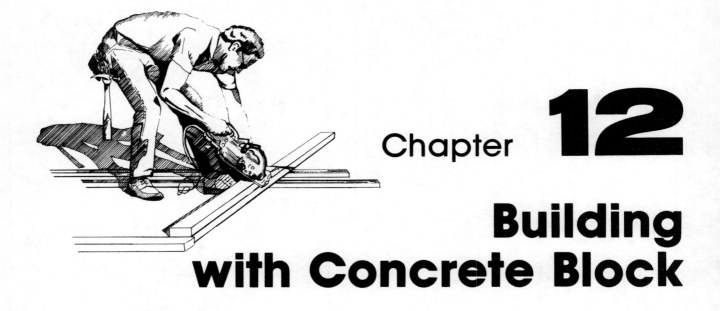

Chapter 12

Building with Concrete Block

As a builder, you'll find many uses for concrete block. In this chapter we'll discuss a few of those uses: basements, apartment buildings, warehouses and passive solar systems.

Basement walls are commonly load-bearing. That makes concrete block ideal for this use. And the basement will stay dry if the block is properly waterproofed.

For many apartment buildings, load-bearing concrete block is the first choice. Properly designed and built, concrete block construction is both attractive and durable, provides good thermal and sound insulation, and is as nearly fireproof as any building material can be.

Many warehouses and commercial buildings have relatively long and high one-story walls. That makes them naturals for concrete block. For the contractor, commercial and industrial buildings are a challenge. It can be a profitable field if you take the time to master what's required. This chapter covers most of the key points you'll need to know about warehouse and similar construction.

While we'll look at concrete block passive solar systems, keep in mind that this is still pretty experimental. Until more research is done, you can't accurately predict the heating and cooling benefits.

Concrete Block Basements

A wet basement does for your reputation about what a leaky roof does for a roofer's. The only way to build a concrete block basement is the *right* way. You really can't afford to skimp here.

Start with the right materials. The concrete block listed below are the best for basement wall construction. They meet the standards of the American Society for Testing and Materials Standards (ASTMS):

- Hollow Load Bearing ASTM C90
- Solid Load Bearing ASTM C145

Mortar

Your basement walls must be able to resist the pressure exerted by the backfill. A strong bond between the blocks is important. So use either high- or moderate-strength mortars. They have a superior bond to concrete block. Figure 12-1 lists the mortar mixes for basement construction. The type S (moderate strength) mortar is especially good for this purpose.

Estimating Materials

Make your job easier and more economical: minimize job-site cutting by taking advantage of full-length and half-length concrete blocks.

Use Figure 12-2 to find the number of blocks and the amount of mortar you need for each 100 square feet of basement wall. The amounts in the table

Mortar type, C 270	Portland cement	Parts by Volume		Sand, damp (loose volume)
		Masonry cement	Hydrated lime	
M (high strength)	1	---	¼	Not less than 2¼ and not more than 3 times the sum of the volumes of the cements and lime.
	1	1 (Type II)		
S (moderate strength)	1	---	¼ to ½	
	½	1 (Type II)	---	
N* (low strength)	1	---	½ to 1¼	
	---	1 (Type II)	---	

*Limited to walls with maximum 5' depth below grade

Mortar mixes for basement construction
Figure 12-1

Actual height of units (inches)	Units per 100 SF	Cu. ft. mortar per 100 SF
7⅝ (modular)	113	8.5
3⅝ (modular)	225	13.5
8 (nonmodular)	103	8.0

Estimating wall materials
Figure 12-2

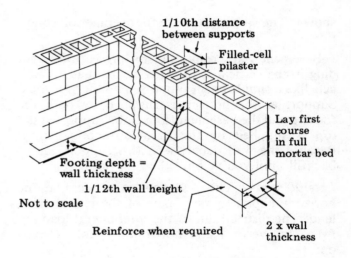

Corner detail for 10'' and 12'' walls
Figure 12-3

assume that mortar joints are 3/8'' thick and that you'll use face-shell mortar bedding. They include a 10% waste allowance.

Pay special attention to concrete block basement walls 10'' or 12'' thick. You want the best running bond pattern for maximum stability and a pleasing interior finish. Figure 12-3 shows details. Use L-corner units, or fill in with bricks. If the soil is unreliable, reinforce the footing.

Designing Basement Walls

Here are a few things to consider when you design concrete block basement walls:

1) The maximum earth pressure that will be exerted against the wall. This varies with the depth of the basement below grade and with the soil type.

2) The weight of the vertical load which the basement wall must support. Heavy loads increase wall stability.

3) The wall thickness required by local code.

4) The length, or height, of the wall between lateral supports.

5) Maximum depths below grade. For residential basement walls of 8'', 10'', and 12'' thick, see Figure 12-4. For comparison, the maximum depths below grade for frame construction are also

Nominal wall thickness	Maximum depth below grade	
	Frame construction	Masonry or masonry veneer construction
Hollow load-bearing:		
8"	5'	5'
10"	6'	7'
12"	7'	7'
Solid load-bearing:		
8"	5'	7'
10"	7'	7'
12"	7'	7'

Based on FHA "MPS" for average soil conditions.

Maximum depth below grade: concrete block and frame walls
Figure 12-4

shown. The values listed are for average soil conditions.

Basement walls resist the soil pressure by bending in the vertical span. This means that the wall acts like a simple beam supported on the two ends. Support at the bottom is supplied by both the footing and the basement floor slab. Support at the top is supplied by the first-floor construction. Because the wall depends on this support, don't backfill until you've built the first floor.

Lateral load— The wall acts as a beam in the horizontal span to carry part of the lateral earth load. The distribution of the *total* lateral load between the vertical and horizontal spans depends on:

1) the wall height and length, and

2) the stiffness of the wall in the vertical and horizontal spans.

The total lateral load is divided equally between horizontal and vertical spans, if the wall's length between supports is no greater than its height. The lateral load is carried entirely in the vertical span when the wall's length between supports is 3½ to 4 times the height. The distribution of the lateral load falls somewhere between these limits for other ratios of wall length to height.

Increase the stability of a basement wall by increasing its stiffness in either the vertical or horizontal spans, or by reducing the length of one span (generally the horizontal span).

Lateral support in the horizontal span is provided by the wall's corners; by the intersecting partition walls; and by the pilasters (or similar supports) built into the wall. You can also use a pilaster to support the building's heavy center beams. A pilaster should project from the wall about 1/12 of the vertical span between supports. And the pilaster's width should be about 1/10 of the horizontal span between supports. Look again at Figure 12-3. A partition wall, used for lateral support, can replace the center beam, thus dividing the basement into work and play areas at the same time.

Although it's not common except in earthquake zones, you can increase wall stiffness in the vertical span by embedding vertical steel in the hollow cores and then filling the cores with grout. Increase wall stiffness in the horizontal span by building a bond beam, or by embedding horizontal steel joint reinforcement in the mortar.

Construction Footings
Support basement walls with cast-in-place concrete footings. Make the footing twice as wide as the wall is thick. Be sure the soil under the footing is firm, and if possible, undisturbed. Always ensure that the footing rests below the frost line. If you must excavate under a footing for a sewer, a water line, or for whatever reason, be sure to backfill the

excavation with compacted stone or gravel.

Install drain tiles above the base of the footing. This avoids a possible wash-out of the soil. Cover any open joints with a strip of roofing felt to keep soil from entering the drain tile. Backfill with coarse gravel or stone to a height of 12'' above the tile.

Walls

Lay the first course of concrete blocks in a full bed of mortar, as in Figure 12-5. Lay the remaining courses with face-shell bedding of horizontal and vertical joints, 3/8'' thick. Tool joints firmly after the mortar has stiffened (thumbprint hard), unless the wall will be plastered or parged.

First course in full mortar bed
Figure 12-5

To ensure even distribution of vertical loads, build the bearing (top) course solidly, as described below:

1) Fill the cores of hollow blocks with mortar or concrete, as shown in Figure 12-6. Put wire screen, or metal lath, in the joint under the cores you're going to fill. This keeps fill out of the cores below. Toenail the joist to the sill, or anchor it with a Trip-L-Grip, or similar, anchor. Make sure the 1/2''-diameter anchor bolts extend at least 15'' into the filled cores. And space them not more than 6' o.c.

2) You can buy special solid-top blocks, as shown in Figure 12-7, for the course supporting the floor joists. Make sure that the joists have a minimum of 3'' of overlap (bearing) on the blocks. Embed one end of the twisted steel plate anchors into the

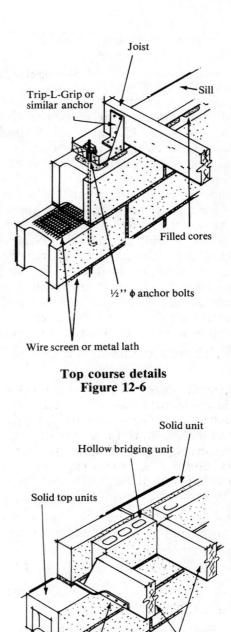

Top course details
Figure 12-6

Solid top
Figure 12-7

horizontal mortar joint. The following section on anchorage gives details of spacings for anchor bolts, floor joists, and metal straps.

3)   As a third alternative, provide a reinforced bond beam as the top unit.

Anchorage

Anchor the sill plates to the wall with 1/2''-diameter bolts (look at Figure 12-6 again), extending at least 15'' into filled cores of the blocks. Space anchor bolts not more than 6' apart. Put one bolt not more than 12'' from each end of the sill plate.

Anchor the ends of floor joists at 6' intervals, at every fourth joist. Look again at Figures 12-6 and 12-7. Also, anchor the joists that are next to, and parallel to, the wall, at intervals of not more than 8'. Figure 12-8 shows how. Be sure that the anchors have a split end embedded in the mortar joint. Or you can bend the end down into a block core filled with mortar. Make sure each anchor is long enough to span at least three joists. Next, nail the anchors to the undersides of the joists. Cross-brace at each anchor and at intermediate spacings, as required.

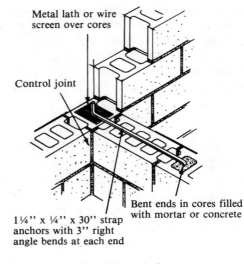

Partition tie-in
Figure 12-9

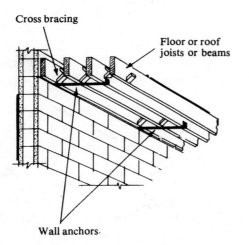

Wall anchors for joists parallel to wall
Figure 12-8

Anchor intersecting walls to basement walls with metal straps, spaced not more than 32'' vertically. The metal straps in Figure 12-9 are 30'' long, 1/4'' thick, and 1¼'' wide. Put metal lath or wire screen over the cores below to support the mortar or concrete fill. Bed the ends of the straps, and anchor them with the fill. Rake out the mortar at the intersection of the two walls, then caulk it to form a control joint. This allows slight longitudinal movement in the building.

Essentially the same techniques as those shown in Figure 12-9 are used to tie non-bearing partitions, unless the partition won't be a lateral support for the basement wall. You can substitute strips of metal lath or galvanized hardware cloth for the steel straps.

Backfilling

Don't backfill the basement walls before the first floor is in place. If you backfill in spots, brace the walls until the first floor is installed. Figure 12-10 shows one way to do this.

Brace walls for backfill
Figure 12-10

Slope the finish grade away from the basement walls for good surface drainage. Always expect some settling of the backfill. Don't operate heavy equipment any closer to the wall than a distance equal to the height of fill.

Waterproofing

Waterproofing is putting a continuous, effective coating from a point above the finish grade to a point below the top of the footing. Figure 12-11 shows how to waterproof a basement wall in well-drained soils. First, clean the outside of the wall. Then, parge with a 1/4" coat of portland cement-sand plaster. You can dampen the wall to get proper curing of the cement plaster. Follow it with a brush coat of a bituminous (tar or asphalt) coating. (Or use two 1/4" coats of portland cement plaster.) Be sure to let the plaster coat cure and dry before the applying the bituminous coat: these coatings need a dry surface for proper adhesion. Eliminate pin holes in the bituminous cover by applying a second coat. Finally, cover the whole wall with a heavy, troweled-on coat of cold, fiber-reinforced asphaltic mastic.

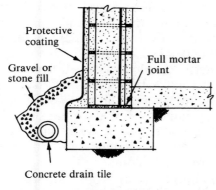

**Waterproofing basement walls in
well-drained soil
Figure 12-11**

Figure 12-12 shows how to waterproof in wet and impermeable soils. Dowel the wall to the footing if the floor might not support the wall laterally (as where a bituminous joint is used between floor and wall). Next, clean the exterior of the wall. Then waterproof with either (1) two 1/4"-thick coats of portland cement plaster, plus two brush coats of bituminous waterproofing, or (2) one 1/4"-thick coat of portland cement plaster,

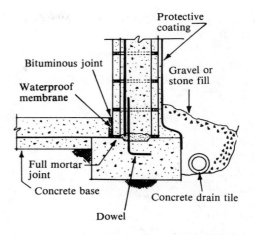

**Waterproofing basement walls in wet or
impermeable soil
Figure 12-12**

plus one heavy troweled-on coat of cold, fiber-reinforced asphaltic mastic.

Condensation

Damp or wet basements are usually attributed to water leaking through the basement walls. But the trouble is often caused by moisture condensing from the air inside the basement.

Condensation occurs whenever the air in the basement has both high temperature and high relative humidity. For example, condensation is common in areas with warm, humid climates. It can also be caused by appliances located in the basement, such as a washer and dryer or the hot water heater. Condensation happens any time the surface temperature of the wall is below the dew-point temperature of the air in the basement.

You can minimize condensation by regulating ventilation in the basement. In general, basement windows should be kept closed whenever the heating system is operating. And they should be kept closed in the spring, until warm weather has raised the temperature of the ground and the basement walls. Be sure to provide proper venting for gas hot water heaters, clothes dryers, and other appliances that release moisture into the air.

Termite Protection

When you're building the basement walls is the time to consider termite protection for the wood portions of the structure. There are several ways to protect the wood in the house from termite attack:

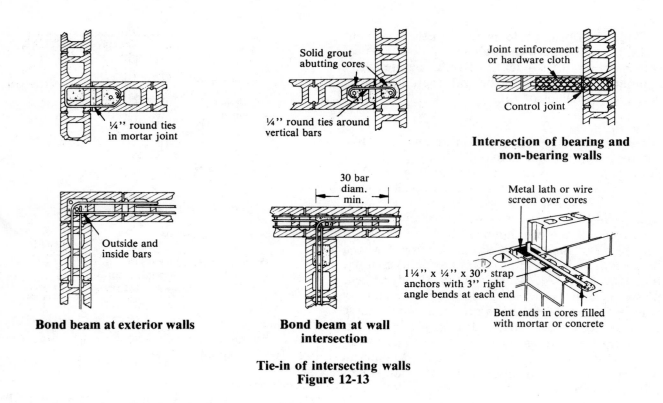

Bond beam at exterior walls   **Bond beam at wall intersection**

¼" round ties in mortar joint

Solid grout abutting cores

¼" round ties around vertical bars

Joint reinforcement or hardware cloth

Control joint

Intersection of bearing and non-bearing walls

Outside and inside bars

30 bar diam. min.

Metal lath or wire screen over cores

1¼" x ¼" x 30" strap anchors with 3" right angle bends at each end

Bent ends in cores filled with mortar or concrete

Tie-in of intersecting walls
Figure 12-13

1) With a special capping course of blocks on the top of the basement wall.

2) With chemical treatment of the soil next to the basement, piers, porches, and under slab-on-grade concrete.

3) Using chemically-treated wood (more on this in the next chapter, The All-Weather Wood Foundation).

4) By installing metal termite shields on top of the foundation walls.

Check your building code to find out which of these methods to use. It depends on the termite hazard in your area.

There are other ways you can reduce the chance of termite attack: You can be sure the site and building have good drainage. You can make sure there's enough flashing at the heads and sills of all openings. You can make sure closed areas have enough ventilation. You can keep the site clean, including removing all waste wood. And you can make sure that all structural wood in the foundation area is inspected.

Concrete Block Apartment Buildings
Concrete block masonry is good for multi-family residential construction. It's strong. It's economical. When properly designed and built, it's attractive and durable and provides thermal, sound, and fire insulation. But attention to construction details is essential in this kind of building.

Modern techniques and designs use walls and floors to make the building stable. The result is taller buildings with thinner walls. For this reason, it's very important that connections between intersecting walls, and between walls and floors, are strong enough to safely transmit the loading expected on the building.

The importance of proper connection details has been shown by the good performance of engineered masonry buildings, particularly in areas where high winds and earthquakes are expected.

The exact construction details depend on the particular combination of materials and conditions. But the following figures are a guide to a few of the important connection details for reinforced and non-reinforced concrete masonry walls. Included are wall intersection details, connection of walls to several commonly-used kinds of floor and roof systems, and details for shear transfer through precast floor slabs.

Figure 12-13 shows how to tie-in intersecting walls. In the bottom left section of the figure (the

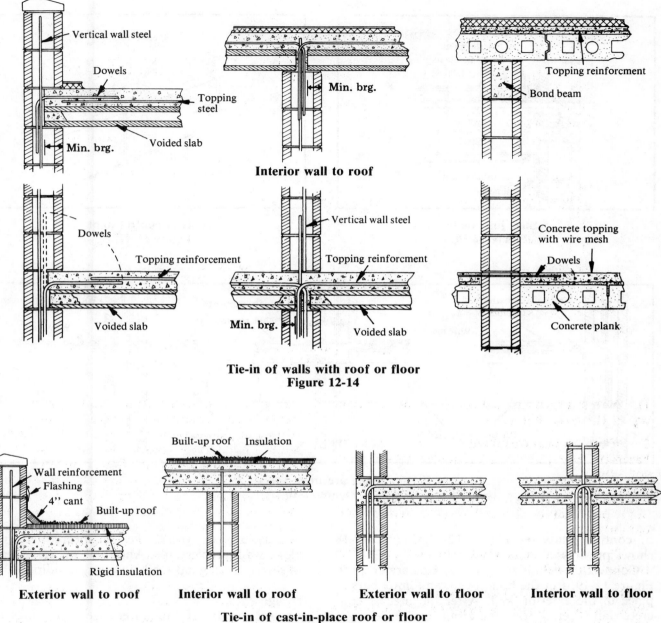

Interior wall to roof

Tie-in of walls with roof or floor
Figure 12-14

Exterior wall to roof   **Interior wall to roof**   **Exterior wall to floor**   **Interior wall to floor**

Tie-in of cast-in-place roof or floor
Figure 12-15

exterior wall bond beam), note that the outside bars extend around the corner of the intersection. And the inside bars extend as far as possible, then bend into the corner core. In the upper right section, note that the joint reinforcement (or hardware cloth) is used on every second course. In the lower right section, the strap anchors shown should have a vertical spacing of not more than 48'' o.c.

Figure 12-14 gives details at intersection of walls and floor or roof. The two figures at the left show how the reinforcing dowels are inserted in the wall, then bent down after the voided slab is in place. In the lower center part of the figure, the vertical steel wall reinforcement is lapped with the reinforcement from the wall below.

Cast-in-place roofs and floors are shown in Figure 12-15.

For steel joist tie-in, see Figure 12-16. For details

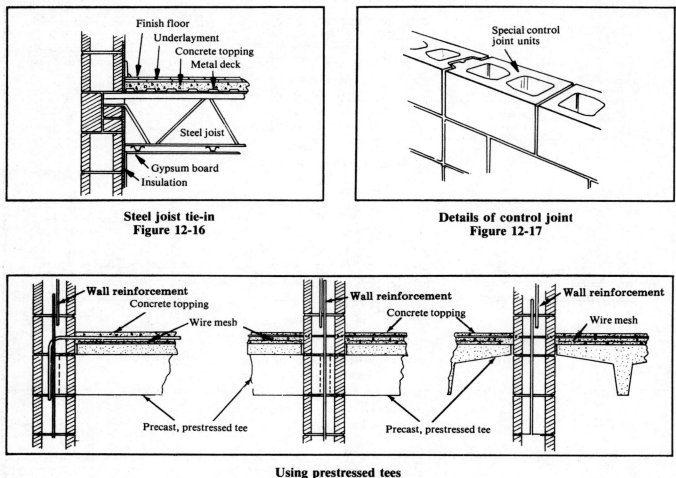

Steel joist tie-in
Figure 12-16

Details of control joint
Figure 12-17

Using prestressed tees
Figure 12-18

of control joints see Figure 12-17. Figure 12-18 shows prestressed tees.

Connection details for precast slabs are given in Figure 12-19. Put the bond beam and lintel bars in place in the exterior wall. After the precast concrete slab is set, vibrate the grout into place.

Electrical outlet box and duct installation is shown in Figure 12-20.

Concrete Block Warehouse Walls

Warehouses and similar buildings (manufacturing plants, machine shops, grocery stores) with relatively long and high one-story walls can be built with engineered concrete block walls. These walls can be used as non-bearing infill walls in a structural frame. Or they can be used as load-bearing walls to carry all the vertical and lateral loads to the foundation. As you'll see, engineered concrete block walls are versatile. And they can be designed to safely carry wind loads in many conditions.

Design
Lateral loads: If the warehouse roof loads are carried by a structural frame, the warehouse walls will only have to support lateral loads from wind pressure or earthquakes. If the total wall height is less than 30', the wind pressure is often given by this formula:

$$P = 0.0033V^2$$

P is wind pressure in pounds per square foot, and V is the peak gust velocity in miles per hour.

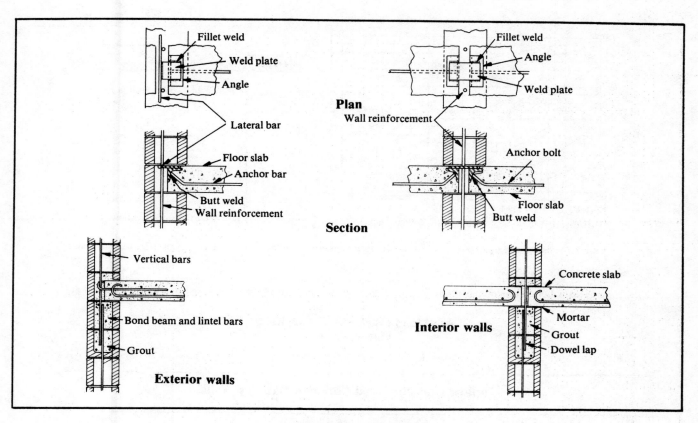

Connecting precast slabs
Figure 12-19

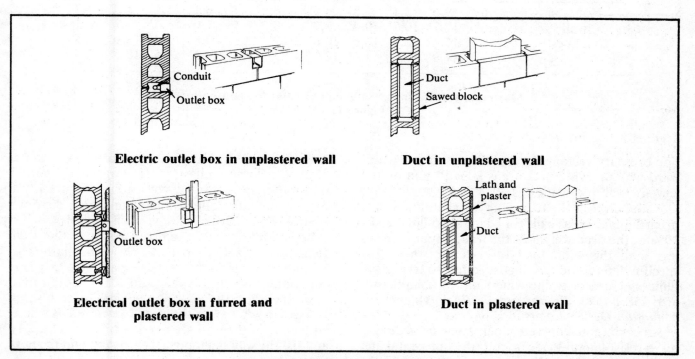

Installing electrical boxes and ducts
Figure 12-20

313

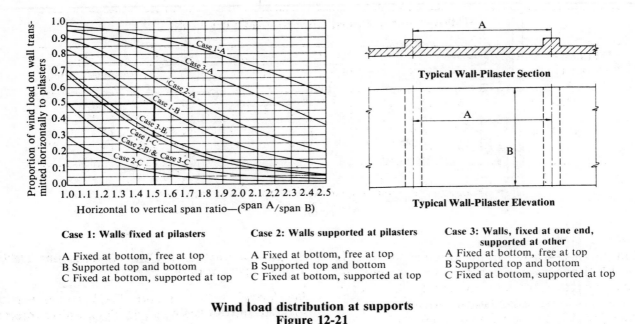

Case 1: Walls fixed at pilasters

A Fixed at bottom, free at top
B Supported top and bottom
C Fixed at bottom, supported at top

Case 2: Walls supported at pilasters

A Fixed at bottom, free at top
B Supported top and bottom
C Fixed at bottom, supported at top

Case 3: Walls, fixed at one end, supported at other

A Fixed at bottom, free at top
B Supported top and bottom
C Fixed at bottom, supported at top

Wind load distribution at supports
Figure 12-21

Hollow Nonreinforced Concrete Masonry Walls

Wind load, w (psf) ➝	10		15		20		25		30	
Mortar type ➝	M,S	N	M,S	N	M,S	N	M,S	N	M,S	N
Nominal 6"	9.6'	8.1'	7.8'	6.6'	6.8'	5.7'	6.1'	5.1'	5.5'	4.7'
wall 8"	12.7'	10.7'	10.4'	8.7'	9.0'	7.5'	8.0'	6.7'	7.3'	6.1'
thickness 10"	15.4'	12.9'	12.6'	10.6'	10.9'	9.1'	9.8'	8.2'	8.9'	7.5'
12"	17.9'	15.0'	14.6'	12.3'	12.7'	10.6'	11.3'	9.5'	10.3'	8.7'

Maximum vertical span, walls supported at top and bottom
Figure 12-22

For ordinary rectangular buildings, on the windward wall the wind pressure is 80% P, and on the leeward wall the suction is 20% P.

About earthquake forces: a typical building code in earthquake areas calls for a lateral design load of 20% of the dead weight of the wall. Usually, this is less than the wind load on exterior walls. For specific lateral design loads, refer to the local building code, or to the American National Standard A58.1-1972, *Building Code Requirements for Minimum Design Loads in Buildings.*

An engineered concrete block wall can be designed to resist lateral loads in both the horizontal and vertical spans, depending on its height, and on the length between structural supports — beams, girders, columns, pilasters. If a wall is supported on more than two sides, you can use Figure 12-21 to determine the approximate wind load distribution to the horizontal and vertical supports.

Let's look at an example. Assume the wall distance (horizontal span A) between pilasters is 30' and the wall height (vertical span B) is 15'. The horizontal-to-vertical span ratio is 1:5. You'll find it on the graph shown at the left in Figure 12-21.

Assume you have a situation like Case 1-B where both ends of the wall are fixed at pilasters (Case 1) and (B) the wall is supported at the top and the bottom.

Hollow Nonreinforced Concrete Masonry Walls

Wind load, w (psf) →		10		15		20		25		30	
Mortar type →		M,S	N	M,S	N	M,S	N	M,S	N	M,S	N
Nominal	6"	13.6'	11.4'	11.1'	9.3'	9.7'	8.1'	8.6'	7.2'	7.8'	6.6'
wall	8"	18.0'	15.1'	14.7'	12.3'	12.7'	10.6'	11.4'	9.5'	10.3'	8.7'
thickness	10"	21.9'	18.3'	17.8'	14.9'	15.5'	12.9'	13.8'	11.6'	12.6'	10.6'
	12"	25.4'	21.2'	20.7'	17.3'	18.0'	15.0'	16.1'	13.4'	14.6'	12.2'

Maximum horizontal span, walls supported at each end
Figure 12-23

Find the Case 1-B line in the graph at the left of Figure 12-21 and locate the point where the line intersects the 1:5 horizontal-to-vertical span ratio line. Here, reading to the left, it also intersects the 0.5 line, giving the proportion of wind load on wall transmitted horizontally to pilasters.

Thus, 50% of the wind load on the wall is absorbed by the pilasters.

As you can see by the graph, walls free at the top require a greater absorption of the wind load by the pilasters.

Vertical loads: Where the dead and live loads of the warehouse roof are carried directly by the walls, the walls must be able to support these vertical loads as well as the lateral loads discussed above.

Design Tables

Nonreinforced walls — Maximum wall spans for engineered, nonreinforced, hollow concrete block walls subjected to uniformly-distributed wind loads are given in Figures 12-22, 12-23, and 12-24. Figure 12-22 covers simply supported (minimum anchorage) walls that span vertically. Figure 12-23 covers simply supported walls that span horizontally. And Figure 12-24 covers walls that are simply supported on all four sides. Design stresses used to develop these tables are based on those given in the National Concrete Masonry Association (NCMA) *Specification for the Design and Construction of Load-Bearing Concrete Masonry* for architectural, or engineering-inspected, nonreinforced construction.

Design tables for nonreinforced concrete block walls subject to vertical loads, and for walls with combined vertical and lateral loading, are contained in *Nonreinforced Concrete Masonry Design Tables,* NCMA-1971.

Reinforced walls — Figure 12-25 shows vertical spans for reinforced concrete block walls subjected to uniformly distributed wind loads. Figure 12-26 shows maximum horizontal spans for the same walls. Design stresses used in these tables are also based on those given in the NCMA *Specification for the Design and Construction of Load-Bearing Concrete Masonry* for engineered reinforced concrete masonry construction.

Allowable vertical loads and combined vertical and lateral loads for reinforced concrete block walls are given in *Reinforced Concrete Masonry Design Tables,* NCMA-1971.

Design examples — This section shows you how to use the design tables. It's not really hard to do. Assume that you're building a one-story concrete block warehouse. The design wind load (w) equals 20 psf. Now, choose non-bearing exterior walls for the following three lateral support conditions. The walls may be either reinforced or nonreinforced.

1) Lateral supports at top and bottom of wall; wall height is 12':

Nonreinforced — From Figure 12-22, find 12" nonreinforced wall (type M or S mortar). The maximum vertical span is 12.7'.

Reinforced — from Figure 12-25, find 6" reinforced wall, with No. 5 bars spaced 48" o.c.; or 8" reinforced wall, with No. 4 bars spaced 48" o.c. The maximum vertical span is 13.3' and 12.6' respectively.

2) Lateral supports (columns or pilasters) spaced 15' o.c. horizontally:

Nonreinforced — from Figure 12-23, find 10" nonreinforced wall (type M or S mortar); or 12" nonreinforced wall (type N mortar). The maximum horizontal span is 15.5' and 15.0' respectively. Reinforced — from Figure 12-26, find 8" reinforced wall, with bond beams spaced 32" o.c., containing 2 No. 4 bars. The maximum horizontal span is 15.7'.

3) Lateral supports on four sides; 12' apart vertically and 20' apart horizontally.

Nonreinforced — from Figure 12-24, find 10" nonreinforced wall (type M or S mortar). The maximum horizontal span is 22.8'.

Second-Generation Passive Solar Systems

The first generation of passive solar systems (Trombe wall, direct gain, and attached solar greenhouse) had two things in common: (1) The heat-absorbing material was also the heat-storage material; and (2) The material was usually located in the collection area. These characteristics limited the building's design, especially the window, room, and even furniture placement.

Second-generation systems— These systems, often called "hybrid," pass the solar-heated air through the cores of concrete blocks to store and distribute the heat. The main benefit of these systems is that the thermal storage mass can be anywhere in the building. It doesn't depend on where the heat is collected.

Hollow Nonreinforced Concrete Masonry Walls

Wall thickness →	6"								8"							
Lateral load →	15 psf		20 psf		25 psf		30 psf		15 psf		20 psf		25 psf		30 psf	
Mortar type →	M,S	N	M,S	N	M,S	N	M,S	N	M,S	N	M,S	N	M,S	N	M,S	N
8'	19.0'	9.6'	10.2'	8.1'	8.8'	7.3'	7.9'	6.7'	*	*	*	13.6'	*	10.0'	12.2'	8.8'
10'	11.5'	9.3'	9.7'	8.3'	8.8'	7.7'	8.1'	7.3'	*	13.2'	14.5'	10.8'	11.6'	9.5'	10.5'	8.8'
12'	11.2'	9.7'	9.9'	8.9'	9.2'	8.2'	8.8'	7.8'	15.8'	12.5'	12.9'	10.8'	11.4'	9.8'	10.6'	9.2'
14'	11.5'	10.2'	10.5'	9.4'	9.8'	8.7'	9.2'	8.1'	14.7'	12.5'	12.9'	11.2'	11.6'	10.4'	11.1'	9.8'
Wall height 16'	12.0'	10.7'	10.9'	9.8'	9.9'	9.1'	9.6'	8.5'	14.9'	12.8'	13.1'	11.7'	12.3'	10.9'	11.5'	10.2'
18'	12.6'	11.0'	11.2'	10.1'	10.8'	9.5'	10.2'	9.2'	14.9'	13.5'	13.7'	12.1'	12.8'	11.3'	12.1'	10.8'
20'	13.0'	11.6'	12.0'	10.6'	11.0'	10.0'	10.6'	9.6'	15.6'	14.0'	14.2'	12.6'	13.0'	11.8'	12.2'	11.0'
22'	13.6'	12.1'	12.1'	11.2'	11.7'	10.5'	11.0'	9.9'	16.2'	14.3'	14.7'	13.0'	13.8'	12.1'	13.2'	11.7'
24'	13.9'	12.5'	12.7'	11.8'	12.0'	11.0'	11.5'	10.8'	16.8'	14.9'	15.4'	13.7'	14.4'	12.7'	13.2'	12.4'

Wall thickness →	10"								12"							
Lateral load →	15 psf		20 psf		25 psf		30 psf		15 psf		20 psf		25 psf		30 psf	
Mortar type →	M,S	N	M,S	N	M,S	N	M,S	N	M,S	N	M,S	N	M,S	N	M,S	N
8'	*	*	*	*	*	*	*	12.8'	*	*	*	*	*	*	*	*
10'	*	*	*	15.0'	22.0'	11.9'	14.2'	10.7'	*	*	*	*	*	17.8'	*	13.2'
12'	*	22.8'	22.8'	13.1'	14.3'	11.5'	12.8'	10.7'	*	*	*	16.8'	21.0'	13.7'	15.8'	12.4'
14'	20.6'	15.1'	15.8'	13.0'	13.9'	11.8'	12.7'	11.1'	*	19.2'	20.6'	15.1'	16.5'	13.4'	14.8'	12.2'
Wall height 16'	18.4'	14.9'	15.5'	13.3'	14.1'	12.3'	13.3'	11.7'	24.0'	17.6'	18.2'	15.0'	16.2'	13.6'	14.7'	12.8'
18'	17.8'	15.3'	15.7'	13.9'	14.6'	13.0'	13.7'	12.1'	21.2'	17.5'	18.0'	15.3'	16.2'	14.2'	14.9'	13.3'
20'	18.0'	15.8'	16.2'	14.4'	15.0'	13.4'	14.2'	12.6'	21.0'	17.4'	18.0'	15.8'	16.4'	14.6'	15.6'	14.0'
22'	18.3'	16.3'	16.5'	14.9'	15.6'	13.9'	14.7'	13.2'	20.9'	17.8'	18.3'	16.5'	17.2'	15.4'	16.1'	14.5'
24'	19.0'	16.8'	17.3'	15.1'	16.1'	14.4'	15.1'	13.7'	21.1'	18.5'	19.0'	16.8'	17.5'	15.8'	16.8'	14.9'

*Unlimited

Maximum horizontal span, walls supported on four sides
Figure 12-24

Hollow Reinforced Concrete Masonry Walls

Nominal wall thickness	Vertical reinforcement	Wind load, w (psf)				
		10	15	20	25	30
6"	No. 4 @ 48"	15.2'	12.4'	10.7'	9.6'	8.8'
	No. 5 @ 48"	18.8'	15.4'	13.3'	11.9'	10.9'
	No. 6 @ 48"	22.2'	18.1'	15.7'	14.0'	12.8'
	No. 5 @ 24"	25.2'	20.6'	17.8'	15.9'	14.6'
	No. 6 @ 24"	26.7'	21.8'	18.9'	16.9'	15.4'
8"	No. 4 @ 48"	17.8'	14.5'	12.6'	11.2'	10.3'
	No. 5 @ 48"	22.1'	18.0'	15.6'	14.0'	12.7'
	No. 6 @ 48"	26.0'	21.3'	18.4'	16.5'	15.0'
	No. 5 @ 24"	30.7'	25.0'	21.7'	19.4'	17.7'
	No. 6 @ 24"	34.4'	28.1'	24.4'	21.8'	19.9'
10"	No. 4 @ 48"	20.0'	16.4'	14.2'	12.7'	11.6'
	No. 5 @ 48"	24.9'	20.3'	17.6'	15.7'	14.4'
	No. 6 @ 48"	29.4'	24.0'	20.8'	18.6'	17.0'
	No. 5 @ 24"	34.6'	28.3'	24.5'	21.9'	20.0'
	No. 6 @ 24"	40.9'	33.4'	29.0'	25.9'	23.6'
12"	No. 4 @ 48"	22.1'	18.0'	15.6'	14.0'	12.8'
	No. 5 @ 48"	27.4'	22.4'	19.4'	17.4'	15.8'
	No. 6 @ 48"	32.4'	26.5'	22.9'	20.5'	18.7'
	No. 5 @ 24"	38.2'	31.2'	27.0'	24.2'	22.1'
	No. 6 @ 24"	45.2'	36.9'	32.0'	28.6'	26.1'

Maximum vertical span, wall vertically reinforced at centerline
Figure 12-25

Hollow Reinforced Concrete Masonry Walls

Nominal wall thickness	Bond beam spacing	Reinforcement per bond beam	Wind load, w (psf)				
			10	15	20	25	30
6"	24" o.c.	1—No. 4	15.8'	12.9'	11.2'	10.0'	9.1'
		1—No. 5	16.8'	13.7'	11.9'	10.6'	9.7'
		1—No. 6	17.6'	14.4'	12.5'	11.1'	10.2'
8"	32" o.c.	2—No. 4	22.2'	18.1'	15.7'	14.0'	12.8'
		2—No. 5	24.0'	19.6'	17.1'	15.2'	13.8'
		2—No. 6	25.3'	20.7'	17.9'	16.0'	14.6'
10"	40" o.c.	2—No. 4	25.6'*	20.9'*	18.1'*	16.2'*	14.8'*
		2—No. 5	28.5'	23.3'	20.1'	18.0'	16.4'
		2—No. 6	30.2'	24.7'	21.4'	19.1'	17.5'
12"	40" o.c.	2—No. 4	29.2'*	23.8'*	20.6'	18.5'	16.9'
		2—No. 5	35.0'	28.6'	24.8'	22.2'	20.2'
		2—No. 6	37.3'	30.4'	26.4'	23.6'	21.5'

*Maximum wall length limited by allowable steel stress
Maximum horizontal span, walls using horizontal bond beams
Figure 12-26

Using Concrete Block Cores

The two most popular uses of block cores in passive solar systems have been the horizontal block bed floor system, and the vertical block plenum wall.

The block bed floor system— is shown in Figure 12-27 in a direct-gain house. Solar energy enters through direct-gain windows. Then the hot air rises. It's collected in the highest point in the building. This heat is drawn off the ceiling by ducts and a small (less than 1 hp) fan. Solar heated air is then blown through the cores, to heat the blocks. After the heat has been removed from the moving air, the air returns to the room through grilles at the end of the block floor. The solar heated blocks warm the room by radiation and by convection. The time-lag of the block floor system causes the heat to be radiated to the room during the night, when it's needed.

Figure 12-28 shows how to build a block bed floor. The blocks are placed on their sides, with the cores lined up. The blocks usually rest on rigid insulation to prevent heat loss to the ground. A thin

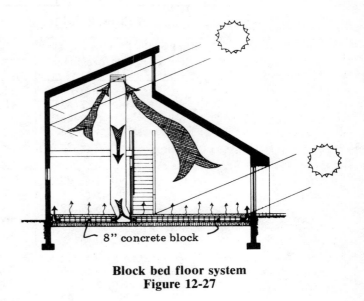

Block bed floor system
Figure 12-27

topping of concrete is poured over the block for a floor surface. Except for placing the blocks (done by laborers), the system is similar to slab-on-grade construction.

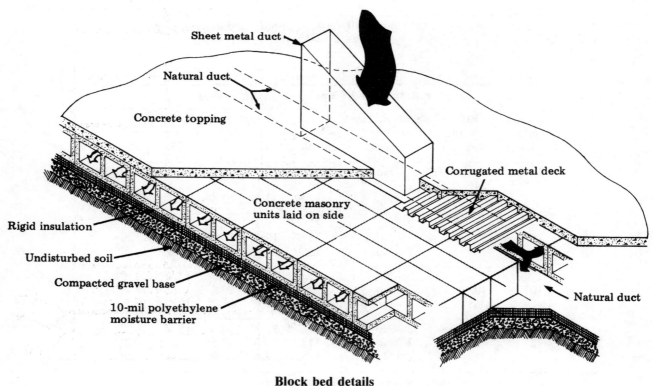

Block bed details
Figure 12-28

The vertical block plenum wall system— This innovative system also uses conventional construction. The solar heated air is passed through the ducts created by the block's cores in the wall. It stores heat during the day for use during the night. Unreinforced, partially-grouted, and surface-bonded walls can all be used. They give equally good results.

In Figure 12-29, the block plenum wall heat-storage system also has a two-story solarium. A standard block thermal-storage wall was added in the solarium because of the large area of south-facing windows. The excess heat generated in the solarium is ducted down the cores of the hollow block wall. So, the plenum wall radiates its heat to the northern rooms in the house, while the solid thermal-storage wall radiates its heat to the southern rooms.

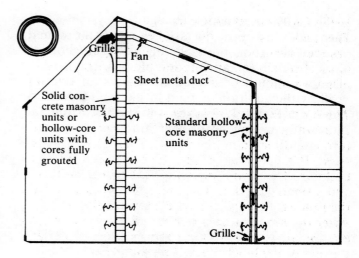

Plenum wall storage system
Figure 12-29

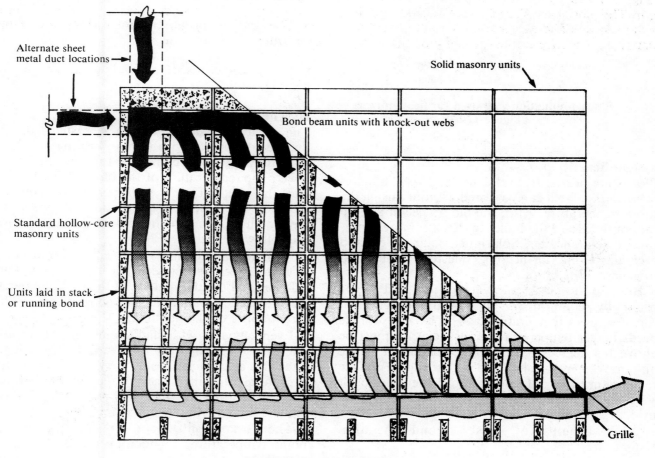

Block plenum wall details
Figure 12-30

The only construction change this system needs is the addition of the ducts. They supply air to, and remove air from, the wall. Figure 12-30 shows a typical detail for an interior block plenum wall. This wall could be altered for use as an exterior wall by the addition of rigid insulation to the outside face. Figure 12-31 shows how a block foundation wall can double for heat storage.

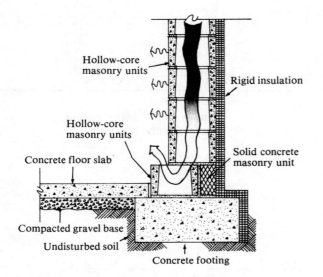

Block foundation wall used for heat storage
Figure 12-31

System Benefits

Blocks are heated from the inside. This eliminates overheating. Heat can be stored without raising the room temperature. Also, the blocks needn't be on the south side of the building. Even remote rooms can be solar heated. Solariums, solar attics, and even active solar air-heating collectors can be used to generate the hot air.

Because the block storage transfers heat to the rooms by radiation, not by forced convection, it can use lower-temperature air. Rock bed storage needs higher temperature air: the exit air must be above 105 degrees F, or people feel chilled by the air movement. The block storage surface temperature has only to be above the thermostat set-point to be comfortable, and to reduce energy use.

Other benefits of these second-generation systems are:

• They're more efficient, because lower temperature differences are used.

• They can be used in commercial, as well as in residential, buildings.

• They're inexpensive, because the storage doubles as a wall or a floor.

• They use standard masonry construction techniques and materials.

• They reduce the need for window insulation at night, because storage systems don't have to face windows.

• Conventional finishes can be used without seriously affecting performance.

• Higher mean radiant temperature allows for lower thermostat setting, and a corresponding energy savings.

• You can adapt current house designs, without changing the floor plan.

• They can be used for cooling, as well as for heating.

Hybrid Cooling

The greatest advantage of these systems is that they're equally effective in both heating and cooling. Blowing cool night air through the cores lowers the temperature of the block. The time lag and the thermal-storage properties of concrete block let the block cool the house by absorbing daytime heat, which is then blown from the cores the following night.

Commercial buildings often require cooling even in winter, because of internal heat generation. Outside air, and the thermal lag of the concrete block, can be used to moderate the temperature in different zones of the building.

Even climates with high humidity can benefit from these systems: the relative humidity doesn't alter the effectiveness of the system. As cooler, very humid air enters the storage, the block raises the temperature of that air, which lowers the relative humidity.

The block doesn't cool down as far as the dew-point temperature, so no moisture condensation occurs. Moisture flows the same direction as heat. Any moisture will flow from the heat source (the block) to the cooler medium (the moving air). Therefore, these systems can be used with earth-cool tubes without high-relative-humidity problems, as long as the air doesn't exit into occupied spaces. Earth-cool tubes (or pipe) are placed about 8' underground, where the temperature remains constant. Air is circulated through the pipe and is used to assist in cooling or heating a building, depending on the season.

The buildings in Figures 12-27 and 12-29 can easily be adapted to passive cooling. Air is ducted from an appropriate cool-air source through the block cores during the summer. This provides radiant cooling. If a closed-loop system is used, the passive system and a mechanical air conditioner can be run at the same time. This way, in theory, the passive system can remove the sensible cooling load, while the mechanical unit removes the latent load.

Of course, much of this is still experimental. But all professional builders should be aware of what's being done to improve the quality of construction.

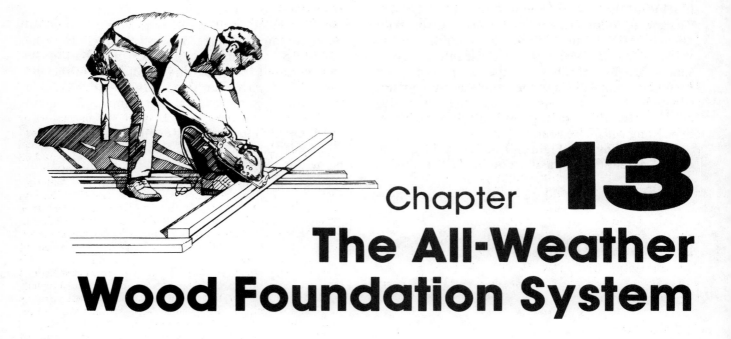

Chapter 13
The All-Weather Wood Foundation System

The All-Weather Wood Foundation (AWWF) uses treated lumber as the building foundation. The lumber industry and the U.S. Forest Service developed it with advice from the Federal Housing Administration and research data from the National Association of Home Builders. The system combines proven construction techniques with good moisture control practices.

It's sturdy— It's a structurally-engineered construction method. A gravel footing and a wood footing plate transfer vertical loads to the soil. The studs and plywood sheathing resist lateral loads of the soil. If the backfill is shallow, use the familiar 2 x 4's on 16'' centers for the AWWF. If the fill is deep, use 2 x 6 studs on 12'' centers. In some situations, you'll use studs that are even larger, or closer together.

It's practical in seismic areas— It won't crack and let water in. All joints between plywood panels are caulked. The whole exterior of a basement (from footing plate to grade level) is covered with a moisture barrier of polyethylene film. Unlike conventional rigid asphaltic dampproofing, this flexible film can adjust to minor movements in the foundation.

It's waterproof— The key to the system is the *continuously-drained gravel pad around and under the footings, and under the floor.* A sump is installed under the floor. (The polyethylene film, and the caulking, are less important in keeping an AWWF basement dry). A dry foundation is assured: water flows easily through the gravel to the sump, where it's pumped or drained away. This method was developed by the NAHB Research Foundation. Thousands of AWWF houses have been built — and not one owner has reported a leaking basement, according to the National Forest Products Association (NFPA).

It can be installed in any weather— That's why it's called the "All-Weather" Wood Foundation. On days when you can't pour concrete because it's too cold, you'll keep building with AWWF. That can add months to your building season. Once you've excavated and installed the gravel, bad weather won't keep you from installing the foundation and framing. You can put foundations up in the coldest weather, in the hottest weather, or in pouring rain.

This chapter will tell you why the AWWF system is an economical and productive foundation. And it will show you how an AWWF is designed. For more information, send for: *All-Weather Wood*

Foundation System — Design, Fabrication and Installation Manual. It's available from the National Forest Products Association, 1619 Massachusetts Avenue NW, Washington, DC 20036.

Differences from Traditional Wood-Frame Construction

The AWWF differs from conventional above-grade wood construction in three ways:

1) The framing and sheathing are stress-graded. They must be carefully engineered to support lateral soil pressures and vertical loads.

2) All wood in contact with, or close to, the ground is pressure-treated with wood preservatives. This protects it against decay and insects. The durability and effectiveness of these preservatives have been proven by decades of U.S. Forest Service research.

3) Moisture-control methods produce dry and comfortable below-grade living space. The most important barrier is the porous gravel envelope surrounding the lower part of the basement. This layer conducts ground water to a sump. Moisture at the upper part of the foundation wall runs down the polyethylene sheeting into the gravel. The result is a dry basement; one easily insulated for comfort and energy conservation.

But note that the AWWF, like all other foundation systems, requires quality workmanship and attention to detail.

The Materials

You have to use the right materials in an AWWF. Let's take them one at a time:

Lumber

Foundation sections can be custom ordered from the mill or built up on the site from standard pieces available from most yards. Use lumber of a species and grade for which allowable unit stresses are listed in the current edition of the *National Design Specification for Wood Construction,* by the NFPA. Every piece of lumber must have an inspection stamp or certificate.

Plywood

Use all-veneer plywood panels, bonded with exterior glue. All plywood must be grademarked by a recognized plywood-inspection agency.

Fasteners

In the AWWF system, the fasteners are important to the strength of the foundation. They're as important as the framing and plywood. Be sure that all nails and staples you use in the foundation comply fully with these AWWF specifications:

The fasteners should be of silicon bronze or copper, or stainless steel types 304 or 316 as defined by the American Iron and Steel Institute classification. Hot-dipped zinc coated steel nails with a minimum average weight of zinc coating of 2.00 oz/ft^2 of surface and that conform to requirements of the current edition of American Society for Testing and Materials Standard A153, Specification for Zinc Coating (Hot-Dip) on Iron and Steel Hardware, shall be permitted in preservative treated wood (i) above grade; (ii) below grade when installed in basement constructions, provided polyethylene sheeting is applied to the exterior of the below grade portions of basement walls, and beneath the joists of wood basement floors; and (iii) below grade when installed in crawl-space constructions situated in well-drained Group 1 soils, provided polyethylene sheeting is applied to the exterior of the below grade portions of exterior walls in accordance with the installation requirements for polyethylene sheeting on basement walls.

Fasteners or fastener materials not otherwise permitted under this Article shall be permitted if adequate comparative test for durability, including the effects associated with the wood treating chemicals, demonstrate performance equal to or greater than the specified fasteners or fastener materials. Hot-dipped zinc coated nails or other coated nails of equivalent performance shall be coated after manufacture to their final form, including pointing, heading, threading or twisting, as applicable. Electro-galvanized nails or staples and hot-dipped zinc coated or other zinc coated staples shall not be permitted.

Framing anchors shall be of zinc coated sheet steel conforming to Grade A, as set forth in the current edition of American Society for Testing and Materials Standard A446, Specification for Steel Sheet Zinc Coated (Galvanized) by the Hot-Dip Process, Physical (Structural) Quality, and having at least the following minimum properties:

Allowable stress in tension	18,000 psi
Yield point	33,000 psi
Ultimate strength	45,000 psi
Elongation in 2 inches	20 percent

The corrosion resistant coating shall be 1.25 oz. pot yield commercial class hot-dipped zinc coating, or 0.625 oz. matte finish hot-dipped zinc coating each side, and may be applied to the steel sheet before the anchor is stamped out. The zinc coating shall conform to the applicable requirements of the current editions of American Society for Testing and Materials Standards A446, A153 and A90, Tests for Weight of Coating on Zinc Coated (Galvanized) Iron or Steel Articles. Nails for use with framing anchors shall conform with the requirements for nails specified elsewhere herein.

Gravel, Sand, or Crushed Stone

The gravel used must be washed and well graded. It should be free from organic, clayey, or silty soils. No stone should be larger than 3/4".

Sand should be coarse, with grains not smaller than 1/16''. It too must be free from organic, clayey, or silty soils.

Crushed stone should be no larger than 1/2''.

Polyethylene Sheeting
Polyethylene sheeting should comply with the current edition of U.S. Department of Commerce Product Standard PS 17, *Polyethylene Sheeting (Construction, Industrial and Agricultural Application)*.

Sealants
Use an adhesive with a good polyethylene-to-wood bond for attaching the polyethylene sheets to the wall. And use a glue with a good polyethylene-to-polyethylene bond for sealing the polyethylene sheets together. Take into account the temperature and moisture-content of the air at the time you'll be applying and using any sealant. Caulking and adhesive materials such as butyl and silicone, which are readily available, have excellent resistance to weather damage. And where the material is protected from the weather, especially from ultraviolet rays, and from drastic changes in temperature and moisture, as is the case in the AWWF system, the caulking and adhesive material can last forever.

Also take into account the temperature and moisture-content conditions when choosing the caulking for joints in the plywood wall sheathing.

Preservative Treatment
Which Wood Is Treated
The following lumber and plywood used in the AWWF system must be preservative-treated:

- All lumber and plywood used in exterior foundation walls (except the upper top plate)

- All interior bearing wall framing and sheathing

- Posts, or other wood supports, used in crawl spaces

- All sleepers, joists, blocking, and plywood subflooring used in the basement floor

- All other plates, framing, and sheathing in the ground or in direct contact with concrete

If a large part of the bottom-story wall is above ground level (as with a house on sloping ground) good engineering practices dictate how much of the wall is considered foundation wall. Some members of the wall, such as the window and door headers, or the top plate, may not need preservative treatment. At the least, have all exterior wall framing lumber and plywood sheathing that will be less than 8'' above finish grade preservative treated.

How Wood is Treated
The lumber and plywood should be pressure-treated as described in the publication *American Wood Preservers Bureau Standard AWPB-FDN, Quality Control Program for Softwood Lumber, Timber and Plywood Pressure Treated with Water Borne Preservatives For Ground Contact Use in Residential and Light Commercial Foundations*.

After treatment, each piece of lumber and plywood is dried to a moisture content of 19% or less. Each piece must also bear the mark of a recognized inspection agency. And the symbol *AWPB-FDN,* and other markings required by that standard, must be on each piece.

If you cut or drill FDN lumber after treatment, field-treat the cut surface with one of the following preservatives: Copper Napthenate, Ammoniacal Copper Arsenate (ACA), Chromated Copper Arsenate (CCA), Fluor Chrome Arsenate Phenol (FCAP), or Acid Copper Chromate (ACC). Brush, dip, or soak the cut area again and again until the wood won't absorb any more preservative.

Prepare Copper Napthenate with a solvent conforming to *AWPA Standard P5.* The solution must contain a minimum of 2% copper. You can also use commercial preparations.

Water-borne preservatives ACA and CCA Types A, B, and C should have a minimum concentration of 3% in solution.

Water-borne preservatives FCAP and ACC can be used for field treatment of material originally treated with CCA and ACA water-borne preservatives. The concentration of FCAP or ACC must be a minimum of 5% in solution.

Designing the AWWF
Like every foundation, the AWWF has to support the intended load. The local building code will tell you what loads to design for. If there's no local code, use the current edition of the American National Standards Institute Standard A58.1, *Building Code Requirements for Minimum Design Loads in Buildings and Other Structures,* (or another nationally-recognized design reference) as a guide.

Soil group	Unified soil classification system symbol	Soil description	Allowable bearing in pounds per square foot with medium compaction or stiffness[3][4]	Drainage Characteristics[2]	Frost heave potential	Volume change potential expansion
Group I Excellent	GW	Well-graded gravels, gravel sand mixtures, little or no fines.	8000	Good	Low	Low
	GP	Poorly graded gravels or gravel sand mixtures, little or no fines.	8000	Good	Low	Low
	SW	Well-graded sands, gravelly sands, little or no fines	6000	Good	Low	Low
	SP	Poorly graded sands or gravelly sands, little or no fines	5000	Good	Low	Low
	GM	Silty gravels, gravel-sand-silt mixtures.	4000	Good	Medium	Low
	SM	Silty sand, sand-silt mixtures.	4000	Good	Medium	Low
Group II Fair to Good	GC	Clayey gravels, gravel-sand-clay mixtures.	4000	Medium	Medium	Low
	SC	Clayey sands, sand-clay mixture.	4000	Medium	Medium	Low
	ML	Inorganic silts and very fine sands, rock flour, silty or clayey fine sands or clayey silts with slight plasticity.	2000	Medium	High	Low
	CL	Inorganic clays of low to medium plasticity, gravelly clays, sands clays, silty clays, lean clays.	2000	Medium	Medium	Medium[1] to Low
Group III Poor	CH	Inorganic clays of high plasticity, fat clays	2000	Poor	Medium	High[1]
	MH	Inorganic silts, micaceous or diatomaceous fine sandy or silty soils, elastic silts	2000	Poor	High	High
Group IV Unsatisfactory	OL	Organic silts and organic silty clays of low plasticity.	400	Poor	Medium	Medium
	OH	Organic clays of medium to high plasticity, organic silts.	-0-	Unsatisfactory	Medium	High
	Pt	Peat and other highly organic soils.	-0-	Unsatisfactory	Medium	High

[1] Dangerous expansion might occur if these two soil types are dry but subject to future wetting.

[2] The percolation rate for good drainage is over 4 inches per hour, medium drainage is 2 to 4 inches per hour, and poor is less than 2 inches per hour.

[3] Building code allowable bearing values may differ from those tabulated.

[4] Allowable bearing value may be increased 25 percent for very compact, coarse grained gravelly or sandy soils or very stiff fine-grained clayey or silty soils. Allowable bearing value shall be decreased 25 percent for loose, coarse-grained gravelly or sandy soils, or soft, fine-grained clayer or silty soils. To determine compactness or stiffness to estimate allowable bearing capacity, measure the number of blows required to drive a 2-inch outside diameter, 1.375-inch inside diameter split-barrel sampler one foot into the soil by dropping a 140 pound hammer through a distance of 30 inches, Select compactness or stiffness value as follows:

Coarse grained gravelly or sandy soils (Unified Soil Classification System types GW, GP, GM, GC, SW, SP, SM, SC.)			Fine grained clayey or silty soils (Unified Soil Classification System types ML, CL, OL, MH, CH, OH.)		
Loose	Medium compaction	Very compact	Soft	Medium stiffness	Very stiff
		(number of blows per foot)			
10 or less	11-50	51 or more	4 or less	5-15	16 or more

Types of soils and their design properties
Figure 13-1

Soil Characteristics

Soils are defined by the Unified Soil Classification System (USCS). Four main soil groups, with their USCS classifications, are shown in Figure 13-1. That table also describes the various soil types.

In Group III soils, CH type backfill should not be compacted when dry. MH type backfill should be well-compacted, to keep surface water from soaking in.

Don't build in Group IV soils unless you take special steps. If you need to build wood foundations on Group IV soils, hire a soils engineer to advise you on the design of the entire soil system.

Soil Loads

When planning a wood foundation, think of the soil outside the foundation wall as a fluid. The soil's lateral (sideways) pressure on the wall is the same as that of a fluid with a given equivalent-fluid weight. The local building code may specify equivalent-fluid weight. Or, it may describe how to determine lateral soil design loads. If it doesn't, an equivalent-fluid weight of 30 pounds per cubic foot is right for most soils. In any of the following cases, however, either hire a qualified soils engineer, or have the local authority approve the installation:

- For buildings on unstable Group III soils.

- If there's a history of problems with foundations in the area.

- For buildings on Group IV soils.

Useful Design Aids

You'll often want to choose the lumber, plywood, or fastenings for an AWWF without doing all the design computations. Figures 13-2 through 13-18 have *pre-engineered* design information. The tables show framing lumber, plywood sheathing, and fastening designs that meet or exceed the minimum requirements of the AWWF. Use them as a guide and quick reference. For each combination of conditions, there are several possible solutions. The tables show only one or two solutions. An engineer can probably find a more economical choice of materials for your specific job.

Allowable Stresses for Species of Lumber

Allowable stresses for the species of lumber you'll use in an AWWF are listed in Figure 13-2. This table describes the species and grade combinations you'll find in Figures 13-3 through 13-9.

Foundation Wall Framing

Look at Figures 13-3 through 13-9 to find the size and grade of lumber to use for an AWWF. The table you use will depend on the design of the foundation: crawl-space walls 2' high are covered by one table; basement walls are covered by another.

Sometimes a crawl space will need end-wall struts to counteract lateral soil pressures. Figure 13-10 gives strut requirements and nailing schedule.

Minimum Footing Plate Size

Look at Figure 13-11 to find the footing plate sizes to use for the loadings, and other conditions, detailed in Figures 13-3 through 13-9.

Foundation Wall Sheathing

Figure 13-12 shows the grades and thicknesses of plywood sheathing to use for different design conditions.

Fastening Requirements

Figure 13-13 gives a general AWWF nailing schedule. Use more fastenings for heavy loads — especially for backfill higher than 4'. The information in Figures 13-13 through 13-16 applies to common wire steel nails. It also applies to other fasteners with equivalent allowable loads.

Figure 13-14 gives the fastener number, size, and spacing for typical top plate to stud, and top plate to top plate connections.

Figure 13-15 gives the fastener requirements for typical floor joist to wall connections.

If joists are parallel to the wall, and backfill is more than 4' high, you'll need to use blocking in the first joist space. Block at intervals of 4' or less. Figure 13-16 gives a nailing schedule for fastening this blocking.

Sheathing— Figure 13-17 gives a general minimum fastener schedule for attaching sheathing to the studs. If there's more than 2' difference in backfill height on opposite sides of the foundation, go to Figure 13-18. It's a nailing schedule for the perpendicular connecting walls. Nail all plywood panel edges in these connecting walls to 2'' (or thicker) framing lumber — unless shear design shows otherwise.

Lumber combination	Minimum allowable design values[1]							Examples of grade/species combinations meeting the minimum design values[4, 5]	
	F_b (Repetitive)		F_v	$F_{c\perp}$	F_c		E		
	2 x 6 2 x 8	2 x 4			2 x 6 2 x 8	2 x 4		Grade	Species
	Psi	Psi	Psi	Psi	Psi	Psi	Psi		
B-1	1700	1950	90	385	1250	1250	1,700,000	No. 1	Douglas Fir-Larch
B-2	1400	1650	90	385	1000	975	1,600,000	No. 2	or
B-3	800	900	90	385	625	575	1,400,000	{ Stud[3] or No. 3 }	Southern Pine[2]
C-1	1400	1600	70	245	975	975	1,400,000	No. 1	Hem-Fir
C-2	1100	1300	70	245	825	775	1,300,000	No. 2	or
C-3	650	725	70	245	525	475	1,100,000	{ Stud[3] or No. 3 }	Northern Pine
D-1	1200	1400	70	235	850	850	1,200,000	No. 1	Ponderosa Pine
D-2	975	1150	70	235	700	675	1,100,000	No. 2	- Sugar Pine
D-3	575	625	70	235	450	400	1,000,000	{ Stud[3] or No. 3 }	

Notes:

1. Minimum allowable design values for lumber shall be as tabulated in the National Design Specification for Wood Construction (NDS) for normal duration of load and dry conditions of use. Increase of design values for lumber manufactured and used at 15 percent maximum moisture content (MC-15) shall not apply. Design value symbols have the same meaning as the symbols in the NDS, as follows:

F_b = stress at extreme fiber in bending, as tabulated for repetitive member uses; psi
F_v = horizontal shear stress; psi
$F_{c\perp}$ = compression perpendicular to grain stress; psi
F_c = compression parallel to grain stress; psi
E = modulus of elasticity

2. Use tabulated values above for Southern Pine surfaced dry (S-Dry) or at 15 percent moisture content (Southern Pine KD).

3. "Stud" grade is not available in 2 x 8 size.

4. When 2" to 4" lumber is designed for use where the moisture content will exceed 19 percent, as in footing plates and crawl spaces, for an extended period of time, the design values shown herein shall be multiplied by the following factors, except for Southern Pine.

2" to 4" thick lumber used where moisture content will exceed 19%					
Extreme fiber in bending "F_b"	Tension parallel to grain "F_t"	Horizontal shear "F_v"	Compression perpendicular to grain "$F_{c\perp}$"	Compression parallel to grain "F_c"	Modulus of elasticity "E"
0.86	0.84	0.97	0.67	0.70	0.97

5. When Southern Pine lumber is surfaced dry or at 15% maximum moisture content (KD) and is used where the moisture content will exceed 19% for an extended period of time, the design values shown in Table 4A of the NDS shall be multiplied by a factor of 0.90 and then by the factors in Footnote 4. The net green size may be used in such designs.

6. Design values given apply only to material identified by the grademark of or certificate of inspection issued by, a grading or inspection bureau or agency recognized as being competent.

Lumber grade and species combinations
Figure 13-2

Figure 13-18 also gives the nailing schedule, and design shear strength, for walls subject to racking loads. No let-in braces are allowed in AWWF walls; whether they're used to counteract racking forces, or for any other purpose.

Building the AWWF
This section shows typical design details useful in preparing architectural drawings. These details have been developed by builders and engineers experienced with All-Weather Wood Foundations.

Height of fill (inches)	Stud size	Stud spacing (inches)	Lumber species and grade		
86	2 x 6	12	B-1	—	—
	2 x 8	16	B-2	C-1*	D-1*
		12	B-2	C-2	D-2
72	2 x 6	16	B-2	C-1*	—
		12	B-2	C-2	D-1
	2 x 8	16	B-3	C-2	D-2
		12	B-3	C-3	D-2
60	2 x 4	12	B-1	—	—
	2 x 6	16	B-2	C-2	D-2
		12	B-3	C-2	D-2
	2 x 8	16	B-3	C-3	D-3
48	2 x 4	16	B-2	C-1	D-1
		12	B-2	C-2	D-2
	2 x 6	16	B-3	C-3	D-3
36	2 x 4	16	B-3	C-3	D-2
		12	B-3	C-3	D-3
24	2 x 4	16	B-3	C-3	D-3

*Where indicated length of end splits or checks at lower end of studs not to exceed width of piece.

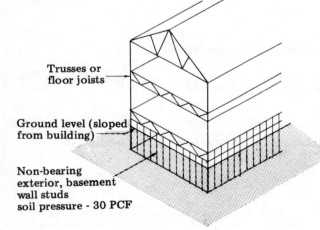

Trusses or floor joists

Ground level (sloped from building)

Non-bearing exterior, basement wall studs soil pressure - 30 PCF

Requirements for non-bearing, exterior, basement wall studs
Figure 13-3

The general notes, below, apply to all the figures.

General Notes to Figures
In Figure 13-19, the fill under the slab is level with the *top* of the footing plate. This is the best method. It allows for the same slab thickness and gives the most bearing against the stud. And it's less likely to leak. Another method is to leave the fill level with the *bottom* of the footing plate.

If prefabricated top and bottom plates for wall panels are not continuous pieces, reinforce butt joints in these members with splice plates so they'll stand up better to handling on the job site.

Supporting brick veneer— Use knee walls to support brick veneer on the foundation footing plate. Figure 13-20 shows how. Use 10'' (nominal) wide wood footing plates if Figure 13-2 calls for 2 x 4 foundation wall studs. Use 12'' wide plates if Figure 13-2 calls for 2 x 6, or 2 x 8 studs. Nail the knee wall according to the schedule in Figure 13-13.

Figure 13-20 shows a 2 x 6 top plate for the knee wall. The top plate edge that is next to the foundation wall is flush with the supporting framing. The opposite edge overhangs the stud framing. (Look at Figure 13-20 again.) Put double studs under any butt joints in the top plate, so each end of the top plate is supported on one stud.

Figure 13-21 shows how to build a crawl-space wall, with a knee wall for supporting brick veneer.

For brick up to 18' high, frame the knee wall with 2 x 4 studs on 16'' centers. Use a 1 x 4 bottom plate, and a 2 x 4 top plate. But, if Figure 13-2 calls for lumber combination "D," and if the brick is more than 16'8'' high, use a double top plate.

The Water Table
Avoid building livable space below the permanent water table. To do this, you need an engineer or an architect to design special waterproofing measures. And you yourself must make sure that the building will comply with the local code.

The Porous Layer
This layer of gravel, sand, or crushed stone must be at least 4'' thick. It goes under basement floor slabs, wood floors, and all wall footings. For basements in Group III soils, use a 6''-thick layer under footings, slabs, and wood floors.

Install a sump to drain the porous layer under a basement (unless the foundation is in a Group I soil — see Figure 13-1 again). Make the sump at least 24'' in diameter, or 20'' square. Extend it at least 24'' below the bottom of the basement floor. Install a pump if a gravity drain to daylight isn't practical.

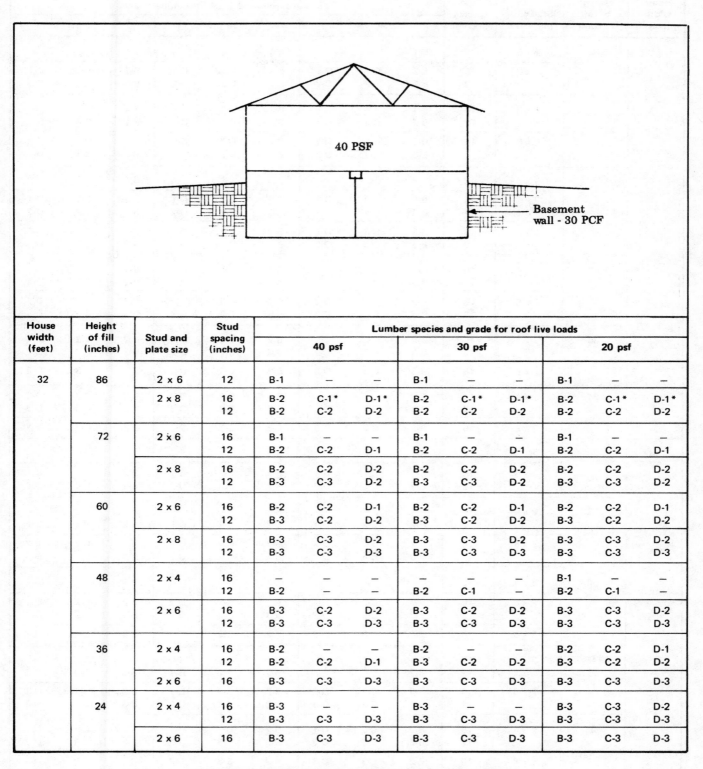

House width (feet)	Height of fill (inches)	Stud and plate size	Stud spacing (inches)	Lumber species and grade for roof live loads								
				40 psf			30 psf			20 psf		
32	86	2 x 6	12	B-1	—	—	B-1	—	—	B-1	—	—
		2 x 8	16	B-2	C-1*	D-1*	B-2	C-1*	D-1*	B-2	C-1*	D-1*
			12	B-2	C-2	D-2	B-2	C-2	D-2	B-2	C-2	D-2
	72	2 x 6	16	B-1	—	—	B-1	—	—	B-1	—	—
			12	B-2	C-2	D-1	B-2	C-2	D-1	B-2	C-2	D-1
		2 x 8	16	B-2	C-2	D-2	B-2	C-2	D-2	B-2	C-2	D-2
			12	B-3	C-3	D-2	B-3	C-3	D-2	B-3	C-3	D-2
	60	2 x 6	16	B-2	C-2	D-1	B-2	C-2	D-1	B-2	C-2	D-1
			12	B-3	C-2	D-2	B-3	C-2	D-2	B-3	C-2	D-2
		2 x 8	16	B-3	C-3	D-2	B-3	C-3	D-2	B-3	C-3	D-2
			12	B-3	C-3	D-3	B-3	C-3	D-3	B-3	C-3	D-3
	48	2 x 4	16	—	—	—	—	—	—	B-1	—	—
			12	B-2	—	—	B-2	C-1	—	B-2	C-1	—
		2 x 6	16	B-3	C-2	D-2	B-3	C-2	D-2	B-3	C-3	D-2
			12	B-3	C-3	D-3	B-3	C-3	D-3	B-3	C-3	D-3
	36	2 x 4	16	B-2	—	—	B-2	—	—	B-2	C-2	D-1
			12	B-2	C-2	D-1	B-3	C-2	D-2	B-3	C-2	D-2
		2 x 6	16	B-3	C-3	D-3	B-3	C-3	D-3	B-3	C-3	D-3
	24	2 x 4	16	B-3	—	—	B-3	—	—	B-3	C-3	D-2
			12	B-3	C-3	D-3	B-3	C-3	D-3	B-3	C-3	D-3
		2 x 6	16	B-3	C-3	D-3	B-3	C-3	D-3	B-3	C-3	D-3

**Minimum framing requirements for 8' basement wall—
one-story with clear-span roof trusses and center-bearing floors
Figure 13-4**

House width (feet)	Height of fill (inches)	Stud and plate size	Stud spacing (inches)	40 psf			30 psf			20 psf		
28	86	2 x 6	12	B-1	—	—	B-1	—	—	B-1	—	—
		2 x 8	16	B-2	C-1*	D-1*	B-2	C-1*	D-1*	B-2	C-1*	D-1*
			12	B-2	C-2	D-2	B-2	C-2	D-2	B-2	C-2	D-2
	72	2 x 6	16	B-1	—	—	B-1	—	—	B-1	—	—
			12	B-2	C-2	D-1	B-2	C-2	D-1	B-2	C-2	D-1
		2 x 8	16	B-2	C-2	D-2	B-2	C-2	D-2	B-2	C-2	D-2
			12	B-3	C-3	D-2	B-3	C-3	D-2	B-3	C-3	D-2
	60	2 x 4	12	—	—	—	—	—	—	B-1	—	—
		2 x 6	16	B-2	C-2	D-1	B-2	C-2	D-1	B-2	C-2	D-1
			12	B-3	C-2	D-2	B-3	C-2	D-2	B-3	C-2	D-2
		2 x 8	16	B-3	C-3	D-2	B-3	C-3	D-2	B-3	C-3	D-2
			12	B-3	C-3	D-3	B-3	C-3	D-3	B-3	C-3	D-3
	48	2 x 4	16	—	—	—	B-1	--	—	B-1	—	—
			12	B-2	C-1	—	B-2	C-1	—	B-2	C-2	D-1
		2 x 6	16	B-3	C-2	D-2	B-3	C-3	D-2	B-3	C-3	D-2
			12	B-3	C-3	D-3	B-3	C-3	D-3	B-3	C-3	D-3
	36	2 x 4	16	B-2	—	—	B-2	C-2	—	B-2	C-2	D-1
			12	B-3	C-2	D-2	B-3	C-2	D-2	B-3	C-2	D-2
		2 x 6	16	B-3	C-3	D-3	B-3	C-3	D-3	B-3	C-3	D-3
	24	2 x 4	16	B-3	—	—	B-3	C-2	D-2	B-3	C-3	D-2
			12	B-3	C-3	D-3	B-3	C-3	D-3	B-3	C-3	D-3
		2 x 6	16	B-3	C-3	D-3	B-3	C-3	D-3	B-3	C-3	D-3
24	86	2 x 6	12	B-1	—	—	B-1	—	—	B-1	—	—
		2 x 8	16	B-2	C-1*	D-1*	B-2	C-1*	D-1*	B-2	C-1*	D-1*
			12	B-2	C-2	D-2	B-2	C-2	D-2	B-2	C-2	D-2
	72	2 x 6	16	B-1	—	—	B-1	—	—	B-1	—	—
			12	B-2	C-2	D-1	B-2	C-2	C-1	B-2	C-2	C-1
		2 x 8	16	B-2	C-2	D-2	B-2	C-2	D-2	B-2	C-2	D-2
			12	B-3	C-3	D-2	B-3	C-3	D-2	B-3	C-3	D-2
	60	2 x 4	12	—	—	—	B-1	—	—	B-1	—	—
		2 x 6	16	B-2	C-2	D-2	B-2	C-2	D-2	B-2	C-2	D-2
			12	B-3	C-2	D-2	B-3	C-2	D-2	B-3	C-2	D-2
		2 x 8	16	B-3	C-3	D-3	B-3	C-3	D-3	B-3	C-3	D-3
	48	2 x 4	16	B-1	—	—	B-1	—	—	B-2	—	—
			12	B-2	C-1	—	B-2	C-2	D-1	B-2	C-2	D-1
		2 x 6	16	B-3	C-3	D-2	B-3	C-3	D-2	B-3	C-3	D-2
			12	B-3	C-3	D-3	B-3	C-3	D-3	B-3	C-3	D-3
	36	2 x 4	16	B-2	C-2	D-1	B-2	C-2	D-1	B-2	C-2	D-2
			12	B-3	C-2	D-2	B-3	C-2	D-2	B-3	C-3	D-2
		2 x 6	16	B-3	C-3	D-3	B-3	C-3	D-3	B-3	C-3	D-3
	24	2 x 4	16	B-3	C-2	D-2	B-3	C-3	D-2	B-3	C-3	D-2
			12	B-3	C-3	D-3	B-3	C-3	D-3	B-3	C-3	D-3
		2 x 6	16	B-3	C-3	D-3	B-3	C-3	D-3	B-3	C-3	D-3

*Length of end splits or checks at lower ends of stud not to exceed the width of the stud.

Minimum framing requirements for 8' basement wall
Figure 13-4 (continued)

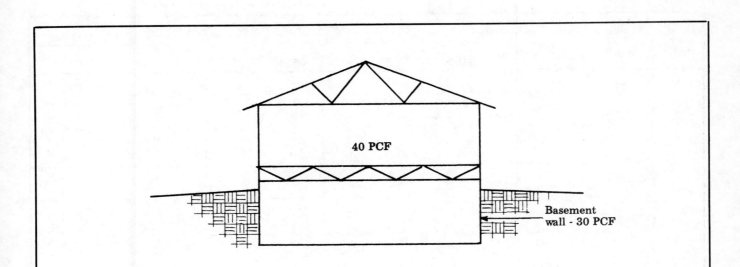

House width (feet)	Height of fill (inches)	Stud and plate size	Stud spacing (inches)	Lumber species and grade for roof live loads								
				40 psf			30 psf			20 psf		
32	86	2 x 6	12	B-1	—	—	B-1	—	—	B-1	—	—
		2 x 8	16	B-2	C-1*	—	B-2	C-1*	—	B-2	C-1*	—
			12	B-2	C-2	D-2	B-2	C-2	D-2	B-2	C-2	D-2
	72	2 x 6	16	B-1	—	—	B-1	—	—	B-1	—	—
			12	B-2	C-1	D-1	B-2	C-1	D-1	B-2	C-1	D-1
		2 x 8	16	B-2	C-2	D-2	B-2	C-2	D-2	B-2	C-2	D-2
			12	B-3	C-3	D-2	B-3	C-3	D-2	B-3	C-3	D-2
	60	2 x 6	16	B-2	C-1	D-1	B-2	C-2	D-1	B-2	C-2	D-1
			12	B-2	C-2	D-2	B-3	C-2	D-2	B-3	C-2	D-2
		2 x 8	16	B-3	C-3	D-2	B-3	C-3	D-2	B-3	C-3	D-2
			12	B-3	C-3	D-3	B-3	C-3	D-3	B-3	C-3	D-3
	48	2 x 4	12	B-1	—	—	B-1	—	—	B-2	—	—
		2 x 6	16	B-3	C-2	D-2	B-3	C-2	D-2	B-3	C-2	D-2
			12	B-3	C-3	D-2	B-3	C-3	D-3	B-3	C-3	D-3
		2 x 8	16	B-3	C-3	D-3	B-3	C-3	D-3	B-3	C-3	D-3
	36	2 x 4	16	B-1	—	—	B-1	—	—	B-2	—	—
			12	B-2	—	—	B-2	C-2	—	B-2	C-2	D-1
		2 x 6	16	B-3	C-3	D-3	B-3	C-3	D-3	B-3	C-3	D-3
	24	2 x 4	16	B-2	—	—	B-2	—	—	B-3	—	—
			12	B-3	—	—	B-3	C-2	—	B-3	C-3	D-2
		2 x 6	16	B-3	C-3	D-3	B-3	C-3	D-3	B-3	C-3	D-3

**Minimum framing requirements for 8' basement wall—
one-story with clear-span roof trusses and clear-span floor trusses
Figure 13-5**

House width (feet)	Height of fill (inches)	Stud and plate size	Stud spacing (inches)	40 psf			30 psf			20 psf		
28	86	2 x 6	12	B-1	—	—	B-1	—	—	B-1	—	—
		2 x 8	16	B-2	C-1*	D-1*	B-2	C-1*	D-1*	B-2	C-1*	D-1*
			12	B-2	C-2	D-2	B-2	C-2	D-2	B-2	C-2	D-2
	72	2 x 6	16	B-1	—	—	B-1	—	—	B-1	—	—
			12	B-2	C-2	D-1	B-2	C-2	D-1	B-2	C-2	D-1
		2 x 8	16	B-2	C-2	D-2	B-2	C-2	D-2	B-2	C-2	D-2
			12	B-3	C-3	D-2	B-3	C-3	D-2	B-3	C-3	D-2
	60	2 x 6	16	B-2	C-2	D-1	B-2	C-2	D-1	B-2	C-2	D-1
			12	B-3	C-2	D-2	B-3	C-2	D-2	B-3	C-2	D-2
		2 x 8	16	B-3	C-3	D-2	B-3	C-3	D-2	B-3	C-3	D-2
			12	B-3	C-3	D-3	B-3	C-3	D-3	B-3	C-3	D-3
	48	2 x 4	12	B-2	—	—	B-2	—	—	B-2	C-1	—
		2 x 6	16	B-3	C-2	D-2	B-3	C-2	D-2	B-3	C-2	D-2
			12	B-3	C-3	D-3	B-3	C-3	D-3	B-3	C-3	D-3
	36	2 x 4	16	B-2	—	—	B-2	—	—	B-2	—	—
			12	B-2	C-2	D-1	B-2	C-2	D-1	B-3	C-2	D-2
		2 x 6	16	B-3	C-3	D-3	B-3	C-3	D-3	B-3	C-3	D-3
	24	2 x 4	16	B-2	—	—	B-3	—	—	B-3	—	—
			12	B-3	C-3	D-2	B-3	C-3	D-2	B-3	C-3	D-3
		2 x 6	16	B-3	C-3	D-3	B-3	C-3	D-3	B-3	C-3	D-3
24	86	2 x 6	12	B-1	—	—	B-1	—	—	B-1	—	—
		2 x 8	16	B-2	C-1*	D-1*	B-2	C-1*	D-1*	B-2	C-1*	D-1*
			12	B-2	C-2	D-2	B-2	C-2	D-2	B-2	C-2	D-2
	72	2 x 6	16	B-1	—	—	B-1	—	—	B-1	—	—
			12	B-2	C-2	D-1	B-2	C-2	D-1	B-2	C-2	D-1
		2 x 8	16	B-2	C-2	D-2	B-2	C-2	D-2	B-2	C-2	D-2
			12	B-3	C-3	D-2	B-3	C-3	D-2	B-3	C-3	D-2
	60	2 x 6	16	B-2	C-2	D-1	B-2	C-2	D-1	B-2	C-2	D-1
			12	B-3	C-2	D-2	B-3	C-2	D-2	B-3	C-2	D-2
		2 x 8	16	B-3	C-3	D-2	B-3	C-3	D-2	B-3	C-3	D-2
			12	B-3	C-3	D-3	B-3	C-3	D-3	B-3	C-3	D-3
	48	2 x 4	16	—	—	—	—	—	—	B-1	—	—
			12	B-2	—	—	B-2	C-1	—	B-2	C-1	—
		2 x 6	16	B-3	C-2	D-2	B-3	C-2	D-2	B-3	C-2	D-2
			12	B-3	C-3	D-3	B-3	C-3	D-3	B-3	C-3	D-3
	36	2 x 4	16	B-2	—	—	B-2	—	—	B-2	C-1	—
			12	B-2	C-2	D-2	B-3	C-2	D-2	B-3	C-2	D-2
		2 x 6	16	B-3	C-3	D-3	B-3	C-3	D-3	B-3	C-3	D-3
	24	2 x 4	16	B-3	—	—	B-3	—	—	B-3	C-2	D-2
			12	B-3	C-3	D-2	B-3	C-3	D-3	B-3	C-3	D-3
		2 x 6	16	B-3	C-3	D-3	B-3	C-3	D-3	B-3	C-3	D-3

*Length of end splits or checks at lower ends of stud not to exceed the width of the stud.

Minimum framing requirements for 8' basement wall
Figure 13-5 (continued)

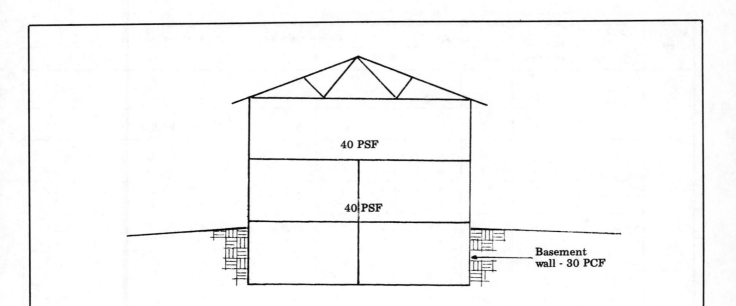

House width (feet)	Height of fill (inches)	Stud and plate size	Stud spacing (inches)	Lumber species and grade for roof live loads								
				40 psf			**30 psf**			**20 psf**		
32	86	2 x 6	12	B-1	—	—	B-1	—	—	B-1	—	—
		2 x 8	16	B-2	C-1*	—	B-2	C-1*	—	B-2	C-1*	—
			12	B-2	C-2	D-2	B-2	C-2	D-2	B-2	C-2	D-2
	72	2 x 6	16	B-1	—	—	B-1	—	—	B-1	—	—
			12	B-2	C-1	D-1	B-2	C-1	D-1	B-2	C-1	D-1
		2 x 8	16	B-2	C-2	D-1	B-2	C-2	D-1	B-2	C-2	D-1
			12	B-3	C-2	D-2	B-3	C-2	D-2	B-3	C-2	D-2
	60	2 x 6	16	B-2	C-1	D-1	B-2	C-1	D-1	B-2	C-1	D-1
			12	B-2	C-2	D-2	B-2	C-2	D-2	B-2	C-2	D-2
		2 x 8	16	B-3	C-2	D-2	B-3	C-3	D-2	B-3	C-3	D-2
			12	B-3	C-3	D-3	B-3	C-3	D-3	B-3	C-3	D-3
	48	2 x 6	16	B-3	C-2	D-2	B-3	C-2	D-2	B-3	C-2	D-2
			12	B-3	C-3	D-2	B-3	C-3	D-3	B-3	C-3	D-3
		2 x 8	16	B-3	C-3	D-3	B-3	C-3	D-3	B-3	C-3	D-3
	36	2 x 6	16	B-3	C-3	D-2	B-3	C-3	D-3	B-3	C-3	D-3
			12	B-3	C-3	D-3	B-3	C-3	D-3	B-3	C-3	D-3
	24	2 x 6	16	B-3	C-3	D-3	B-3	C-3	D-3	B-3	C-3	D-3

**Minimum framing requirements for 8' basement wall—
two-story with clear-span roof trusses and center-bearing floors
Figure 13-6**

House width (feet)	Height of fill (inches)	Stud and plate size	Stud spacing (inches)	40 psf			30 psf			20 psf		
28	86	2 x 6	12	B-1	—	—	B-1	—	—	B-1	—	—
		2 x 8	16	B-2	C-1*	—	B-2	C-1*	—	B-2	C-1*	—
			12	B-2	C-2	D-2	B-2	C-2	D-2	B-2	C-2	D-2
	72	2 x 6	16	B-1	—	—	B-1	—	—	B-1	—	—
			12	B-2	C-1	D-1	B-2	C-1	D-1	B-2	C-1	D-1
		2 x 8	16	B-2	C-2	D-2	B-2	C-2	D-2	B-2	C-2	D-2
			12	B-3	C-3	D-2	B-3	C-3	D-2	B-3	C-3	D-2
	60	2 x 6	16	B-2	C-1	D-1	B-2	C-2	D-1	B-2	C-2	D-1
			12	B-3	C-2	D-2	B-3	C-2	D-2	B-3	C-2	D-2
		2 x 8	16	B-3	C-3	D-2	B-3	C-3	D-2	B-3	C-3	D-2
			12	B-3	C-3	D-3	B-3	C-3	D-3	B-3	C-3	D-3
	48	2 x 6	16	B-3	C-2	D-2	B-3	C-2	D-2	B-3	C-2	D-2
			12	B-3	C-3	D-3	B-3	C-3	D-3	B-3	C-3	D-3
	36	2 x 6	16	B-3	C-3	D-3	B-3	C-3	D-3	B-3	C-3	D-3
	24	2 x 6	16	B-3	C-3	D-3	B-3	C-3	D-3	B-3	C-3	D-3
24	86	2 x 6	12	B-1	—	—	B-1	—	—	B-1	—	—
		2 x 8	16	B-2	C-1*	—	B-2	C-1*	—	B-2	C-1*	—
			12	B-2	C-2	D-2	B-2	C-2	D-2	B-2	C-2	D-2
	72	2 x 6	16	B-1	—	—	B-1	—	—	B-1	—	—
			12	B-2	C-2	D-1	B-2	C-2	D-1	B-2	C-2	D-1
		2 x 8	16	B-2	C-2	D-2	B-2	C-2	D-2	B-2	C-2	D-2
			12	B-3	C-3	D-2	B-3	C-3	D-2	B-3	C-3	D-2
	60	2 x 6	16	B-2	C-2	D-1	B-2	C-2	D-1	B-2	C-2	D-1
			12	B-3	C-2	D-2	B-3	C-2	D-2	B-3	C-2	D-2
		2 x 8	16	B-3	C-3	D-2	B-3	C-3	D-2	B-3	C-3	D-2
			12	B-3	C-3	D-3	B-3	C-3	D-3	B-3	C-3	D-3
	48	2 x 6	16	B-3	C-2	D-2	B-3	C-2	D-2	B-3	C-2	D-2
			12	B-3	C-3	D-3	B-3	C-3	D-3	B-3	C-3	D-3
	36	2 x 6	16	B-3	C-3	D-3	B-3	C-3	D-3	B-3	C-3	D-3
	24	2 x 6	16	B-3	C-3	D-3	B-3	C-3	D-3	B-3	C-3	D-3

*Length of end splits or checks at lower ends of stud not to exceed the width of the stud.

Minimum framing requirements for 8' basement wall
Figure 13-6 (continued)

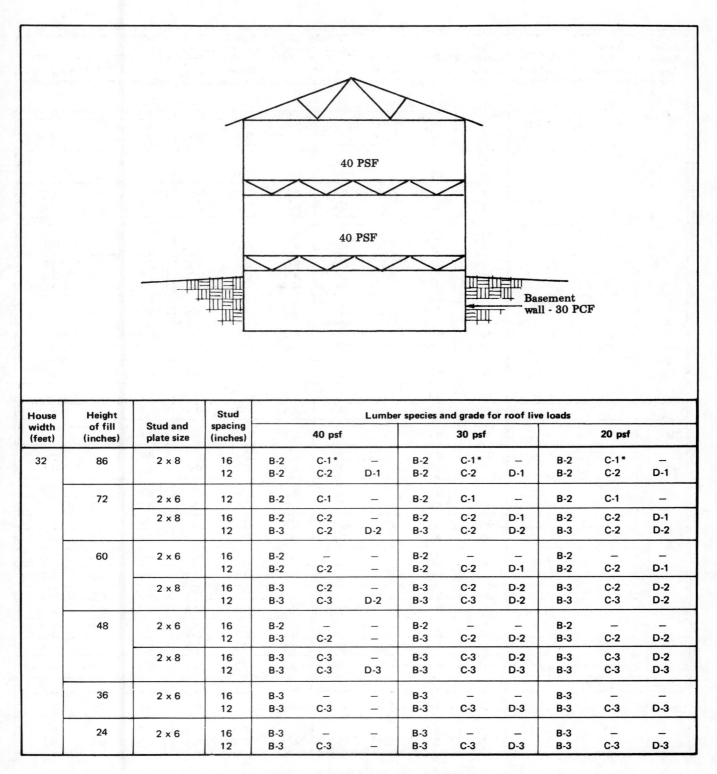

House width (feet)	Height of fill (inches)	Stud and plate size	Stud spacing (inches)	Lumber species and grade for roof live loads								
				40 psf			30 psf			20 psf		
32	86	2 x 8	16	B-2	C-1*	—	B-2	C-1*	—	B-2	C-1*	—
			12	B-2	C-2	D-1	B-2	C-2	D-1	B-2	C-2	D-1
	72	2 x 6	12	B-2	C-1	—	B-2	C-1	—	B-2	C-1	—
		2 x 8	16	B-2	C-2	—	B-2	C-2	D-1	B-2	C-2	D-1
			12	B-3	C-2	D-2	B-3	C-2	D-2	B-3	C-2	D-2
	60	2 x 6	16	B-2	—	—	B-2	—	—	B-2	—	—
			12	B-2	C-2	—	B-2	C-2	D-1	B-2	C-2	D-1
		2 x 8	16	B-3	C-2	—	B-3	C-2	D-2	B-3	C-2	D-2
			12	B-3	C-3	D-2	B-3	C-3	D-2	B-3	C-3	D-2
	48	2 x 6	16	B-2	—	—	B-2	—	—	B-2	—	—
			12	B-3	C-2	—	B-3	C-2	D-2	B-3	C-2	D-2
		2 x 8	16	B-3	C-3	—	B-3	C-3	D-2	B-3	C-3	D-2
			12	B-3	C-3	D-3	B-3	C-3	D-3	B-3	C-3	D-3
	36	2 x 6	16	B-3	—	—	B-3	—	—	B-3	—	—
			12	B-3	C-3	—	B-3	C-3	D-3	B-3	C-3	D-3
	24	2 x 6	16	B-3	—	—	B-3	—	—	B-3	—	—
			12	B-3	C-3	—	B-3	C-3	D-3	B-3	C-3	D-3

**Minimum framing requirements for 8' basement wall—
two-story with clear-span roof trusses and clear-span floor trusses
Figure 13-7**

House width (feet)	Height of fill (inches)	Stud and plate size	Stud spacing (inches)	Lumber species and grade for roof live loads 40 psf			30 psf			20 psf		
28	86	2 x 6	12	B-1	—	—	B-1	—	—	B-1	—	—
		2 x 8	16	B-2	C-1*	—	B-2	C-1*	—	B-2	C-1*	—
			12	B-2	C-2	D-1	B-2	C-2	D-1	B-2	C-2	D-1
	72	2 x 6	16	B-1	—	—	B-1	—	—	B-1	—	—
			12	B-2	C-1	—	B-2	C-1	—	B-2	C-1	—
		2 x 8	16	B-2	C-2	D-1	B-2	C-2	D-1	B-2	C-2	D-1
			12	B-3	C-2	D-2	B-3	C-2	D-2	B-3	C-2	D-2
	60	2 x 6	16	B-2	—	—	B-2	—	—	B-2	—	—
			12	B-2	C-2	D-2	B-2	C-2	D-2	B-2	C-2	D-2
		2 x 8	16	B-3	C-2	D-2	B-3	C-2	D-2	B-3	C-2	D-2
			12	B-3	C-3	D-3	B-3	C-3	D-3	B-3	C-3	D-3
	48	2 x 6	16	B-2	—	—	B-2	—	—	B-2	—	—
			12	B-3	C-2	D-2	B-3	C-3	D-2	B-3	C-3	D-2
		2 x 8	16	B-3	C-3	D-3	B-3	C-3	D-3	B-3	C-3	D-3
	36	2 x 6	16	B-3	—	—	B-3	—	—	B-3	—	—
			12	B-3	C-3	D-3	B-3	C-3	D-3	B-3	C-3	D-3
	24	2 x 6	16	B-3	—	—	B-3	—	—	B-3	—	—
			12	B-3	C-3	D-3	B-3	C-3	D-3	B-3	C-3	D-3
24	86	2 x 6	12	B-1	—	—	B-1	—	—	B-1	—	—
		2 x 8	16	B-2	C-1*	—	B-2	C-1*	—	B-2	C-1*	—
			12	B-2	C-2	D-2	B-2	C-2	D-2	B-2	C-2	D-2
	72	2 x 6	16	B-1	—	—	B-1	—	—	B-1	—	—
			12	B-2	C-1	—	B-2	C-1	—	B-2	C-1	—
		2 x 8	16	B-2	C-2	D-2	B-2	C-2	D-2	B-2	C-2	D-2
			12	B-3	C-2	D-2	B-3	C-2	D-2	B-3	C-2	D-2
	60	2 x 6	16	B-2	—	—	B-2	C-1	—	B-2	C-1	—
			12	B-2	C-2	D-2	B-2	C-2	D-2	B-2	C-2	D-2
		2 x 8	16	B-3	C-2	D-2	B-3	C-2	D-2	B-3	C-2	D-2
			12	B-3	C-3	D-3	B-3	C-3	D-3	B-3	C-3	D-3
	48	2 x 6	16	B-2	—	—	B-2	C-2	D-2	B-2	C-2	D-2
			12	B-3	C-3	D-2	B-3	C-3	D-2	B-3	C-3	D-2
		2 x 8	16	B-3	C-3	D-3	B-3	C-3	D-3	B-3	C-3	D-3
	36	2 x 6	16	B-3	—	—	B-3	C-3	D-2	B-3	C-3	D-2
			12	B-3	C-3	D-3	B-3	C-3	D-3	B-3	C-3	D-3
	24	2 x 6	16	B-3	—	—	B-3	C-3	D-3	B-3	C-3	D-3
			12	B-3	C-3	D-3	B-3	C-3	D-3	B-3	C-3	D-3

*Length of end splits or checks at lower ends of stud not to exceed the width of the stud.

Minimum framing requirements for 8' basement wall
Figure 13-7 (continued)

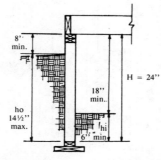

House width (feet)	Condition outside fill inside fill (inches)	Stud and plate size	Stud spacing (inches)	Lumber species and grade for roof live load								
				40 psf			30 psf			20 psf		
32	2 Story Trussed Floors	2 x 6	16	B-3	C-2	D-2	B-3	C-2	D-2	B-3	C-2	D-2
			12	B-3	C-3	D-3	B-3	C-3	D-3	B-3	C-3	D-3
28	14.5 Outside	2 x 6	16	B-3	C-2	D-2	B-3	C-2	D-2	B-3	C-3	D-2
			12	B-3	C-3	D-3	B-3	C-3	D-3	B-3	C-3	D-3
24	6 Inside	2 x 6	16	B-3	C-3	D-2	B-3	C-3	D-3	B-3	C-3	D-3
			12	B-3	C-3	D-3	B-3	C-3	D-3	B-3	C-3	D-3
32	2 Story Joisted Floors	2 x 6	16	B-3	C-3	D-3	B-3	C-3	D-3	B-3	C-3	D-3
			12	B-3	C-3	D-3	B-3	C-3	D-3	B-3	C-3	D-3
28	14.5 Outside	2 x 6	16	B-3	C-3	D-3	B-3	C-3	D-3	B-3	C-3	D-3
			12	B-3	C-3	D-3	B-3	C-3	D-3	B-3·	C-3	D-3
24	6 Inside	2 x 6	16	B-3	C-3	D-3	B-3	C-3	D-3	B-3	C-3	D-3
			12	B-3	C-3	D-3	B-3	C-3	D-3	B-3	C-3	D-3
32	1 Story Trussed Floors	2 x 6	16	B-3	C-3	D-3	B-3	C-3	D-3	B-3	C-3	D-3
			12	B-3	C-3	D-3	B-3	C-3	D-3	B-3	C-3	D-3
		2 x 4	16	B-2	C-2	D-2	B-2	C-2	D-2	B-3	C-2	D-2
			12	B-3	C-2	D-2	B-3	C-3	D-2	B-3	C-3	D-3
28	14.5 Outside	2 x 6	16	B-3	C-3	D-3	B-3	C-3	D-3	B-3	C-3	D-3
			12	B-3	C-3	D-3	B-3	C-3	D-3	B-3	C-3	D-3
		2 x 4	16	B-3	C-2	D-2	B-3	C-2	D-2	B-3	C-3	D-2
			12	B-3	C-3	D-2	B-3	C-3	D-3	B-3	C-3	D-3
24	6 Inside	2 x 6	16	B-3	C-3	D-3	B-3	C-3	D-3	B-3	C-3	D-3
			12	B-3	C-3	D-3	B-3	C-3	D-3	B-3	C-3	D-3
		2 x 4	16	B-3	C-2	D-2	B-3	C-3	D-2	B-3	C-3	D-2
			12	B-3	C-3	D-3	B-3	C-3	D-3	B-3	C-3	D-3
32	1 Story Joisted Floors	2 x 6	16	B-3	C-3	D-3	B-3	C-3	D-3	B-3	C-3	D-3
			12	B-3	C-3	D-3	B-3	C-3	D-3	B-3	C-3	D-3
		2 x 4	16	B-3	C-2	D-2	B-3	C-3	D-2	B-3	C-3	D-3
			12	B-3	C-3	D-3	B-3	C-3	D-3	B-3	C-3	D-3
28	14.5 Outside	2 x 6	16	B-3	C-3	D-3	B-3	C-3	D-3	B-3	C-3	D-3
			12	B-3	C-3	D-3	B-3	C-3	D-3	B-3	C-3	D-3
		2 x 4	16	B-3	C-3	D-2	B-3	C-3	D-2	B-3	C-3	D-3
			12	B-3	C-3	D-3	B-3	C-3	D-3	B-3	C-3	D-3
24	6 Inside	2 x 6	16	B-3	C-3	D-3	B-3	C-3	D-3	B-3	C-3	D-3
			12	B-3	C-3	D-3	B-3	C-3	D-3	B-3	C-3	D-3
		2 x 4	16	B-3	C-3	D-2	B-3	C-3	D-3	B-3	C-3	D-3
			12	B-3	C-3	D-3	B-3	C-3	D-3	B-3	C-3	D-3

Minimum framing requirements for 2' crawl space wall
Figure 13-8

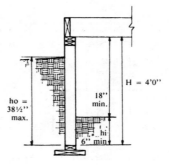

House width (feet)	Condition outside fill inside fill (inches)	Stud and plate size	Stud spacing (inches)	Lumber species and grade for roof live load								
				40 psf			30 psf			20 psf		
32	2 Story Trussed Floors	2 x 6	16	B-2	C-2	D-1	B-2	C-2	D-2	B-2	C-2	D-2
			12	B-3	C-2	D-2	B-3	C-2	D-2	B-3	C-2	D-2
28	38.5 Outside	2 x 6	16	B-2	C-2	D-2	B-3	C-2	D-2	B-3	C-2	D-2
			12	B-3	C-2	D-2	B-3	C-3	D-2	B-3	C-3	D-3
24	6 Inside	2 x 6	16	B-3	C-2	D-2	B-3	C-2	D-2	B-3	C-2	D-2
			12	B-3	C-3	D-3	B-3	C-3	D-3	B-3	C-3	D-3
32	2 Story Joisted Floors	2 x 6	16	B-3	C-2	D-2	B-3	C-2	D-2	B-3	C-2	D-2
			12	B-3	C-3	D-3	B-3	C-3	D-3	B-3	C-3	D-3
28	38.5 Outside	2 x 6	16	B-3	C-2	D-2	B-3	C-3	D-2	B-3	C-3	D-2
			12	B-3	C-3	D-3	B-3	C-3	D-3	B-3	C-3	D-3
24	6 Inside	2 x 6	16	B-3	C-3	D-2	B-3	C-3	D-3	B-3	C-3	D-3
			12	B-3	C-3	D-3	B-3	C-3	D-3	B-3	C-3	D-3
32	1 Story Trussed Floors	2 x 6	16	B-3	C-2	D-2	B-3	C-2	D-2	B-3	C-3	D-2
			12	B-3	C-3	D-3	B-3	C-3	D-3	B-3	C-3	D-3
		2 x 4	16	B-2	C-2	—	B-2	C-2	—	B-2	C-2	D-1
			12	B-2	C-2	D-1	B-2	C-2	D-2	B-2	C-2	D-2
28	38.5 Outside	2 x 6	16	B-3	C-2	D-2	B-3	C-3	D-2	B-3	C-3	D-3
			12	B-3	C-3	D-3	B-3	C-3	D-3	B-3	C-3	D-3
		2 x 4	16	B-2	C-2	—	B-2	C-2	D-1	B-2	C-2	D-1
			12	B-2	C-2	D-2	B-2	C-2	D-2	B-3	C-2	D-2
24	6 Inside	2 x 6	16	B-3	C-3	D-2	B-3	C-3	D-3	B-3	C-3	D-3
			12	B-3	C-3	D-3	B-3	C-3	D-3	B-3	C-3	D-3
		2 x 4	16	B-2	C-1	D-1	B-2	C-2	D-1	B-2	C-2	D-1
			12	B-2	C-2	D-2	B-3	C-2	D-2	B-3	C-2	D-2
32	1 Story Joisted Floor	2 x 6	16	B-3	C-3	D-2	B-3	C-3	D-3	B-3	C-3	D-3
			12	B-3	C-3	D-3	B-3	C-3	D-3	B-3	C-3	D-3
		2 x 4	16	B-2	C-1	D-1	B-2	C-2	D-1	B-2	C-2	D-2
			12	B-2	C-2	D-2	B-3	C-2	D-2	B-3	C-2	D-2
28	38.5 Outside	2 x 6	16	B-3	C-3	D-3	B-3	C-3	D-3	B-3	C-3	D-3
			12	B-3	C-3	D-3	B-3	C-3	D-3	B-3	C-3	D-3
		2 x 4	16	B-2	C-2	D-1	B-2	C-2	D-1	B-2	C-2	D-2
			12	B-3	C-2	D-2	B-3	C-2	D-2	B-3	C-2	D-2
24	6 Inside	2 x 6	16	B-3	C-3	D-3	B-3	C-3	D-3	B-3	C-3	D-3
			12	B-3	C-3	D-3	B-3	C-3	D-3	B-3	C-3	D-3
		2 x 4	16	B-2	C-2	D-1	B-2	C-2	D-2	B-2	C-2	D-2
			12	B-3	C-2	D-2	B-3	C-2	D-2	B-3	C-3	D-2

Minimum framing requirements for 4' crawl space wall[1, 2]
Figure 13-9

House width (feet)	Condition outside fill inside fill (inches)	Stud and plate size	Stud spacing (inches)	Lumber species and grade for roof live load								
				40 psf			30 psf			20 psf		
32	2 Story Trussed Floors	2 x 6	16	B-2	C-2	D-2	B-2	C-2	D-2	B-2	C-2	D-2
			12	B-3	C-2	D-2	B-3	C-2	D-2	B-3	C-3	D-2
28	38.5 Outside	2 x 6	16	B-2	C-2	D-2	B-3	C-2	D-2	B-3	C-2	D-2
			12	B-3	C-3	D-2	B-3	C-3	D-2	B-3	C-3	D-3
24	18 Inside	2 x 6	16	B-3	C-2	D-2	B-3	C-2	D-2	B-3	C-2	D-2
			12	B-3	C-3	D-3	B-3	C-3	D-3	B-3	C-3	D-3
32	2 Story Joisted Floors	2 x 6	16	B-3	C-2	D-2	B-3	C-2	D-2	B-3	C-3	D-2
			12	B-3	C-3	D-3	B-3	C-3	D-3	B-3	C-3	D-3
28	38.5 Outside	2 x 6	16	B-3	C-2	D-2	B-3	C-3	D-2	B-3	C-3	D-3
			12	B-3	C-3	D-3	B-3	C-3	D-3	B-3	C-3	D-3
24	18 Inside	2 x 6	16	B-3	C-3	D-2	B-3	C-3	D-3	B-3	C-3	D-3
			12	B-3	C-3	D-3	B-3	C-3	D-3	B-3	C-3	D-3
32	1 Story Trussed Floors	2 x 6	16	B-3	C-2	D-2	B-3	C-2	D-2	B-3	C-3	D-2
			12	B-3	C-3	D-3	B-3	C-3	D-3	B-3	C-3	D-3
		2 x 4	16	B-2	C-1	—	B-2	C-1	D-1	B-2	C-2	D-1
			12	B-2	C-2	D-2	B-2	C-2	D-2	B-2	C-2	D-2
28	38.5 Outside	2 x 6	16	B-3	C-2	D-2	B-3	C-3	D-2	B-3	C-3	D-3
			12	B-3	C-3	D-3	B-3	C-3	D-3	B-3	C-3	D-3
		2 x 4	16	B-2	C-1	D-1	B-2	C-2	D-1	B-2	C-2	D-1
			12	B-2	C-2	D-2	B-2	C-2	D-2	B-3	C-2	D-2
24	18 Inside	2 x 6	16	B-3	C-3	D-3	B-3	C-3	D-3	B-3	C-3	D-3
			12	B-3	C-3	D-3	B-3	C-3	D-3	B-3	C-3	D-3
		2 x 4	16	B-2	C-2	D-1	B-2	C-2	D-1	B-2	C-2	D-2
			12	B-3	C-2	D-2	B-3	C-2	D-2	B-3	C-2	D-2
32	1 Story Joisted Floors	2 x 6	16	B-3	C-3	D-3	B-3	C-3	D-3	B-3	C-3	D-3
			12	B-3	C-3	D-3	B-3	C-3	D-3	B-3	C-3	D-3
		2 x 4	16	B-2	C-2	D-1	B-2	C-2	D-1	B-2	C-2	D-2
			12	B-2	C-2	D-2	B-3	C-2	D-2	B-3	C-2	D-2
28	38.5 Outside	2 x 6	16	B-3	C-3	D-3	B-3	C-3	D-3	B-3	C-3	D-3
			12	B-3	C-3	D-3	B-3	C-3	D-3	B-3	C-3	D-3
		2 x 4	16	B-2	C-2	D-1	B-2	C-2	D-2	B-2	C-2	D-2
			12	B-3	C-2	D-2	B-3	C-2	D-2	B-3	C-3	D-2
24	18 Inside	2 x 6	16	B-3	C-3	D-3	B-3	C-3	D-3	B-3	C-3	D-3
			12	B-3	C-3	D-3	B-3	C-3	D-3	B-3	C-3	D-3
		2 x 4	16	B-2	C-2	D-2	B-2	C-2	D-2	B-3	C-2	D-2
			12	B-3	C-2	D-2	B-3	C-3	D-2	B-3	C-3	D-3

Minimum framing requirements for 4' crawl space wall[1, 2]
Figure 13-9 (continued)

House width (feet)	Condition outside fill / inside fill (inches)	Stud and plate size	Stud spacing (inches)	Lumber species and grade for roof live load 40 psf			30 psf			20 psf		
32	2 Story Trussed Floors	2 x 6	16	B-2	C-2	D-2	B-3	C-2	D-2	B-3	C-2	D-2
			12	B-3	C-2	D-2	B-3	C-3	D-2	B-3	C-3	D-3
28	38.5 Outside	2 x 6	16	B-3	C-2	D-2	B-3	C-2	D-2	B-3	C-2	D-2
			12	B-3	C-3	D-3	B-3	C-3	D-3	B-3	C-3	D-3
24	30 Inside	2 x 6	16	B-3	C-2	D-2	B-3	C-2	D-2	B-3	C-3	D-2
			12	B-3	C-3	D-3	B-3	C-3	D-3	B-3	C-3	D-3
32	2 Story Joisted Floors	2 x 6	16	B-3	C-3	D-2	B-3	C-3	D-3	B-3	C-3	D-3
			12	B-3	C-3	D-3	B-3	C-3	D-3	B-3	C-3	D-3
28	38.5 Outside	2 x 6	16	B-3	C-3	D-3	B-3	C-3	D-3	B-3	C-3	D-3
			12	B-3	C-3	D-3	B-3	C-3	D-3	B-3	C-3	D-3
24	30 Inside	2 x 6	16	B-3	C-3	D-3	B-3	C-3	D-3	B-3	C-3	D-3
			12	B-3	C-3	D-3	B-3	C-3	D-3	B-3	C-3	D-3
32	1 Story Trussed Floors	2 x 6	16	B-3	C-3	D-2	B-3	C-3	D-3	B-3	C-3	D-3
			12	B-3	C-3	D-3	B-3	C-3	D-3	B-3	C-3	D-3
		2 x 4	16	B-2	C-2	D-1	B-2	C-2	D-1	B-2	C-2	D-2
			12	B-2	C-2	D-2	B-3	C-2	D-2	B-3	C-2	D-2
28	38.5 Outside	2 x 6	16	B-3	C-3	D-3	B-3	C-3	D-3	B-3	C-3	D-3
			12	B-3	C-3	D-3	B-3	C-3	D-3	B-3	C-3	D-3
		2 x 4	16	B-2	C-2	D-1	B-2	C-2	D-2	B-2	C-2	D-2
			12	B-3	C-2	D-2	B-3	C-2	D-2	B-3	C-3	D-2
24	30 Inside	2 x 6	16	B-3	C-3	D-3	B-3	C-3	D-3	B-3	C-3	D-3
			12	B-3	C-3	D-3	B-3	C-3	D-3	B-3	C-3	D-3
		2 x 4	16	B-2	C-2	D-2	B-2	C-2	D-2	B-3	C-2	D-2
			12	B-3	C-3	D-2	B-3	C-3	D-2	B-3	C-3	D-3
32	1 Story Joisted Floors	2 x 6	16	B-3	C-3	D-3	B-3	C-3	D-3	B-3	C-3	D-3
			12	B-3	C-3	D-3	B-3	C-3	D-3	B-3	C-3	D-3
		2 x 4	16	B-2	C-2	D-2	B-2	C-2	D-2	B-3	C-2	D-2
			12	B-3	C-2	D-2	B-3	C-3	D-2	B-3	C-3	D-3
28	38.5 Outside	2 x 6	16	B-3	C-3	D-3	B-3	C-3	D-3	B-3	C-3	D-3
			12	B-3	C-3	D-3	B-3	C-3	D-3	B-3	C-3	D-3
		2 x 4	16	B-2	C-2	D-2	B-3	C-2	D-2	B-3	C-2	D-2
			12	B-3	C-3	D-2	B-3	C-3	D-3	B-3	C-3	D-3
24	30 Inside	2 x 6	16	B-3	C-3	D-3	B-3	C-3	D-3	B-3	C-3	D-3
			12	B-3	C-3	D-3	B-3	C-3	D-3	B-3	C-3	D-3
		2 x 4	16	B-3	C-2	D-2	B-3	C-2	D-2	B-3	C-3	D-2
			12	B-3	C-3	D-3	B-3	C-3	D-3	B-3	C-3	D-3

[1] For inside fill heights not tabulated, use framing tables for lower fill height.

[2] See Figure 13-10 for strut requirements for non-bearing walls.

Minimum framing requirements for 4' crawl space wall
Figure 13-9 (continued)

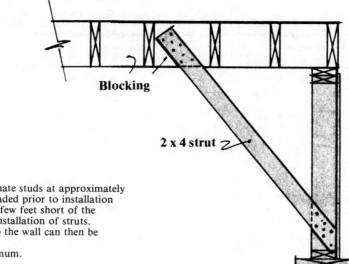

Blocking

2 x 4 strut

(Minimum species grade D-3)
Where required install on alternate studs at approximately
45° angle. Inside fill may be graded prior to installation
of struts if the fill is stopped a few feet short of the
exterior end wall to allow for installation of struts.
Final inside grading adjacent to the wall can then be
done by hand shoveling.
Outside fill height 38½'' maximum.

▓▓▓ **Pressure treated wood**

Strut nailing schedule[1, 2, 3]

Inside fill height (inches)	Conventional construction					Modular mobile construction (any number of stories)		
	Stud spacing (inches)	Number of stories	Species			Stud spacing (inches)	Species	
			B	C, D			B	C, D
6	12	1	3-10d	3-10d		12	6-10d	8-10d
		2	3-10d	3-10d				
	16	1	3-10d	3-10d		16	7-10d	10-10d
		2	3-10d	3-10d				
18	12	1	3-10d	3-10d		12	5-10d	6-10d
		2	Not required					
	16	1	3-10d	3-10d		16	6-10d	8-10d
		2	Not required					
30	12	1	Not required			12	3-10d	3-10d
		2	Not required					
	16	1	Not required			16	3-10d	3-10d
		2	Not required					

[1] Use species of strut or stud, whichever is limiting.

[2] All fasteners for struts are 10d common nails.

[3] Nail and strut requirements are based on soil load of 30 pounds per cubic foot.

Strut requirements for non-bearing 4' crawl space end wall
Figure 13-10

For house constructions and sizes in Figures 13-4 through 13-9
Floor loads: 1st floor-LL + DL = 50psf; 2nd floor-LL + DL = 50psf
(LL = live load; DL = dead load)

Joisted Floors (center bearing under 1st or 1st and 2nd floors)

Loading		House width (feet)	2 x 4 wall plate	2 x 6 wall plate	2 x 8 wall plate
Roof 40 psf LL 10 psf DL Ceiling 10 psf	2 Story	32		2x10 + (1) ½x10	2x10
		28		2x10 + (1) ½x10	2x10
		24		2x8	2x8
	1 Story	32	2x8 + (1) ½x8	2x8	2x8
		28	2x8 + (1) ½x8	2x8	2x8
		24	2x6	2x6	2x8
Roof 30 psf LL 10 psf DL Ceiling 10 psf	2 Story	32		2x10 + (1) ½x10	2x10
		28		2x8	2x8
		24		2x8	2x8
	1 Story	32	2x8 + (1) ½x8	2x8	2x8
		28	2x6	2x6	2x8
		24	2x6	2x6	2x8
Roof 20 psf LL 10 psf DL Ceiling 10 psf	2 Story	32		2x8	2x8
		28		2x8	2x8
		24		2x6	2x8
	1 Story	32	2x6	2x6	2x8
		28	2x6	2x6	2x8
		24	2x6	2x6	2x8

Trussed Floors (no center bearing)

Loading		House width (feet)	2 x 4 wall plate	2 x 6 wall plate	2 x 8 wall plate
Roof 40 psf LL 10 psf DL Ceiling 10 psf	2 Story	32		2x12 + (2) 5/8x12	2x12 + (1) 5/8x12
		28		2x12 + (2) 5/8x12	2x12 + (1) 5/8x12
		24		2x10 + (1) ½x10	2x10
	1 Story	32	2x10 + (2) 5/8x10	2x10 + (1) ½x10	2x10
		28	2x8 + (1) ½x8	2x8	2x8
		24	2x8 + (1) ½x8	2x8	2x8
Roof 30 psf LL 10 psf DL Ceiling 10 psf	2 Story	32		2x12 + (2) 5/8x12	2x12 + (1) 5/8x12
		28		2x12 + (2) 5/8x12	2x12 + (1) 5/8x12
		24		2x10 + (1) ½x10	2x10
	1 Story	32	2x8 + (1) 5/8x8	2x8	2x8
		28	2x8 + (1) ½x8	2x8	2x8
		24	2x8 + (1) ½x8	2x8	2x8
Roof 20 psf LL 10 psf DL Ceiling 10 psf	2 Story	32		2x12 + (2) 5/8x12	2x12 + (1) 5/8x12
		28		2x10 + (1) ½x10	2x10
		24		2x10 + (1) ½x10	2x10
	1 Story	32	2x8 + (1) ½x8	2x8	2x8
		28	2x8 + (1) ½x8	2x8	2x8
		24	2x6	2x6	2x8

[1] Where 2x10+(1)½x10 and similar configurations are designated, 2x10 indicates lumber footing plate to be used, (1) indicates number of plywood strips to be used, and ½x10 indicates thickness and width of plywood. Use continuous treated plywood strips with face grain perpendicular to footing; minimum grade C-D (exterior glue). Use plywood of same width as footing plate and fasten with two 6d galvanized nails spaced 16inches on center.

[2] Footing plate shall not be less than species combination "D" from Figure 13-2.

Minimum footing plate size
Figure 13-11

Height of fill (inches)	Stud spacing (inches)	Face grain across studs[2]			Face grain parallel to studs		
		Grade[3]	Minimum thickness[1]	Identification index	Grade[3]	Minimum thickness[1,4]	Identification index
24	12	B	15/32	32/16	A	15/32	32/16
					B	15/32[5]	32/16
	16	B	15/32	32/16	A	15/32[5]	32/16
					B	19/32[5](4,5 ply)	40/20
36	12	B	15/32	32/16	A	15/32	32/16
					B	15/32[5](4,5 ply)	32/16
					B	19/32 (4,5 ply)	40/20
	16	B	15/32[5]	32/16	A	19/32	40/20
					B	23/32	48/24
48	12	B	15/32	32/16	A	15/32[5]	32/16
					B	19/32[5](4,5 ply)	40/20
	16	B	19/32	40/20	A	19/32[5]	40/20
					A	23/32	48/24
60	12	B	15/32	32/16	A	19/32	40/20
					B	19/32[5](5 ply)	40/20
					B	23/32	48/24
	16	B	19/32[5]	40/20	A	23/32[5]	48/24
72	12	B	15/32[5]	32/16	A	19/32	40/20
					B	23/32[5]	48/24
	16	B	23/32[5]	48/24	–	–	–
86	12	B	19/32	40/20	A	19/32[5]	40/20
					A	23/32	48/24
	16	B	23/32[5]	48/24	–	–	–

1. Minimum thickness $^{15}/_{32}$ inch except crawl space sheathing may be ⅜ inch for face grain across studs 16 inches on center and maximum 2 foot depth of unequal fill.

2. Minimum 2 inch blocking between studs required at all horizontal panel joints more than 4 feet below adjacent ground level.

3. Plywood shall be of the following minimum grades:

   (i)   U.S. Product Standard PS 1, Construction and Industrial Plywood, grades marked
   A. Structural I C-D
   B. C-D (Exterior glue)

   (ii)   Performance rated all veneer plywood grades marked
   A. Structural I
   B. C-D (Exterior glue)
   which meet all U.S. Product Standard PS 1 requirements except thickness and which have been evaluated under approved performance specifications.

   (iii)   Where a major portion of the wall is exposed above ground and a better appearance is desired, the following U.S. Product Standard PS 1 Exterior grades are suitable:
   A. Structural I A-C, Structural I B-C or Structural I C-C (Plugged)
   B. A-C Exterior Group I, B-C Exterior Group I, C-C (Plugged) Exterior Group 1 or MDO Exterior Group 1

4. When face grain is parallel to studs, all veneer plywood panels of the required thickness, grade and identification index may be of any construction permitted (see "plywood" in text).

5. For this fill height, thickness and grade combination, panels which are continuous over less than three spans (across less than three stud spacings) require blocking 16 inches above the bottom plate. Offset adjacent blocks and fasten through studs with two 16d corrosion resistant nails at each end.

Plywood grade and thickness for foundation construction
Figure 13-12

Applicable to minimal soil pressures
Heavy loads require more or larger fasteners
See Figures 13-14, 13-15 and 13-16

Joint description	Minimum nail size	Number or spacing
Bottom plate to footing plate — Face nail	10d	12" oc
Bottom plate to stud-End nail — 2" plate	16d	2
— 1" plate	8d	2
Top plate to stud — End nail minimum (See Figure 13-14)	16d	2
Upper top plate to top plate — Face nail minimum (See Figure 13-14)	10d	8" oc
(No overlap of plywood)		
Header joist to upper top plate — Toe nail minimum (See Figure 13-15)	8d	16" oc
Joist to upper top plate — Toe nail minimum (See Figure 13-15)	8d / 10d	3 } / 2 }
End joist to plate (joists parallel to wall) — Toe nail minimum (See Figure 13-15)	8d	4" oc
Window header support studs to window sill — End nail minimum	16d	2
Window sill to studs under — End nail minimum (See Figure 13-14)	16d	2
Window header to stud — End nail	16d	4
Knee wall top plate to studs — End nail	16d	2
Knee wall bottom plate to studs — End nail	8d	2
Knee wall top plate to foundation wall — Toe nail	16d	1 per stud
Knee wall stud over 5' long to foundation wall stud — Toe nail at mid-height of stud	16d	2 per stud
Knee wall bottom plate to footing plate — Face nail	8d	2 per stud space
Window, door or beam pocket header support stud to stud — Face nail	10d	24" oc
Corner posts - stud to stud — Face nail	16d	16" oc

Note: Nailing schedules in Figures 13-14, 13-15 and 13-16 apply only to standard common wire nails of the following minimum sizes: 6d-2" x 0.113"; 8d-2.5" x 0.131"; 10d-3" x 0.148"; 16d-3.5" x 0.162"; 20d-4" x 0.192". To provide equivalent structural joint capacity with shorter and/or thinner nails, additional nails will be required.

General nailing schedule
Figure 13-13

For lumber species, grade, size and spacing combinations in Figure 13-2

Height of fill (inch)	Treated lumber species	End-nail treated top plate to treated studs		Face-nail untreated top plate to treated top plate			
				No overlap of plywood		3/4" plywood overlap	
		Nail size[2]	Number per joint	Nail size[2]	Spacing (inch)	Nail size[2]	Spacing (inch)
24	All	16d	2	10d	8	10d	16
48	All	16d	2	10d	8	10d	16
72	B	16d[3]	3	10d	6	10d	8
	C,D	16d[3]	4	10d	4[4]	10d	4[4]
86	B	20d[3]	3	10d	3[4]	10d	4[4]
	C,D	20d[3]	4	10d	2[4]	10d	3[4]

[1] Based on 30 pcf equivalent-fluid weight soil pressure and dry lumber.

[2] Common wire nails; hot-dipped zinc coated steel.

[3] Alternatively, may use "U" type framing anchor or hanger having a minimum load capacity (live plus dead load, normal duration) of 340 pounds in species combination "B" (see Figure 13-2).

[4] Alternatively, two nails 2-1/2 inches apart across the grain at twice the spacing indicated may be used.

Nailing schedule:
top plate to stud, and plate to plate connections
Figure 13-14

Height of fill (inch)	Joist spacing (inch)	Joists perpendicular to wall					Joists parallel to wall					
		Toe-nail header joist to plate		Toe-nail each joist to plate		Framing anchor, each joist to plate	Blocking between joists, spacing (inch)	Toe-nail end joist to plate		Toe-nail blocking to plate		Framing anchor each block to plate
		Nail size	Spacing (inch)	Nail size	No. per joist			Nail size	Spacing (inch)	Nail size	No. per block	
48 or less	16	8d	16	8d	3	none	No blocking	8d	4	none		none
		10d	16	10d	2	none						
	24	8d	8	8d	3	none						
		10d	8	10d	2	none						
72	16	8d	8	8d	3	none	48	8d	4	8d	3	none
		10d	8	10d	2	none		10d	4	10d	2	none
		8d	16	none		1		10d	6	10d	4	none
	24	10d	8	10d	3	none		8d	6	none		1
		8d	16	none		1						
86	16	8d	8	none		1	24	8d	4	none		1
	24	8d	4	none		1						

[1] Based on 30 pcf equivalent-fluid weight soil pressure and dry lumber. Untreated top plate not less than species combination "D" from Figure 13-2.

[2] Toe-nails driven at angle of approximately 30° with the piece and started approximately one-third the length of the nail from the end or edge of the piece.

[3] See Figure 13-16 for additional spacing requirements for blocking, and for subfloor to blocking nailing schedule.

[4] Framing anchors shall have a minimum load capacity (live load plus dead load, normal duration) of 320 pounds per joist. If plate or joist is species combination "C" or "D", then rated load capacity of anchors when installed in species combination "B" shall be not less than 395 pounds per joist.

[5] Common wire steel nails.

Nailing schedule: floor joists to wall connections
Figure 13-15

Height of fill (inch)	Species of joist and blocking lumber[4]					
	"B"			"C" or "D"		
	Block spacing (inch)	Nails per block[1]		Block spacing (inch)	Nails per block[1]	
		6d	8d		6d	8d
60	48	4	3	48	6	4
72	48	8	6	48	11	8
	24	3	2	24	5	3
86	24	7	5	24	9	7

[1] Common wire steel nails. Nails shall be spaced 2 inches on centers or more; where block length requires, nails may be in two rows.

[2] See Figure 13-15 for additional requirements for block spacing and nailing.

[3] Based on 30 pcf equivalent-fluid weight soil pressure and dry lumber.

[4] See Figure 13-2 for minimum properties of lumber species combinations.

Nailing schedule: subfloor to end wall blocking
Figure 13-16

Backfill condition on opposing walls	Panel location	Fastener size and spacing
Equal	Edge	8d com. nail . . . . 6″ o.c. 16ga X 1½″ staple . . . . 4″ o.c.
	Intermediate	8d com. nail . . . . 12″ o.c. 16ga X 1½″ staple . . . . 8″ o.c.
Unequal— long walls (4ft. differential)	Edge	8d com. nail . . . . 6″ o.c. 16ga X 1½″ staple . . . . 5″ o.c.
	Intermediate	8d com. nail . . . . 12″ o.c. 16ga X 1½″ staple . . . . 12″ o.c.
Unequal— short walls (4ft differential)	Edge	8d com. nail . . . . 6″ o.c. (blocking required)[3, 4] 16ga X 1½″ staple . . . . 4″ o.c. required)[3] (blocking required)[3]
	Intermediate	8d com. nail . . . . 12″ o.c. 16ga X 1½″ staple . . . . 8″ o.c.[2]

[1] In crawl space, provide a fastener within 1½″ of the bottom of each stud.

[2] Use this schedule for higher backfill on one long wall than on opposing wall. Use "short wall" schedule for long walls when backfill is higher on one short wall than on opposing wall. Maintain this schedule for a minimum length equal to one half the length of the short wall. The "long wall" schedule applies to the remaining long walls and all short walls.

[3] All plywood panel edges shall be fastened to blocking or framing of 2 inch nominal lumber.

[4] Blocking can be omitted if nailing at intermediate supports is spaced 6″ on center.

[5] For fastening plywood blocking to end walls, see Figure 13-15 and 13-16.

Minimum fastener schedule: sheathing to wall framing
Figure 13-17

Loading	Plywood[1, 2, 3]			Nail spacing at panel edges (inches)[5]				
	Minimum thickness (inches)	Grade	Nail size[4]	All panel edges supported				One or more panel edge not supported
				6	4	3	2	6
Wind or	15/32	All grades	8d	260	380	490	640	180
wind plus	15/32	All grades	10d	310	460	600[6]	770[6]	190
unequal	15/32	Structural 1 grades	10d	340	510	665[6]	870[6]	215
backfill	19/32, 23/32	All grades	10d	340	510	665[6]	870[6]	215
Unequal	15/32	All grades	8d	175	255	330	435	120
backfill	15/32	All grades	10d	210	310	405[6]	520[6]	130
	15/32	Structural 1 grades	10d	230	345	450[6]	590[6]	145
	19/32, 23/32	All grades	10d	230	345	450[6]	590[6]	145

1   Values apply to all plywood conforming to the requirements set forth in this report, except plywood made with Group 5 species.

2   Values shown are for studs of species/grade combinations B-1, B-2, B-3 from Figure 13-2. Use 81% of these design shears for combinations C-1, C-2, C-3, D-1, D-2 and D-3.

3   Design shears for ¾″ plywood used in crawl space construction shall be the same as the values for 8d nails provided studs are spaced no more than 16″ o.c.

4   Nails to be hot dipped zinc coated steel; or copper, silicone bronze or stainless steel Types 304 or 316 having equivalent lateral load capacity. Values apply to standard common wire nails or box nails of the following minimum sizes: 8d - 2.5″ x 0.113″ (0.281″ head); 10d - 3″ x 0.128″ (0.312″ head).

5   All panel edges backed with 2″ (nominal) or wider framing, except where noted. Space nails at 12″ o.c. along intermediate framing members.

6   Framing at panel edges shall be 3″ nominal or thicker and nails shall be staggered where nails are spaced 2″ o.c., and where 10d nails having penetration into framing of more than 1⅝″ are spaced 3″ o.c. *Exception:* Unless otherwise required, 2″ nominal framing may be used where full nailing surface width is available and nails are staggered.

Wood foundation walls sheathed with ¹⁵⁄₃₂″ or thicker plywood:
nailing schedule and design shear strength (pounds per foot of wall)
Figure 13-18

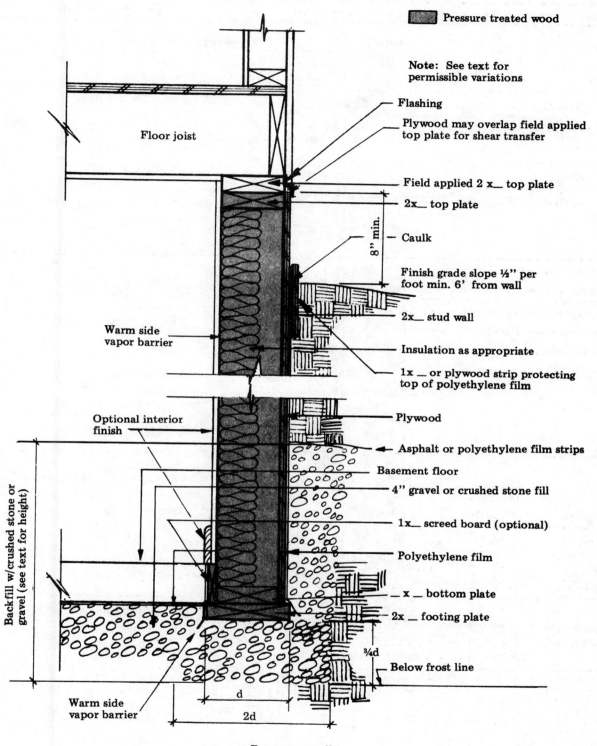

Pressure treated wood

Note: See text for permissible variations

Flashing

Plywood may overlap field applied top plate for shear transfer

Field applied 2 x__ top plate

2x__ top plate

8" min.

Caulk

Finish grade slope ½" per foot min. 6' from wall

2x__ stud wall

Insulation as appropriate

1x __ or plywood strip protecting top of polyethylene film

Plywood

Asphalt or polyethylene film strips

Basement floor

4" gravel or crushed stone fill

1x__ screed board (optional)

Polyethylene film

x __ bottom plate

2x __ footing plate

¾d

Below frost line

Floor joist

Warm side vapor barrier

Optional interior finish

Backfill w/crushed stone or gravel (see text for height)

Warm side vapor barrier

d

2d

Basement wall
Figure 13-19

347

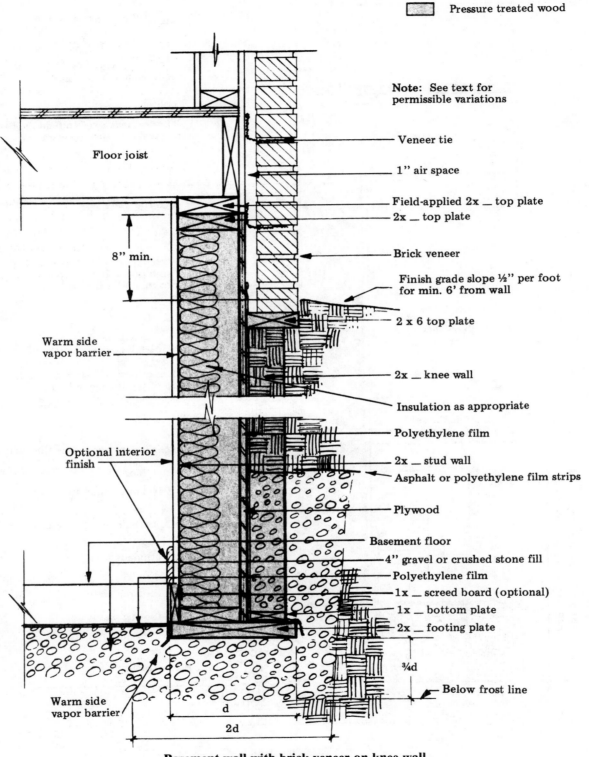

Pressure treated wood

Note: See text for permissible variations

Floor joist

Veneer tie

1" air space

Field-applied 2x __ top plate
2x __ top plate

Brick veneer

8" min.

Finish grade slope ½" per foot for min. 6' from wall

2 x 6 top plate

Warm side vapor barrier

2x __ knee wall

Insulation as appropriate

Polyethylene film

Optional interior finish

2x __ stud wall
Asphalt or polyethylene film strips

Plywood

Basement floor

4" gravel or crushed stone fill
Polyethylene film
1x __ screed board (optional)
1x __ bottom plate
2x __ footing plate

¾d

Below frost line

Warm side vapor barrier

d

2d

Basement wall with brick veneer on knee wall
Figure 13-20

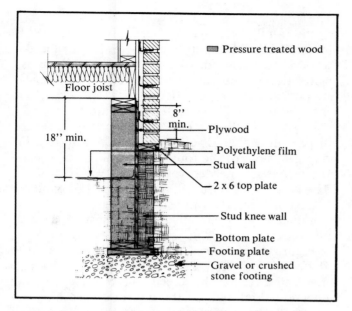

**Crawl space wall with brick veneer on knee wall
Figure 13-21**

The Footing

The AWWF uses a composite footing: a wood footing plate supported by a layer of gravel, coarse sand, or crushed stone. Figures 13-19 through 13-22 show several kinds of AWWF foundation walls. Figure 13-23 shows a basement bearing partition wall.

The width of the footing plate depends on how much pressure the layer of gravel, sand or crushed stone (GSCS) can handle. Look back to Figure 13-1: the allowable bearing pressure on uncompacted GSCS is at least 3,000 pounds per square foot. Sometimes the footing plate is wider than the bottom wall plate, as in Figures 13-19 through 13-23. If it is, be sure the stress across the grain of the bottom of the footing plate is less than 1/3 the allowable shear stress for the footing plate. You can use plywood strips to reinforce the lumber footing plate.

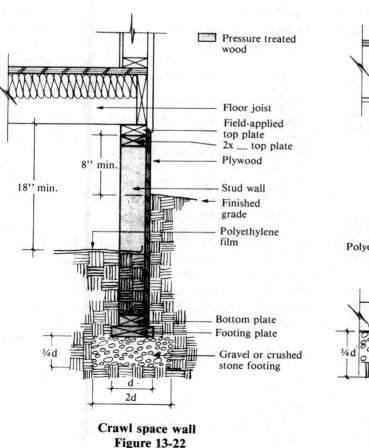

**Crawl space wall
Figure 13-22**

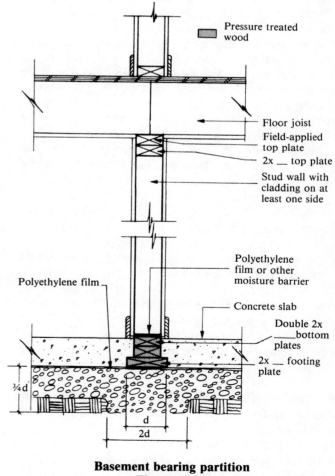

**Basement bearing partition
Figure 13-23**

The thickness and width of the GSCS depends on the soil. But the GSCS footing should never be less than twice the width of the footing plate. See (2d) at the bottom of Figure 13-19. And the GSCS thickness should never be less than 3/4 the width of (d). The GSCS is kept from spreading by the backfill, the granular fill, the undisturbed soil, or the foundation wall.

The bottom of the wood footing plate should extend down to the frost line, unless the GSCS footing does. Be sure the GSCS has its positively-drained sump at or below the frost line. This isn't necessary in Group I soils where the permanent water table is below the frost line. A footing can be connected to a sump by a trench or a pipe. The footing should drain to whichever is highest: the bottom of the sump, the bottom of the trench, or the bottom of the pipe.

The bottom of a wood footing plate for a crawl space wall may be above the frost line. Be sure to cover the GSCS outside the wall with strips of plastic sheeting, or something similar, to keep soil out of the GSCS.

If a wood footing plate is close to finished grade (such as when a deep GSCS footing goes to the frost line) protect the GSCS footing from surface erosion or mechanical disturbance. Install gutters and drains and slope the ground around the edge to protect the footing. Temporary barriers, such as shrubbery or trees, add extra protection.

Follow your local code when building posts and piers or crawl space footings. These footings can be of treated wood, treated wood and gravel, or precast concrete.

See Figure 13-22 for one way to build a crawl space bearing partition.

You can use a continuous concrete footing (instead of a wood and gravel footing) with a wood foundation basement. Pour the concrete over a 4'' thick layer of GSCS. Arrange this layer so water drains from the backfill to the porous layer under the slab. Or, provide drainage across the concrete footing. Use transverse pipes or drain tiles embedded in the concrete every 6 feet around the foundation.

Stepped footing— Figure 13-24 shows how to build a stepped footing for a wood foundation. First, build the full basement wall and the extended footing plate. Build the support frame. Cover it with plywood sheathing. Then place the gravel or crushed stone around the support frame. Finally, install the crawl space, or the elevated wall.

The Floor

Apply 6-mil polyethylene sheeting over the GSCS layer. Then, either pour a concrete slab, or build a wood basement floor over the sheeting. If you use wood floors, you can put the sheeting *over* wood sleepers supporting the floor joists. But don't put sheeting under the wood footing plate.

If you install a wood basement floor where there's more than 2' difference in depth-of-fill on opposite sides of the foundation wall, some special provisions are needed to keep the walls in place. The best way is to install treated wood or concrete beams between the opposite footing plates.

Concrete slab basement floors must be at least 3'' thick.

Figure 13-19 shows screed boards between the floor slab and the wall framing. They're optional. You can leave out the screed boards and extend the concrete floor slab to the plywood sheathing between studs. The screed makes the floor-to-wall joint extra strong and rigid. Use this method if walls are subject to heavy racking loads.

The Walls

In basement walls, use caulking compound (butyl or silicone) to seal the full length of all joints between the plywood panels. Caulk any unbacked panel joints while you're fastening the panels to the framing.

Before you backfill, cover the exterior of the below-grade part of basement walls with 6-mil polyethylene sheeting. Lap all joints in the sheeting at least 6''. Then glue the seam with a butyl or silicone adhesive. The sheeting should extend down to the bottom of the wood footing plate. But don't let it overlap or extend into the gravel footing.

Glue the top edge of the sheeting to the sheathing. Caulk the full length of this joint. Then cover the joint with a strip of treated lumber or plywood. This wood strip should extend several inches above and below the finish grade. It protects the sheeting from exposure to light and from mechanical damage at or near grade. Instead of the wood strip, you could use asbestos-cement board, brick, stucco, or other covering that fits the design of the house.

The Backfill

In most cases you'll want to place gravel or crushed rock backfill in the space between the side of the basement excavation and the basement. This backfill should extend to at least half the height of

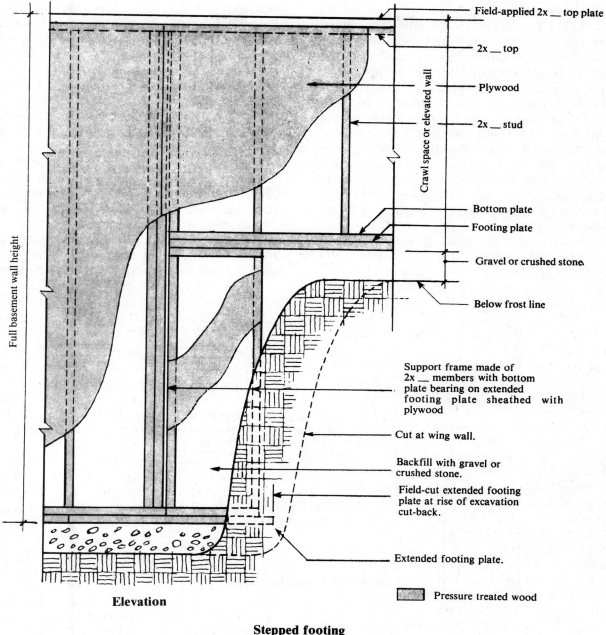

Field-applied 2x __ top plate

2x __ top

Plywood

2x __ stud

Crawl space or elevated wall

Bottom plate

Footing plate

Gravel or crushed stone

Below frost line

Support frame made of 2x __ members with bottom plate bearing on extended footing plate sheathed with plywood

Cut at wing wall.

Backfill with gravel or crushed stone.

Field-cut extended footing plate at rise of excavation cut-back.

Extended footing plate.

Pressure treated wood

Full basement wall height

Elevation

Stepped footing
Figure 13-24

the excavation. For basements in Group I soils (or in any soil if you can get the approval of your inspector), the fill may have to go only 1' above the footing.

If the backfill is deeper than 4' and the joists are parallel to the wall, use blocking. Figure 13-25 shows how to put blocking in the first joist space. Spacing shouldn't be more than 48" intervals. (Figure 13-16 gave nailing requirements for fastening this blocking.)

Cover the top of the fill with strips of either 6-mil polyethylene sheeting or 30-pound asphalt paper. Overlap the strips so water can seep through without letting soil pass through. An alternative is to use perforated sheeting or some other filtering cover that will keep the soil above the backfill.

Slope the ground away from the building. The slope should be at least 1/2" per foot for at least 6'. Provide drainage to prevent accumulation of surface water.

The Insulation

Insulate wood foundations with insulation either between the studs or outside the foundation wall.

Between-stud insulation can be used with or without a vented air space between the insulation and the foundation wall. With a vented airspace, run the insulation and vapor barrier all the way down to the bottom plate. Without a vented airspace, end the insulation and vapor barrier about 1' below the outside grade. There, fold an extension of the vapor barrier to make a tab. Attach the tab to the foundation wall. This separates the insulated and non-insulated parts of the stud cavity. You can install more insulation, *without* vapor barrier, below this level.

When you install any insulation between the studs below grade, leave at least 2'' between the end of the insulation and the bottom plate.

Figure 13-26 compares thermal resistance values of concrete block and AWWF systems. You can see that the AWWF system provides a warmer basement.

Clear Span Trusses

Figure 13-27 shows the top chord bearing attachment and blocking details for clear span trusses.

Figure 13-28 shows end-wall blocking details for clear span trusses.

Comparative Cost Study

AWWF can save you money. This means more profit for you. The NAHB Research Foundation did a cost study of two test houses. One was in Washington, DC, the other in nearby Maryland. One had a concrete block foundation, the other an AWWF. The study compared in-place material and

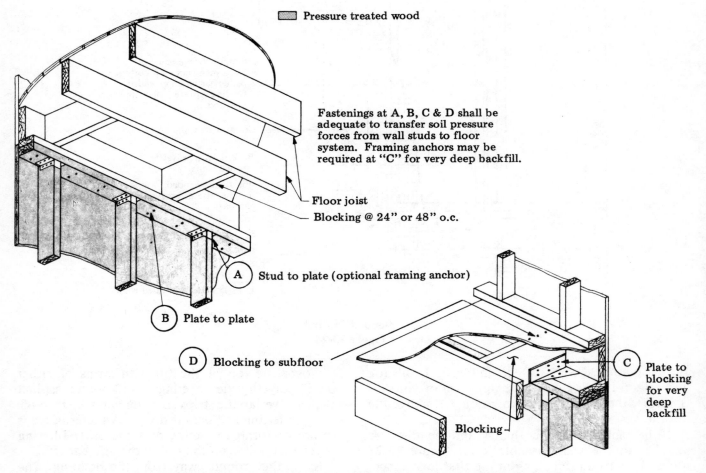

Pressure treated wood

Fastenings at A, B, C & D shall be adequate to transfer soil pressure forces from wall studs to floor system. Framing anchors may be required at "C" for very deep backfill.

Floor joist
Blocking @ 24'' or 48'' o.c.

(A) Stud to plate (optional framing anchor)

(B) Plate to plate

(D) Blocking to subfloor

(C) Plate to blocking for very deep backfill

Blocking

Fastening foundation end walls to floor system for deep backfill (more than 4')
Figure 13-25

Wall element	Thermal resistance (R)	8" concrete wall			8" block wall			AWWF wall insulation & drywall
		No interior finish	Furring & drywall	Furring insulation & drywall	No interior finish	Furring & drywall	Furring insulation & drywall	
Outside air film	0.17	0.17	0.17	0.17	0.17	0.17	0.17	0.17
8" poured concrete	0.88	0.88	0.88	0.88	--	--	--	--
8" cinder block	1.72	--	--	--	1.72	1.72	1.72	--
½" plywood	0.62	--	--	--	--	--	--	0.62
¾" air space	1.01	--	1.01	--	--	1.01	--	--
R-4 insulation batts in ¾" cavity	3.00	--	--	3.00	--	--	3.00	--
R-11 insulation batts in 3½" cavity	11.00	--	--	--	--	--	--	11.00
½" gypsum drywall	0.45	--	0.45	0.45	--	0.45	0.45	0.45
Inside air film	0.68	0.68	0.68	0.68	0.68	0.68	0.68	0.68
Total thermal resistance of wall, $R_T = R_1 + R_2 + R_3$		1.73	3.19	5.18	2.57	4.03	6.02	12.92
Coefficient of heat transmission for wall, $U = 1/R$		0.58	0.31	0.19	0.39	0.25	0.17	0.08

Thermal resistance values of foundation constructions
Figure 13-26

labor costs for the two houses. The result showed a cost savings of 10 percent with AWWF.

The cost savings, and the livability, of an AWWF basement have persuaded many builders to try AWWF. And in crawl space construction, AWWF means low cost and fast installation. Since it takes only a single trade (carpenters), some builders can finish an AWWF in one day.

Here are some of the benefits of the AWWF:

• Foundations are drier because drainage is better.

• Because the foundation area is drier, your tradesmen have better working conditions.

• You can buy post bearing plates and partition bearing plates to speed up construction. This means you can finish the roof before you have to pour any concrete to hold up a partition or a beam.

• You can spread gravel in the basement before setting the walls, thus eliminating an expensive wheelbarrow job.

• In areas with a deep frostline, you need deep footings for crawl space and slab-on-grade construction. This means a deep — and expensive — wall. With AWWF, you can use gravel for the footing. Just fill up the footing from the frostline with gravel, and use a normal-height AWWF wall.

• Except where the frostline is involved, you don't need to dig deep footings at all.

• With known soil type, soil conditions, and backfill, the AWWF system provides more shear and lateral strength. It acts as a beam-type construction that trusses the load across poor soil. Properly stress-graded lumber resists lateral pressure better than masonry. (Masonry has no more lateral strength than the mortar joint has.)

• You don't need to use drain tile.

• You can build the foundation in any weather; there's no concrete to freeze or to crack.

• Brick veneer won't crack (as with some concrete foundations) because the footing won't break.

• You don't need a termite shield. The wood treatment keeps termites and other insects out of the wood.

• No waterproofing is needed, but you must use polyethylene and caulk on basement foundations.

• It's easy to install water lines and electric wiring through the walls.

• Fuse boxes and switches can be recessed into the walls.

• No furring is needed: you finish the walls just like the upstairs stud walls.

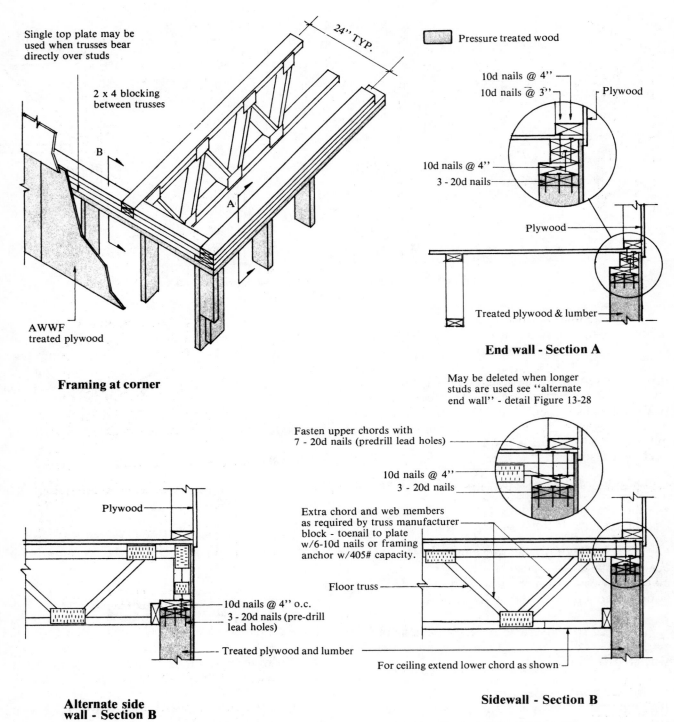

Single top plate may be used when trusses bear directly over studs

2 x 4 blocking between trusses

B

A

24" TYP.

AWWF treated plywood

Framing at corner

Pressure treated wood

10d nails @ 4"
10d nails @ 3"
Plywood

10d nails @ 4"
3 - 20d nails

Plywood

Treated plywood & lumber

End wall - Section A

May be deleted when longer studs are used see "alternate end wall" - detail Figure 13-28

Fasten upper chords with 7 - 20d nails (predrill lead holes)

10d nails @ 4"
3 - 20d nails

Extra chord and web members as required by truss manufacturer block - toenail to plate w/6-10d nails or framing anchor w/405# capacity.

Floor truss

For ceiling extend lower chord as shown

Sidewall - Section B

Plywood

10d nails @ 4" o.c.
3 - 20d nails (pre-drill lead holes)

Treated plywood and lumber

Alternate side wall - Section B

Note: Fastener schedule shown for 7' fill. fasteners may be reduced for lesser fill heights. For 4' fill, fasteners may be reduced 50%

Clear-span trusses — top chord bearing attachment and blocking details
Figure 13-27

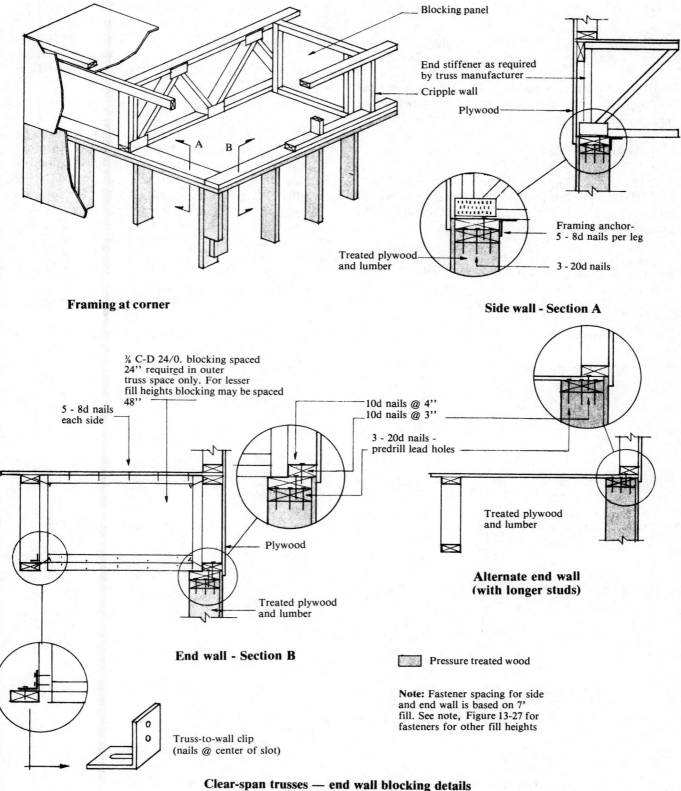

Blocking panel

End stiffener as required by truss manufacturer

Cripple wall

Plywood

Framing anchor- 5 - 8d nails per leg

3 - 20d nails

Treated plywood and lumber

Framing at corner

Side wall - Section A

⅜ C-D 24/0. blocking spaced 24" required in outer truss space only. For lesser fill heights blocking may be spaced 48"

5 - 8d nails each side

10d nails @ 4"
10d nails @ 3"

3 - 20d nails - predrill lead holes

Plywood

Treated plywood and lumber

End wall - Section B

Alternate end wall (with longer studs)

Treated plywood and lumber

 Pressure treated wood

Note: Fastener spacing for side and end wall is based on 7' fill. See note, Figure 13-27 for fasteners for other fill heights

Truss-to-wall clip (nails @ center of slot)

Clear-span trusses — end wall blocking details
Figure 13-28

- It's easy to fix any mistakes in construction.

- The foundation makes it easy to add wood stoops, siding, decks, porches, or stairways.

- The AWWF system is fast. An average-size home of 30' x 50' can be set up and ready for the floor system in less than half a day.

- Scheduling is easy with this system.

- The AWWF foundation is energy-efficient. Actual HUD documentation of two homes, one wood and one masonry, shows a 20% heat savings with the wood basement, split-foyer home.

- Basements can be used as living quarters, instead of merely as storage or utility areas, if you provide windows and doors for ventilation.

- The AWWF basement provides a little extra square footage in the house. You can expect to gain 1' for the front and back walls, and 1' for the side walls. So, on a 30' by 50' house, you gain 80 square feet. This extra space makes the house, theoretically, worth $1500 more.

Summary

Thousands of wood foundations have been built in the U.S. and Canada during the last five years. Many code agencies allow the wood foundation system, when it's built according to NFPA Technical Report No. 7, *All-Weather Wood Foundation System, Basic Requirements*. (This report is periodically updated by the National Forest Products Association.) The All-Weather Wood Foundation System is accepted by HUD/FHA, and the Farmers Home Administration, as well as the national model code organizations.

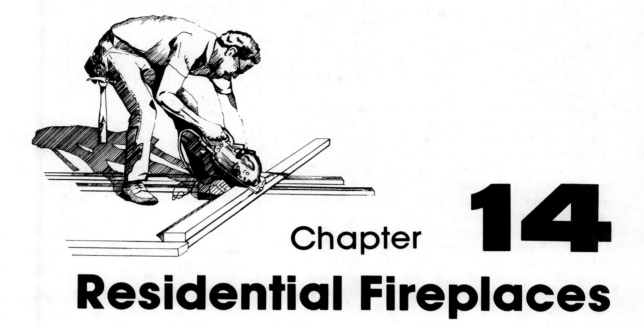

Chapter 14
Residential Fireplaces

Your homes will sell faster if they include the physical and emotional warmth of a fireplace. Your customers expect to see a fireplace in all but the most modestly priced homes. In more expensive homes, an extra fireplace in the master bedroom makes a nice selling point.

There are many types of fireplaces. In addition to the conventional masonry fireplace, there are prefabricated built-ins and freestanding units. The prefab fireplace with its metal chimney has grown in popularity over the last few years. It's probably the most common fireplace installed in new and remodeled homes now. And for good reason. The units are economical and highly efficient.

Because of the upsurge in demand for fireplaces in new single-family homes (30% in the early 1970's to nearly 77% in the early 1980's), a fireplace can add several thousand dollars to the value of a home.

Most manufacturers of prefabricated fireplaces include installation instructions with their units. If you need additional installation instructions for prefab units, they're included in *Manual of Professional Remodeling,* published by Craftsman Book Company. For details, look at the order form at the end of this book.

This chapter deals with information that you may not be able to find in other sources — how to build an efficient conventional masonry fireplace. You'll find the detailed instructions you need to do the job right.

There are three parts to a good fireplace: the right design, the right materials, and good construction techniques. This chapter covers all three. The information applies to almost any residential fireplace. Of course, as in all construction, you must follow your local building code.

Residential Fireplace Design

For years, fireplaces were used for decoration, not for heating. Recently, though, people are again heating with fireplaces. This means that fireplaces are used at high operating temperatures. They're used often, and for long periods of time. Fireplaces must be designed and built to withstand this heavier use.

New ideas have been developed to make conventional fireplaces more efficient. And some successful concepts that weren't used with decorative fireplaces are being rediscovered.

The combustion process is complex. Rates of fuel consumption vary widely. This makes it difficult to test fireplace designs. Very little laboratory testing has been done. Test methods that cover all possible variables are just now being developed. Most of the designs in this chapter are based on concepts that have worked in the past. Fireplace design is not an exact science: it's an art, to be applied with judgment.

We'll discuss the design of successful wood-burning fireplaces. And we'll introduce some ways to increase their efficiency.

Every energy-efficient fireplace includes an exterior air supply. The air supply has to be the right size and in the right place. You have to make sure the dampers fit tightly. Adding glass screens or similar devices can increase efficiency. Regardless of how good the design is, however, good workmanship is critical. We'll talk more about energy-efficient fireplaces later in the chapter.

Common Fireplace Features

Fireplaces are for burning wood. They're not designed or built for fuels that burn at higher temperatures than wood. Fireplaces perform two main functions: they contain a fire safely, and they heat the house. All fireplace styles have several common construction features.

The fireplace must not be a safety hazard. Your local building code will have specific requirements. Some general requirements, found in many building codes, are:

• Firestop all spaces between masonry fireplaces and wood or other combustible materials. Use a 1'' thickness of a noncompressible, noncombustible material, like fibrous insulation, in these spaces.

• No combustible material is permitted within 6'' of the fireplace opening.

• Combustible material within 12'' of the fireplace opening can only project 1/8'' (or less) for each inch of distance from the opening. For instance, wood trim that is 3/4'' thick must be at least 6'' from the opening.

Fill any nonfunctional empty areas in the fireplace — from the foundation through the chimney — solidly, with masonry mortared in place. (The functional empty areas are the air passageway, the ashpit, the 1'' airspace between the combustion chamber and surrounding brickwork, and the smoke chamber.) Tool all exposed mortar joints. Concave jointing is best. Filling these areas makes the fireplace work better and helps it last longer. Filling also makes it less likely to leak in the rain.

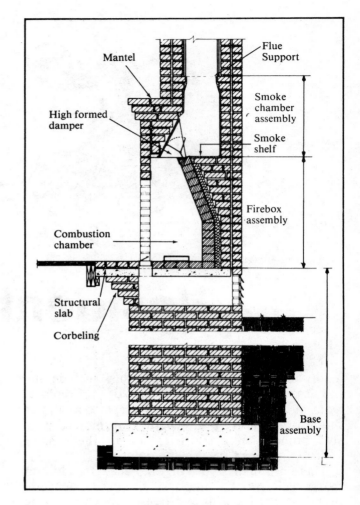

Typical fireplace section
Figure 14-1

Figure 14-1 shows a typical fireplace section with the parts labeled.

Kinds of Fireplaces

There are several popular residential fireplace designs and many variations of each style. The basic principles, however, are the same.

Single-Face Fireplaces

Single-face fireplaces have been used and developed over the centuries. What we know today about proper opening, damper, and flue sizes is the result of trial and error experiments.

Single-face fireplaces can be efficient heating units. The more brick masonry surface that can be

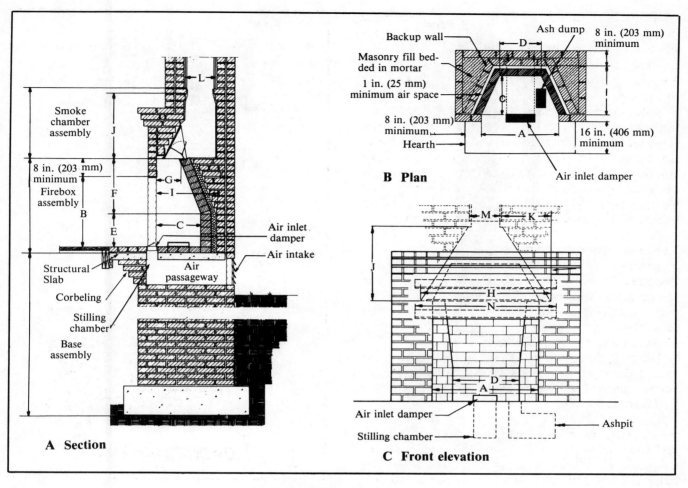

Single-face fireplace
Figure 14-2

exposed to the fire, the better. It means more radiated and reflected heat. The mass of the brick in the combustion chamber also stores more heat, for radiation after the fire is out. Figure 14-2 shows a section, a plan, and an elevation of a single-face fireplace.

Rumford fireplaces— The Rumford fireplace is a single-face fireplace. It has widely-splayed sides, a shallow back, and a high opening. The Rumford is being re-examined in the search for more energy-efficient fireplaces.

Most of the heat from a fireplace is radiated heat. The more surface there is to radiate heat, the better. A fireplace with a shallow firebox and flared sides and back radiates the most heat. Add an outside air supply and a glass screen at the open-

ing for maximum all-round energy performance. It will give the most radiated heat, with the least air infiltration.

Figure 14-3 shows a plan, a section, and an elevation of a Rumford fireplace, and how it's built. Figure 14-4 gives the sizes of the dimensions marked with letters in Figures 14-2 and 14-3.

Multi-Face Fireplaces
Although we often think of it as contemporary, the multi-face fireplace is really an old design. Consider, for example, the corner fireplace with its two adjacent open sides. It's been used in Scandinavia for centuries.

Multi-face fireplaces may have adjacent, opposite, three, or even all faces open. But be careful. Some multi-face fireplaces present design problems

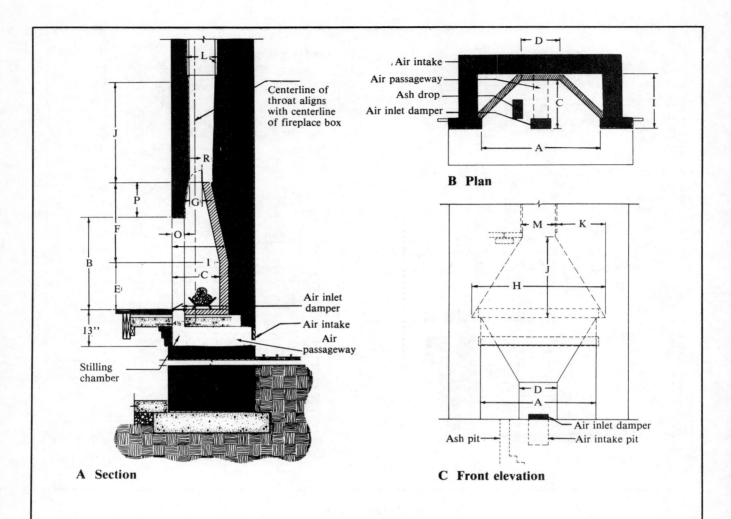

A Section

B Plan

C Front elevation

For maximum heat radiation from a Rumford Fireplace:

1) The width of the firebox (D) must equal the depth (C).

2) The vertical portion of the firebox (E) must equal the width (D).

3) Thickness of the firebox (I minus C) should be at least 2¼ inches.

4) Area of fireplace opening (A x B) must not exceed ten times the flue opening area.

5) The width of the fireplace opening (A), and its height (B) should each be two to three times the depth of the firebox (C).

6) Opening height (B) should not be larger than width (A).

7) The throat (G) should be no less than three or more than four inches.

8) Centerline of the throat must align with the centerline of the firebox base.

9) The smokeshelf (R) should be four inches wide.

10) The width of the lintel (O) should be no less than four nor more than five inches.

11) Vertical distance from lintel to throat (P) must be at least 12 inches.

12) A flat plate damper is required at the throat, and must open towards the smokeshelf.

Rumford fireplace with exterior air supply system
Figure 14-3

Single-face fireplace dimensions (inches[a,d])													
Finished fireplace opening							Rough brickwork				Flue size[b]		Steel angle[c]
A	B	C	D	E	F	G	H	I	J	K	L	M	N
24	24	16	11	14	18	8¾	32	21	19	10	8 × 12		A-36
26	24	16	13	14	18	8¾	34	21	21	11	8 × 12		A-36
28	24	16	15	14	18	8¾	36	21	21	12	8 × 12		A-36
30	29	16	17	14	23	8¾	38	21	24	13	12 × 12		A-36
32	29	16	19	14	23	8¾	40	21	24	14	12 × 12		A-42
36	29	16	23	14	23	8¾	44	21	27	16	12 × 12		A-42
40	29	16	27	14	23	8¾	48	21	29	16	12 × 16		A-48
42	32	16	29	14	26	8¾	50	21	32	17	16 × 16		B-48
48	32	18	33	14	26	8¾	56	23	37	20	16 × 16		B-54
54	37	20	37	16	29	13	68	25	45	26	16 × 16		B-60
60	37	22	42	16	29	13	72	27	45	26	16 × 20		B-66
60	40	22	42	16	31	13	72	27	45	26	16 × 20		B-66
72	40	22	54	16	31	13	84	27	56	32	20 × 20		C-84
84	40	24	64	20	28	13	96	29	61	36	20 × 24		C-96
96	40	24	76	20	28	13	108	29	75	42	20 × 24		C-108

[a]Adopted from The Donley Brothers Company, *Book of Successful Fireplaces—How to Build Them*

[b]Flue sizes conform to modular dimensional system.

[c]Angle sizes: A—3 × 3 × 3/16 in., B—3½ × 3 × ¼ in., C—5 × 3½ × 5/16 in.

[d]SI conversion: mm = in. × 25.4

**Single-face fireplace dimensions
Figure 14-4**

which can result in insufficient draw, external draft requirements and excessive smoking.

Brick for Fireplaces
Most building codes require you use solid bricks for fireplaces. Facing brick should conform to ASTM C 216. Building brick should conform to ASTM C 62. Or, use hollow brick (ASTM C 652) with vertical reinforcement. Use grade SW brick in exposed areas. It's more durable. It might be more economical to use Grade MW or NW brick in areas not exposed to weather and dirt.

You can use facing brick, stone, tile, or marble to face the fireplace.

Residential Fireplace Construction
Though the design and construction information here is for single-face fireplaces, all fireplaces have the same basic parts. These are the base assembly, the firebox assembly, and the smoke chamber assembly. Figure 14-1 shows how they combine to make a fireplace.

Base Assembly
The base assembly has two parts: the foundation and the hearth support. Figure 14-5 shows them. No rule says you must always use both of these components. In slab-on-grade construction, for example, the slab can act as both foundation and hearth support.

Foundation: The foundation supports the whole fireplace and chimney assembly. It must be heavy enough to carry the load. (Figure 14-5 shows a typical foundation.) Footings support either foundation walls or a structural slab. Most building codes, however, don't let you support other parts of the building on the fireplace/chimney assembly. It's not a structural element.

When designing the foundations, consider soil

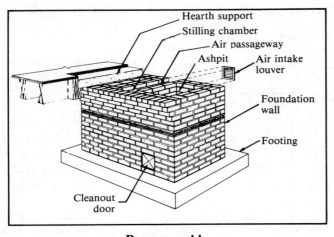

Base assembly
Figure 14-5

type and condition. Undisturbed or well-compacted soil will usually be adequate. But some types of soil, or some conditions of the soil, may need more analysis. Watch out for:

- Inorganic clays of high plasticity
- Fat clays
- Inorganic silty soils
- Elastic silts
- Organic silts or silty clays of low plasticity
- Organic clays of medium to high plasticity
- Peat and other highly organic soils

Footings— Make them of concrete at least 12" thick. Extend them at least 6" past the fireplace walls on each side. Footings should go below the frost line. Footings in a basement, or other area that won't freeze, needn't penetrate the frost line. Build footings on undisturbed or properly compacted soil. You may have to build special foundations for oversized chimneys.

Foundation walls— Build foundation walls of masonry or concrete. Most building codes call for an 8" minimum thickness. In brick foundation walls, solidly fill the voids (except the ashpit and external air ducts). Use brick bedded in mortar.

Structural slab— When the fireplace is on a slab-on-grade, you'll need to thicken the slab under the fireplace. It has to support all loads from the fireplace and chimney.

Hearth Support: There are several ways to support the hearth. These include: *corbeled brickwork,* a *structural slab,* or *cantilevered, reinforced brick masonry.*

Codes for corbeling usually limit the individual maximum projection. No unit can project more than half its height. No unit can project more than a third its thickness. When corbeling from walls, limit the *overall* horizontal projection to half the wall thickness, unless reinforced. Your local code and foundation design may also limit the maximum horizontal projections, both overall and individual. A hearth supported by corbeling and a structural slab is shown in Figure 14-2.

Firebox
The firebox assembly consists of the firebox opening, the fireplace hearth, the combustion chamber, the throat, and the smoke shelf. Figure 14-6 shows a typical firebox.

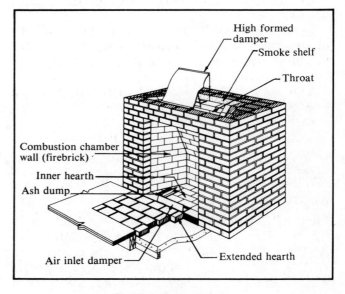

Firebox assembly
Figure 14-6

Firebox Opening: The size and shape of the firebox opening determine the design of the rest of the fireplace components. Look back to Figure 14-4. It shows fireplace opening sizes to use for the most attractive and most efficient fireplace. You can vary these dimensions slightly, to allow for regular brick coursing.

Sizing the opening— The fireplace's size should complement the room's size. Both size and location are important, both for appearance and for good operation. With a too-small opening, a fireplace might not warm the room, even though it works perfectly. With a too-large opening, a properly-made fire might be too hot for the room. It also would need a larger flue area. And it would use a lot of interior air, unless exterior combustion air was provided.

A fireplace in the 12'-long wall of a 12' x 20' room should have a 32'' to 36'' wide opening. If it's in the 20' wall, a 36'' to 40'' wide opening is appropriate. A 36' wall can handle a 48'' or 50''-wide fireplace.

On a square-foot basis, a 30''-wide fireplace will serve 300 square feet of floor area.

Support above the fireplace opening— The brickwork above the fireplace opening must be supported. There are several ways to do it. You can use *brick arches, reinforced brick masonry lintels,* or *steel angle lintels.*

Brick arches don't usually need steel reinforcement. And they're attractive. When figuring the height of an opening, use the maximum height of the arch soffit.

You can use reinforced brick masonry lintels, instead of precast lintels. They have a brick soffit, without any exposed steel angle.

Steel angle lintels are most often used. For this reason, Figure 14-4 gives steel angle dimensions for fireplaces. Here are some general rules: Steel angle lintels should be at least 1/4'' thick. They should have a horizontal leg at least 3½'' long, when used with a (nominal) 4''-thick brick face wall. And they should have a horizontal leg of 2½'', when used with a (nominal) 3''-thick face wall. The minimum bearing length is 4''. Use corrosion-resistant steel (ASTM A 36) for lintels. The maximum clear spans might be limited by the fire-protection requirements in some building codes.

Fireplace Hearth: The hearth consists of two parts: the inner hearth, and the extended hearth. The hearth may be raised or flush with the floor. A fireplace hearth flush with the floor was shown in Figure 14-6.

Inner hearth— The inner hearth is the floor of the combustion chamber. The fire is above it. The inner hearth should be of noncombustible material at least 4'' thick. This 4'' includes both the inner hearth and the noncombustible hearth support.

Extended hearth— The extended hearth is the part that projects into the room. It should also be noncombustible. A reinforced brick cantilever makes an attractive extended hearth.

Most building codes require the extended hearth to project at least 8'' on each side of the opening and 16'' in front of it. If the opening's area is 6 square feet or more, some building codes require the hearth to project 12'' on each side of the opening and 20'' in front of it.

Combustion Chamber: The shape and depth of the combustion chamber are important. They determine the draft and the air needed for combustion. They also determine how much heat is reflected into the room. Figure 14-7 shows the combustion chamber and smoke chamber. Look back at Figures 14-2 and 14-3 for plan views of the combustion chamber. Then, look at Figure 14-4 for recommended dimensions. You can change these dimensions *slightly.* But don't make any drastic changes. These dimensions are based on fireplace designs that work.

The sides of the combustion chamber are vertical. So is the lower part of the back. Above the vertical part in the back, the brick slopes forward toward

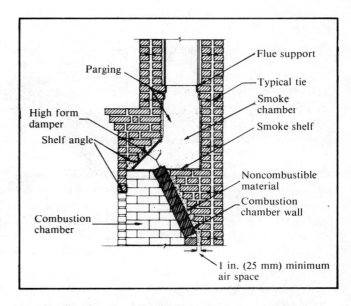

Combustion chamber and smoke chamber
Figure 14-7

the opening. The top of this slope will support the damper. For maximum heating of the room, make this sloped part flat, not concave. If it's concave, the heat reflects back into the fire, not into the room. Splayed, rather than vertical, sides also reflect more heat into the room.

Laying out the combustion chamber— The combustion chamber *must* be laid out right. It's inside the firebox assembly but *isolated* from it. Look at Figure 14-8. It shows one way to locate the combustion chamber. First, choose the dimensions from Figure 14-4. Then, locate the front wall (facing) of the fireplace. Line A-A marks the inside face position. Next, find Point I. It's at the center of line A-A. Now, square from Point I back into the combustion chamber. Mark the chamber's depth. This is Point II. Line B-B connects Points I and II. From Point II, square two perpendiculars to line B-B, one in each direction. The length of each perpendicular is half the width of the rear chamber wall. At the ends of these perpendiculars are Points c and d. Now, from Point I on line A-A, measure half the fireplace opening dimension, in each direction. You've now located Points a and b. Finally, connect the four points (a, b, c and d) for the outline of the inside face of the combustion chamber.

Building the combustion chamber— The combustion chamber should be of brick 4" thick or thicker. Firebrick (ASTM C 64) is best. It's the most resistant to heat and temperature changes. You can lay it with any face exposed, but it's best as a stretcher course. Use fireclay mortar (ASTM C 105). Make mortar joints no thicker than 3/16". The mortar only has to be thick enough to fill irregularities in the firebrick. Use the "pick and dip" method for thin mortar joints: Dip the brick into a soupy mix of fireclay mortar, then immediately put it in place.

Alternatively, build the combustion chamber of Grade SW brick (ASTM C 62, or ASTM C 216). Use Type N or Type O, portland cement-lime mortar (ASTM C 270, or BIA Designation M1-72). These high-lime mortars are more heat-resistant than high-portland cement mortars. (You can also use these mortars with the firebrick discussed in the last paragraph.) Because it's not as durable as firebrick, lay grade SW brick as a stretcher course. Again, make the mortar joints no more than 3/16" thick. They'll be less likely to crack and deteriorate.

Building the Firebox: You can build the surrounding brickwork while you're building the combustion chamber or after the combustion chamber is

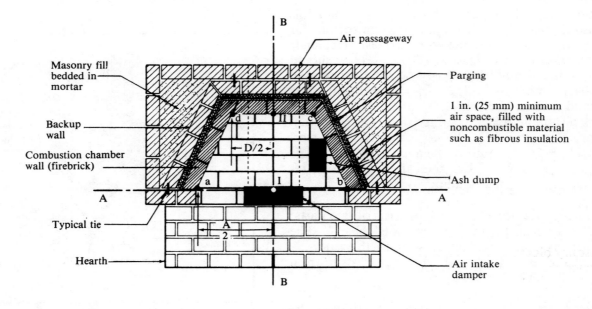

Sizing the combustion chamber
Figure 14-8

completed. Either way, build a backup wall. This is a full course of masonry surrounding the combustion chamber. Look at Figure 14-8 again. Make this wall at least 4'' thick to support the smoke chamber and chimney above it. You might need to build a thicker wall to support a higher chimney.

Leave at least a 1'' airspace between the combustion chamber wall and the backup wall. This is expansion space for the heated combustion chamber. Wrap a compressible, noncombustible material (fibrous insulation is good) around the combustion chamber. This assures that the 1'' air space won't be filled with mortar droppings. The air space isolates the combustion chamber, reducing stress from thermal expansion.

Use wall ties at all intersections where the wall isn't masonry bonded, except in the combustion chamber walls. Use wire ties (ASTM A 82 or ASTM A 185), at least 9 gauge, and corrosion-resistant. Space the ties no more than 16'' apart vertically. Embed the ties at least 2'' into the bed joints of the brick masonry.

Horizontal joint reinforcement is a good idea. Use it at the corners of adjacent wythes and in the wythe (backup wall) surrounding the combustion chamber walls. This reinforcement will reduce cracking.

Wait 30 days after construction to build a fire. The mortar needs moisture to cure; fires built too soon dry out the masonry too fast.

Throat: Because it affects the draft, design the fireplace throat carefully. Put it at least 8'' above the highest point of the fireplace opening. Again, Figure 14-4 gives the dimensions. Details are shown in Figures 14-2 and 14-3.

Damper: Use a metal damper, as wide as the opening. Put it in the throat. The valve plate should open toward the back of the fireplace. This plate, when open, will deflect downdrafts.

There are many damper designs. The high formed damper is a good choice: It extends the throat. It also forms a critical part of the smoke chamber. This damper style also lessens the chance of masonry blocking the damper's valve plate. A high formed damper is shown in the fireplace throat, in Figure 14-7.

Don't embed the damper in mortar. Just seat it on a thin setting bed. Spread a mortar bed. Use the same mortar as in the combustion chamber. Make

the bed just thick enough for a seat. This seat also seals against gas and smoke leakage.

The damper assembly shouldn't touch any masonry, other than what it bears on. So make sure it's seated. Then, wrap it with a compressible, noncombustible material. See Figure 14-9. This material provides space for thermal expansion and space for damper movement during fireplace use.

Smoke Shelf: The smoke shelf is directly under the flue. Figure 14-7 shows its location. Design the smoke shelf to provide a uniform flow of air. It should also cause any downdrafts to eddy, then drift upward. The smoke shelf is level and on an even horizontal plane with the base of the damper. It can be flat or curved to blend with the rear wall of the smoke chamber. See Figure 14-10. Figure 14-4 gives recommended dimensions.

Smoke Chamber Assembly
The smoke chamber assembly forms the chimney flue support, as shown in Figures 14-1 and 14-11. The smoke chamber is above the throat and smoke shelf. It supports the flue lining on all sides. Note Figure 14-1. The back wall of the chamber is vertical. The side walls slope evenly toward the center. The front wall above the throat is also sloped. Figure 14-4 gives recommended dimensions for the smoke chamber.

Building the Smoke Chamber: A steel lintel, *not* the damper, supports the front wall above the throat. Put a noncombustible, compressible material between the masonry and the damper and steel lintel. Figure 14-9 shows how. This lets both the damper and the lintel expand.

Give the smoke chamber a smooth slope. For the required size, corbel each course of brick. Parge the inside of the smoke chamber. This reduces friction and prevents smoke leakage.

Corbeling in the smoke chamber— Starting at the smoke shelf, corbel in the front and sides of the smoke chamber. Build the rear wall vertically. This ensures that the entire edge of the flue liner will be supported.

The usual limitations for corbeling walls don't apply in the smoke chamber. Here, the fireplace's configuration determines the limitations. The corbels are laterally supported by adjacent masonry. The maximum corbel for each unit is: the horizon-

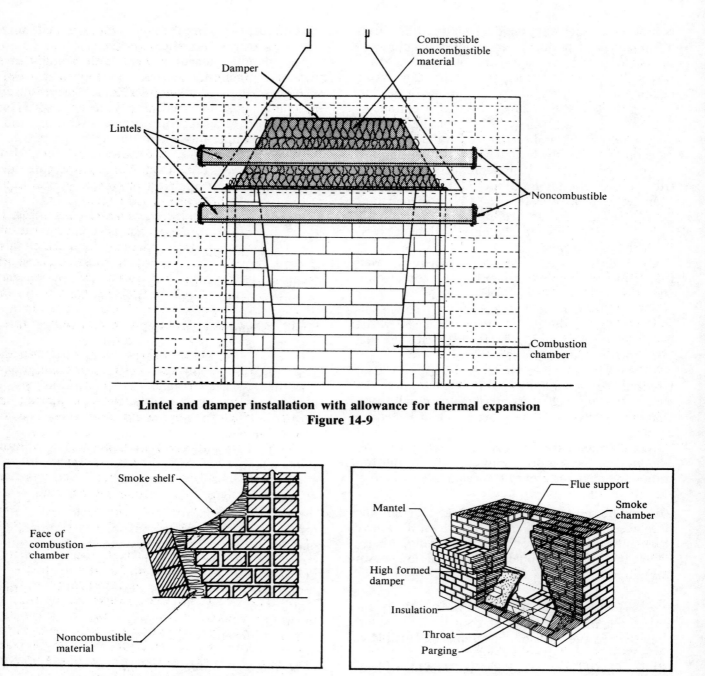

Lintel and damper installation with allowance for thermal expansion
Figure 14-9

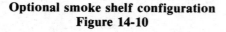

Optional smoke shelf configuration
Figure 14-10

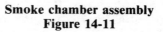

Smoke chamber assembly
Figure 14-11

tal distance to be corbeled, divided by the number of courses from the bottom of the flue liner to the first corbeled course.

Lay the two last courses before the flue liner as headers. Cut them so they totally support the flue liner. But make sure they don't obstruct the flue liner opening.

Chimney Flue: Both flue area and chimney height affect proper draft. Figure 14-4 shows recommended flue areas. The next chapter, Residential Chimneys, discusses flue area and chimney height in detail.

Energy-Efficient Fireplaces

People want fireplaces in the homes they buy. A fireplace makes a home more attractive. And it can be used for supplemental heating. So the owners not only save on utility bills, they use fewer nonrenewable resources. Make conventional fireplaces more energy-efficient with a few modifications. Of course, when any fireplace is in use, the mechanical heating system should be turned down — or even off.

Locating the Fireplace

For the most efficient heating, put the fireplace completely inside the house. The mass of the fireplace stores heat. The masonry then radiates the heat, long after the fire has gone out. For even heating of the living areas, put the fireplace near the center of the house. Cold spots, in areas away from the fire, are kept to a minimum. And heat from an air-circulating fireplace can easily be vented into other areas.

Outside Air Supply

An outside air supply system makes a fireplace more efficient. It reduces the amount of heated air the fire draws from the house for combustion and draft. The outside air can be drawn straight from the outside, or from unheated areas of the building. The next section discusses possible locations. As mentioned before, leaving out this system results in a conventional fireplace design.

Building the Outside Air Supply System

There are many ways to lay out and build the air passageway discussed in this chapter. If you make any changes from the suggested details, keep in mind the components' purpose. And, of course, follow the manufacturer's directions for installing a specific component.

There are many ways to bring the outside air into the firebox area: one example is shown in Figure 14-2 A. The three basic parts, required by any method, are: the *intake,* the *passageway,* and the *inlet.* Install tight-fitting inlet dampers and tight-closing intake louvers. This keeps cold air from seeping in when the fireplace isn't in use.

The intake— Locate the intake on an outside wall, or in the back of the fireplace. Or, put it in a crawl space or other unheated area. Don't put it in a garage; most building codes won't allow it. Cover the intake with a screen-backed, closeable louver. Use one that can be operated from inside the house.

The passageway— A passageway, or duct, connects the intake to the inlet. The size and construction details can vary. Sizes anywhere from 6 to 55 square inches have been used successfully.

Build the passageway into the base assembly. If this isn't practical (due to fireplace configuration, or exterior grade level) change the fireplace by raising the hearth. You could also channel a ductwork passageway between floor joists. The passageway could also enter above the fireplace base, then connect to inlets in the sides of the firebox. In any case, insulate the passageway to reduce heat loss.

The inlet— The inlet brings outside air into the firebox. Put it in the sides or floor of the combustion chamber, in front of the grate. Don't put it toward the back of the combustion chamber: updrafts through the inlet could blow ashes into the room.

Fit the inlet with a damper to control the volume and direction of air. Proper operation of this damper is critical to the fireplace's performance. This is because cold outside air expands when brought into the fireplace. This expansion could produce more air than the fire needs: the extra cold air goes out into the room.

The air inlet damper's area, as shown in Figure 14-2, can range from 6 to 55 square inches. It depends on the other parts of the air supply system. You can use corbeling to get proper size or location of the air inlet. If you do, limit the maximum horizontal projection to half the distance from the ashpit face to the exterior of the fireplace assembly. Limit the maximum projection for an individual unit to either half the height of the unit or one-third the bed depth.

Sometimes air moving quickly through the inlet causes a large temperature increase in the combustion chamber. These higher temperature can literally burn up the grates and inlet dampers. To help slow down the air through the inlet, build a *stilling chamber.* This is a space before the inlet. See Figure 14-2 A.

The Ash Drop and Ashpit

Figure 14-12 shows a plan view of an air passageway and ashpit. If there's enough room to work, put the ashpit cleanout door inside the house, preferably in the basement. Use a metal ashpit cleanout door. And be sure it fits tightly, to reduce air infiltration.

The ash drop and ashpit are often left out of the fireplace's design. There's no space for an ashpit in slab-on-grade construction. Sometimes the designer or owner just doesn't think it's necessary.

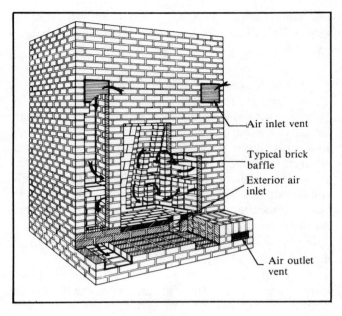

Air-circulating fireplace (Brick-O-Lator)
Figure 14-13

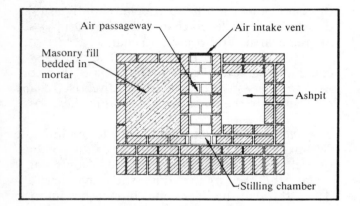

Air passageway and ashpit plan
Figure 14-12

Glass Fireplace Screens

Install glass screens on *all* fireplaces: both conventional ones and those with an outside air supply. A glass screen keeps heated air from escaping up the chimney while allowing smoke and fumes to escape. Seal the glass around the edges. Use tightfitting doors and vents so the fireplace doesn't let in cold air when not in use.

These screens are closed when the fire is out or smoldering and before it's safe to close the damper. Don't close glass screens while a large fire is burning. High temperatures, due to increased air velocities through the intakes, can cause problems.

Air-Circulating Fireplaces

Room air can be circulated through brick baffles behind the combustion chamber. This delivers more heat to the room — or to other areas of the house. One example of an air-circulating fireplace is the *Brick-O-Lator*[TM]. It's shown in Figure 14-13 and can be very efficient. It combines the basic

principles of a fireplace with those of a forced air system. Air-circulating fireplaces make a good supplementary heating system. Brick radiates heat long after the fire's out.

Summary

Prefabricated fireplaces arrive on the job ready to install. You don't have to worry about throat size, damper, smoke chamber and the other variables that go into a masonry fireplace.

Prefab units save time and money, and they're attractive. You can give them a masonry front just as in conventional fireplaces. The units that include a blower to circulate the warm air are efficient, as well. Many builders still build conventional masonry fireplaces in their better, more traditional homes, especially in the East. But use care in selecting masons to build a masonry fireplace. Every mason isn't a fireplace expert. Poor workmanship can cause serious problems, even in a well-designed fireplace. And it's hard to correct construction faults once they've been made.

The information and suggestions in this chapter are based on the experience of this author and the technical recommendations of the Brick Institute of America (BIA). Use the dimensions and materials suggested and you'll build quality fireplaces that will give years of dependable service.

Chapter **15**

Residential Chimneys

Some residential chimneys serve fireplaces; others serve appliances. Both types are made of similar materials and must conform to the same building code requirements. And both must do the same job — channel smoke and gases away from the building while creating a draft.

A draft is important for combustion. Heated air creates a draft because warm air is lighter than cool air. As the warm air rises up the chimney, it's replaced by cool air rich in oxygen. Without this fresh air and oxygen, the fire would die.

Even though the fireplace and appliance chimney may convey different types of gas and smoke at different velocities, they both have to work without interfering with the burning process. They must also release the smoke at a safe *height* and *location* and keep gases from escaping into the house.

This chapter explains what you need to know to design and build any common type of residential chimney. Like fireplace design, chimney design is based on experience — what worked in the past. Chimney height and flue area are the two critical factors in chimney design.

Chimney Design

When designing fireplace and appliance chimneys, be sure the height both complies with the code and is correct for the flue size so there's enough draft.

Building Code Requirements

Though these vary from one area to another, there are several requirements that apply just about anywhere. They include:

• For fire safety, the minimum chimney height is the *greater* of the following: 3' above the highest point where the chimney comes through the roof, or 2' higher than any part of the structure, or adjoining structures within 10' of the chimney. Figure 15-1 illustrates these points.

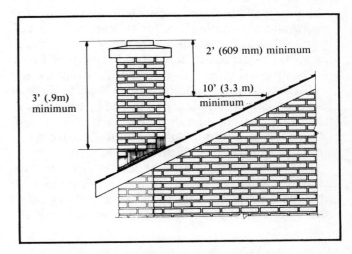

**Building code dimension requirements
Figure 15-1**

• Make chimney walls at least 4'' (nominal) thick. If no flue liner is used, make the walls at least 8'' thick.

• Neither chimney nor flue liner can change size or shape within 6'' of floor components, ceiling components, or rafters.

• The chimney must be at least 2'' from combustible material, unless the chimney is entirely outside the structure. A 1'' clearance is acceptable in that case.

• Firestop the spaces between a chimney and any combustible material with at least a 1'' thickness of noncombustible material.

• Seal all exterior spaces between the chimney and adjacent components. This is usually done by flashing and caulking.

• Do not corbel masonry chimneys more than 6'' from a wall or foundation. Don't corbel a chimney from a wall or foundation that's less than 12'' thick, unless it projects equally on each side of the wall. The exception is on the second story of a two-story house. There, the corbeling of the chimney, or of the exterior of the enclosing wall, may equal the wall thickness. Don't project corbeling more than 1'' for each course of brick projected.

Multiple Flues

Flues may be sloped for several reasons. They may be sloped to join with other flues, to discharge into a common flue. Or they may slope because the chimney has to be in a particular location. The maximum allowable slope is 30 degrees from vertical. When you join flues, size the common flue to handle the combined flow from the smaller flues.

Don't use the same flue for both a fireplace and a furnace. Most building codes won't allow it anyway. Connecting more than one heating unit to the same flue cuts the draft for each unit in half. Remember this rule: Don't combine flues from different systems, or from different fuels.

But you *can* combine separate flues in one chimney. Just be sure to meet minimum wall thickness requirements. And lay a full wythe of brick, bonded to the chimney walls, between the flues.

Chimneys as Structural Supports

Sometimes, you'll want to use the chimney as a

structural element. You can do it, and still stay within most building codes. Keep the normal chimney wall thickness, then add a structural wall around the chimney. This support wall may be built as part of the chimney wall as shown in Figure 15-2.

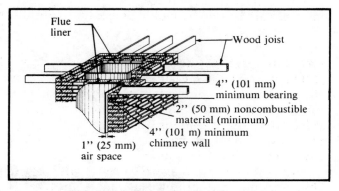

Chimney used as structural support
Figure 15-2

Most building codes require a minimum bearing of 4''. When all building code requirements are followed, the minimum wall thickness of a lined chimney used as a structural component is 10''. This 10'' includes: a 4'' brick chimney wall, 2'' of noncombustible material (brick), and a 4'' bearing length (brick). Figure 15-2 shows how. The minimum wall thickness for an unlined chimney is 14''. The only difference from the lined chimney is that the chimney wall must be 8'' thick.

Designing Fireplace Chimneys

The height of a residential fireplace chimney depends on:

• The size of the fireplace opening

• The size of the flue liner

• The size of the damper throat

• Whether the chimney has a chimney cap

As a general rule, appearance rather than performance determines the size of the fireplace opening. The size of the fireplace opening is the basis for the design of the rest of the chimney.

The American Society of Heating, Refrigerating,

and Air-Conditioning Engineers (ASHRAE) says that a frontal face velocity (draft) at the fireplace opening of 0.80 feet per second is enough to keep smoke and gases from discharging into living areas. This is a minimum velocity. It's what you would expect when a fire is started. As the fire generates heat, the draft increases.

You can see how fireplace opening size and flue liner size are related by looking back to Figure 14-4 in Chapter 14.

Calculating Height, Based on Draft

Chimney height is measured from the top of the fireplace opening. Fire safety considerations determine the minimum heights in building codes. Draft calculations have nothing to do with it. Always meet these heights, even if draft calculations show the chimney could be shorter.

This section shows how to calculate the shortest chimney that will produce adequate draft. You only have to know the opening and flue sizes. Remember, it's the *minimum*. Code may require a taller chimney.

For these calculations, you'll use Figure 14-4 to find some values and Figure 15-3 to find other values. The numbers in the equations are rounded off to the nearest hundredth.

Nominal size (inches)	Inside perimeter, P_1 (feet)	Equivalent diameter, D (inches)	Minimum area, A_F (square feet)
8 × 12	2.7	9	0.42
12 × 12	3.0	10	0.56
12 × 16	3.6	12	0.78
16 × 16	4.2	14	1.08
16 × 20	4.8	16	1.38
20 × 20	5.3	18	1.78
20 × 24	5.9	20	2.16
24 × 24	6.5	22	2.64

Flue liner dimensions
Figure 15-3

First, we'll describe the steps to follow. Then we'll take you through an actual example.

Step 1— Go to Figure 14-4 in Chapter 14. Choose the fireplace opening dimensions A and B. Also choose the corresponding flue liner size, L and M (L and M are used later in the calculations). Now,

use Equation 1: Divide the fireplace opening area, in inches (A x B), by 144 (the number of square inches in a square foot). The answer is the *Fireplace Opening Area, Ao.*

$$A_o = \frac{A \text{ in.} \times B \text{ in.}}{144"}$$

Equation 1

Step 2— Now you need to use Equation 2 to find the *Flue Friction Coefficient, Kt.* But before using Equation 2, you must find K1, K2, and K3.

In residential applications, the *Flue Gas Velocity, K1,* always equals 1.0. It's the friction loss when the moving air in the chimney accelerates to Flue Gas Velocity.

Now find the *Inlet Loss Coefficient, K2.* You'll have to pick a preliminary damper size. At this point, you only need to decide whether the damper throat area will be equal to the flue area, or twice the flue area. If the damper throat area equals the flue area, K2 equals 2.5. If the damper throat area is twice the flue area, K2 equals 1.0.

Now it's time to decide whether to use a rain cap. And how far above the top of the chimney will it be? Choose the *Termination Coefficient, K3,* using the following information: If no rain cap is used, K3 equals 0.0. If the rain cap is at a distance of D/2 (See Figure 15-3 for D) above the end of the flue liner, K3 can vary from 0.0 to 4.0. You can get this information from the manufacturer.

Now that you've found K1, K2, and K3, use Equation 2 to figure the *Flue Friction Coefficient, Kt.*

$$K_T = K_1 + K_2 + K_3$$

Equation 2

Step 3— Remember, in Step 1 you chose the flue liner size from Figure 14-4. Using the L and M you found, look at Figure 15-3 to find the *Minimum Flue Area, Af,* and the *Inside Perimeter, P1.* Use Equation 3. Divide Af by P1, to find the *Hydraulic Radius, RH.*

$$R_H = \frac{A_F}{P_1}$$

Equation 3

Step 4— Now use Equation 4. Calculate the *Minimum Height, H,* needed for adequate draft. Equation 4 is the general equation for chimneys with rectangular flues.

Measure this height from the lintel above the fireplace opening. To find the height from the floor of the combustion chamber, just add the opening height. Of course, you still have to follow the building code. So make the chimney height equal to whichever is greater: the value you just calculated, or the building code requirements.

$$ H = \frac{K_T}{\left[\frac{5A_F}{A_o}\right]^2 - \left[\frac{0.0083}{R_H}\right]} $$

Equation 4

An Example:
Now you know how to calculate chimney height for proper draft. Let's actually figure one out.

Step 1— Look at Figure 14-4, Chapter 14. Let's choose fireplace opening dimensions of 30'' wide and 29'' high. And choose a flue liner size of 12'' by 12''. Use Equation 1 to calculate Ao:

$$ A_o = \frac{30'' \times 29''}{144''} = 6.04 \text{ sq. ft.} $$

Step 2— We know that K1 always equals 0.0 in residential applications. Now, let's decide that the damper throat area is twice the flue area. This makes K2 equal to 1.0. And let's decide that there's no rain cap, so K3 is equal to 0.0. Now, use Equation 2 to find Kt:

$$ K_T = 1.0 + 1.0 + 0.0 = 2.0 $$

Step 3— Now, look at Figure 15-3. Find Af: it's 0.56. Find P1: it's 3.0. Now, solve Equation 3 for *RH, Hydraulic Radius*:

$$ R_H = \frac{0.56 \text{ sq. ft.}}{3.0'} = 0.19' $$

Step 4— Finally, substitute all these values into Equation 4, and calculate the *Minimum Height, H* for adequate draft.

$$ H = \frac{2.0}{\left[\frac{5(0.56)}{6.04}\right]^2 - \left[\frac{0.0083}{0.19}\right]} = 11.68' $$

Designing Appliance Chimneys
An appliance chimney can vent one appliance, as in Figure 15-4A. Or it can vent two or more appliances, as in Figure 15-4B. When designing these chimneys, you'll usually know both the input rating and the system configuration. Typical design criteria are shown in Figures 15-5A and 15-5B. Again, building code requirements are the *minimum* heights for fire safety, so follow them.

Chimney Construction
This section describes the materials to use for residential chimneys. It also has some hints for good construction. Fireplace and appliance chimneys have identical functions, so their materials and constructions are similar. Building code construction requirements are identical.

Materials
Don't skimp on material quality in chimney work. Chimneys have to withstand not only the effects of the weather, but also the heat from the smoke and gases going through them. Below are the kinds of materials you should use.

Brick— At least part of the chimney will be exposed to the weather. For durability, use brick that conforms to ASTM C 216, Grade SW; or to ASTM C 62, Grade SW.

Mortar— To handle both weathering and temperature changes, use Type N portland cement-lime mortar for the chimney. Type S portland cement-lime mortar is better for chimneys subject to wind loads over 25 psf, or to earthquake loads. Use Type M portland cement-lime mortar where the chimney touches the ground.

The mortar used to bed the flue liner must stand up under high temperatures. Use fireclay mortar here. An acceptable substitute, though, is Type N portland cement-lime mortar.

Flue liners— Flue liners should conform to ASTM C 315. Inspect them thoroughly just before installation. Look for cracks and any other damage that might let smoke and flue gas leak out.

Flashing— Corrosion-resistant sheet metal flashing is specified by most codes. Use quality materials that will last: flashing replacement is expensive and troublesome.

Chimney caps— Use a prefabricated chimney cap. It's more durable than a cast-in-place cap. It's also easier to make water-resistant. If you do use a cast-in-place cap, make it the same shape as the prefabricated cap. The thickened sides and overhangs mean less water will leak in.

Rain caps— Rain caps vary. They can be sophisticated turbine-type metal caps or simple slabs set above the top of the flue liner. Before specifying a manufactured rain cap, get informa-tion from the manufacturer about its effect on the gas flow through the chimney. If the cap is metal, be sure it's corrosion-resistant.

Sealants— Sometimes caulking is used to correct or to hide poor workmanship. But it really is an important step in chimney-building. Always use a good grade of polysulfide, butyl, or silicone rubber sealant. Don't use oil-based sealants. Apply caulking carefully like a professional. Seal all joints. Proper priming and backing rope are a must with any kind of caulking.

Ties and reinforcement— Use corrosion-resistant metal ties in chimney construction. Reinforcing steel used in chimneys should conform to the following ASTM Standards:

- Welded wire: ASTM A 185

- Steel bar: ASTM A 615, ASTM A 616, or ASTM A 617

- Wire: ASTM A 82

Building Fireplace Chimneys

A chimney runs from the base of the first flue liner to the top of the last flue liner, or to any rain cap above it.

Attach single-wythe chimneys to the house. Use

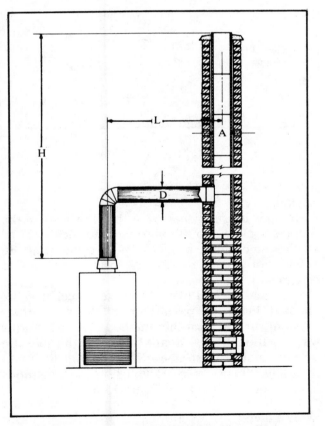

Masonry chimney serving a single appliance
Figure 15-4A

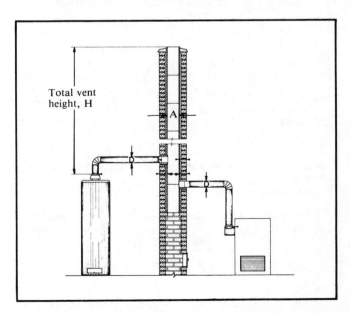

Masonry chimney serving two or more appliances
Figure 15-4B

Height H, (feet)	Lateral L, (feet)	Minimum internal area of chimney, A, (square feet)				
		0.26	0.35	0.47	0.66	0.92
6	2	130	180	247	400	580
	5	118	164	230	375	560
8	2	145	197	265	445	650
	5	133	182	246	422	638
	10	123	169	233	400	598
10	2	161	220	297	490	722
	5	147	203	276	465	710
	10	137	189	261	441	665
	15	125	175	246	421	634
15	2	178	249	335	560	840
	5	163	230	312	531	825
	10	151	214	294	504	774
	15	138	198	278	481	738
	20	128	184	261	459	706
20	2	200	273	374	625	950
	5	183	252	348	594	930
	10	170	235	330	562	875
	15	156	217	311	536	835
	20	144	202	292	510	800
30	2	215	302	420	715	1110
	5	196	279	391	680	1090
	10	182	260	370	644	1020
	15	168	240	349	615	975
	20	155	223	327	585	932
	30	NR*	182	281	544	865
50	2	250	350	475	810	1240
	5	228	321	442	770	1220
	10	212	301	420	728	1140
	15	195	278	395	695	1090
	20	180	258	370	660	1040
	30	NR*	NR*	318	610	970
		6	7	8	10	12
		Single-wall vent connector diameter, D, inches				

SI conversions: W = Btu/h × 0.293; m = ft × 0.3048; mm = in. × 25.4; mm² = in.² × 645
*Not recommended.

Capacity of masonry chimney serving a single appliance (maximum appliance input rating, thousands of Btu/h)
Figure 15-5A

corrosion-resistant metal ties, spaced no more than 24" o.c. If multi-wythe chimneys aren't masonry bonded, bond them together with metal wire ties.

Total vent height H, (feet)	Minimum internal area of chimney, A, (square feet)			
	0.26	0.35	0.54	0.79
6	102	142	245	NR
8	118	162	277	405
10	129	175	300	450
15	150	210	360	540
20	170	240	415	640
30	195	275	490	740
50	NR*	325	600	910

SI conversions: W = Btu/h × 0.293; m = ft × 0.3048; mm = in. × 25.4.
*Not recommended.

Capacity of masonry chimney serving two or more appliances (combined appliance input rating, thousands of Btu/h)
Figure 15-5B

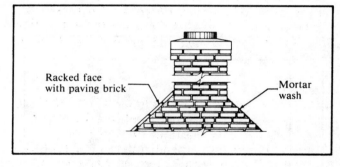

Racked face with paving brick — Mortar wash

Racking a chimney
Figure 15-6

Racking

A chimney is usually narrower than the body of the fireplace below. You'll often need to rack back to get the dimensions, or location, that you want. Be careful not to rack too far. No brick cores should be exposed.

Put a setting bed over the racked face, then set uncored brick or paving brick for a weather-resistant surface. Another method is to use mortar washes, though they might not be as durable. Be sure the mortar wash doesn't bridge over the rack. It should fill each step individually. Both methods of racking are shown in Figure 15-6.

Flue Liners

Support the first flue liner around its entire bottom edge with masonry. (The previous chapter,

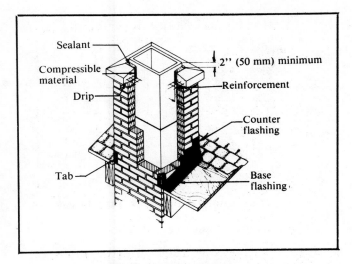

**Section and flashing
Figure 15-7**

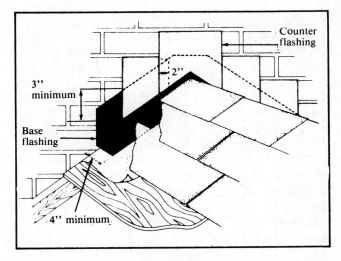

**Flashing detail
Figure 15-8**

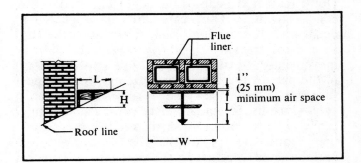

**Chimney cricket framing
Figure 15-9A**

Roof pitch ratio	H
1 to 1	½ of W
2 to 3	⅓ of W
1 to 2	¼ of W
1 to 3	⅙ of W
1 to 4	⅛ of W

**Cricket dimensions
Figure 15-9B**

Residential Fireplaces, gave details.) Bed the liner in mortar. Cut the joints flush. Smooth the joint on the inside of the flue. Parge the outsides of the joints with lots of mortar. Set each section of flue liner before you finish the previous section of chimney brickwork.

Flashing

Install base flashing and counter (step) flashing where the chimney and roof meet. See Figure 15-7. First, put the base flashing on the faces of the chimney. Put them perpendicular to the roofline, with tabs at each corner. Extend the flashing at least 4" up the chimney face, and along the roof.

Next, put counter flashing over the base flashing. Figure 15-8 details how to do it. Insert the flashing 3/4" to 1" into a mortar joint and mortar it solidly into the joint. Lap at least 3" of counter flashing over the base flashing. If you install the flashing in sections, lap the flashing higher up the roofline at least 2" over the flashing below. Seal all joints in the base flashing and the counter flashing thoroughly. Seal the underside of the flashing extending onto the roof.

Cricket

Use a cricket if the chimney doesn't intersect the ridgeline and if it measures over 30" parallel to the ridgeline. Figure 15-9A shows how to build a cricket. The size of the cricket is based on the chimney's measurement parallel to the ridgeline and on the roof pitch. Figure 15-9B gives dimensions.

Flash and counter-flash the intersection of the cricket and the chimney. Do it just like a normal chimney-roof intersection. Extend the flashing at the roofline to at least 4" under the roofing material.

Chimney Caps

Extend the flue liner at least 2" above the top of the chimney cap. See Figure 15-10. Thoroughly prime, back, and seal where the cap and flue liner meet. This makes it less likely to leak. If the cap doesn't overhang the chimney face, corbel the last two courses of the chimney brickwork out to form a drip. This means less water will run down the face of the chimney when it rains.

There are, as mentioned before, two kinds of chimney caps: prefabricated and cast-in-place. Prefabricated caps are best. Set them in place on a mortar bed. Leave a bond break between the brickwork and the setting bed. This allows the cap to move a little, without damaging the brickwork. Section B of Figure 15-10 shows a typical prefabricated cap.

Make cast-in-place caps match the shape and minimum dimensions shown in Figure 15-10 B. Don't feather the cap to the edge. This makes the edge too thin. Thin caps are more likely to deteriorate. Use enough reinforcement in the cap to control cracking from shrinkage and thermal movements. You might need extra reinforcement in the part of the cap that overhangs the chimney face.

Use flashing for cast-in-place caps. This flashing can also be the bond break material. Waterproofing requirements aren't the same as for prefabricated caps since the concrete is sure to shrink as it cures. Section A of Figure 15-10 shows one way to form a cast-in-place chimney cap. Extend flashing under and at the back of the cap between the flue liner and cap. Seal the flashing joint to the flue liner and caulk between flashing and cap as shown in A.

Building Appliance Chimneys

Appliance chimneys are similar to fireplace chimneys. Only three parts are different: the foundation, the cleanout door and the thimble. Suggestions for building these parts follow. Otherwise, follow the directions already given for building fireplace chimneys.

Foundation

The foundation supports the chimney. It must be big enough to support all likely loads.

Consider the soil condition and type when you design the foundation. Undisturbed or well-compacted soil will usually be adequate. Other soil conditions need more analysis. Figure 13-1,

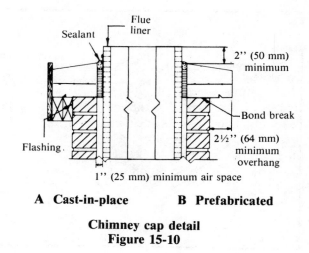

A Cast-in-place   B Prefabricated

Chimney cap detail
Figure 15-10

Chapter 13, gives types of soils and their design properties. Group III and IV soils are poor or unsatisfactory for heavy chimney loads.

Building codes usually require the foundation to be at least 12" thick and to extend at least 6" past each face of the masonry bearing on it. It should also penetrate the frost line. This makes it less likely that the foundation will heave when ground is frozen.

Cleanout Door

Appliances burning natural gas don't need a cleanout door. Other fuels, though, produce ashes. These ashes must periodically be cleaned out from the bottom of the chimney. Your main concern in sizing and locating the cleanout door is how easy it

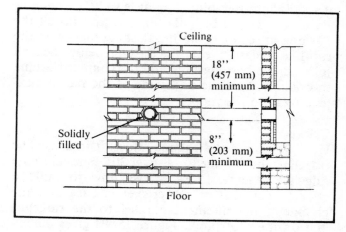

Thimble detail
Figure 15-11

Residential Chimneys

is to use. It's nice to have the cleanout door inside the house. Use a ferrous metal cleanout door, and make it as airtight possible.

Thimble

The thimble is the lined opening through the chimney wall. It receives the smoke pipe connector. Look at Figure 15-11 for an example.

Build the thimble into the chimney where the pipe connector enters it. Set the thimble flush with the inside face of the flue liner. Set it at least 18" below the ceiling. Be sure at least 8" of flue liner is below the thimble's lowest point. Make it airtight with either boiler putty or asbestos cement.

Safety Is the Key

Fire safety is the most important consideration on every chimney you install. Build them with quality materials and quality workmanship. Remember, not every mason is a chimney master. It's the general contractor's responsibility to deliver the safe and durable chimney the owner bargained for.

377

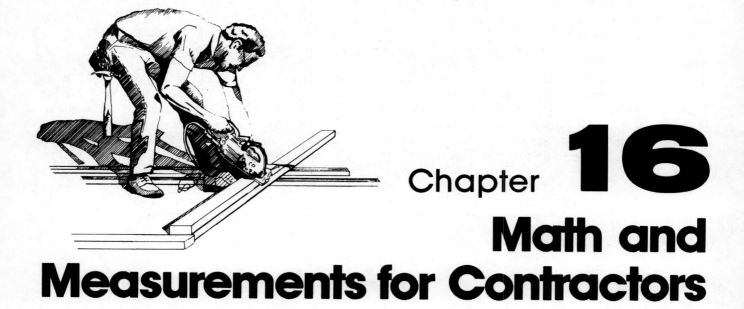

Chapter 16

Math and Measurements for Contractors

All construction contractors make calculations. Math is a tool of the trade. Carpenters use math to find lengths of lumber. The contractor uses math to find the quantity of materials and the cost of construction. A working knowledge of practical construction math distinguishes a craftsman from a common laborer.

Most people have forgotten a good deal of the math they learned in school — probably because they don't use it regularly. This chapter is a reference and a review. It covers most of the math needed in construction.

Everyone working in construction should know how to use the math in this chapter. If you have trouble with the concepts here, spend an hour or two to review the key points. It will be time well spent. Knowledge of practical construction math will save you time, trouble, mistakes and money. A builder who knows practical math will have no trouble figuring costs, laying out the building, calculating cabinet widths and heights, laying out the stairways, and locating correct positions for doors and windows.

I hope that a lot of the math review that follows seems too easy. If so, congratulations. If not, here's your chance to pick up some important skills. Remember, it's very easy to make a mistake. And it can be a real shock when a trivial math error costs you the entire profit in a job. It's worth your time to review these pages.

Using Feet and Inches
Adding Feet and Inches
Obviously, every addition problem must involve like numbers. But feet and inches aren't alike. The following example shows how to add feet and inches so the result is correct, in feet and inches: Add 23'7'', 14'3'', and 8'6'':

$$\begin{array}{r} 23'7'' \\ 14'3'' \\ \underline{8'6''} \\ 46'4'' \end{array}$$

Adding 7'' plus 3'' plus 6'' gives 16''. But, since there are only 12'' in a foot, you have 1'4''. Write the 4'', and add the 1' to the next column.

To find the total length of something when you know the lengths of its parts simply add them all together. Look at Figure 16-1. It shows some windows in a brick wall. And it shows the widths of the windows and the widths of the wall between the windows. To find the total width of the wall, add

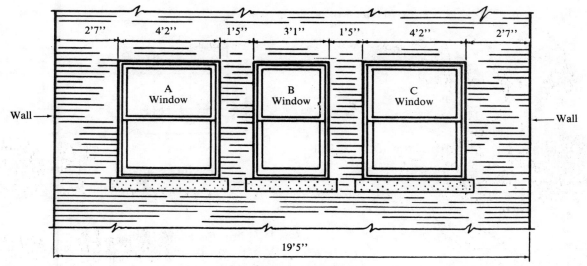

Finding total width
Figure 16-1

all the separate widths together as shown below. This is simple math but if you add wrong, you'll be making another trip to the lumber yard to replace the lumber you spoiled.

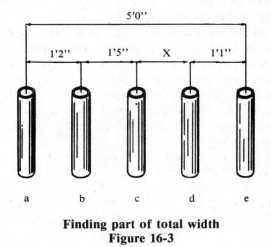

Subtracting feet and inches
Figure 16-2

2'7"
4'2"
1'5"
3'1"
1'5"
4'2"
2'7"
―――
19'5"

Subtracting Feet and Inches
There are three floor joists in Figure 16-2 (*A, B,* and *C*). You want to find the width, center to center, of *B-C*. This is *X*. The total width, center to center, of *A-C* is 4'6''. And the width of *A-B*, center to center, is 3'6''. Now, just subtract 3'6'' from 4'6'': this gives you 1'0'' for *X,* the width, center to center, of *B-C*.

Figure 16-3 shows five parallel pipes: *a, b, c, d* and *e.* The total distance, center to center, between the outer pipes *(a-e)* is 5'0''. You know all the spacing between pipes, except the space marked *X, c-d.* To find *X* , first add all the known spacings between pipes, then subtract this total from the

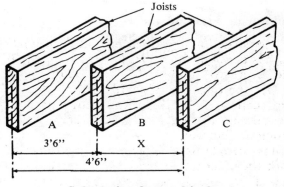

Finding part of total width
Figure 16-3

overall dimension, 5'0''. The sum of the known spacings is:

$$
\begin{array}{r}
1'2'' \\
1'5'' \\
\underline{1'1''} \\
3'8''
\end{array}
$$

Now, subtract this 3'8'' from the total dimension:

$$
\begin{array}{r}
4'12'' \\
\underline{-3'\ 8''} \\
1'\ 4''\ \text{(spacing X)}
\end{array}
$$

Here's the trick. You need something to subtract the 8'' from. So, take 12'' from the 5'0'', and put it in the inch column. This gives you 4'12'', from which you can subtract the 3'8''.

Multiplying Feet and Inches

When you multiply dimensions containing both feet and inches, multiply the feet and inches separately, just as you add and subtract them separately. Say you have a space of 3'2'' measured off four times along the length of a board, and you want to find the total length of spaces marked off:

$$
\begin{array}{r}
3'2'' \\
\underline{\times\ 4} \\
12'8''
\end{array}
$$

If the inches in the answer are more than 12, change them to feet and inches. Figure 16-4 is an example: the seven circular holes in the board have six spaces, each 1'3'' o.c. You want to find *X*, the dimension, center to center, between the end holes.

$$
\begin{array}{r}
1'\ 3'' \\
\underline{\times\ 6} \\
6'18''
\end{array}
$$

Here you see that 3'' multiplied by 6 gives 18''. Since 18'' is equal to 1'6'', drop the 18'' in the result and add 1'6'' to the 6' of the result. The final answer is 7'6''.

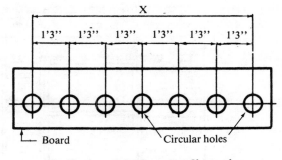

Finding center-to-center dimension
Figure 16-4

Dividing Feet and Inches

When you divide dimensions containing both feet and inches, once again, divide the feet and inches separately. For example, divide 12'6'' by 3:

$$
3\overline{\smash{\big)}12'6''} \quad = 4'2''
$$

First, divide the 12', to get 4'. Then, divide the 6'', to get 2''. The result is 4'2''. This is a simple example. Often, the number you're dividing by won't go evenly into the number you want to divide. There's an easy way to handle this problem, though. Just change the whole dimension to inches, then divide it. You can see at once that 7'6'' can't be equally divided by 3. So, first convert the dimension to inches: multiply the feet by 12'', then add the inches. Now, you're dividing 90'' by 3. It's clear that the answer is 30'', or 2'6''.

This method avoids the use of fractions. And it's so simple that you can often do it mentally. For example, divide 3'6'' by 6. Here 3' equals 36'', which, when added to the 6'' of the dimension, gives 42'' total. You now have: 42'' divided by 6, which equals 7''.

The rule is: Reduce each length to inches, then divide.

Using Fractions

First, some terms you should know. Consider the fractions 2/3 and 3/4. The top (in this book, first) number is the *numerator* (the 2 and the 3). And the bottom number is the *denominator* (3 and 4).

Fractions can be either *proper* or *improper*. The

difference is easy to see: a proper fraction is always less than one. An improper fraction is always equal to, or more than, one. Some improper fractions: 18/16, 10/4, 8/7. We use improper fractions when changing a *mixed number*, like 1¼, to a fraction for calculation purposes. Mixed numbers and improper fractions are closely related. If you know one, you can easily figure out the other.

To change mixed numbers to improper fractions, first multiply the whole number by the denominator. Next, add the numerator to this result. Finally, write the whole amount over the denominator. Here's an example of changing 3⅞ to an improper fraction:

$$3 \times 8 = 24 \text{ and } 24 + 7 = 31, \text{ so } 3\text{-}7/8 = 31/8.$$

To change improper fractions to mixed numbers, first divide by the denominator. The number of times it will go is the whole-number part of the mixed number. Then, put any remainder over the denominator, to get the fraction part of the mixed number. For example, change 20/16 to a mixed number:

Dividing 20 by 16 gives a quotient of 1 and a remainder of 4.

Thus $\dfrac{20}{16}$ = 1-4/16 = 1-1/4.

Reducing fractions— A fraction is "reduced to its lowest terms" when you can no longer divide the numerator and denominator by the same number. Thus, 7/8 is already reduced to its lowest terms, since 7 and 8 can't be divided evenly by the same number. But 2/4 is not reduced to its lowest terms, since both numbers (2 and 4) can be divided by 2. If you divide both numbers by 2, as you should, you get 1/2. When you write fractions, use reduced form. Reducing to the lowest terms doesn't change the value of the fraction, of course. Let's change 12/16 to its lowest terms:

$$\frac{12 \div 2 = 6}{16 \div 2 = 8} \quad \frac{6 \div 2 = 3}{8 \div 2 = 4} \text{ or } \frac{12 \div 4 = 3}{16 \div 4 = 4}$$

Changing fractions to higher terms— In many problems, you'll want to change fractions to higher terms. You may have several fractions of different values, like 3/8, 5/16, and 3/16. You need to get them all into 16ths before you can do anything with them. Do it like this: multiply both the numerator and denominator by the same number. To change 3/8 to sixteenths:

$$\frac{3 \times 2 = 6}{8 \times 2 = 16}$$

How do you know which number to multiply by? It's easy: see how many times the denominator of the fraction you have goes into the denominator of the fraction you want to change it to (8 goes into 16 twice, so multiply by 2).

Adding Fractions

Suppose you want to find the combined thickness of two boards. One of them is 3/4" thick, and the other is 7/8" thick. If you could measure them, you'd see at once that the combined thickness is 1⅝". But you might not have the boards handy, so you figure it on paper. First, find a common denominator for the fractions. Then add the fractions. Let's add 1/4" and 7/8":

$$\begin{aligned} 1/4" \times 2/2 &= 2/8" \\ 7/8" &= 7/8" \\ \hline 9/8" &= 1\text{-}1/8" \end{aligned}$$

Change 1/4" to eighths by multiplying both parts by 2. The other fraction is already in eighths, so now add the numerators and put the answer over the common denominator: 9/8. Change 9/8 to the mixed number 1⅛".

Finding the Lowest Common Denominator (LCD)

As you can see from the previous example, you can't add fractions unless their denominators are alike. (The same applies to subtracting fractions.) Change fractions with different denominators to new fractions with the same common denominator, called the Lowest Common Denominator (LCD). It's the smallest denominator you can use for all the fractions. For example, add 4¼, 3/4 and 7/8.

$$
\begin{array}{rcl}
4\text{-}1/4 & = & 4\text{-}\ 2/8 \\
3/4 & = & 6/8 \\
7/8 & = & 7/8 \\
\hline
& & 4\text{-}15/8 = 5\text{-}7/8
\end{array}
$$

The LCD is often obvious when you look at the fractions. In this problem, it's easy to see that the LCD equals 8. So, change all the fractions. Then, add all the numerators and put them over the LCD: 15/8. Now, change this to a mixed number, 1⅞. Finally, add 1⅞ to 4 for the answer.

If you're using fractions of an inch — halves, quarters, eighths, sixteenths, thirty-seconds, and sixty-fourths — it's easy to find the LCD. It's the largest denominator that occurs in any group of fractions of an inch. So, to find the LCD of 7/8, 9/16, 3/4, 11/32, 1/4, 1/2, and 7/16:

$$
\begin{array}{rclcl}
7/8 & x & 4/4 & = & 28/32 \\
9/16 & x & 2/2 & = & 18/32 \\
3/4 & x & 8/8 & = & 24/32 \\
11/32 & x & & = & 11/32 \\
1/4 & x & 8/8 & = & 8/32 \\
1/2 & x & 16/16 & = & 16/32 \\
7/16 & x & 2/2 & = & 14/32
\end{array}
$$

You can see that the largest denominator is 32. Now, take each fraction in turn, and mentally divide its denominator into 32, to see what number to multiply it by. Finally, multiply both parts by this number.

Finding the LCD when it's not obvious— Sometimes you need to find the LCD of several fractions but can't do it by just looking at them. In this case, first put the denominators in a row, separated by dashes. Then write, on the left of this row, the smallest number (except 1) that will divide into two or more of the denominators. (Look at the first line in the example that follows.) Draw a line under the row. Do the division, writing the answers below the line. Bring down unchanged any number you can't divide evenly (like the 15).

Treat the second row of figures just like the first. Repeat these steps until you can only divide one of the numbers below the line. Finally, multiply the

number at the left of each row, and all the numbers in the last row. Now you have the LCD. Look at this example of finding the LCD of 4/15, 3/20, 7/30, and 9/10:

$$
\begin{array}{l}
\begin{array}{rcccc}
2) & 15 & - 20 & - 30 & - 10 \\
\hline
3) & 15 & - 10 & - 15 & - 5 \\
\hline
5) & 5 & - 10 & - 5 & - 5 \\
\hline
& 1 & - 2 & - 1 & - 1
\end{array} \\
\text{LCD} = 2 \times 3 \times 5 \times 1 \times 2 \times 1 \times 1 = 60
\end{array}
$$

Subtracting Fractions

To subtract fractions, first change them to their LCD, just as when adding fractions. For example, subtract 7¾ from 15⁹⁄₁₆:

$$
\begin{array}{rclclcl}
15\text{-}9/16 & = & 15\text{-}\ 9/16 & = & 14\text{-}25/16 \\
-7\text{-}3/4 & = & -7\text{-}12/16 & = & -7\text{-}12/16 \\
\hline
& & & & 7\text{-}13/16
\end{array}
$$

The LCD is obviously 16. So 7¾ is changed to 7-12/16 and subtracted from 15⁹⁄₁₆. But there's a problem: 12/16 can't be subtracted from 9/16. Fix it by borrowing 1 (16/16) from the 15. Add it to the 9/16, making 25/16. Last, subtract 7 from 14 (instead of 15); for an answer of 7¹³⁄₁₆.

Multiplying Fractions

Multiplying a fraction by a whole number— Just multiply the numerator by the whole number, then divide this answer by the denominator. For example, multiply 3/8 by 2:

$$
\text{This is written } \frac{3}{8} \text{ x } 2
$$

$$
\frac{3}{8} \text{ x } 2 = \frac{3 \times 2}{8} = \frac{6}{8} = \frac{3}{4}
$$

Multiplying two fractions together— It's easy. Just multiply the numerators together for the new numerator. And multiply the denominators together for the new denominator. Let's say we're multiplying 3/4 by 7/8:

$$
\frac{3}{4} \text{ x } \frac{7}{8} = \frac{21}{32}
$$

Multiplying several numbers with fractions— Now, imagine that you want to multiply 15 by 2¼ by 2/3:

$$\frac{15}{1} \times \frac{\overset{3}{\cancel{9}}}{\underset{2}{\cancel{4}}} \times \frac{\overset{1}{\cancel{2}}}{\underset{1}{\cancel{3}}} = \frac{45}{2} = 22\text{-}1/2$$

First, change the mixed number to an improper fraction (2¼ equals 9/4). Next, find the LCD. The cancellation method is quick: By trial, we find that 2 goes evenly into both 2 and 4. This leaves 1 and 2. Cross out the 2 and 4, and write the 1 and 2 by them, as in the example. Again by trial, we find that 3 goes evenly into both 3 and 9. This leaves 1 and 3. After canceling, the numbers to multiply in the numerators are 15 x 3 x 1 = 45. And the numbers left in the denominators are 1 x 2 x 1 = 2.

Multiplying feet and inches containing fractions— To multiply mixed numbers with feet and inches (when figuring areas, for instance) first change all the numbers to inches. Then change the inches to improper fractions. Finally, multiply them in the normal way. Let's say we're multiplying 3'2¾'' by 5'7½'':

```
  3'2-3/4"        5'7-1/2"
x12"            x12"
-------         -------
  36"             60"
+ 2-3/4"        + 7-1/2"
-------         -------
 38-3/4"         67-1/2"
```

38-3/4" = 155/4" 67-1/2" = 135/2"

38-3/4" x 67-1/2" = 155/4 x 135/2 =

$$\frac{20925}{8} = 2615\text{-}5/8 \text{ square inches}$$

Dividing Fractions

There's one rule you need to remember when dividing one fraction by another:

Invert (turn upside down) the number you're dividing by, then proceed as in multiplication. Say you want to divide 5¼ by 3/4:

$$\frac{21}{4} \div \frac{3}{4} = \frac{\overset{7}{\cancel{21}}}{\underset{1}{\cancel{4}}} \times \frac{\overset{1}{\cancel{4}}}{\underset{1}{\cancel{3}}} = \frac{7}{1} = 7$$

Change the mixed number, 5¼, to the fraction 21/4. Thus, 21/4 divided by 3/4 equals 21/4 times 4/3. By cancellation you get 7/1, or 7.

Dividing whole numbers by fractions— To divide a whole number (4 or 6) by a fraction, first convert the whole number into an improper fraction with a denominator of 1. Thus, 4 becomes 4/1; 6 becomes 6/1. Next, invert the fraction. Finally, multiply the numerators and denominators as usual. For example, divide 8 by 3/4:

$$8 \div \frac{3}{4} = \frac{8}{1} \div \frac{3}{4} = \frac{8}{1} \times \frac{4}{3} = 10\text{-}2/3$$

Change 8 to 8/1. Then, 8/1 divided by 3/4 becomes 8/1 times 4/3, which gives the answer 32/3, or 10⅔.

Dividing a mixed number by a mixed number— First, change both numbers to improper fractions. Then, invert the number you're dividing by, and multiply the numerators and denominators as before. As an example, divide 2½ by 1⅓. 2½ changed to an improper fraction is 5/2. 1⅓ changed to an improper fraction is 4/3. Inverted, it's 3/4. Therefore, 5/2 times 3/4 equals 15/8, or 1⅞.

Dividing feet and inches containing fractions— Let's divide 11'2¾'' by 3'2½'': First change each number to inches. Then change them to fractions. Invert the second number, and multiply:

```
  11'2¾"          3'2½"
x12"            x12"
-------         -------
  22              36"
  11            + 2½"
-------         -------
 132"            38½"
+ 2¾"
-------
134¾"
```

$$134\tfrac{3}{4}'' = \frac{134 \times 4 + 3}{4} = \frac{539}{4}$$

$$38\tfrac{1}{2} = \frac{38 \times 2 + 1}{2} = \frac{77}{2}$$

$$\frac{539}{4} \div \frac{77}{2} = \frac{\overset{}{\cancel{539}}}{\underset{2}{\cancel{4}}} \times \frac{\overset{1}{\cancel{2}}}{\underset{1}{\cancel{77}}} = \frac{7}{2} = 3\tfrac{1}{2}$$

Using Decimals

You can use decimal fractions to simplify your calculations. Common fractions are important to the craftsman, architect, or builder, but decimals make estimating and engineering calculations much faster.

As you move to the right from the decimal point, the value of each figure is 1/10 that of the figure to its left. It's just like U.S. money: a penny is worth 1/10 of a dime.

Location of the decimal point— The decimal point's location is important. A decimal fraction is multiplied by 10 for each place the decimal point moves to the right. And it's divided by 10 for each place the decimal point moves to the left. For example, 1.75 becomes 17.5 by moving the point one place toward the right (because 1.75 x 10 equals 17.5) And it becomes 0.175, when the point is moved one place toward the left (1.75 divided by 10 equals 0.175). *Adding* zeros after a decimal doesn't affect its value. So 0.5 is the same as 0.50, or 0.500, since 5/10 is the same as 50/100, or 500/1000, when reduced.

You may find it helpful to write decimal fractions with a zero to the left of the decimal point, to make clear where the decimal point is located.

Adding and Subtracting Decimals

The only thing you need to remember when adding and subtracting decimals is to always write them so the decimal points are lined up.

Multiplying Decimals

When multiplying decimals, ignore the decimal point until you've done the multiplication. Then point off, from right to left in the answer, as many decimal places as there are in all the numbers multiplied. Let's multiply 8.57 by 2.8:

$$
\begin{array}{r}
8.57 \\
\times\ 2.8 \\
\hline
6856 \\
1714 \\
\hline
23.996
\end{array}
$$

In this example, there are two decimal places in the first number and one in the second, giving us three decimal places in the answer.

Now, multiply 0.0094 by 4.7:

$$
\begin{array}{r}
.0094\ \text{(4 decimal places)} \\
\times\ 4.7\ \text{(1 decimal place)} \\
\hline
658 \\
376 \\
\hline
.04418\ \text{(5 decimal places)}
\end{array}
$$

In this example, we have to make the answer contain five decimal places. So we put a zero between the decimal point and the 4.

One application of multiplying decimals is in estimating lumber. Framing lumber is sold by the 1,000 board feet, so 1,000 is the unit used when estimating lumber. Thus, 14,895 board feet equals 14.895 MBF (thousand board feet).

For example, suppose we have 150 pieces of 16-foot 2 x 8, which contains 3,200 board feet. This is the same as

$$
3\frac{200}{1000} = 3\frac{2}{\cdot 10} = 3.2
$$

thousandths, when stated in decimals. Find the cost of this lumber, at $95.00 per 1,000 BF, by multiplying: $95.00 by 3.2. This equals $304.00.

Here's the rule for finding how much a given number of board feet of lumber costs: *Multiply the total number of board feet by the cost per 1,000, and move the decimal place three places to the left.*

In ordinary construction calculations, three decimal places are plenty, and usually two are enough. Therefore, if your multiplication answer has four or five decimal places, just drop all but two. There's no practical reason to use more.

Dividing Decimals

Dividing decimals is as easy as multiplying them: just forget the decimal point until you've done the division. Then, subtract the number of decimal places in the *divisor* (number you divided by) from the decimal places in the *dividend* (number that was divided). What's left is the number of decimal places in the answer:

$$
\begin{array}{r}
3.23 \\
25.4\ )\overline{82.042} \\
76\ 2 \\
\hline
5\ 84 \\
5\ 08 \\
\hline
762 \\
762 \\
\hline\hline
\end{array}
$$

In the first example, there is one decimal place in the divisor, and three in the dividend. Therefore, there are two decimal places in the answer.

Let's divide 15 by 0.625:

$$
\begin{array}{r}
24 \\
.625\,)\overline{15.000} \\
\underline{12\ 50} \\
2\ 500 \\
\underline{2\ 500}
\end{array}
$$

If there are more decimal places in the divisor than in the dividend, add zeros to the decimal part of the dividend. This does not change the value. In this example, we had to add enough zeros to make three decimal places in the dividend. Since there are also three decimal places in the divisor, the answer has no decimal places.

Changing fractions to decimals— Change a fraction to a decimal by dividing the numerator by the denominator. Add enough zeroes after the decimal in the numerator to give the number of decimal places you want in the answer. Change 3/8 to a decimal:

$$
\begin{array}{r}
.375 \\
8\,)\overline{3.000} \\
\underline{2\ 4} \\
60 \\
\underline{56} \\
40 \\
\underline{40}
\end{array}
$$

Changing decimals to fractions— Change a decimal to a fraction by first writing it as a fraction, then reducing the fraction to its lowest terms (or to a mixed number) as shown here:

$$.05 = \frac{5}{100} = \frac{1}{20}$$

$$.625 = \frac{625}{1000} = \frac{5}{8}$$

$$1.25 = 1\frac{25}{100} = 1\text{-}1/4$$

Changing inches to decimal parts of a foot— You'll often need to change inches to decimal parts of a foot. It makes calculating excavations, cubic yards of concrete, or the square feet in floor, walls, and roofs easier. Consider a room 12'5'' wide. We want to change the inches to decimals of a foot to make calculations easier. Do it this way: First, 12'5'' equals 12 and 5/12 feet. Now, change the fraction 5/12 to a decimal:

$$
\begin{array}{r}
.4166 \\
12\,)\overline{5.0000} \\
\underline{4\ 8} \\
20 \\
\underline{12} \\
80 \\
\underline{72} \\
80
\end{array}
$$

Now you see that 12'5'' equals 12.4166 feet. Round it off to 12.417 feet.

You don't really need to calculate, though. Take a look at the conversion chart in Figure 16-5. To find the decimal equivalent of 1/2'' (0.0147) look in the column headed 0'', opposite 1/2 in the "Inch" column. Or, find that 2¼'' equals 0.1875 feet by looking in the 2'' column, opposite 1/4 in the "inch" column.

You can also use Figure 16-5 to change decimal parts of a foot into inches. For example, you've calculated that a rafter has to be 22.317 feet long. Look for 0.3177 in the table. It's in the column headed 3'', and opposite 13/16 in the "inch" column. You'll use 0.317, not 0.3177, for practical work, but the values are so close that it's no problem. The rafter's length is 22'3¹³⁄₁₆''.

Using Percentages

Percentage is simply a decimal fraction with a denominator of 100. But we say "percent," instead of writing the 100. When we say "five percent," we mean 5/100, or 0.05. They're all the same.

Changing percents to decimals— It's easy to change percents to decimals. Write the number, leaving off the percent sign. Then move the decimal point two places to the *left*. Examples:

INCH	0"	1"	2"	3"	4"	5"	6"	7"	8"	9"	10"	11"
0	0	.0833	.1667	.2500	.3333	.4167	.5000	.5833	.6667	.7500	.8333	.9167
1/16	.0052	.0885	.1719	.2552	.3385	.4219	.5052	.5885	.6719	.7552	.8385	.9219
1/8	.0104	.0937	.1771	.2604	.3437	.4271	.5104	.5937	.6771	.7604	.8437	.9271
3/16	.0156	.0990	.1823	.2656	.3490	.4323	.5156	.5990	.6823	.7656	.8490	.9323
1/4	.0208	.1042	.1875	.2708	.3542	.4375	.5208	.6042	.6875	.7708	.8542	.9375
5/16	.0260	.1094	.1927	.2760	.3594	.4427	.5260	.6094	.6927	.7760	.8594	.9427
3/8	.0312	.1146	.1979	.2812	.3646	.4479	.5312	.6146	.6979	.7812	.8646	.9479
7/16	.0365	.1198	.2031	.2865	.3698	.4531	.5365	.6198	.7031	.7865	.8698	.9531
1/2	.0417	.1250	.2083	.2917	.3750	.4583	.5417	.6250	.7083	.7917	.8750	.9583
9/16	.0469	.1302	.2135	.2969	.3802	.4635	.5469	.6302	.7135	.7969	.8802	.9635
5/8	.0521	.1354	.2188	.3021	.3854	.4688	.5521	.6354	.7188	.8021	.8854	.9688
11/16	.0573	.1406	.2240	.3073	.3906	.4740	.5573	.6406	.7240	.8073	.8906	.9740
3/4	.0625	.1458	.2292	.3125	.3958	.4792	.5625	.6458	.7292	.8125	.8958	.9792
13/16	.0677	.1510	.2344	.3177	.4010	.4844	.5677	.6510	.7344	.8177	.9010	.9844
7/8	.0729	.1562	.2396	.3229	.4062	.4896	.5729	.6562	.7396	.8229	.9062	.9896
15/16	.0781	.1615	.2448	.3281	.4115	.4948	.5781	.6615	.7448	.8281	.9115	.9948
1												1.0000

Decimals of a foot/inches conversion chart
Figure 16-5

3%	=	.03
85%	=	.85
2½%	=	.02½ or .025
125%	=	1.25
.2%	=	.002

Changing decimals to percents— This is the reverse of changing percents to decimals. Write the number. Then move the decimal two places to the *right*. If it's now at the end of the number, don't write it. Last, write the percent sign. Examples:

.07	=	7%
.15	=	15%
1.35	=	135%
.5	=	50%

Changing fractions to percents— First, divide the numerator by the denominator, for the decimal fraction. Find the answer to two decimal places. Next, rewrite the answer, omitting the decimal point. Last, write the percent sign. For example, change 18/25 to percent:

$$\begin{array}{r} .72 \\ 25\overline{)18.00} \\ 17\ 5 \\ \hline 50 \\ 50 \\ \hline \end{array}$$

First, divide 18 by 25 to find the decimal fraction, 0.72. Now change the decimal to a percent:

move the decimal point two places to the right, and add the percent sign. The answer is 72%.

Common Uses of Percent

In construction work: You add a certain amount to the estimated cost of a building, for profit. You also have many items of overhead (rent, taxes, bookkeeping costs, phone calls, and more) which must be paid for, although they're not directly related to the cost of a particular building. So, you add a certain percent to the estimated cost of a building to cover these items.

In shop work: The main use of percentage is to describe the amount of loss or gain; or to state the amount that's used or unused, good or bad, finished or unfinished. Sometimes you hear a statement like: "Two out of five of these tiles are bad." In this case, we can talk in terms of percentages by saying: "Forty percent of these tiles are bad." This is because "two out of five" is the same as:

$$2/5 = \frac{2 \times 20}{5 \times 20} = \frac{40}{100} = .40 = 40\%$$

First, find the number to multiply by that will make the denominator equal 100. In this example, it's 20. Next, multiply both numbers by it. Turn the resulting fraction into a decimal. Last, move the decimal point two places to the right, and add the percent sign.

In specifications: There are specs for the portions

of various parts needed to make a mixture. Proportions of solders for sheet metal work; or of brasses, bronzes, and other alloys, are given by percent. For example, brass is about 65% copper and 35% zinc. Then, in 100 pounds of brass, there are 65 pounds of copper and 35 pounds of zinc. Suppose, however, you want 15 pounds of brass, not 100 pounds. Then, the amount of copper and zinc needed is:

.65 x 15 = 9.75 or 9-3/4 lb. **copper needed**
.35 x 15 = 5.25 or 5-1/4 lb. **zinc needed**

In figuring discounts: It's customary on millwork, hardware, and many other building materials, for dealers to give contractors a certain percentage off the list price. This lets the dealer print catalog prices that will be usable, even if the prices of the materials change from time to time. The price to the contractor is changed by giving a different discount. As an example, find the total cost of 1,500 pounds of 16d nails, at 65 cents per pound list price, with a 12% discount:

1500 x   .65 = $975.00 **List price of nails**
$975.00 x   .12 = $117.00 **Discount on nails**
$975.00 − $117.00 = $858.00 **Cost of nails**

In describing grades or slopes: These are usually given in percent. That is, in the percentage of the rise in relation to the length of the distance traveled. So, a rise of 1' in 100' of horizontal distance is a 1% grade. A rise of 4' in 100' is a 4% grade. Look at Figure 16-6. The top of a 1¼% grade is 56.2' higher than the bottom. How long must the horizontal distance (L) be to obtain this grade?

Since, for every 100' of horizontal distance the rise is 1¼', we can get the number of 100' lengths corresponding to the total rise of 56.2' by finding the number of times 56.2 is larger than 1.25. There are then, $\frac{56.2}{1.25}$ = 44.96 hundred foot lengths; or the total horizontal length "L" is 44.96 x 100 = 4496'.

Decimals in Surveying

A surveyor uses a steel tape divided into feet and decimal parts of a foot. In surveying, decimal parts of a foot are used instead of inches and fractions of

an inch. For example, 50'9'' is given as 50.75 feet, since 9'' is 9/12, or 3/4, or 0.75 foot. Likewise, in finding elevations, a graduated rod, divided into feet and decimal parts of a foot, is used. These measurements hardly ever need to be closer than thousandths of a foot. More often, the nearest hundredth of a foot is accurate enough.

Changing decimals of a foot to inches— After construction measurements are laid out, you'll often need to change the decimals of a foot to inches. Just multiply the decimal part by 12 (inches in a foot). For example, change 125.75 feet into feet and inches:

First multiply the decimal .75 by 12; 0.75 x 12 = 9''. Now add this to the feet: 125' + 9'' = 125'9''.

Here's a trick: *Remember that 0.01' equals about 1/8''.* By keeping this in mind, you can easily change decimals of a foot to inches, and vice versa. For example, 0.04' is 4 x 1/8'', which equals 4/8, or 1/2''. And 0.1' is 10 x 1/8'', which equals 10/8, or 1¼''.

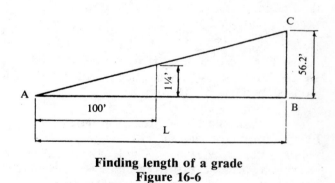

Finding length of a grade
Figure 16-6

Decimals in Elevations

Surveyors refer elevations to some base line, or point of zero elevation. This may be located at the nearest manhole, on a curb near the building, or in part of a nearby building. The plan of the grounds and building shows the height of the grade from the base line.

The blueprints of large structures often show elevations with a plus sign (+) or a minus sign (−) in front of the figures. The plus sign means the elevation is above the base line. The minus sign means that it's below the base line. These heights are given in feet and decimals.

Look at the foundation shown in Figure 16-7. The elevation is marked + 3.48. This means the top is 3.48', or 3'5¾'', above the base line. Figure 16-7 also shows that grade is 9'' below the top of the foundation. This 9'' is 9/12', or 3/4', which equals 0.75'. The elevation of the grade line, therefore, is marked 2.73' (which equals 3.48 minus 0.75).

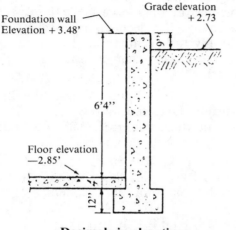

Decimals in elevations
Figure 16-7

Similarly, the elevation of the top of the basement floor is marked − 2.85'. This shows the number of feet below the base line. If we add 3.48' to 2.85', we get 6.33', or 6'4''. This checks with the dimensions shown on the drawing.

Construction Measurements

Every contractor works with lengths, areas, or volumes nearly every day. The length of rafters to be cut; the area of roof to be shingled; the surface area to be painted; the volume of earth to be excavated. You must be able to calculate these items quickly and accurately. Of course, you now use a hand-held calculator instead of figuring with pencil and paper. So this section is a review of methods for solving common problems. It's not, however, a complete textbook. You won't, for example, learn how to calculate the volume of a pyramid — that's something for which you'd use the reference tables in a mathematics book.

Angles

Angles are formed wherever two lines meet. The size of an angle depends on the size of the opening, not on the length of the sides.

Angles and arcs are measured in degrees (one degree is one 360th of a circle). Finer measurements are minutes (60ths of a degree, shown as: '), and seconds (60ths of a minute, shown as: ''). Use a protractor, like the one in Figure 16-8, to lay out or to measure angles in the shop or on paper.

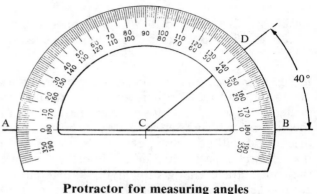

Protractor for measuring angles
Figure 16-8

Right angles— A right angle measures 90 degrees. It's formed by two lines that are perpendicular to each other. In construction, you'll usually use the steel square to lay out right angles. (An acute angle measures less than 90 degrees; and an obtuse angle measures more than 90 degrees.)

In structural steel work, and on roof elevations, angles are given in terms of the rise in twelve inches. To find the angle, first lay out a line 12'' long. Then mark off the rise, at right angles to it. Consider a rise of 5⅛'', as shown in Figure 16-9.

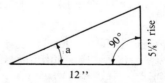

Finding the angle when the rise is known
Figure 16-9

There's the base of 12'', and the rise of 5⅛''. Connect the ends of these lines, then find the angle (a) by measuring with a protractor. Use this easy method to lay out any angle that's given in terms of a base length of 12''.

Contractors often use the 6-8-10 method to mark

off a right angle. Mark a distance 6' from the corner on one building line, like (b) in Figure 16-10. Next, mark a distance 8' from the corner on the other building line: (a) in Figure 16-10. If the distance (c) measures 10', you know the corner is a right angle. If it doesn't measure 10', then recheck the distances (a) and (b) until it does. This is a good method for staking out lots and square corners.

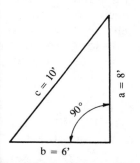

**Using the 6-8-10 method to mark off a right angle
Figure 16-10**

Finding the angles in a triangle— If you know two angles of a triangle, it's easy to find the third one. Just remember that *the sum of the three angles in a triangle is 180 degrees.* All you have to do is add the two angles you know, then subtract the answer from 180 degrees.

If one angle of a triangle is 90 degrees, and the other is 22 degrees 40 minutes, what's the third angle? First, add the known angles, then subtract from 180:

$$\begin{array}{ll} 90°\ 00' & 180°\ 00'' = 179°\ 60' \\ +22°\ 40' & -112°\ 40'' = 112°\ 40' \\ \hline 112°\ 40' & 67°\ 20' \end{array}$$

But there's a problem in the subtraction: you can't subtract 40 minutes from the 00 minutes of the 180 degrees. So, take away a degree (60 minutes) from the degrees column (leaving 179 degrees), and add it to the minutes column. Now you can subtract.

Circles

Sometimes you'll need to find the *circumference* — distance around the outside — of a circle. Or the *radius* — distance from the center point to the circumference. Or the *diameter* — distance through the center, with both ends at the circumference. The diameter is always twice the radius. (Or, the radius is always half the diameter.) And the circumference of a circle equals 3.1416 times the diameter. Find the circumference of a circle with a diameter of 6'':

Circumference = 3.1416 times the diameter. Therefore circumference = 3.1416 x 6 = 18.8496, or 18.85''

And you can reverse the calculation, if you need to find the diameter or the radius. Divide the circumference by 3.1416, and you have the diameter. Half the diameter is the radius.

To find the diameter and radius of a circle when the circumference is 12.5664'':

$$\text{Diameter} = \frac{\text{Circumference}}{3.1416} = \frac{12.5664}{3.1416} = 4''$$

$$\text{Radius} = \frac{4}{2} = 2''$$

Area Calculations

Some of the calculations you, as a contractor, have to make involve measurements of surface area, or of cubical contents. You'll use linear feet, square feet, square yards, squares containing 100 square feet, cubic feet, cubic yards, and gallons. And these are often changed further, to thousands of brick, board feet of lumber, and so on. But don't let the simplicity of these calculations encourage you to be careless. The price of a mistake is high.

A gallon of paint covers 350 SF, and you plan to paint a warehouse wall 12' x 189'. You need to know how many gallons of paint to buy to do the job.

Before you can contract to install roofing shingles on a house, you must know how to figure the roof area.

How many square yards of carpet will be required for that spec house you hope to build?

You're bidding on a house. The footing and concrete slab require some number of cubic yards of plant mix. Can you convert the size of footings and slab to cubic yards?

How many bricks will you need for the job? The house plan gives the measurements of the exterior walls. All you have to do is figure out how many bricks it will take.

Following is some review to help you make these calculations quickly and accurately.

A "square foot" isn't always one foot square: an area 2' by 6'' also contains a square foot. And note the relative areas of the square inch, the square foot, and the square yard. Since there are 12 inches in a foot, then the area of a 1-foot square is 12'' x 12'', or 144 square inches. And the area of a 3-foot square — a square yard — is 3' x 3', or 9 square feet. The following conversion table is a handy reference:

144 square inches	= 1 square foot
9 square feet	= 1 square yard
100 square feet	= 1 square

Areas of Squares and Rectangles

Calculating the area of a square or a rectangle is simple: multiply the length by the width. And you can find the perimeter (P) by adding twice the length to twice the width. For example, a bungalow measures 26'6'' by 18'6''. What are the distance around it, and the floor area? (In the calculation, P is perimeter, L is length on one side, W is width of the building, and A is floor area.)

$P = 2L + 2W$
$L = 26'6'' = 26.5'$
$W = 18'6'' = 18.5'$
$P = 2 \times 26.5 + 2 \times 18.5$
$P = 53 + 37 = 90'$

The perimeter equals two times the length, plus two times the width, of the building.

$A = 26.5 \times 18.5 = 490.25$ square feet.
The bungalow is shaped like a rectangle, therefore the floor area, in square feet equals the length times the width.

Of course, in the case of a square, every side is equal, so you end up multiplying the length of one side by itself — "squaring" it. So, to find the area of a square with 4' sides:

$A = S^2$ (A = Area and S = length of one side)
$S = 4'$ (Area = the square of one side, or S x S.)
$A = 4' \times 4'$ Therefore, $A = 4^2 = 4 \times 4 = 16$ square feet.
$A = 16$ square feet. If the length of the side "S" is in inches, then the area will be in square inches.

A small increase in the length of the sides of a square causes a large increase in the area. A 3' x 3' square has an area of nine square feet; adding one foot on each side to make a 4' x 4' square gives it an area of 16 square feet. Nearly double the area.

Square root— Occasionally, to find the lengths of rafters, braces, and stair stringers, you'll need to find the square root of a number. The square root of a number is the unknown number which, when multiplied by itself, equals the given number.

We won't tell you how to do square roots by hand: it takes too long, and it's too easy to make an error. Just about every calculator has a square-root function. Use it.

Areas of Parallelograms

Occasionally, you'll need to figure the area of a parallelogram; for ductwork or awning work, for instance. Look at Figure 16-11; it shows a four-sided figure. The opposite sides are equal and parallel: a parallelogram. A rectangle is also parallelogram: one with four 90-degree interior angles. Just as in a rectangle, a parallelogram's four interior angles add up to 360 degrees. But, the formula for finding the area of a parallelogram like this is no longer length times the width.

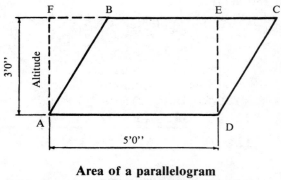

Area of a parallelogram
Figure 16-11

When angle "A" is less than 90 degrees, then the height is less than the length of side A-B. So, use the height A-F (or *altitude*) instead. A-F is the distance measured at right angles to the length, or *base*, A-D. Since the triangles AFB and DEC are equal, the parallelogram ABCD is equal to the rectangle AFED. So, *the area of a parallelogram equals the base times the height.* In Figure 16-11, the area is 5' x 3', or 15 square feet.

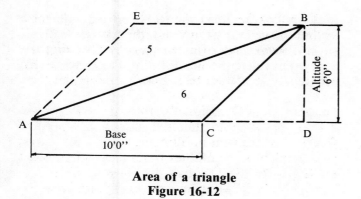

Area of a triangle
Figure 16-12

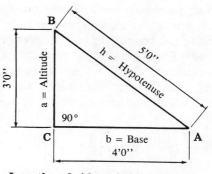

Lengths of sides of right triangle
Figure 16-13

Areas of Triangles

The area of any triangle is equal to *1/2 of the base times the height.* Look at Figure 16-12. It shows a parallelogram cut into two equal triangles, marked (5) and (6). Since ABC, of (6), equals one-half the parallelogram AEBC, its area equals one-half the product of its base times its height.

Length of Sides of a Triangle

You need to be able to figure out the length for a piece of stock to make a base, or the length of stock for the stringer of a stairway. These calculations involve triangles. This section will show you how to figure the lengths of the sides of different kinds of triangles.

Triangles with One 90-degree Angle

Look at Figure 16-13. It shows a right triangle, which has one 90-degree angle. The slanting line A-B is the *hypotenuse.* It illustrates a useful rule: *In any right triangle, the square of the hypotenuse is equal to the sum of the squares of the other two sides.*

$$h^2 = a^2 + b^2$$

But remember that this rule applies *only* to right triangles. Other triangles must be figured out differently.

But, if you know two sides of a right triangle, you can use this rule to find the third side. Just remember that the formula can be written three ways:

(1) $h^2 = a^2 + b^2$
(2) $a^2 = h^2 - b^2$
(3) $b^2 = h^2 - a^2$

This is useful for accurately laying out right angles on a construction job. In fact, this is why the 6-8-10 rule we discussed earlier works.

Let's find the hypotenuse of a right triangle whose height (a) is 15', and base (b) is 36':

$$h^2 = a^2 + b^2$$
$$\text{then } h = \sqrt{a^2 + b^2}$$
$$h = \sqrt{15^2 + 36^2}$$
$$h = \sqrt{225 + 1296}$$
$$h = \sqrt{1521}$$
$$h = 39'$$

Here's another fact to remember: Since (h) squared equals (a) squared plus (b) squared, then (h) equals the *square root of (a) squared plus (b) squared.*

Triangles with Two Sides Equal

If a triangle has two equal sides, it also has the angles opposite these sides equal. It's called an isosceles triangle. In Figure 16-14, the angles at A

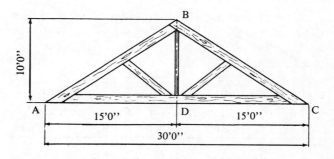

Lengths of sides of isosceles triangle
Figure 16-14

and C are equal. Many roof trusses and gable ends have this shape. If we draw line BD from the peak of the roof to the base, we divide one large triangle ABC into two equal small right triangles, BAD and BCD. The line AD is now the hypotenuse of the right triangle BAD, which has a height of 10' and a base of 15'. Now we can calculate the length of AB:

```
Square of altitude = 10  x  10  =   100
Square of base     = 15  x  15  = +225
                                   325
   AB  = √325  =  18.027'  =  18'5/16"
```

Triangles with All Three Sides Equal

If a triangle has all three sides equal, it also has all three angles equal. It's called an *equilateral* triangle. See Figure 16-15. We only need to know

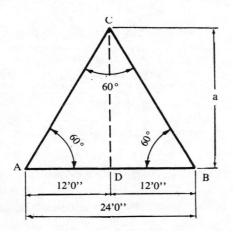

Lengths of sides of equilateral triangle
Figure 16-15

the length of one side, say AB. Then our problem is to find the height, CD. Again, the triangle is divided into two equal right triangles by CD, and the base (b) of the half-triangle becomes 12', while (h), or AC, equals 24'. Now we use the formula:

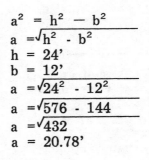

$$a^2 = h^2 - b^2$$
$$a = \sqrt{h^2 - b^2}$$
$$h = 24'$$
$$b = 12'$$
$$a = \sqrt{24^2 - 12^2}$$
$$a = \sqrt{576 - 144}$$
$$a = \sqrt{432}$$
$$a = 20.78'$$

Areas of Sides or Ends of Buildings

The sides and ends of most buildings are combinations of triangular and rectangular surfaces. So to find their areas, divide the surface into squares, rectangles, triangles, and parallelograms. Next figure the separate areas. Finally, add them together to find the total area. Say we want to find the area of the end elevation of the building in Figure 16-16. The figure shows how the area has been cut into two parts, along line BD. Now let's do the calculation:

```
Area of triangle      = 0.5 x 32' x 8' = 128 SF
Area of rectangle     = 32 x 9        =+288 SF
Total gross area of
  surface                              = 416 SF
Deduct area of door   = 7' x 3'       = - 21 SF
Net area of surface                    = 395 SF
```

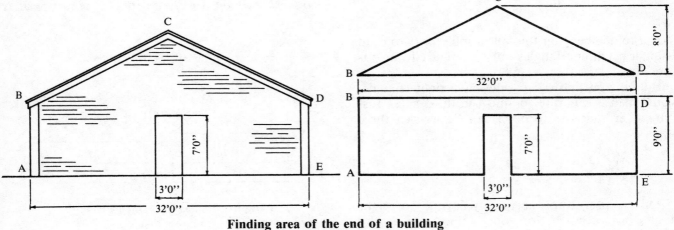

Finding area of the end of a building
Figure 16-16

Surface Areas

A rectangle like a room, or like the one in Figure 16-17, has six sides. Each side is rectangular in shape. And opposite sides are equal. Let's figure out the wall area of a room 15' long, 12' wide, and 8'6'' high. Ignore doors and windows for now. Let's also figure out the total area. In the following equations, WA is wall area, L is length, W is width, H is height, and TA is total area:

$$WA = 2 \times (L + W) \times H$$
$$L = 15', W = 12', H = 8.5'$$
$$WA = 2 \times (15 + 12) \times 8.5$$
$$WA = 2 \times 27 \times 8.5$$
$$WA = 54 \times 8.5 = 459 \text{ SF}$$
$$TA = WA + 2 \times (L \times W)$$
$$TA = 459 + 2 \times (15 \times 12)$$
$$TA = 459 + 360$$
$$TA = 819 \text{ SF}$$

To find the total area, add the ceiling and floor areas to the wall area. They're both rectangles, and they're equal. So figure the area for one (L x W), double it, and add the answer to WA. Now you have total area of the room.

Volume Calculations

You'll often need to figure out volumes. You'll use cubic inches for small objects. Use cubic feet for figuring the space occupied by buildings, or the amount of air needed for heating or ventilating. Use cubic yards for figuring the amount of concrete needed for foundations and walls, or how much dirt will be excavated. Since volume is a measure of contents, you'll deal with length, width, and height. Remember to use the units of cubic measure after the figures, to distinguish them from surface area or linear measurements.

Volumes of Solids

To find volume, multiply the length by the width by the height (or depth). When you increase the three dimensions of a solid, the volume increases at a great rate. Doubling the sides of a 1'' cube increases the volume to 2 x 2 x 2, or 8 times. Increasing the sides further, to 3'', increases the volume to 3 x 3 x 3, or 27 times.

Converting Between Cubic Inches, Feet, and Yards

It's easy to convert between cubic inches, cubic

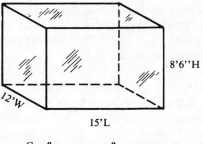

Surface area of a room
Figure 16-17

feet, and cubic yards. Here are the conversion factors:

- To convert cubic inches into cubic feet, divide the cubic inches by 1,728.

- To convert cubic feet into cubic inches, multiply the cubic feet by 1,728.

- To convert cubic feet into cubic yards, divide the cubic feet by 27.

- To convert cubic yards into cubic feet, multiply the cubic yards by 27.

- To convert cubic inches into cubic yards, divide the cubic inches by 46,656.

- To convert cubic yards into cubic inches, multiply the cubic yards by 46,656.

When do you need to use these conversion factors? Let's say we have to excavate for a basement. The building is 45' long, 30' wide, and the excavation is 6' deep. You'll need to know the amount of earth to be removed, in cubic yards. The contents, in cubic feet, is 45 x 30 x 6, which is 8,100 cubic feet. Since there are 27 cubic feet to one cubic yard, we have:

$$\frac{8,100}{27} = 300 \text{ cubic yards}$$

Here's another example: A 20'' by 40'' brick pier is 90'' tall. Find the volume in cubic feet.

Volume in cubic inches = 40'' x 20'' x 90'' = 72,000 cubic inches

Volume in cubic feet = $\frac{72,000}{1,728}$ = 41.67 cubic feet

In reality, you'd probably solve this problem faster by converting the dimensions to feet and decimal parts of a foot, then multiplying these to find the volume in cubic feet.

Volumes of Liquids

Liquid volumes are measured in units like gill, pint, quart, gallon, and barrel. The gallon is the basic unit, and in the U.S. it equals 231 cubic inches. Canada's British Imperial gallon, however, equals 277 cubic inches. Be careful. And don't estimate in terms of barrels: there isn't a standard size (oil barrels are 42 gallons, gasoline barrels are 50 to 52 gallons), so everyone must know exactly what kind of barrel you mean. Stick to gallons.

The cubic foot contains 1,728 cubic inches, and the gallon has 231 cubic inches. Therefore, the equivalent of a cubic foot in gallons is:

$$\text{Cubic foot} = \frac{1,728}{231} = 7.48 \text{ gallons.}$$

Here are some conversion factors:

- To convert cubic feet into gallons, multiply the cubic feet by 7.48.

- To convert gallons into cubic feet, divide the gallons by 7.48.

- To convert cubic inches into gallons, divide the cubic inches by 231.

- To convert gallons into cubic inches, multiply the gallons by 231.

For example, say we have a tank with an inside length of 10', a width of 5', and a depth of 6'. Find the filled capacity, in gallons:

Cubic contents = 10' x 5' x 6' = 300 cubic feet
Contents in gallons = 300 x 7.48 = 2,244 gallons

To find the number of gallons in a tank containing 57,800 cubic inches:

$$\frac{57,800}{231} = 250.2 \text{ gallons}$$

Weights of Materials

Each material used in construction has a fixed weight per cubic foot, or per cubic inch. If you know this weight, you can figure the weight of any volume of material: Multiply the unit weight by the volume in cubic units.

To be exact, you should determine each material's weight per cubic unit separately, since the unit weights will vary slightly depending on type, method of manufacture, etc. But average figures are plenty close enough for ordinary construction work. Figure 16-18 shows typical average values of weights per cubic yard for the most common materials.

Asphalt	2700
Brick, common	3375
Brick, pressed	4050
Clay	3300
Clay and gravel (dry)	3400
Clay and gravel (wet)	4300
Concrete	3710
Concrete, lightweight	2970
Earth, loam (dry)	3000
Earth and gravel	3240
Granite	4500
Granite masonry	4200
Gravel (dry)	3100
Gravel (wet)	3600
Hemlock dry	675
Limestone	4200
Maple (dry)	1325
Mud (dry)	2500
Mud (wet)	3250
Oak, white (dry)	1296
Pine, white (dry)	675
Pine yellow (dry)	920
Pine yellow, southern	1215
Rock, crushed	4000
Rock, well blasted	4000
Sand (dry)	3250
Sand (wet)	3400
Sand and gravel	3100
Sandstone	4140
Shale	2800
Slag	1890
Trap rock	5057

**Typical building material weight
in pounds per cubic yard
Figure 16-18**

Class of soil	Bearing value in tons per square foot
Sand, clean and dry	2.0
Sand, compact, not drained	4.5
Clay, deep beds, always dry	4.5
Clay, average, moist	3.5
Clay, soft	1.0
Gravel, well banded with coarse sand	4.5
Hard rock, thick layers, undisturbed	200.0
Soft rock, limestone strata	50.0
Soft rock, limestone broken	15.0

Safe bearing values of soils
Figure 16-19

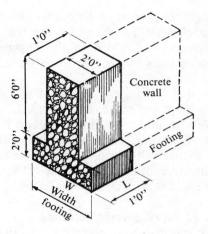

Figuring width of footing
Figure 16-20

Bearing of Soils

Figure 16-19 shows the pressure, in tons per square foot, that various classes of soils will support safely. Let's say you need to know how wide (W) to make the footing for the foundation wall in Figure 16-20. It carries a load of four tons per foot of length. To this, add the weight of the foundation wall and of the footing itself. First, calculate the weight of one foot (L) of the wall and footing. The footing is three feet wide. It will rest on dry sand:

Weight of wall = 2' x 6' x 1' x 150 lbs	1,800 lbs.
Weight of footing = 2' x 3' x 1' x 150	900 lbs.
Total weight of wall and footing = 1,800 + 900 = 2,700 lbs.	1.35 tons
Added load carried	4.00 tons
Total weight and load per linear foot (L)	5.35 tons

Since it will rest on clean, dry sand, the required area, at the safe bearing value of two tons per square foot, is:

$$\text{Area} = \text{width (W)} = \frac{5.35}{2} = 2.675 \text{ square feet}$$

Since you've figured the weight per foot of length, the area of 2.675 square feet is the same as the width (W). So, the footing width is 2.675 feet. To convert this to inches, multiply by 12: 12 x 2.675 equals 32.1''. In such rough work, you can ignore the one-tenth inch, so (W) equals 32''.

Calculating Lumber Amounts

Lumber is sold by the *board foot,* which is just a special application of cubic measure. The board foot is more convenient than the cubic foot, since most boards are comparatively thin.

A board foot is the volume of a piece of lumber one foot square and one inch thick, or its equivalent. The cubic contents of a board foot is equal to 12'' x 12'' x 1'', or 144 cubic inches. It's also equal to 1/12 of a cubic foot. These facts make calculations simple. The general rule is:

Multiply the width in inches by the thickness in inches; then multiply this answer by the length in feet. Divide by 12 to find the number of board feet.

For example, find the board feet in a plank 12'' wide, 4'' thick, and 16' long:

$$\text{BF} = \frac{12'' \text{ x } 4'' \text{ x } 16'}{12} = 64 \text{ board feet}$$

When making calculations, remember that a board less than 1'' thick is calculated as if it were 1'' thick. No allowance is made for the amount lost in planing or finishing. Pieces thicker than 1'' are figured in proportion to the increased thickness.

Matched flooring— T&G flooring is cut from boards of nominal size, and the actual width is further reduced by the length of the tongue. When calculating the number of pieces to use in a floor, use the face width. The loss of width from shaping can be from 15% to 33% of the nominal width. Add to this about 5% for waste from cutting and fitting.

Shiplap— Shiplap also has loss of width, due to the lap. Make allowance for finish on two edges, and for waste in cutting. The loss in size is about 7%, while the waste is from 5% to 15%. This gives a total allowance of from 12% to 25% over the amount required for the nominal size.

Estimating Flooring and Sheathing

When estimating the quantity of flooring or sheathing needed to cover a given area, figure the number of board feet (directly from the area). Then add a waste allowance, according to the kind of lumber used.

For example, say you're going to lay a floor in a room 12' by 18'. You're using 1 x 4 T&G pine lumber. Find the number of board feet required for the job. Allow 26% for waste:

Area of room = 12'0" x 18'0" = 216 square feet
Allowance = 216 x .26 = 56.16 square feet
Total = area + allowance = 216 + 56.16 = 272.16 square feet

Since the nominal thickness of the flooring is 1'', then the board feet will be the same as the number of square feet.

A Final Example

Now, look at Figure 16-21. It shows a building with a pitched roof. The sides and ends will be covered with 1 x 6 T&G sheathing. The roof will be covered with 1'' sheathing, then with shingles laid 5'' to the weather. You need to find the quantity of sheathing, in board feet, and the number of shingles that will be needed. For now, ignore window and door openings.

First, figure the sheathing requirement. This means you have to find the area of the two ends, and the two sides. The side EFGB and the side opposite are rectangles. Each side has an area of 12' x 30', or 360 square feet. So, you have an area of 720 square feet for both sides.

The ends contain the roof triangle, ABC, with a base of 16' and altitude of 4'. The area of one triangle is 0.5 height x base (4 x 16'), or 32 square feet. The lower rectangle on the end, ADEB, is 12'

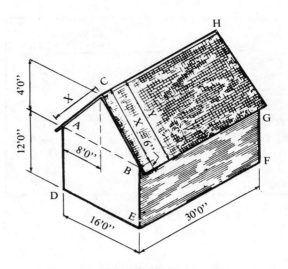

Estimating sheathing and roofing for a building
Figure 16-21

x 16', or 192 square feet. So, the total area of one end is equal to the triangle plus the rectangle: 32 + 192, or 224 square feet. And both ends are: 2 x 224, or 448 square feet.

The total wall area covered by sheathing, therefore, is:

Area of sides	720 square feet
Area of ends	+448 square feet
Total	1,168 square feet

Since the sheathing is 1'' thick, the area in square feet is also the actual quantity in board feet. But you must add a waste allowance. Let's make it about 23%. So, 1,168 x 0.23 equals 269 board feet. Add this to the actual amount: 1,168 plus 269, or 1,437 board feet. That's how much sheathing to buy for the building. Next, find the roof area. The rafter length, X, is the hypotenuse of the right-angle triangle whose height is 4', and whose base is 8' (or 1/2 the building width of 16'). The length X is equal to the square root of the sum of the height squared plus the base squared:

$$x = \sqrt{4^2 + 8^2} = \sqrt{16 + 64} = \sqrt{80} = 8.94$$

The eaves overhang by 6'', so the total length of X is 8.94 + 0.5, or 9.44 feet. Assume that the roof has a 1' overhang at each end. The total length then becomes 30' + 2', or 32'. The area of one side of the roof is 9.44' x 32', or 302 square feet. The whole roof area is 2 x 302, or 604 square feet.

For the sheathing, add 23% waste to 604: 604 x 0.213 is 139. And 604 + 139 gives a total of 743 board feet of sheathing.

For the shingles, use roofing measure: one ''square'' equals 100 square feet. So you need 6.04 squares to cover the roof. From a shingle table you find that about 800 shingles are needed pe. square when they're laid 5'' to the weather. So the number of shingles needed becomes 6.04 x 800, or 4,832 shingles. This completes your estimate.

Summary

It is said that ''The contractor who can figure, can figure on success.'' Likewise, a carpenter or builder who can't apply this chapter's basic mathematics is shorting himself. He might learn —

too late — that he, in reality, did learn too late!

Contracting is an art. Construction is a science — putting exactly-measured parts in exact places. The art becomes possible only after you master the science. Put another way, if you can't measure a board, or figure out how much concrete it takes to fill a hole, you're in the wrong business.

Math is elusive. If you don't use it, you lose it. Many high school graduates have trouble balancing their checkbooks — because they stopped using what they learned as students.

If this chapter appeared foggy when you read it, go back and study it again. A surveyor may no longer string out a steel tape, but instead use computerized equipment to do the measuring. The measurement is made nonetheless. The measurement is the key. Once you know how to do it, you eliminate a lot of the problems that can keep you from becoming a successful builder.

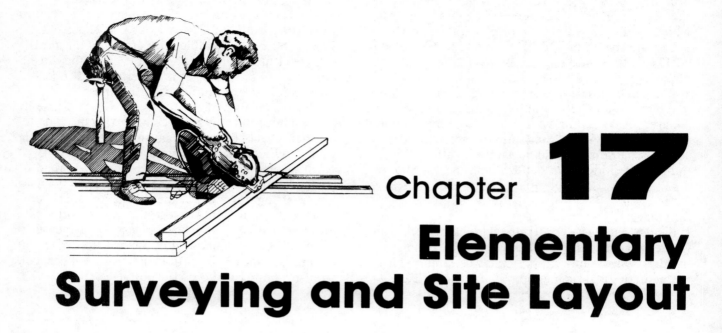

Chapter **17**

Elementary Surveying and Site Layout

In 1785 the first U.S. ordinance for the division of public lands was enacted. The basic principles in that law are still being followed. Under the law, public lands were divided into townships with sides six miles long. Each township was subdivided into 36 square mile sections. Of course, few townships are perfectly square.

Figure 17-1 shows a typical square township. Each square mile section contains 640 acres and is further subdivided into quarter and quarter-quarter (1/16) sections. Plot A is a quarter-quarter section designated as the NE 1/4 of SE 1/4 of Section 8. Plot B is located at the E 1/2 of NW 1/4 of Section 22.

Surveying always begins from an *initial point* located on a north-south line called a *principal meridian*. Figure 17-2 shows how this is done.

Division of Townships and Ranges

Townships which extend east and west of the principal meridian are called *tiers. Tiers are numbered from the surveying base line* as shown in Figure 17-2. Townships extending north and south of the base line are called ranges. They are numbered from the principal meridian. Find Plot A in Figure 17-2. It is three tiers south of seven ranges east of the initial point on the principal meridian. So it's designed T3S, R7E.

"Quarter section corners" are found at the intersections of boundary lines.

6	5	4	3	2	1
7	8 Ⓐ	9	10	11	12
18	17	16	15	14	13
19	20	21	B 22	23	24
30	29	28	27	26	25
31	32	33	34	35	36

⟵ 6 miles ⟶

Township with 36 sections
Figure 17-1

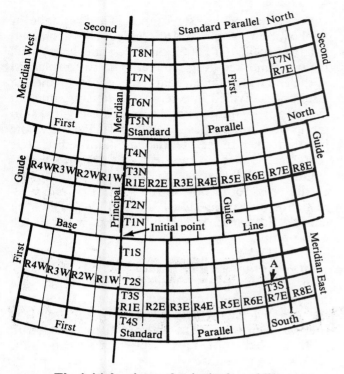

The initial point and principal meridian
Figure 17-2

Monuments are used to mark corners. They range from official markers to ordinary stones and trees. When monuments are built from a pile of materials, they are called *memorials*. Monuments may be numbered or lettered with a wide range of abbreviations. Examples are B.T. (bearing tree), S.M.C. (special meander corner), W.P. (witness point), and C.P. (concrete post). Disagreements about coordinates or corners must be resolved in court.

Never lay out a job until you know exactly where the plot lines are. Markers are sometimes moved. Many builders have discovered too late that part of the building they just finished is on a neighbor's property. There's not a lot you can do about it but hope and pay. *Have a professional surveyor establish the lines.*

Project Layout
Project layout is the transferring of measurements on plans and survey maps to the ground at the job site. In large projects, this work is usually done from drawings prepared by a civil engineer. The site is divided into a set of triangles or squares. The proposed buildings are then laid out along the lines forming the squares or triangles, as shown in Figure 17-3.

Figure 17-4 shows a typical layout for a house, giving specific measurements. Note that the property line in this figure begins at the sidewalk. The monuments (Mon) are shown at the lot corners.

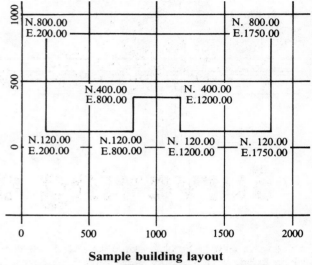

Sample building layout
Figure 17-3

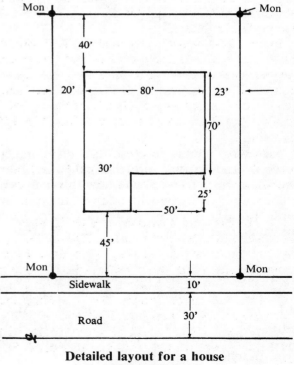

Detailed layout for a house
Figure 17-4

Your plot may show I.P. (iron pins) as the corner monuments.

A street layout is shown in Figure 17-5. Distances are marked from a reference point called a *station* in 50-foot increments called *plusses*. For example, the second manhole from the station is marked as 1 plus 50. This means that it's 150 feet away from the station. The entire length of the street is marked off in this manner.

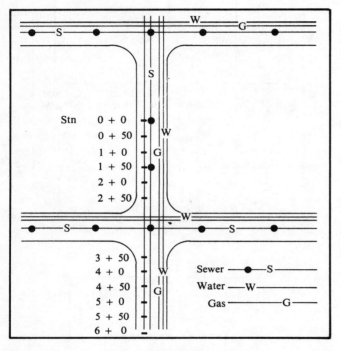

Typical street layout
Figure 17-5

Surveying is an exact science. A competent surveyor is a trained professional. This chapter doesn't attempt to teach you how to survey. What it does is acquaint you with a few of the principles, and highlight the importance of exact measurements in your site and building layout.

Measuring Tapes

There are three kinds of tapes: cloth, metallic, and steel. Don't use cloth tapes for construction. They stretch too easily, especially when wet. Metallic tapes have a fine metal thread wrapped around each cloth thread. But don't use them, either. They also stretch.

For accuracy, use steel tapes. They're made of steel ribbon, about 5/16'' wide. They come in lengths of 50', 100', 200', or 300'. Some heavy steel tapes are marked every 10 feet. The 10-foot length at the end of the tape is marked every foot, and the last foot is divided into tenths. Others are marked every foot — these have the end foot divided into tenths and hundredths. Light tapes have the whole length marked in feet, tenths, and hundredths. The marks are etched on the steel.

On many light tapes, the zero point is at the end of the brass loop, not on the steel. This is convenient for measuring, but it's a bad idea. The loop can get flattened; this changes the position of the zero point.

Using Measuring Tapes

Your measurements must be accurate, so use measuring tapes properly. Keep them wound on the reel. Clean them at the end of each day. If they've been wet, clean and dry them. Next, rub with an oiled cloth. Last, wipe with a clean cloth.

When you measure with an unsupported tape, its weight pulls it into a curve. This means your measurement isn't accurate, because you're measuring the length of the curve. A curve, of course, is longer than the distance between the two points. You must eliminate this deflection with supports.

To measure a distance longer than the tape — on a floor slab, for instance, stretch a chalk line along the distance to be measured. Snap it, so it marks the slab. Next, mark some partial distances. Make each one a bit shorter than the tape. Measure these distances. Then, measure from the last mark to the end of the chalk line. Finally, find the total distance by adding up all the measurements. Add carefully.

Surveying Instruments

It's possible to lay out an entire building with a hand level. Every job site, though, should have a builder's level. Use it for checking lines and elevations. If you know how to use it, a builder's level will more than pay for itself in time saved and accuracy gained.

Surveyors measure in feet and tenths. But architects, and your workmen, use feet and inches. So change survey elevations on a building to read in feet and inches. For example, to change an elevation of 10.302', multiply 0.302 by 12''. This equals 3.625''. Now, use a conversion table (like the one in Chapter 16) to find that 3.625'' equals 3⅝''. Change the mark to 10'3⅝''.

The Builder's Level

Builder's levels are used for finding differences in elevation quickly and accurately. A builder's level has three main parts: the *telescope,* the *leveling vial,* and the *circle.* The person using the level is the *levelman.*

The telescope— This is a precision optical instrument. It has a clear, magnified image. Crosshairs help you center the object you're viewing. The amount of magnification is known as the *power.* An object seems 18 times closer than normal when viewed through an 18x telescope. A fixed-lens telescope only has one power; a zoom lens telescope has variable powers. Zoom lenses come in various ranges. Less expensive telescopes have a lower power; use them over shorter distances.

The leveling vial— This important part of the builder's level is similar to a spirit level. It's a glass tube, with the inside upper surface ground to a curve. Figure 17-6 shows this clearly. A liquid fills the tube, with enough space left to form a bubble.

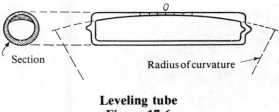

Leveling tube
Figure 17-6

The leveling vial has a scale. It's either a fastened-on metal scale, or one etched into the glass of the tube. The zero point is at the center of the tube. Scale graduations are numbered, both ways from the zero point. The bubble is centered when its ends fall on the same marking, on opposite sides of the zero point. This means the tube is horizontal.

The radius of the curve determines how sensitive the level is. A long radius means the bubble travels further with each movement of the end of the tube. Therefore, it takes longer to center the bubble each time the tube is moved. The longer it takes to center the bubble, the more sensitive the level is.

The bubble's length changes with the temperature. This is because the tube's size increases as the glass expands with the heat. It doesn't affect the readings, though.

The circle— The telescope rotates on this plate, which is marked in degrees. The more precise models have vernier scales, which subdivide each degree into minutes, and each minute into seconds.

Buy the most precise level you can afford. You want accuracy, even on small jobs. And, as your skill increases, you'll use it for more kinds of jobs.

The dumpy level— This level is popular with architects, superintendents, and contractors. One make is shown in Figure 17-7. The uprights, carrying the telescope, are firmly fastened to the bar. The level tube, mounted on top of the bar, is easy to adjust. All these parts are rigidly fastened together. Because it has few moving parts, the dumpy level can stand rough treatment. It doesn't get out of adjustment easily.

The transit level— This level can be raised and lowered vertically. And it can be locked at zero vertical reading, then used horizontally. Figure 17-8 shows a transit level. This is a versatile, and popular, instrument.

The automatic level— This self-leveling instrument is also popular. Rough leveling is done with a small, circular bubble. For more precise leveling, a suspended compensator keeps the instrument level. It's very accurate, since the instrument isn't affected by wind or traffic.

The Theodolite

This is another surveying instrument. It has an "N" on its dial which always points to the object sighted rather than north. So, if you sight an object and the needle points to the "W" part of the dial, the object must be west of north to the degree indicated by the needle. Such an object would be given a "bearing" of N 50 degrees W, if the needle pointed to 50. You can use this procedure to get a bearing on any object. Use the theodolite to measure vertical and horizontal angles.

The Leveling Rod

The leveling rod is an important part of a leveling outfit. It's used to measure elevation differences. This is the difference between the level's line of

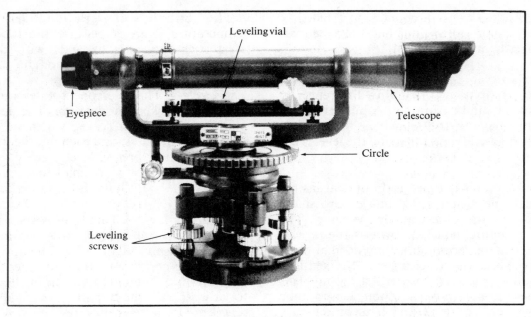

Dumpy level
Figure 17-7

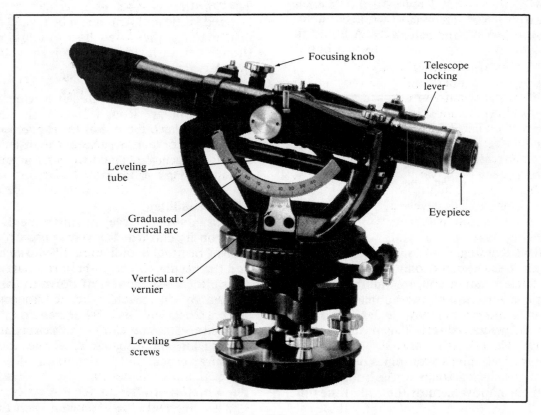

Transit level
Figure 17-8

sight, and the point where the rod is held. There are two types of rods: self-reading and target. Self-reading rods can be read, through the telescope, by the levelman. Target rods can be read only by the rodman. They're mainly used when the levelman can't get a reading because the rod is too far from the telescope.

There are also combination rods, like the popular Philadelphia combination. It can be a single strip of wood. Or two or more strips can slide over each other to make an extension rod. When extended, the parts are clamped in position. The most common length is 12' to 13' long.

Section A of Figure 17-9 shows part of a one-piece rod, 12.5' long. Its whole length is graduated into feet, tenths, and hundredths. The zero point is at the bottom of the rod. The gradations are painted black, and are 0.01' wide. They are spaced 0.01' apart. Therefore, each alternate space — black or white — is 0.01' high. The large numbers mark the feet, and are painted red. The small numbers, marking tenths of a foot, are painted black.

For construction work, I recommend an extension Philadelphia rod. It's divided into feet, inches, and eighths of inches, and extends from 7' to 13'.

Using the Builder's Level

Be sure your level is properly adjusted. And be sure it's set up correctly when you use it, or you won't get accurate sights. The most important part of setting up the instrument is leveling it.

The manual that came with your builder's level has step-by-step instructions for setup and leveling. Because procedures vary, we won't go into detail here. The basic idea, though, is for the bubble to stay in the center of the leveling vial, no matter which way the telescope is turned.

Taking a Rod Reading

The rodman holds the rod on station. You may want the rodman facing you, standing behind the rod. Or, you might have him stand beside the rod. It's absolutely necessary that he holds the rod plumb. The levelman directs the rodman which way to move the rod, using arm signals, until it's parallel to the telescope's vertical cross-hair. Vertical lines on nearby buildings can help the rodman judge when the rod is plumb.

Using a self-reading rod— Now, at the point where the horizontal crosshair bisects the rod, read the

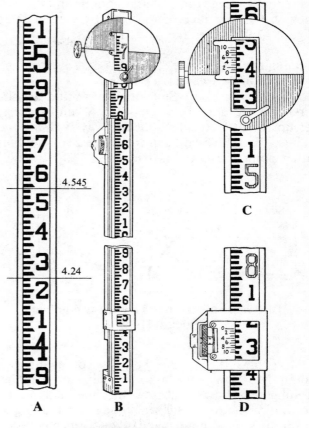

Philadelphia leveling rods
Figure 17-9

feet and tenths. Approximate the hundredths. (Of course, with a construction rod, you'd read the feet, inches and eighth-inches.) For example, the reading of the lower wire in Section A of Figure 17-9, shows 4.24'. By estimation, the upper wire reads 4.545'.

Using a target rod— Say you're using a rod like the one in Section B of Figure 17-9, and the reading will be less than 7'. Signal the rodman to move the target up or down on the rod. When the telescope's horizontal crosshairs bisect the target, the rodman clamps it to the rod. He then takes the reading from the vernier scale on the target. Section C of Figure 17-9, for example, shows a reading of 5.386'.

For readings higher than 7', the rodman clamps the centerline of the target (the zero point of the target vernier) at the 7' mark. Then he raises the rear strip of the rod until the crosshair bisects the target. He clamps the two strips together. Then he takes the reading from the graduations and the vernier on the rear face of the rod. In Section D of Figure 17-9, the reading is 8.238'.

Finding the Difference in Elevation Between Two Points

Look at Figure 17-10. Say you want to find the difference in elevation between points M and N. First, set up the instrument about midway between the points. Place it so you can easily see a rod held on either point. Level the instrument.

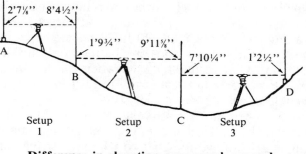

Difference in elevation over rough ground
Figure 17-11

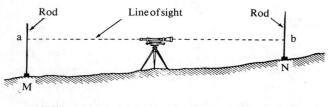

Difference in elevation between two points
Figure 17-10

Next, have the rodman hold the rod at the first point, M. Now, take a rod reading at *a*. (This reading tells how far below the line-of-sight the bottom of the rod is.) Then, take a rod reading at *b*, on the second point, N. The difference between the two readings is the difference in elevation between the two points.

Finding the Difference in Elevation over Rough Ground

Sometimes the ground is irregular, or there's a large elevation difference. In this case, use several set-ups to find elevations. And use a plus-and-minus reading system to calculate the final elevation difference.

The plus sights are from the setup, back toward the starting point. Plus sights are known elevations, in relation to the starting point. And minus sights are from the setup, away from the starting point. Minus sights are unknown elevations, in relation to the starting point — until you take a back-sight. Look at Figure 17-11 to see what we mean.

When you've sighted, and recorded, the last point, add all the plus sights. Then add all the minus sights. Finally, subtract the smaller number from the larger. A minus sum means that the end point is lower than the starting point. And a plus sum means the end point is higher than the starting point.

Set up	+ Sight	- Sight
1	2' 7-1/8"	8' 4-1/2"
2	1' 9-3/4"	9'11-5/8"
3	7'10-1/4"	1' 2-1/2"
	+12' 3-1/8"	-19' 6-5/8"

In Figure 17-11, the difference in elevation between A and D is 12'3⅛" minus 19'6⅝", or −7'3½". The minus sign means that D is 7'3½" below A.

Making Construction Stakes the Same Height

When staking out a building, you'll want to make a line of stakes all the same height. You can use the level to line up any number of stakes. See Figure 17-12. First, set up the instrument. Then have the

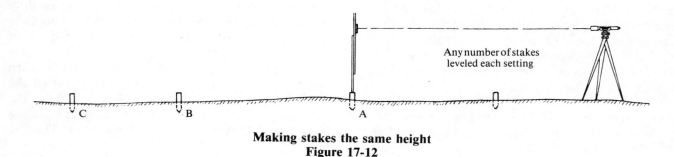

Making stakes the same height
Figure 17-12

rod placed on A, whose height you know. Take a reading on it.

Next, have the rod moved to the next stake, B. Have the rodman raise or lower the rod until you get the original reading. Finally, have him adjust the stake, so it touches the bottom of the rod. Repeat for all the stakes. The top of each stake is now the same level as the original, stake A.

The Bench Mark

Before you can lay out a building, establish a *bench mark*. Use it for checking first-floor heights, grade lines, and so on. Take all measurements from this bench mark.

Start with a permanent object. This could be a post driven into the ground. Put a permanent mark on it, at a known height above a certain part of the surroundings. You could use the crown of a road, the concrete curbing, or the sill of a nearby building. Pick something that lets you put the mark at a convenient height.

Here's how to use the bench mark during excavation: The blueprint shows the finish grade 1'8'' below the finish floor line and 6'8'' above the top of the footing. By adding 1'8'' and 6'8'', you get 8'4'' from the footing top to the finish floor line (bench mark).

Assume that at one corner of the excavation, the ground level is 3'4'' below the bench mark. You subtract 3'4'' from 8'4'', getting 5'. Add the 5' to the thickness of the footing to find the depth of the excavation required at this point.

The Datum Plane

On the working drawings, an architect may assume an imaginary level surface. This surface may be 100' below the actual level planned for a part of the building (the top of the foundation, the top of the first floor joists, or the top of the finished first floor.) All points on this assumed level surface have an elevation of zero. This assumed surface is the *datum plane.*

Sometimes a city or town establishes a datum plane to use as a reference point in measuring depths and elevations. In some areas mean sea level is used as the datum plane.

On the drawings, the building's elevations are written as if they were measured from the datum plane. For example, Figure 17-13 shows the elevations of the tops of joists for different floors. The elevation of the first-floor joists is marked 100'0''. The architect has measured it from a datum plane, 100' below the first-floor joists.

The grade-line elevation is marked 98'3''. It's 98'3'' above the datum plane. To find the height of the first-floor joists above grade, subtract 98'3'' from 100'. The first-floor joists are 1'9'' above grade.

Likewise, the footing elevation is marked 91'4⅜''. To find how far below grade the top of the footing is, subtract 91'4⅜'' from 98'3''. The top of the footing is 6'10⅝'' below grade. To find the distance between the top of the first-floor joists and the top of the second-floor joists, subtract 100'0'' from 109'3 5/16''. The answer is 9'3 5/16''. Figure out other elevations the same way.

Examine the Site

When you're familiar with the project, go look at the site. Look for any natural obstacles on the site. And look for the kinds of markers you can use. Make a sketch and take notes. If you'll need a plan for a temporary building, storage shed, or access road, note the best sites. Taking notes on the site will help you avoid unnecessary problems when actually doing the layout.

Now you have enough information to make up a layout schedule. Decide how many people you need. Decide how much time it will all take. Decide, on the basis of the contract schedule and construction methods, when to do the various stages of the layout.

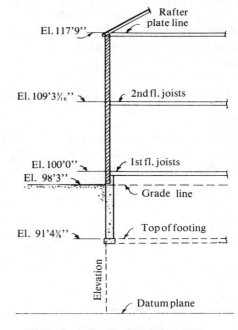

Elevations from the datum plane
Figure 17-13

Stages of Layout Work

Layout work can be divided into three main categories: rough setting out, first-stage accurate setting out, and second-stage accurate setting out. The order in which these stages are done varies from job to job. So does the extent of each stage. The order described here is only an example.

Once you've drawn up the layout schedule, make a detailed plan of each main stage. But be flexible: as construction proceeds, you're better able to decide what kind of layout is most suitable.

Setting Out Buildings

Be sure your preliminary layout is ready when needed. Station points and building corners must be set out before actual construction begins. Once excavation has begun, setting-out is much harder.

Baseline grid— Make sure the surveyor doing the setting out marks the plan with the approximate position of the baseline (or setting-out) grid that will be used. If building frontages are parallel to each other, a baseline grid parallel to them is used. Points of intersection between the frontages are also marked. Look back to Figures 17-2 and 17-3.

Establish Reference Points and Bench Marks

The corner points will disappear during excavation. Be sure the surveyor ties in their positions, so you can locate them again. The lines that mark the outside shape of the building must extend beyond the building lines. Mark the reference points on these extensions. See Figure 17-14. The corner points are where these lines intersect.

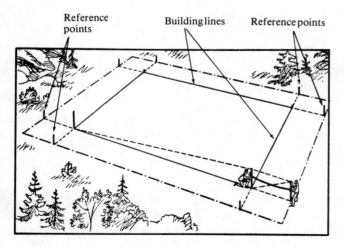

Rough layout for excavation
Figure 17-14

Make sure you have enough reference points. Have them positioned so that, during construction, you can still see them. Reference points must be permanent. Figure 17-15 shows suitable reference points. If rock isn't available, set the marks at a frost-free depth.

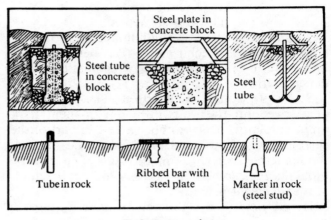

Reference points
Figure 17-15

Establish some bench marks. You'll use them in leveling later on. See Figure 17-16. Keep the layout plan in mind when choosing the positions of these marks.

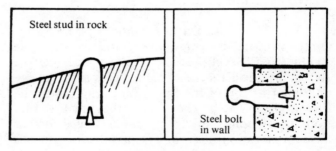

Bench marks
Figure 17-16

First-Stage Accurate Layout

This stage of the layout involves setting up for excavation, leveling the foundation, and forming for the foundation.

Setting Up Batterboards

Erect batterboards around the site. They make foundation setting-out, and excavation, faster. Use corner boards at the corners. Use single boards and auxiliary boards along the sides. Put the boards 6' to 10' from the outside line of the building. See Figure 17-17.

First, set out *profile stakes*. Drive them into firm ground and brace them with angled boards. In rock, attach them to bars set into drilled holes. Connect the tops of the stakes with horizontal boards. These must be the same height.

Be sure the lines are straight. With horizontal boards around the entire building, you can set out all points at the same time, with a tape, from one bench mark. The tape is supported along its entire length. Also use the lines to establish elevations such as excavation and footing depths.

Mark points; either with a cut, or with nails driven into the boards. The lower left section of Figure 17-17 shows the two methods. Mark the points with the same number that's on the layout drawing.

If you used a baseline grid, mark the baselines. Use wires stretched between the marks on the profile. You could fix the wire to the mark with two nails, but it's safer to make a cut in the board. The cut will guide the wire. Use an ordinary plumb bob to plumb the intersections of the wires.

Leveling the Bottom of the Excavation

The bottom of the excavation must be at the right elevation, and reasonably even. Use a traveler, between fixed sight rails, for this. The tops of the sight rails, and of the traveler, must be horizontal. The plane between the sight rails is called the *sight plane*. See Figure 17-18. Make the sight rails so the sight plane is 3' above finished level. Always mark the height on the sight rails. Make the traveler the same as the sight rails. Paint the sight rails and traveler with contrasting colors to make sighting easier.

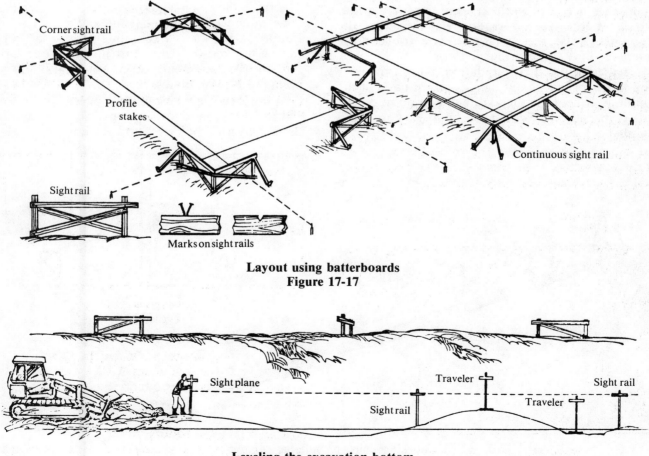

Layout using batterboards
Figure 17-17

Leveling the excavation bottom
Figure 17-18

Now, sight along the top of one sight rail until it lines up with a second sight rail. This establishes the sight plane. Then, put the traveler in the line of sight. Its position shows you whether the ground at that point is too high, too low, or just right.

Forming for the Foundation

Once excavation is complete, start forming for the foundation. Align the formwork along the wires you stretched on the profiles. The formwork could get knocked out of place during work. So check its position with a builder's level, or even better, with a theodolite before and after pouring the concrete.

Set up the theodolite over a baseline or over some other line that's parallel to the side of the formwork. Sight along this line. Measure the distance between the line-of-sight and the edges of the formwork. This way, you can check the position of the formwork. See Figure 17-19.

Second-Stage Accurate Layout

Once the foundation formwork is removed, mark your reference lines on the sides of the foundation. Do this before formwork and walls obstruct the sight lines. Use these marks if you transfer the reference lines up to a higher floor.

Next, on top of the foundation, set out the lines you'll use for setting out dimensions of parts of the building. Set up the theodolite over a point on the line to be set out. See Figure 17-20. Try to avoid parallel offset.

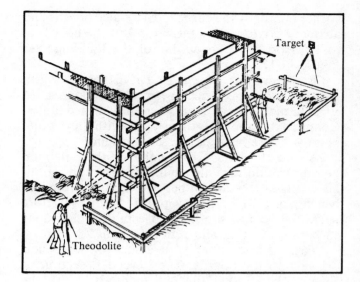

Checking formwork
Figure 17-19

Scratch the concrete to mark the line. Then mark the scratch with a spot of paint so it's easy to find. Important points — like points of intersection — can be marked on plates cast into the foundation slab or the basement roof. Drill marks for the points into these plates.

At an early stage, establish a bench mark, as in Figure 17-21, to use for leveling inside the building. Use an inverted bolt cast into the concrete.

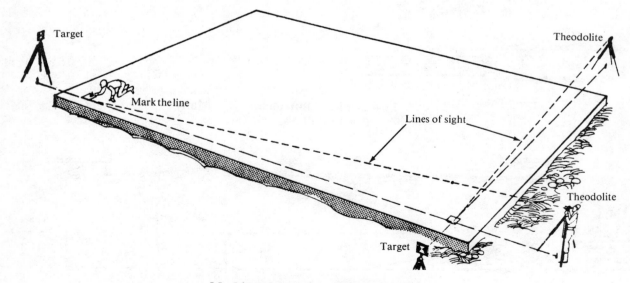

Marking slabs for building components
Figure 17-20

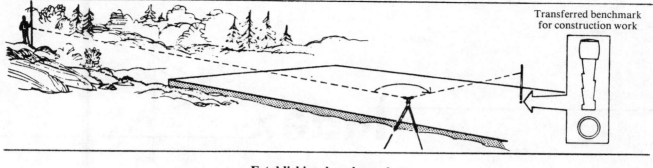

Establishing bench marks
Figure 17-21

Layout and Plumbing of Formwork

If there's a cast-in-place building part like a stair-case, set out and mark baselines on the foundation slab. Use these for setting out points for the form-work. Before you begin, though, agree with the foreman, or the carpenter in charge of formwork, on which points to set out. You'll avoid misunderstandings that might cause hold-ups or faulty formwork.

Once formwork is erected, set out a convenient level line on it. Mark it with nails at a few places. Drive the nails through the timber; they'll mark the level on the inside of the form. Always write on the timber to note what level this represents. Now it's easy to see where the form must be cut back, or what level the concrete must be cast to. Figure 17-22 shows the bench mark being transferred to floor level in a building. The point marked by the bolt will be flush with the top of the slab.

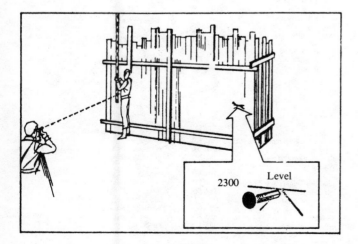

Establishing a level line on formwork
Figure 17-22

Before starting to pour concrete, do some checking. Check the positions of the forms, and the distances between walls. And check that any recesses or holes are in the right place.

The Importance of Layout

In the building trades, knowledge is profit. And being able to determine specifically where a structure is to be located is the first priority. Whether you build one house at a time on lots already surveyed and plotted, or many houses on undeveloped land, the layout procedure is the same.

Few builders are surveyors or have the civil engineering background necessary to establish property boundaries. Leave work like that to competent surveyors. But anyone can establish bench marks and use these marks for building layout.

In reasonably level terrain, where you'll just build one or two houses, you may just pour a slab on grade and let the top surface become the bench mark.

While the builder's level, the transit and the theodolite are useful instruments in surveying and establishing various lines and elevations, they're not essential. Many builders get along nicely without them, using a carpenter's steel square and a 4' level. If you're building a multiple house project, such instruments are a good investment.

Job layout is doing first things first, with a schedule that lets work progress smoothly and accurately according to some definite plan. Naturally, completed work must be checked to make sure it's right.

This completes Volume 1. If you've digested the information in this manual, you're ready to tackle the management and estimating subjects in Volume 2. It's an important part of what every successful construction contractor needs to know.

Index

Other Practical References

Cost Records for Construction Estimating

How to organize and use cost information from jobs just completed to make more accurate estimates in the future. Explains how to keep the cost records you need to reflect the time spent on each part of the job. Shows the best way to track costs for sitework, footing, foundations, framing, interior finish, siding and trim, masonry, and subcontract expense. Provides sample forms. **208 pages, 8½ x 11, $15.75**

Contractor's Year-Round Tax Guide

How to set up and run your construction business to minimize taxes: corporate tax strategy and how to use it to your advantage, and what you should be aware of in contracts with others. Covers tax shelters for builders, write-offs and investments that will reduce your taxes, accounting methods that are best for contractors, and what the I.R.S. allows and what it often questions. **192 pages, 8½ x 11, $16.50**

Wood-Frame House Construction

From the layout of the outer walls, excavation and formwork, to finish carpentry, and painting, every step of construction is covered in detail with clear illustrations and explanations. Everything the builder needs to know about framing, roofing, siding, insulation and vapor barrier, interior finishing, floor coverings, and stairs. . . complete step by step "how to" information on what goes into building a frame house. **240 pages, 8½ x 11, $11.25. Revised edition**

Residential Electrical Design

Explains what every builder needs to know about designing electrical systems for residential construction. Shows how to draw up an electrical plan from the blueprints, including the service entrance, grounding, lighting requirements for kitchen, bedroom and bath and how to lay them out. Explains how to plan electrical heating systems and what equipment you'll need, how to plan outdoor lighting, and much more. If you are a builder who ever has to plan an electrical system, you should have this book. **194 pages, 8½ x 11, $11.50**

Roof Framing

The only book ever published that explains in detail how to cut and frame gable, hip, Dutch, Tudor, gambrel, shed and gazebo roofs. Shows how to use an inexpensive hand-held calculator to calculate lengths of common, hip, valley and jack rafters — including rafters on irregular roofs that would stump even experienced roof framers. Includes over 200 illustrations and pictures of each roof that the author describes. Anyone who knows the procedures and methods described in this practical manual should have no trouble making a good living as a master roof framer. **480 pages, 5½ x 8½, $19.50**

Manual of Professional Remodeling

This is the practical manual of professional remodeling written by an experienced and successful remodeling contractor. Shows how to evaluate a job and avoid 30-minute jobs that take all day, what to fix and what to leave alone, and what to watch for in dealing with subcontractors. Includes chapters on calculating space requirements, repairing structural defects, remodeling kitchens, baths, walls and ceilings, doors and windows, floors, roofs, installing fireplaces and chimneys (including built-ins), skylights, and exterior siding. Includes blank forms, checklists, sample contracts, and proposals you can copy and use. **400 pages, 8½ x 11, $18.75**

Spec Builder's Guide

Explains how to plan and build a home, control your construction costs, and then sell the house at a price that earns a decent return on the time and money you've invested. Includes professional tips to ensure success as a spec builder: how government statistics help you judge the housing market, cutting costs at every opportunity without sacrificing quality, and taking advantage of construction cycles. Every chapter includes checklists, diagrams, charts, figures, and estimating tables. **448 pages, 8½ x 11, $24.00**

National Construction Estimator

Current building costs in dollars and cents for residential, commercial and industrial construction. Prices for every commonly used building material, and the proper labor cost associated with installation of the material. Everything figured out to give you the "in place" cost in seconds. Many time-saving rules of thumb, waste and coverage factors and estimating tables are included. **512 pages, 8½ x 11, $16.00. Revised annually.**

Remodeler's Handbook

The complete manual of home improvement contracting: Planning the job, estimating costs, doing the work, running your company and making profits. Pages of sample forms, contracts, documents, clear illustrations and examples. Chapters on evaluating the work, rehabilitation, kitchens, bathrooms, adding living area, re-flooring, re-siding, re-roofing, replacing windows and doors, installing new wall and ceiling cover, re-painting, upgrading insulation, combating moisture damage, estimating, selling your services, and bookkeeping for remodelers. **416 pages, 8½ x 11, $18.50**

Construction Superintending

Explains what the "super" should do during every job phase from taking bids to project completion on both heavy and light construction: excavation, foundations, pilings, steelwork, concrete and masonry, carpentry, plumbing, and electrical. Explains scheduling, preparing estimates, record keeping, dealing with subcontractors, and change orders. Includes the charts, forms, and established guidelines every superintendent needs. **240 pages, 8½ x 11, $22.00**

Plumbers Handbook Revised

This new edition shows what will and what will not pass inspection in drainage, vent, and waste piping, septic tanks, water supply, fire protection, and gas piping systems. All tables, standards, and specifications are completely up-to-date with recent changes in the plumbing code. Covers common layouts for residential work, how to size piping, selecting and hanging fixtures, practical recommendations and trade tips. This book is the approved reference for the plumbing contractors exam in many states. **240 pages, 8½ x 11, $16.75**

Computers: The Builder's New Tool

Shows how to avoid costly mistakes and find the right computer system for your needs. Takes you step-by-step through each important decision, from selecting the software to getting your equipment set up and operating. Filled with examples, checklists and illustrations, including case histories describing experiences other contractors have had. If you're thinking about putting a computer in your construction office, you should read this book before buying anything. **192 pages, 8½ x 11, $17.75**

Builder's Guide to Accounting

Explains how to set up and operate the record systems best for your business: simplified payroll and tax record keeping plus quick ways to make forecasts, spot trends, prepare estimates, record sales, receivables, checks and costs, and control losses. Loaded with charts, diagrams, blank forms and examples to help you create the strong financial base your business needs. **304 pages, 8½ x 11, $12.50**

How to Sell Remodeling

Proven, effective sales methods for repair and remodeling contractors: finding qualified leads, making the sales call, identifying what your prospects really need, pricing the job, arranging financing, and closing the sale. Explains how to organize and staff a sales team, how to bring in the work to keep your crews busy and your business growing, and much more. Includes blank forms, tables, and charts. **240 pages, 8½ x 11, $17.50**

Berger Building Cost File

Labor and material costs needed to estimate major projects: shopping centers and stores, hospitals, educational facilities, office complexes, industrial and institutional buildings, and housing projects. All cost estimates show both the manhours required and the typical crew needed so you can figure the price and schedule the work quickly and easily. **344 pages, 8½ x 11, $30.00**

Blueprint Reading for the Building Trades

How to read and understand construction documents, blueprints, and schedules. Includes layouts of structural, mechanical and electrical drawings, how to interpret sectional views, how to follow diagrams; plumbing, HVAC and schematics, and common problems experienced in interpreting construction specifications. This book is your course for understanding and following construction documents. **192 pages, 5½ x 8½, $11.25**

Rough Carpentry

All rough carpentry is covered in detail: sills, girders, columns, joists, sheathing, ceiling, roof and wall framing, roof trusses, dormers, bay windows, furring and grounds, stairs and insulation. Many of the 24 chapters explain practical code approved methods for saving lumber and time without sacrificing quality. Chapters on columns, headers, rafters, joists and girders show how to use simple engineering principles to select the right lumber dimension for whatever species and grade you are using. **288 pages, 8½ x 11, $14.50**

Estimating Home Building Costs

Estimate every phase of residential construction from site costs to the profit margin you should include in your bid. Shows how to keep track of manhours and make accurate labor cost estimates for footings, foundations, framing and sheathing finishes, electrical, plumbing and more. Explains the work being estimated and provides sample cost estimate worksheets with complete instructions for each job phase. **320 pages, 5½ x 8½, $14.00**

Paint Contractor's Manual

How to start and run a profitable paint contracting company: getting set up and organized to handle volume work, avoiding the mistakes most painters make, getting top production from your crews and the most value from your advertising dollar. Shows how to estimate all prep and painting. Loaded with manhour estimates, sample forms, contracts, charts, tables and examples you can use. **224 pages, 8½ x 11, $19.25**

Estimating Tables for Home Building

Produce accurate estimates in minutes for nearly any home or multi-family dwelling. This handy manual has the tables you need to find the quantity of materials and labor for most residential construction. Includes overhead and profit, how to develop unit costs for labor and materials and how to be sure you've considered every cost in the job. **336 pages, 8½ x 11, $21.50**